FREE Test Taking Tips DVD Offer

To help us better serve you, we have developed a Test Taking Tips DVD that we would like to give you for FREE. **This DVD covers world-class test taking tips that you can use to be even more successful when you are taking your test.**

All that we ask is that you email us your feedback about your study guide. Please let us know what you thought about it – whether that is good, bad or indifferent.

To get your **FREE Test Taking Tips DVD**, email freedvd@studyguideteam.com with "FREE DVD" in the subject line and the following information in the body of the email:

a. The title of your study guide.

b. Your product rating on a scale of 1-5, with 5 being the highest rating.

c. Your feedback about the study guide. What did you think of it?

d. Your full name and shipping address to send your free DVD.

If you have any questions or concerns, please don't hesitate to contact us at freedvd@studyguideteam.com.

Thanks again!

GED Preparation Canada

Study Guide and Practice Test Questions
[Book Includes Detailed Answer Explanations]

Joshua Rueda

Written and edited by TPB Publishing.

ISBN 13: 9781637750018
ISBN 10: 1637750013

Table of Contents

Quick Overview

As you draw closer to taking your exam, effective preparation becomes more and more important. Thankfully, you have this study guide to help you get ready. Use this guide to help keep your studying on track and refer to it often.

This study guide contains several key sections that will help you be successful on your exam. The guide contains tips for what you should do the night before and the day of the test. Also included are test-taking tips. Knowing the right information is not always enough. Many well-prepared test takers struggle with exams. These tips will help equip you to accurately read, assess, and answer test questions.

A large part of the guide is devoted to showing you what content to expect on the exam and to helping you better understand that content. In this guide are practice test questions so that you can see how well you have grasped the content. Then, answer explanations are provided so that you can understand why you missed certain questions.

Don't try to cram the night before you take your exam. This is not a wise strategy for a few reasons. First, your retention of the information will be low. Your time would be better used by reviewing information you already know rather than trying to learn a lot of new information. Second, you will likely become stressed as you try to gain a large amount of knowledge in a short amount of time. Third, you will be depriving yourself of sleep. So be sure to go to bed at a reasonable time the night before. Being well-rested helps you focus and remain calm.

Be sure to eat a substantial breakfast the morning of the exam. If you are taking the exam in the afternoon, be sure to have a good lunch as well. Being hungry is distracting and can make it difficult to focus. You have hopefully spent lots of time preparing for the exam. Don't let an empty stomach get in the way of success!

When travelling to the testing center, leave earlier than needed. That way, you have a buffer in case you experience any delays. This will help you remain calm and will keep you from missing your appointment time at the testing center.

Be sure to pace yourself during the exam. Don't try to rush through the exam. There is no need to risk performing poorly on the exam just so you can leave the testing center early. Allow yourself to use all of the allotted time if needed.

Remain positive while taking the exam even if you feel like you are performing poorly. Thinking about the content you should have mastered will not help you perform better on the exam.

Once the exam is complete, take some time to relax. Even if you feel that you need to take the exam again, you will be well served by some down time before you begin studying again. It's often easier to convince yourself to study if you know that it will come with a reward!

Test-Taking Strategies

1. Predicting the Answer

When you feel confident in your preparation for a multiple-choice test, try predicting the answer before reading the answer choices. This is especially useful on questions that test objective factual knowledge. By predicting the answer before reading the available choices, you eliminate the possibility that you will be distracted or led astray by an incorrect answer choice. You will feel more confident in your selection if you read the question, predict the answer, and then find your prediction among the answer choices. After using this strategy, be sure to still read all of the answer choices carefully and completely. If you feel unprepared, you should not attempt to predict the answers. This would be a waste of time and an opportunity for your mind to wander in the wrong direction.

2. Reading the Whole Question

Too often, test takers scan a multiple-choice question, recognize a few familiar words, and immediately jump to the answer choices. Test authors are aware of this common impatience, and they will sometimes prey upon it. For instance, a test author might subtly turn the question into a negative, or he or she might redirect the focus of the question right at the end. The only way to avoid falling into these traps is to read the entirety of the question carefully before reading the answer choices.

3. Looking for Wrong Answers

Long and complicated multiple-choice questions can be intimidating. One way to simplify a difficult multiple-choice question is to eliminate all of the answer choices that are clearly wrong. In most sets of answers, there will be at least one selection that can be dismissed right away. If the test is administered on paper, the test taker could draw a line through it to indicate that it may be ignored; otherwise, the test taker will have to perform this operation mentally or on scratch paper. In either case, once the obviously incorrect answers have been eliminated, the remaining choices may be considered. Sometimes identifying the clearly wrong answers will give the test taker some information about the correct answer. For instance, if one of the remaining answer choices is a direct opposite of one of the eliminated answer choices, it may well be the correct answer. The opposite of obviously wrong is obviously right! Of course, this is not always the case. Some answers are obviously incorrect simply because they are irrelevant to the question being asked. Still, identifying and eliminating some incorrect answer choices is a good way to simplify a multiple-choice question.

4. Don't Overanalyze

Anxious test takers often overanalyze questions. When you are nervous, your brain will often run wild, causing you to make associations and discover clues that don't actually exist. If you feel that this may be a problem for you, do whatever you can to slow down during the test. Try taking a deep breath or counting to ten. As you read and consider the question, restrict yourself to the particular words used by the author. Avoid thought tangents about what the author *really* meant, or what he or she was *trying* to say. The only things that matter on a multiple-choice test are the words that are actually in the question. You must avoid reading too much into a multiple-choice question, or supposing that the writer meant something other than what he or she wrote.

5. No Need for Panic

It is wise to learn as many strategies as possible before taking a multiple-choice test, but it is likely that you will come across a few questions for which you simply don't know the answer. In this situation, avoid panicking. Because most multiple-choice tests include dozens of questions, the relative value of a single wrong answer is small. As much as possible, you should compartmentalize each question on a multiple-choice test. In other words, you should not allow your feelings about one question to affect your success on the others. When you find a question that you either don't understand or don't know how to answer, just take a deep breath and do your best. Read the entire question slowly and carefully. Try rephrasing the question a couple of different ways. Then, read all of the answer choices carefully. After eliminating obviously wrong answers, make a selection and move on to the next question.

6. Confusing Answer Choices

When working on a difficult multiple-choice question, there may be a tendency to focus on the answer choices that are the easiest to understand. Many people, whether consciously or not, gravitate to the answer choices that require the least concentration, knowledge, and memory. This is a mistake. When you come across an answer choice that is confusing, you should give it extra attention. A question might be confusing because you do not know the subject matter to which it refers. If this is the case, don't eliminate the answer before you have affirmatively settled on another. When you come across an answer choice of this type, set it aside as you look at the remaining choices. If you can confidently assert that one of the other choices is correct, you can leave the confusing answer aside. Otherwise, you will need to take a moment to try to better understand the confusing answer choice. Rephrasing is one way to tease out the sense of a confusing answer choice.

7. Your First Instinct

Many people struggle with multiple-choice tests because they overthink the questions. If you have studied sufficiently for the test, you should be prepared to trust your first instinct once you have carefully and completely read the question and all of the answer choices. There is a great deal of research suggesting that the mind can come to the correct conclusion very quickly once it has obtained all of the relevant information. At times, it may seem to you as if your intuition is working faster even than your reasoning mind. This may in fact be true. The knowledge you obtain while studying may be retrieved from your subconscious before you have a chance to work out the associations that support it. Verify your instinct by working out the reasons that it should be trusted.

8. Key Words

Many test takers struggle with multiple-choice questions because they have poor reading comprehension skills. Quickly reading and understanding a multiple-choice question requires a mixture of skill and experience. To help with this, try jotting down a few key words and phrases on a piece of scrap paper. Doing this concentrates the process of reading and forces the mind to weigh the relative importance of the question's parts. In selecting words and phrases to write down, the test taker thinks about the question more deeply and carefully. This is especially true for multiple-choice questions that are preceded by a long prompt.

9. Subtle Negatives

One of the oldest tricks in the multiple-choice test writer's book is to subtly reverse the meaning of a question with a word like *not* or *except*. If you are not paying attention to each word in the question, you can easily be led astray by this trick. For instance, a common question format is, "Which of the following is...?" Obviously, if the question instead is, "Which of the following is not...?," then the answer will be quite different. Even worse, the test makers are aware of the potential for this mistake and will include one answer choice that would be correct if the question were not negated or reversed. A test taker who misses the reversal will find what he or she believes to be a correct answer and will be so confident that he or she will fail to reread the question and discover the original error. The only way to avoid this is to practice a wide variety of multiple-choice questions and to pay close attention to each and every word.

10. Reading Every Answer Choice

It may seem obvious, but you should always read every one of the answer choices! Too many test takers fall into the habit of scanning the question and assuming that they understand the question because they recognize a few key words. From there, they pick the first answer choice that answers the question they believe they have read. Test takers who read all of the answer choices might discover that one of the latter answer choices is actually *more* correct. Moreover, reading all of the answer choices can remind you of facts related to the question that can help you arrive at the correct answer. Sometimes, a misstatement or incorrect detail in one of the latter answer choices will trigger your memory of the subject and will enable you to find the right answer. Failing to read all of the answer choices is like not reading all of the items on a restaurant menu: you might miss out on the perfect choice.

11. Spot the Hedges

One of the keys to success on multiple-choice tests is paying close attention to every word. This is never truer than with words like almost, most, some, and sometimes. These words are called "hedges" because they indicate that a statement is not totally true or not true in every place and time. An absolute statement will contain no hedges, but in many subjects, the answers are not always straightforward or absolute. There are always exceptions to the rules in these subjects. For this reason, you should favor those multiple-choice questions that contain hedging language. The presence of qualifying words indicates that the author is taking special care with his or her words, which is certainly important when composing the right answer. After all, there are many ways to be wrong, but there is only one way to be right! For this reason, it is wise to avoid answers that are absolute when taking a multiple-choice test. An absolute answer is one that says things are either all one way or all another. They often include words like *every*, *always*, *best*, and *never*. If you are taking a multiple-choice test in a subject that doesn't lend itself to absolute answers, be on your guard if you see any of these words.

12. Long Answers

In many subject areas, the answers are not simple. As already mentioned, the right answer often requires hedges. Another common feature of the answers to a complex or subjective question are qualifying clauses, which are groups of words that subtly modify the meaning of the sentence. If the question or answer choice describes a rule to which there are exceptions or the subject matter is complicated, ambiguous, or confusing, the correct answer will require many words in order to be expressed clearly and accurately. In essence, you should not be deterred by answer choices that seem excessively long. Oftentimes, the author of the text will not be able to write the correct answer without offering some qualifications and modifications. Your job is to read the answer choices thoroughly and completely and to select the one that most accurately and precisely answers the question.

13. Restating to Understand

Sometimes, a question on a multiple-choice test is difficult not because of what it asks but because of how it is written. If this is the case, restate the question or answer choice in different words. This process serves a couple of important purposes. First, it forces you to concentrate on the core of the question. In order to rephrase the question accurately, you have to understand it well. Rephrasing the question will concentrate your mind on the key words and ideas. Second, it will present the information to your mind in a fresh way. This process may trigger your memory and render some useful scrap of information picked up while studying.

14. True Statements

Sometimes an answer choice will be true in itself, but it does not answer the question. This is one of the main reasons why it is essential to read the question carefully and completely before proceeding to the answer choices. Too often, test takers skip ahead to the answer choices and look for true statements. Having found one of these, they are content to select it without reference to the question above. Obviously, this provides an easy way for test makers to play tricks. The savvy test taker will always read the entire question before turning to the answer choices. Then, having settled on a correct answer choice, he or she will refer to the original question and ensure that the selected answer is relevant. The mistake of choosing a correct-but-irrelevant answer choice is especially common on questions related to specific pieces of objective knowledge. A prepared test taker will have a wealth of factual knowledge at his or her disposal, and should not be careless in its application.

15. No Patterns

One of the more dangerous ideas that circulates about multiple-choice tests is that the correct answers tend to fall into patterns. These erroneous ideas range from a belief that B and C are the most common right answers, to the idea that an unprepared test-taker should answer "A-B-A-C-A-D-A-B-A." It cannot be emphasized enough that pattern-seeking of this type is exactly the WRONG way to approach a multiple-choice test. To begin with, it is highly unlikely that the test maker will plot the correct answers according to some predetermined pattern. The questions are scrambled and delivered in a random order. Furthermore, even if the test maker was following a pattern in the assignation of correct answers, there is no reason why the test taker would know which pattern he or she was using. Any attempt to discern a pattern in the answer choices is a waste of time and a distraction from the real work of taking the test. A test taker would be much better served by extra preparation before the test than by reliance on a pattern in the answers.

FREE DVD OFFER

Don't forget that doing well on your exam includes both understanding the test content and understanding how to use what you know to do well on the test. We offer a completely FREE Test Taking Tips DVD that covers world class test taking tips that you can use to be even more successful when you are taking your test.

All that we ask is that you email us your feedback about your study guide. To get your **FREE Test Taking Tips DVD**, email freedvd@studyguideteam.com with "FREE DVD" in the subject line and the following information in the body of the email:

- The title of your study guide.
- Your product rating on a scale of 1-5, with 5 being the highest rating.
- Your feedback about the study guide. What did you think of it?
- Your full name and shipping address to send your free DVD.

Introduction to the GED Exam

Function of the Test

The Canada General Educational Development (GED) Exam is for individuals who are at least eighteen years old who did not receive a high school diploma and would like to earn certification proving they have high-school level skills. Those who pass the GED Exam can use their credentials to become employed, to gain licensing, to be eligible for promotion, among other educational and vocational opportunities. The GED is offered in the United States and Canada.

Test Administration

For those taking the Canada GED, each province has separate rules as to who can sit for the exam. For example, only Ontario residents are able to take the GED Exam in Ontario. However, each province is different, so test takers should check with their own provinces for the rules. The GED is offered in person at an official GED testing center. If you wish to take the computer-based exam, you may register online at ged.com. If you wish to take the paper exam, you may register at a local testing center, which can be found on the website as well.

If individuals do not pass one of the GED sections, they only need to retake that section. Test takers may take the failed subject two additional times; if they fail both times, they have to wait sixty days before their next retest. Test takers requiring accommodations should go to the GED website to begin the process of requesting them. While accommodations are provided, there are multiple steps to be taken before getting approved.

Test Format

The GED is made up of multiple-choice questions and includes an essay section in the writing portion. Students are allowed to use a calculator during part I of the math exam but not part II. Calculators are provided at the testing centers.

Subjects on the GED Exam are math, reading, writing, social studies, and science. The Canada GED is similar to the U.S. version, but the Canada GED covers Canadian history and government. The GED Exam is given over a two-day period with short breaks between the subjects. Below is a table of the subjects which corresponds to the test length and number of questions, followed by the subjects and their content areas.

Subject	Time	Number of Questions
Mathematics	1 hour 30 minutes	50 multiple-choice
Reading	1 hour 5 minutes	40 multiple-choice
Writing	Part 1: 1 hour 15 minutes Part 2: 45 minutes	Part 1: 50 multiple-choice Part 2: 1 essay
Social Studies	1 hour 10 minutes	50 multiple-choice
Science	1 hour 20 minutes	50 multiple-choice

Mathematics:

- 20–30% Number operations and number sense
- 20–30% Measurement and geometry
- 20–30% Data analysis, statistics, and probability
- 20–30% Algebra, functions, and patterns

Reading:

- 75% Literary texts, including poetry, drama, prose fiction before 1920, prose fiction between 1920 and 1960, and prose fiction after 1960
- 25% Nonfiction texts, including nonfiction prose, critical review of visual and performing arts, and workplace and community documents

Writing Part I:

- 30% Sentence structure
- 30% Usage
- 25% Mechanics
- 15% Organization

Writing Part II:

- Opinion essay

Social Studies:

- 40% History
- 25% Civics and government
- 20% Economics
- 15% Geography

Science:

- 45% Life science
- 35% Physical science (physics and chemistry)
- 20% Earth and space science

Scoring

The GED passing scale is 450 on each of the subject areas. The score scale for the GED Exam is from 200 to 800. Scoring for the essay is between a 1 and a 4, with 1 being inadequate and 4 being effective. Two readers will grade the essay, with the final grade being an average between the two.

Study Prep Plan for the GED Exam

1 **Schedule -** Use one of our study schedules below or come up with one of your own.

2 **Relax -** Test anxiety can hurt even the best students. There are many ways to reduce stress. Find the one that works best for you.

3 **Execute -** Once you have a good plan in place, be sure to stick to it.

One Week Study Schedule

Day	Subject
Day 1	Mathematics
Day 2	Reading
Day 3	Writing
Day 4	Social Studies
Day 5	Science
Day 6	Practice Test
Day 7	Take Your Exam!

Two Week Study Schedule

Day	Subject	Day	Subject
Day 1	Mathematics	Day 8	Practice Questions
Day 2	Practice Questions	Day 9	Science
Day 3	Reading	Day 10	Practice Questions
Day 4	Practice Questions	Day 11	Practice Test (Mathematics - Writing)
Day 5	Writing	Day 12	Practice Test (Social Studies - Science)
Day 6	Practice Questions	Day 13	Review Answer Explanations
Day 7	Social Studies	Day 14	Take Your Exam!

One Month Study Schedule

Day 1	Number Operations and Number Sense	Day 11	Organization	Day 21	Life Science
Day 2	Measurement and Geometry	Day 12	Mechanics	Day 22	Earth and Space Science
Day 3	Data Analysis, Statistics, and Probability	Day 13	Practice Questions	Day 23	Practice Questions
Day 4	Algebra, Functions, and Patterns	Day 14	World History	Day 24	Practice Test - Mathematics
Day 5	Practice Questions	Day 15	Canadian History	Day 25	Practice Test - Reading
Day 6	Putting Events in Order	Day 16	Geography	Day 26	Practice Test - Writing
Day 7	Assessing Whether an Argument is Valid	Day 17	Civics and Government	Day 27	Practice Test - Social Studies
Day 8	Literary Text	Day 18	Economics	Day 28	Practice Test - Science
Day 9	Nonfiction Text	Day 19	Practice Questions	Day 29	Review Answer Explanations
Day 10	Practice Questions	Day 20	Physical Science (Physics and Chemistry)	Day 30	Take Your Exam!

Mathematics

Number Operations and Number Sense

Fractions and Decimals in Order

Rational numbers are those that can be written as a fraction or ratio. Within the set of rational numbers, several subsets exist that are referenced throughout the mathematics topics. Counting numbers are the first numbers learned as a child. Counting numbers consist of 1, 2, 3, 4, and so on. Whole numbers include all counting numbers and zero (0, 1, 2, 3, 4, ...). Integers include counting numbers, their opposites, and zero (..., -3, -2, -1, 0, 1, 2 ,3, ...). Rational numbers are inclusive of integers, fractions, and decimals that terminate, or end (1.7, 0.04213) or repeat ($0.136\overline{5}$).

Placing numbers in an order in which they are listed from smallest to largest is known as **ordering**. Ordering numbers properly can help in the comparison of different quantities of items.

When comparing two numbers to determine if they are equal or if one is greater than the other, it is best to look at the digit furthest to the left of the decimal place (or the first value of the decomposed numbers). If this first digit of each number being compared is equal in place value, then move one digit to the right to conduct a similar comparison. Continue this process until it can be determined that both numbers are equal or a difference is found, showing that one number is greater than the other. If a number is greater than the other number it is being compared to, a symbol such as > (greater than) or < (less than) can be utilized to show this comparison. It is important to remember that the "open mouth" of the symbol should be nearest the larger number.

For example:

1,023,100 compared to 1,023,000

First, compare the digit farthest to the left. Both are decomposed to 1,000,000, so this place is equal.

Next, move one place to right on both numbers being compared. This number is zero for both numbers, so move on to the next number to the right. The first number decomposes to 20,000, while the second decomposes to 20,000. These numbers are also equal, so move one more place to the right. The first number decomposes to 3,000, as does the second number, so they are equal again. Moving one place to the right, the first number decomposes to 100, while the second number is zero. Since 100 is greater than zero, the first number is greater than the second. This is expressed using the greater than symbol:

1,023,100 > 1,023,000 because 1,023,100 is greater than 1,023,000 (Note that the "open mouth" of the symbol is nearest to 1,023,100).

Notice the > symbol in the above comparison. When values are the same, the equals sign (=) is used. However, when values are unequal, or an **inequality** exists, the relationship is denoted by various inequality symbols. These symbols describe in what way the values are unequal. A value could be greater than (>); less than (<); greater than or equal to (≥); or less than or equal to (≤) another value.

The statement "five times a number added to forty is more than sixty-five" can be expressed as $5x + 40 > 65$. Common words and phrases that express inequalities are:

Symbol	Phrase
$<$	is under, is below, smaller than, beneath
$>$	is above, is over, bigger than, exceeds
$\leq$	no more than, at most, maximum
$\geq$	no less than, at least, minimum

Another way to compare whole numbers with many digits is to use place value. In each number to be compared, it is necessary to find the highest place value in which the numbers differ and to compare the value within that place value. For example, $4{,}523{,}345 < 4{,}532{,}456$ because of the values in the ten thousands place.

Comparing and Ordering Decimals

To compare decimals and order them by their value, utilize a method similar to that of ordering large numbers.

The main difference is where the comparison will start. Assuming that any numbers to left of the decimal point are equal, the next numbers to be compared are those immediately to the right of the decimal point. If those are equal, then move on to compare the values in the next decimal place to the right.

For example:

Which number is greater, 12.35 or 12.38?

Check that the values to the left of the decimal point are equal:

$$12 = 12$$

Next, compare the values of the decimal place to the right of the decimal:

$$12.3 = 12.3$$

Those are also equal in value.

Finally, compare the value of the numbers in the next decimal place to the right on both numbers:

$$12.3\mathbf{5} \text{ and } 12.3\mathbf{8}$$

Here the 5 is less than the 8, so the final way to express this inequality is:

$$12.35 < 12.38$$

Comparing decimals is regularly exemplified with money because the "cents" portion of money ends in the hundredths place. When paying for gasoline or meals in restaurants, and even in bank accounts, if enough errors are made when calculating numbers to the hundredths place, they can add up to dollars and larger amounts of money over time.

Number lines can also be used to compare decimals. Tick marks can be placed within two whole numbers on the number line that represent tenths, hundredths, etc. Each number being compared can then be plotted. The value farthest to the right on the number line is the largest.

Comparing Fractions

To compare fractions with either the same **numerator** (top number) or same **denominator** (bottom number), it is easiest to visualize the fractions with a model.

For example, which is larger, $\frac{1}{3}$ or $\frac{1}{4}$? Both numbers have the same numerator, but a different denominator. In order to demonstrate the difference, shade the amounts on a pie chart split into the number of pieces represented by the denominator.

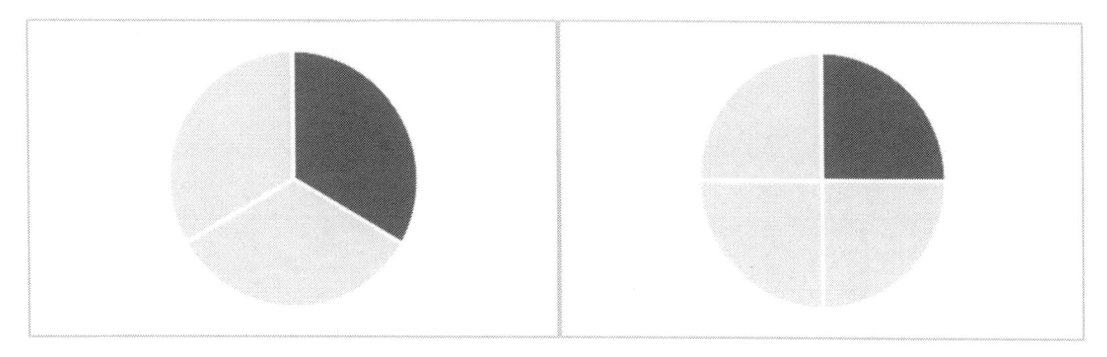

The first pie chart represents $\frac{1}{3}$, a larger shaded portion, and is therefore a larger fraction than the second pie chart representing $\frac{1}{4}$.

If two fractions have the same denominator (or are split into the same number of pieces), the fraction with the larger numerator is the larger fraction, as seen below in the comparison of $\frac{1}{3}$ and $\frac{2}{3}$:

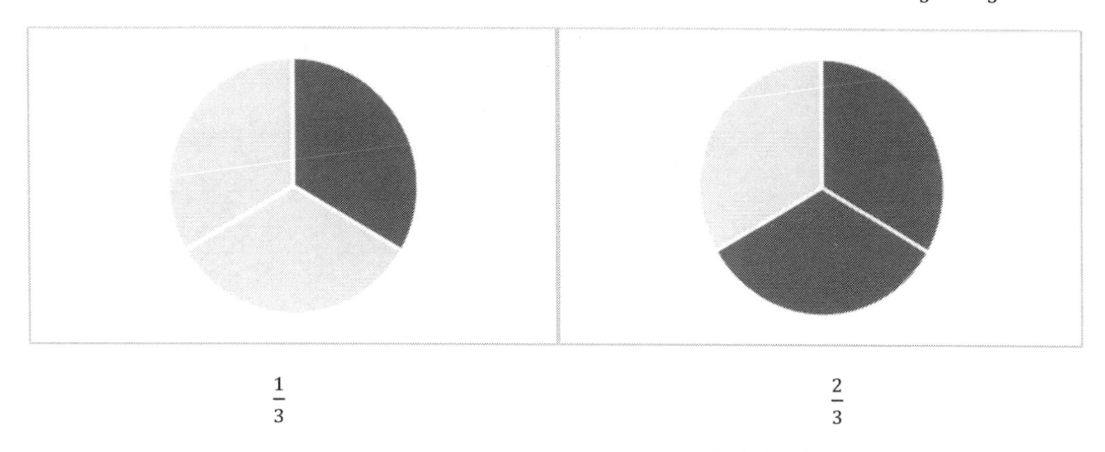

$\frac{1}{3}$ $\frac{2}{3}$

As mentioned, a **unit fraction** is one in which the numerator is 1 ($\frac{1}{2}, \frac{1}{3}, \frac{1}{8}, \frac{1}{20}$, etc.). The denominator indicates the number of equal pieces that the whole is divided into. The greater the number of pieces, the smaller each piece will be. Therefore, the greater the denominator of a unit fraction, the smaller it is in value. Unit fractions can also be compared by converting them to decimals. For example, $\frac{1}{2}$ = 0.5, $\frac{1}{3}$ = $0.\bar{3}$, $\frac{1}{8}$ = 0.125, $\frac{1}{20} = 0.05$, etc.

Comparing two fractions with different denominators can be difficult if attempting to guess at how much each represents. Using a number line, blocks, or just finding a common denominator with which to compare the two fractions makes this task easier.

For example, compare the fractions $\frac{3}{4}$ and $\frac{5}{8}$.

The number line method of comparison involves splitting one number line evenly into 4 sections, and the second number line evenly into 8 sections total, as follows:

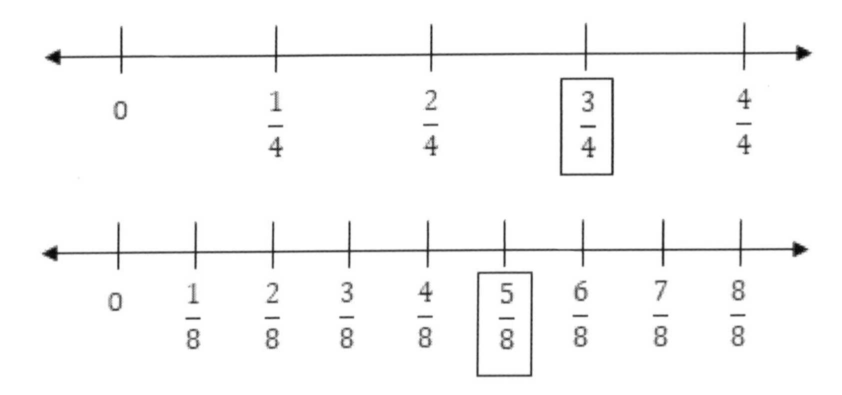

Here it can be observed that $\frac{3}{4}$ is greater than $\frac{5}{8}$, so the comparison is written as $\frac{3}{4} > \frac{5}{8}$.

This could also be shown by finding a common denominator for both fractions, so that they could be compared. First, list out factors of 4: 4, **8**, 12, 16.

Then, list out factors of 8: **8**, 16, 24.

Both share a common factor of 8, so they can be written in terms of 8 portions. In order for $\frac{3}{4}$ to be written in terms of 8, both the numerator and denominator must be multiplied by 2, thus forming the new fraction $\frac{6}{8}$. Now the two fractions can be compared.

Because both have the same denominator, the numerator will show the comparison.

$$\frac{6}{8} > \frac{5}{8}$$

Ordering Numbers

Whether the question asks to order the numbers from greatest to least or least to greatest, the crux of the question is the same—convert the numbers into a common format. Generally, it's easiest to write the numbers as whole numbers and decimals so they can be placed on a number line. Follow these examples to understand this strategy.

1) Order the following rational numbers from greatest to least:

$$\sqrt{36}, 0.65, 78\%, \frac{3}{4}, 7, 90\%, \frac{5}{2}$$

Of the seven numbers, the whole number (7) and decimal (0.65) are already in an accessible form, so concentrate on the other five.

First, the square root of 36 equals 6. (If the test asks for the root of a non-perfect root, determine which two whole numbers the root lies between.) Next, convert the percentages to decimals. A percentage means "per hundred," so this conversion requires moving the decimal point two places to the left, leaving 0.78 and 0.9. Lastly, evaluate the fractions: $\frac{3}{4} = \frac{75}{100} = 0.75$; $\frac{5}{2} = 2\frac{1}{2} = 2.5$

Now, the only step left is to list the numbers in the request order:

$$7, \sqrt{36}, \frac{5}{2}, 90\%, 78\%, \frac{3}{4}, 0.65$$

2) Order the following rational numbers from least to greatest:

$$2.5, \sqrt{9}, -10.5, 0.853, 175\%, \sqrt{4}, \frac{4}{5}$$

$$\sqrt{9} = 3$$

$$175\% = 1.75$$

$$\sqrt{4} = 2$$

$$\frac{4}{5} = 0.8$$

From least to greatest, the answer is: $-10.5, \frac{4}{5}, 0.853, 175\%, \sqrt{4}, 2.5, \sqrt{9}$

It is not possible to give similar relationships between two complex numbers $a + ib$ and $c + id$. This is because the real numbers cannot be identified with the complex numbers, and there is no form of comparison between the two. However, given any polynomial equation, its solutions can be solved in the complex field. If the zeros are real, they can be written as $a + i \times 0$; if they are complex, they can be written as $a + ib$; and if they are imaginary, they can be written as ib.

Multiples and Factors

Multiples of a given number are found by taking that number and multiplying it by any other whole number. For example, 3 is a factor of 6, 9, and 12. Therefore, 6, 9, and 12 are multiples of 3. The multiples of any number are an infinite list. For example, the multiples of 5 are 5, 10, 15, 20, and so on. This list continues without end. A list of multiples is used in finding the **least common multiple**, or LCM, for fractions when a common denominator is needed. The denominators are written down and their multiples listed until a common number is found in both lists. This common number is the LCM.

The **factors** of a number are all integers that can be multiplied by another integer to produce the given number. For example, 2 is multiplied by 3 to produce 6. Therefore, 2 and 3 are both factors of 6. Similarly, $1 \times 6 = 6$ and $2 \times 3 = 6$, so 1, 2, 3, and 6 are all factors of 6. Another way to explain a factor is to say that a given number divides evenly by each of its factors to produce an integer. For example, 6 does not divide evenly by 5. Therefore, 5 is not a factor of 6.

Prime factorization breaks down each factor of a whole number until only prime numbers remain. All composite numbers can be factored into prime numbers. For example, the prime factors of 12 are 2, 2, and 3 ($2 \times 2 \times 3 = 12$). To produce the prime factors of a number, the number is factored, and any composite numbers are continuously factored until the result is the product of prime factors only. A **factor tree**, such as the one below, is helpful when exploring this concept.

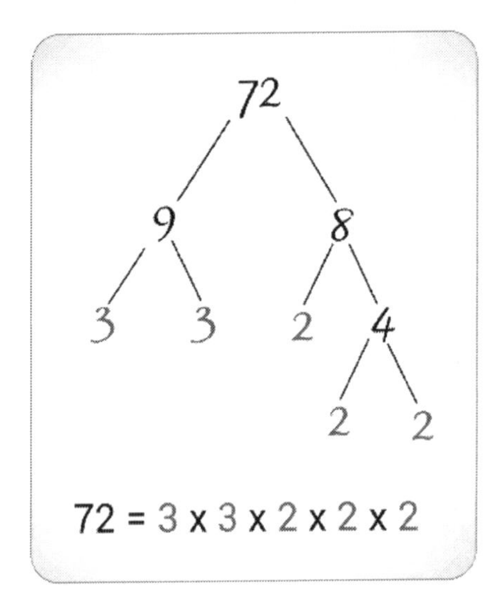

Simplifying Exponents

Exponents are used in mathematics to express a number or variable multiplied by itself a certain number of times. For example, x^3 means x is multiplied by itself three times. In this expression, x is called the **base,** and 3 is the **exponent.** Exponents can be used in more complex problems when they contain fractions and negative numbers.

Fractional exponents can be explained by looking first at the inverse of exponents, which are **roots.** Given the expression x^2, the square root can be taken, $\sqrt{x^2}$, cancelling out the 2 and leaving *x* by itself, if *x* is positive. Cancellation occurs because $\sqrt{x}$ can be written with exponents, instead of roots, as $x^{\frac{1}{2}}$. The numerator of 1 is the exponent, and the denominator of 2 is called the root (which is why it's referred to as **square root**). Taking the square root of x^2 is the same as raising it to the $\frac{1}{2}$ power. Written out in mathematical form, it takes the following progression:

$$\sqrt{x^2} = (x^2)^{\frac{1}{2}} = x$$

From properties of exponents, $2 \times \frac{1}{2} = 1$ is the actual exponent of *x*. Another example can be seen with $x^{\frac{4}{7}}$. The variable *x*, raised to four-sevenths, is equal to the seventh root of *x* to the fourth power: $\sqrt[7]{x^4}$. In general,

$$x^{\frac{1}{n}} = \sqrt[n]{x}$$

and

$$x^{\frac{m}{n}} = \sqrt[n]{x^m}$$

Negative exponents also involve fractions. Whereas y^3 can also be rewritten as $\frac{y^3}{1}$, y^{-3} can be rewritten as $\frac{1}{y^3}$. A negative exponent means the exponential expression must be moved to the opposite spot in a fraction to make the exponent positive. If the negative appears in the numerator, it moves to the denominator. If the negative appears in the denominator, it is moved to the numerator. In general, $a^{-n} = \frac{1}{a^n}$, and a^{-n} and a^n are reciprocals.

Take, for example, the following expression:

$$\frac{a^{-4}b^2}{c^{-5}}$$

Since *a* is raised to the negative fourth power, it can be moved to the denominator. Since *c* is raised to the negative fifth power, it can be moved to the numerator. The *b* variable is raised to the positive second power, so it does not move.

The simplified expression is as follows:

$$\frac{b^2c^5}{a^4}$$

In mathematical expressions containing exponents and other operations, the order of operations must be followed. **PEMDAS** states that exponents are calculated after any parenthesis and grouping symbols, but before any multiplication, division, addition, and subtraction.

There are a few rules for working with exponents. For any numbers a, b, m, n, the following hold true:

$$a^1 = a$$

$$1^a = 1$$

$$a^0, = 1$$

$$a^m \times a^n = a^{m+n}$$

$$a^m \div a^n = a^{m-n}$$

$$(a^m)^n = a^{m \times n}$$

$$(a \times b)^m = a^m \times b^m$$

$$(a \div b)^m = a^m \div b^m$$

Any number, including a fraction, can be an exponent. The same rules apply.

Distance Between Numbers on a Number Line

Aside from zero, numbers can be either positive or negative. The sign for a positive number is the plus sign or the + symbol, while the sign for a negative number is the minus sign or the – symbol. If a number has no designation, then it's assumed to be positive.

Both positive and negative numbers are valued according to their distance from zero. Both +3 and -3 can be considered using the following number line:

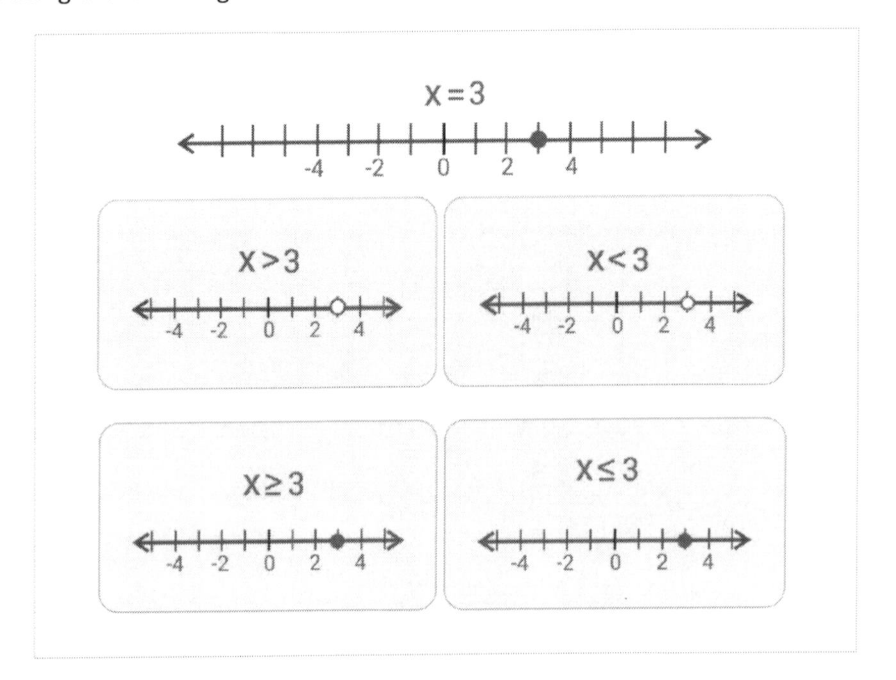

Both 3 and -3 are three spaces from zero. The distance from zero is called its **absolute value**. Thus, both -3 and 3 have an absolute value of 3 since they're both three spaces away from zero.

An absolute number is written by placing | | around the number. So, $|3|$ and $|-3|$ both equal 3, as that's their common absolute value.

Implications for Addition and Subtraction

For addition, if all numbers are either positive or negative, they are simply added together. For example, $4 + 4 = 8$ and $-4 + -4 = -8$. However, things get tricky when some of the numbers are negative and some are positive.

For example, with $6 + (-4)$, the first step is to take the absolute values of the numbers, which are 6 and 4. Second, the smaller value is subtracted from the larger. The equation becomes $6 - 4 = 2$. Third, the sign of the original larger number is placed on the sum. Here, 6 is the larger number, and it's positive, so the sum is 2.

Here's an example where the negative number has a larger absolute value: $(-6) + 4$. The first two steps are the same as the example above. However, on the third step, the negative sign must be placed on the sum, because the absolute value of (-6) is greater than 4. Thus, $-6 + 4 = -2$.

The absolute value of numbers implies that subtraction can be thought of as flipping the sign of the number following the subtraction sign and simply adding the two numbers. This means that subtracting a negative number will, in fact, be adding the positive absolute value of the negative number.

Here are some examples:

$$-6 - 4 = -6 + -4 = -10$$

$$3 - -6 = 3 + 6 = 9$$

$$-3 - 2 = -3 + -2 = -5$$

Implications for Multiplication and Division

For multiplication and division, if both numbers are positive, then the product or quotient is always positive. If both numbers are negative, then the product or quotient is also positive. However, if the numbers have opposite signs, the product or quotient is always negative.

Simply put, the product in multiplication and quotient in division is always positive, unless the numbers have opposing signs, in which case it's negative. Here are some examples:

$$(-6) \times (-5) = 30$$

$$(-50) \div 10 = -5$$

$$8 \times |-7| = 56$$

$$(-48) \div (-6) = 8$$

If there are more than two numbers in a multiplication or division problem, then whether the product or quotient is positive or negative depends on the number of negative numbers in the problem. If there is an odd number of negatives, then the product or quotient is negative. If there is an even number of negative numbers, then the result is positive.

Here are some examples:

$$(-6) \times 5 \times (-2) \times (-4) = -240$$

$$(-6) \times 5 \times 2 \times (-4) = 240$$

Whole Numbers, Fractions, and Decimal Problems

Operations

Addition combines two quantities together. With whole numbers, this is taking two sets of things and merging them into one, then counting the result. For example, $4 + 3 = 7$. When adding numbers, the order does not matter: $3 + 4 = 7$, also. Longer lists of whole numbers can also be added together. The result of adding numbers is called the **sum.**

With fractions, the number on top is the **numerator**, and the number on the bottom is the **denominator**. To add fractions, the denominator must be the same—a **common denominator**. To find a common denominator, the existing numbers on the bottom must be considered, and the lowest number they will both multiply into must be determined. Consider the following equation:

$$\frac{1}{3} + \frac{5}{6} = ?$$

The numbers 3 and 6 both multiply into 6. Three can be multiplied by 2, and 6 can be multiplied by 1. The top and bottom of each fraction must be multiplied by the same number. Then, the numerators are added together to get a new numerator. The following equation is the result:

$$\frac{1}{3}+\frac{5}{6}=\frac{2}{6}+\frac{5}{6}=\frac{7}{6}$$

Subtraction is taking one quantity away from another, so it is the opposite of addition. The expression $4-3$ means taking 3 away from 4. So, $4-3=1$. In this case, the order matters, since it entails taking one quantity away from the other, rather than just putting two quantities together. The result of subtraction is also called the **difference**.

To subtract fractions, the denominator must be the same. Then, subtract the numerators together to get a new numerator. Here is an example:

$$\frac{1}{3}-\frac{5}{6}=\frac{2}{6}-\frac{5}{6}=\frac{-3}{6}=-\frac{1}{2}$$

Multiplication is a kind of repeated addition. The expression 4×5 is taking four sets, each of them having five things in them, and putting them all together. That means $4 \times 5 = 5+5+5+5 = 20$. As with addition, the order of the numbers does not matter. The result of a multiplication problem is called the **product**.

To multiply fractions, the numerators are multiplied to get the new numerator, and the denominators are multiplied to get the new denominator:

$$\frac{1}{3}\times\frac{5}{6}=\frac{1\times 5}{3\times 6}=\frac{5}{18}$$

When multiplying fractions, common factors can cancel or divide into one another, when factors appear in the numerator of one fraction and the denominator of the other fraction. Here is an example:

$$\frac{1}{3}\times\frac{9}{8}=\frac{1}{1}\times\frac{3}{8}=1\times\frac{3}{8}=\frac{3}{8}$$

The numbers 3 and 9 have a common factor of 3, so that factor can be divided out.

Division is the opposite of multiplication. With whole numbers, it means splitting up one number into sets of equal size. For example, $16 \div 8$ is the number of sets of eight things that can be made out of sixteen things. Thus, $16 \div 8 = 2$. As with subtraction, the order of the numbers will make a difference, here. The answer to a division problem is called the **quotient,** while the number in front of the division sign is called the **dividend** and the number behind the division sign is called the **divisor.**

To divide fractions, the first fraction must be multiplied with the reciprocal of the second fraction. The **reciprocal** of the fraction $\frac{x}{y}$ is the fraction $\frac{y}{x}$. Here is an example:

$$\frac{1}{3}\div\frac{5}{6}=\frac{1}{3}\times\frac{6}{5}=\frac{6}{15}=\frac{2}{5}$$

The value of a fraction does not change if multiplying or dividing both the numerator and the denominator by the same number (other than 0).

In other words:

$$\frac{x}{y} = \frac{a \times x}{a \times y} = \frac{x \div a}{y \div a}$$

as long as *a* is not 0.

This means that $\frac{2}{5} = \frac{4}{10}$, for example. If *x* and *y* are integers that have no common factors, then the fraction is said to be **simplified**. This means $\frac{2}{5}$ is simplified, but $\frac{4}{10}$ is not.

Often when working with fractions, the fractions need to be rewritten so that they all share a single denominator—this is called finding a **common denominator** for the fractions. Using two fractions, $\frac{a}{b}$ and $\frac{c}{d}$, the numerator and denominator of the left fraction can be multiplied by *d*, while the numerator and denominator of the right fraction can be multiplied by *b*. This provides the fractions $\frac{a \times d}{b \times d}$ and $\frac{c \times b}{d \times b}$ with the common denominator $b \times d$.

A fraction whose numerator is smaller than its denominator is called a **proper fraction**. A fraction whose numerator is bigger than its denominator is called an **improper fraction**. These numbers can be rewritten as a combination of integers and fractions, called a **mixed number**. For example:

$$\frac{6}{5} = \frac{5}{5} + \frac{1}{5} = 1 + \frac{1}{5}$$

and can be written as $1\frac{1}{5}$.

Estimation is finding a value that is close to a solution but is not the exact answer. For example, if there are values in the thousands to be multiplied, then each value can be estimated to the nearest thousand and the calculation performed. This value provides an approximate solution that can be determined very quickly.

Sometimes when multiplying numbers, the result can be estimated by **rounding**. For example, to estimate the value of 11.2×2.01, each number can be rounded to the nearest integer. This will yield a result of 22.

The **decimal system** is a way of writing out numbers that uses ten different numerals: 0, 1, 2, 3, 4, 5, 6, 7, 8, and 9. This is also called a "base ten" or "base 10" system. Other bases are also used. For example, computers work with a base of 2. This means they only use the numerals 0 and 1.

The decimal place denotes how far to the right of the decimal point a numeral is. The first digit to the right of the decimal point is in the tenths place. The next is the hundredths. The third is the thousandths.

So, 3.142 has a 1 in the tenths place, a 4 in the hundredths place, and a 2 in the thousandths place.

The **decimal point** is a period used to separate the **ones** place from the **tenths** place when writing out a number as a decimal.

A **decimal number** is a number written out with a decimal point instead of as a fraction, for example, 1.25 instead of $\frac{5}{4}$. Depending on the situation, it can sometimes be easier to work with fractions and sometimes easier to work with decimal numbers.

A decimal number is **terminating** if it stops at some point. It is called **repeating** if it never stops but repeats a pattern over and over. It is important to note that every rational number can be written as a terminating decimal or as a repeating decimal.

To add decimal numbers, each number in the column needs to be lined up by the decimal point. For each number being added, the zeros to the right of the last number need to be filled in so that each of the numbers has the same number of places to the right of the decimal. Then, the columns can be added together.

Here is an example of $2.45 + 1.3 + 8.891$ written in column form:

$$\begin{array}{r} 2.450 \\ 1.300 \\ \underline{+\,8.891} \end{array}$$

Zeros have been added in the columns so that each number has the same number of places to the right of the decimal.

Added together, the correct answer is 12.641:

$$\begin{array}{r} 2.450 \\ 1.300 \\ \underline{+\,8.891} \\ 12.641 \end{array}$$

Subtracting decimal numbers is the same process as adding decimals. Here is $7.89 - 4.235$ written in column form:

$$\begin{array}{r} 7.890 \\ \underline{-\,4.235} \\ 3.655 \end{array}$$

A zero has been added in the column so that each number has the same number of places to the right of the decimal.

Decimals can be multiplied as if there were no decimal points in the problem. For example, 0.5×1.25 can be rewritten and multiplied as 5×125, which equals 625.

The final answer will have the same number of decimal points as the total number of decimal places in the problem. The first number has one decimal place, and the second number has two decimal places. Therefore, the final answer will contain three decimal places:

$$0.5 \times 1.25 = 0.625$$

Dividing a decimal by a whole number entails using long division first by ignoring the decimal point. Then, the decimal point is moved the number of places given in the problem.

For example, $6.8 \div 4$ can be rewritten as $68 \div 4$, which is 17. There is one non-zero integer to the right of the decimal point, so the final solution would have one decimal place to the right of the solution. In this case, the solution is 1.7.

Dividing a decimal by another decimal requires changing the divisor to a whole number by moving its decimal point. The decimal place of the dividend should be moved by the same number of places as the divisor. Then, the problem is the same as dividing a decimal by a whole number.

For example, $5.72 \div 1.1$ has a divisor with one decimal point in the denominator. The expression can be rewritten as $57.2 \div 11$ by moving each number one decimal place to the right to eliminate the decimal. The long division can be completed as $572 \div 11$ with a result of 52. Since there is one non-zero integer to the right of the decimal point in the problem, the final solution is 5.2.

In another example, $8 \div 0.16$ has a divisor with two decimal points in the denominator. The expression can be rewritten as $800 \div 16$ by moving each number two decimal places to the right to eliminate the decimal in the divisor. The long division can be completed with a result of 50.

Properties of Operations

Properties of operations exist to make calculations easier and solve problems for missing values. The following table summarizes commonly used properties of real numbers.

Property	Addition	Multiplication
Commutative	$a + b = b + a$	$a \times b = b \times a$
Associative	$(a + b) + c = a + (b + c)$	$(a \times b) \times c = a \times (bc)$
Identity	$a + 0 = a; 0 + a = a$	$a \times 1 = a; 1 \times a = a$
Inverse	$a + (-a) = 0$	$a \times \frac{1}{a} = 1; a \neq 0$
Distributive	$a(b + c) = ab + ac$	

The **commutative property of addition** states that the order in which numbers are added does not change the sum. Similarly, the **commutative property of multiplication** states that the order in which numbers are multiplied does not change the product. The **associative property of addition** and **multiplication** state that the grouping of numbers being added or multiplied does not change the sum or product, respectively. The commutative and associative properties are useful for performing calculations. For example, $(47 + 25) + 3$ is equivalent to $(47 + 3) + 25$, which is easier to calculate.

The **identity property of addition** states that adding zero to any number does not change its value. The **identity property of multiplication** states that multiplying a number by one does not change its value. The **inverse property of addition** states that the sum of a number and its opposite equals zero. Opposites are numbers that are the same with different signs (ex. 5 and -5; $\frac{1}{2}$ and -$\frac{1}{2}$). The **inverse property of multiplication** states that the product of a number (other than zero) and its reciprocal equals one. **Reciprocal numbers** have numerators and denominators that are inverted (ex. $\frac{2}{5}$ and $\frac{5}{2}$). Inverse properties are useful for canceling quantities to find missing values (see algebra content). For example, $a + 7 = 12$ is solved by adding the inverse of $7(-7)$ to both sides in order to isolate *a*.

The **distributive property states** that multiplying a sum (or difference) by a number produces the same result as multiplying each value in the sum (or difference) by the number and adding (or subtracting) the products. Consider the following scenario: You are buying three tickets for a baseball game. Each ticket costs \$18. You are also charged a fee of \$2 per ticket for purchasing the tickets online. The cost is calculated: $3 \times 18 + 3 \times 2$. Using the distributive property, the cost can also be calculated $3(18 + 2)$.

Conversions

To convert a fraction to a decimal, the numerator is divided by the denominator. For example, $\frac{3}{8}$ can be converted to a decimal by dividing 3 by 8 ($\frac{3}{8} = 0.375$). To convert a decimal to a fraction, the decimal point is dropped, and the value is written as the numerator. The denominator is the place value farthest to the right with a digit other than zero. For example, to convert .48 to a fraction, the numerator is 48 and the denominator is 100 (the digit 8 is in the hundredths place). Therefore, .48 = $\frac{48}{100}$. Fractions should be written in the simplest form, or reduced. To reduce a fraction, the numerator and denominator are divided by the largest common factor. In the previous example, 48 and 100 are both divisible by 4. Dividing the numerator and denominator by 4 results in a reduced fraction of $\frac{12}{25}$.

To convert a decimal to a percent, the number is multiplied by 100. To convert .13 to a percent, .13 is multiplied by 100 to get 13 percent. To convert a fraction to a percent, the fraction is converted to a decimal and then multiplied by 100. For example, $\frac{1}{5}$ = .20 and .20 multiplied by 100 produces 20 percent.

To convert a percent to a decimal, the value is divided by 100. For example, 125 percent is equal to 1.25 ($\frac{125}{100}$). To convert a percent to a fraction, the percent sign is dropped, and the value is written as the numerator with a denominator of 100. For example, 80% = $\frac{80}{100}$. This fraction can be reduced ($\frac{80}{100}$ = $\frac{4}{5}$).

Fraction Word Problems

One painter can paint a designated room in 6 hours, and a second painter can paint the same room in 5 hours. How long will it take them to paint the room if they work together?

The first painter paints $\frac{1}{6}$ of the room in an hour, and the second painter paints $\frac{1}{5}$ of the room in an hour.

Together, they can paint $\frac{1}{x}$ of the room in an hour. The equation is the sum of the painter's rate equal to the total job or $\frac{1}{6} + \frac{1}{5} = \frac{1}{x}$.

The equation can be solved by multiplying all terms by a common denominator of $30x$ with a result of $5x + 6x = 30$.

The left side can be added together to get $11x$, and then divide by 11 for a solution of $\frac{30}{11}$ or about 2.73 hours.

Squares, Square Roots, Cubes, and Cube Roots

A **root** is a different way to write an exponent when the exponent is the reciprocal of a whole number. We use the **radical** symbol to write this in the following way: $\sqrt[n]{a} = a^{\frac{1}{n}}$. This quantity is called the *n-th* **root** of *a*. The *n* is called the **index** of the radical.

Note that if the *n*-th root of *a* is multiplied by itself *n* times, the result will just be *a*. If no number *n* is written by the radical, it is assumed that *n* is 2: $\sqrt{5} = 5^{\frac{1}{2}}$. The special case of the 2nd root is called the **square root,** and the third root is called the **cube root**.

A **perfect square** is a whole number that is the square of another whole number. For example, 16 and 64 are perfect squares because 16 is the square of 4, and 64 is the square of 8.

Undefined Expressions

Expressions can be undefined when they involve dividing by zero or having a zero denominator. In simple fractions, the numerator and denominator can be nearly any integer. However, the denominator of a fraction can never be zero, because dividing by zero is a function which is undefined. Trying to take the square root of a negative number also yields an undefined result.

Unit Rates

Unit rate word problems will ask you to calculate the rate or quantity of something in a different value. For example, a problem might say that a car drove a certain number of miles in a certain number of minutes and then ask how many miles per hour the car was traveling. These questions involve solving proportions. Consider the following examples:

1) Alexandra made $96 during the first 3 hours of her shift as a temporary worker at a law office. She will continue to earn money at this rate until she finishes in 5 more hours. How much does Alexandra make per hour? How much will Alexandra have made at the end of the day?

This problem can be solved in two ways. The first is to set up a proportion, as the rate of pay is constant. The second is to determine her hourly rate, multiply the 5 hours by that rate, and then add the $96.

To set up a proportion, put the money already earned over the hours already worked on one side of an equation. The other side has x over 8 hours (the total hours worked in the day). It looks like this: $\frac{96}{3} = \frac{x}{8}$. Now, cross-multiply to get $768 = 3x$. To get x, divide by 3, which leaves $x = 256$. Alternatively, as *x* is the numerator of one of the proportions, multiplying by its denominator will reduce the solution by one step. Thus, Alexandra will make $256 at the end of the day. To calculate her hourly rate, divide the total by 8, giving $32 per hour.

Alternatively, it is possible to figure out the hourly rate by dividing $96 by 3 hours to get $32 per hour. Now her total pay can be figured by multiplying $32 per hour by 8 hours, which comes out to $256.

2) Jonathan is reading a novel. So far, he has read 215 of the 335 total pages. It takes Jonathan 25 minutes to read 10 pages, and the rate is constant. How long does it take Jonathan to read one page? How much longer will it take him to finish the novel? Express the answer in time.

To calculate how long it takes Jonathan to read one page, divide the 25 minutes by 10 pages to determine the page per minute rate. Thus, it takes 2.5 minutes to read one page.

Jonathan must read 120 more pages to complete the novel. (This is calculated by subtracting the pages already read from the total.) Now, multiply his rate per page by the number of pages. Thus, $120 \div 2.5 = 300$. Expressed in time, 300 minutes is equal to 5 hours.

3) At a hotel, $\frac{4}{5}$ of the 120 rooms are booked for Saturday. On Sunday, $\frac{3}{4}$ of the rooms are booked. On which day are more of the rooms booked, and by how many more?

The first step is to calculate the number of rooms booked for each day. Do this by multiplying the fraction of the rooms booked by the total number of rooms.

Saturday: $\frac{4}{5} \times 120 = \frac{4}{5} \times \frac{120}{1} = \frac{480}{5} = 96$ rooms

Sunday: $\frac{3}{4} \times 120 = \frac{3}{4} \times \frac{120}{1} = \frac{360}{4} = 90$ rooms

Thus, more rooms were booked on Saturday by 6 rooms.

4) In a veterinary hospital, the veterinarian-to-pet ratio is 1:9. The ratio is always constant. If there are 45 pets in the hospital, how many veterinarians are currently in the veterinary hospital?

Set up a proportion to solve for the number of veterinarians: $\frac{1}{9} = \frac{x}{45}$

Cross-multiplying results in $9x = 45$, which works out to 5 veterinarians.

Alternatively, as there are always 9 times as many pets as veterinarians, is it possible to divide the number of pets (45) by 9. This also arrives at the correct answer of 5 veterinarians.

5) At a general practice law firm, 30% of the lawyers work solely on tort cases. If 9 lawyers work solely on tort cases, how many lawyers work at the firm?

First, solve for the total number of lawyers working at the firm, which will be represented here with x. The problem states that 9 lawyers work solely on torts cases, and they make up 30% of the total lawyers at the firm. Thus, 30% multiplied by the total, x, will equal 9. Written as equation, this is: $30\% \times x = 9$.

It's easier to deal with the equation after converting the percentage to a decimal, leaving $0.3x = 9$. Thus, $x = \frac{9}{0.3} = 30$ lawyers working at the firm.

6) Xavier was hospitalized with pneumonia. He was originally given 35mg of antibiotics. Later, after his condition continued to worsen, Xavier's dosage was increased to 60mg. What was the percent increase of the antibiotics? Round the percentage to the nearest tenth.

An increase or decrease in percentage can be calculated by dividing the difference in amounts by the original amount and multiplying by 100. Written as an equation, the formula is:

$$\frac{new\ quantity\ - old\ quantity}{old\ quantity} \times 100$$

Here, the question states that the dosage was increased from 35mg to 60mg, so these are plugged into the formula to find the percentage increase.

$$\frac{60\ -\ 35}{35} \times 100 = \frac{25}{35} \times 100 = .7142 \times 100 = 71.4\%$$

Objects at Scale

Scale drawings are used in designs to model the actual measurements of a real-world object. For example, the blueprint of a house might indicate that it is drawn at a scale of 3 inches to 8 feet. Given one value and asked to determine the width of the house, a proportion should be set up to solve the problem. Given the scale of 3in:8ft and a blueprint width of 1 ft (12 in.), to find the actual width of the building, the proportion $\frac{3}{8} = \frac{12}{x}$ should be used. This results in an actual width of 32 ft.

The ratio between two similar geometric figures is called the **scale factor**. For example, a problem may depict two similar triangles, A and B. The scale factor from the smaller triangle A to the larger triangle B is given as 2 because the length of the corresponding side of the larger triangle, 16, is twice the corresponding side on the smaller triangle, 8. This scale factor can also be used to find the value of a missing side, x, in triangle A. Since the scale factor from the smaller triangle (A) to larger one (B) is 2, the larger corresponding side in triangle B (given as 25) can be divided by 2 to find the missing side in A (x = 12.5). The scale factor can also be represented in the equation $2A = B$ because two times the lengths of A gives the corresponding lengths of B. This is the idea behind similar triangles.

Multiple-Step Problems that Use Ratios, Proportions, and Percentages

Solving Real-World Problems Involving Ratios and Rates of Change

Ratios are used to show the relationship between two quantities. The ratio of oranges to apples in the grocery store may be 3 to 2. That means that for every 3 oranges, there are 2 apples. This comparison can be expanded to represent the actual number of oranges and apples. Another example may be the number of boys to girls in a math class. If the ratio of boys to girls is given as 2 to 5, that means there are 2 boys to every 5 girls in the class. Ratios can also be compared if the units in each ratio are the same. The ratio of boys to girls in the math class can be compared to the ratio of boys to girls in a science class by stating which ratio is higher and which is lower.

Rates are used to compare two quantities with different units. **Unit rates** are the simplest form of rate. With unit rates, the denominator in the comparison of two units is one. For example, if someone can type at a rate of 1000 words in 5 minutes, then their unit rate for typing is $\frac{1000}{5} = 200$ words in one minute or 200 words per minute. Any rate can be converted into a unit rate by dividing to make the denominator one. 1000 words in 5 minutes has been converted into the unit rate of 200 words per minute.

Ratios and rates can be used together to convert rates into different units. For example, if someone is driving 50 kilometers per hour, that rate can be converted into miles per hour by using a ratio known as the **conversion factor**. Since the given value contains kilometers and the final answer needs to be in miles, the ratio relating miles to kilometers needs to be used. There are 0.62 miles in 1 kilometer. This, written as a ratio and in fraction form, is $\frac{0.62\ miles}{1\ km}$. To convert 50km/hour into miles per hour, the following conversion needs to be set up:

$$\frac{50\ km}{hour} \times \frac{0.62\ miles}{1\ km} = 31\ miles\ per\ hour$$

When dealing with word problems, there is no fixed series of steps to follow, but there are some general guidelines to use. It is important that the quantity to be found is identified. Then, it can be determined how the given values can be used and manipulated to find the final answer.

Example: Jana wants to travel to visit Alice, who lives one hundred and fifty miles away. If she can drive at fifty miles per hour, how long will her trip take?

The quantity to find is the *time* of the trip. The time of a trip is given by the distance to travel divided by the speed to be traveled. The problem determines that the distance is one hundred and fifty miles, while the speed is fifty miles per hour. Thus, 150 divided by 50 is $150 \div 50 = 3$. Because *miles* and *miles per hour* are the units being divided, the miles cancel out. The result is 3 hours.

Example: Bernard wishes to paint a wall that measures twenty feet wide by eight feet high. It costs ten cents to paint one square foot. How much money will Bernard need for paint?

The final quantity to compute is the *cost* to paint the wall. This will be ten cents ($0.10) for each square foot of area needed to paint. The area to be painted is unknown, but the dimensions of the wall are given; thus, it can be calculated.

The dimensions of the wall are 20 feet wide and 8 feet high. Since the area of a rectangle is length multiplied by width, the area of the wall is $8 \times 20 = 160\ square\ feet$. Multiplying 0.1×160 yields $16 as the cost of the paint.

Solving Real-World Problems Involving Proportions

Much like a scale factor can be written using an equation like $2A = B$, a **relationship** is represented by the equation $Y = kX$. X and Y are proportional because as values of X increase, the values of Y also increase. A relationship that is inversely proportional can be represented by the equation $Y = \frac{k}{X}$, where the value of Y decreases as the value of x increases and vice versa.

Proportional reasoning can be used to solve problems involving ratios, percentages, and averages. Ratios can be used in setting up proportions and solving them to find unknowns. For example, if a student completes an average of 10 pages of math homework in 3 nights, how long would it take the student to complete 22 pages? Both ratios can be written as fractions. The second ratio would contain the unknown.

The following proportion represents this problem, where x is the unknown number of nights:

$$\frac{10\ pages}{3\ nights} = \frac{22\ pages}{x\ nights}$$

Solving this proportion entails cross-multiplying and results in the following equation: $10x = 22 \times 3$. Simplifying and solving for x results in the exact solution: $x = 6.6\ nights$. The result would be rounded up to 7 because the homework would actually be completed on the 7th night.

The following problem uses ratios involving percentages:

If 20% of the class is girls and 30 students are in the class, how many girls are in the class?

To set up this problem, it is helpful to use the common proportion: $\frac{\%}{100} = \frac{is}{of}$. Within the proportion, % is the percentage of girls, 100 is the total percentage of the class, *is* is the number of girls, and *of* is the total number of students in the class. Most percentage problems can be written using this language. To solve this problem, the proportion should be set up as $\frac{20}{100} = \frac{x}{30}$, and then solved for x. Cross-multiplying results in the equation $20 \times 30 = 100x$, which results in the solution $x = 6$. There are 6 girls in the class.

Problems involving volume, length, and other units can also be solved using ratios. For example, a problem may ask for the volume of a cone to be found that has a radius:

$$r = 7m$$

and a height:

$$h = 16m$$

Referring to the formulas provided on the test, the volume of a cone is given as: $V = \pi r^2 \frac{h}{3}$, where r is the radius, and h is the height. Plugging $r = 7$ and $h = 16$ into the formula, the following is obtained:

$$V = \pi(7^2)\frac{16}{3}$$

Therefore, the volume of the cone is found to be $821m^3$. Sometimes, answers in different units are sought. If this problem wanted the answer in liters, 821m^3 would need to be converted. Using the equivalence statement $1\text{m}^3 = 1000\text{L}$, the following ratio would be used to solve for liters:

$$821\text{m}^3 \times \frac{1000L}{1m^3}$$

Cubic meters in the numerator and denominator cancel each other out, and the answer is converted to 821,000 liters, or 8.21×10^5 L.

Other conversions can also be made between different given and final units. If the temperature in a pool is 30°C, what is the temperature of the pool in degrees Fahrenheit? To convert these units, an equation is used relating Celsius to Fahrenheit. The following equation is used:

$$T_{°F} = 1.8T_{°C} + 32$$

Plugging in the given temperature and solving the equation for T yields the result:

$$T_{°F} = 1.8(30) + 32 = 86°F$$

Units in both the metric system and U.S. customary system are widely used.

Here are some more examples of how to solve for proportions:

1) $\frac{75\%}{90\%} = \frac{25\%}{x}$

To solve for x, the fractions must be cross multiplied: ($75\%x = 90\% \times 25\%$). To make things easier, let's convert the percentages to decimals: ($0.9 \times 0.25 = 0.225 = 0.75x$). To get rid of x's co-efficient, each side must be divided by that same coefficient to get the answer $x = 0.3$. The question could ask for the answer as a percentage or fraction in lowest terms, which are 30% and $\frac{3}{10}$, respectively.

2) $\frac{x}{12} = \frac{30}{96}$

Cross-multiply: $96x = 30 \times 12$

Multiply: $96x = 360$

Divide: $x = 360 \div 96$

Answer: $x = 3.75$

3) $\frac{0.5}{3} = \frac{x}{6}$

Cross-multiply: $3x = 0.5 \times 6$

Multiply: $3x = 3$

Divide: $x = 3 \div 3$

Answer: $x = 1$

You may have noticed there's a faster way to arrive at the answer. If there is an obvious operation being performed on the proportion, the same operation can be used on the other side of the proportion to solve for x. For example, in the first practice problem, 75% became 25% when divided by 3, and upon doing the same to 90%, the correct answer of 30% would have been found with much less legwork. However, these questions aren't always so intuitive, so it's a good idea to work through the steps, even if the answer seems apparent from the outset.

Solving Real-World Problems Involving Percentages

Questions dealing with percentages can be difficult when they are phrased as word problems. These word problems almost always come in three varieties. The first type will ask to find what percentage of some number will equal another number. The second asks to determine what number is some percentage of another given number. The third will ask what number another number is a given percentage of.

One of the most important parts of correctly answering percentage word problems is to identify the numerator and the denominator. This fraction can then be converted into a percentage, as described above.

The following word problem shows how to make this conversion: A department store carries several different types of footwear. The store is currently selling 8 athletic shoes, 7 dress shoes, and 5 sandals. What percentage of the store's footwear are sandals?

First, calculate what serves as the **whole**, as this will be the denominator. How many total pieces of footwear does the store sell? The store sells 20 different types ($8\ athletic + 7\ dress + 5\ sandals$). Second, what footwear type is the question specifically asking about? Sandals. Thus, 5 is the numerator. Third, the resultant fraction must be expressed as a percentage. The first two steps indicate that $\frac{5}{20}$ of the footwear pieces are sandals. This fraction must now be converted into a percentage:

$$\frac{5}{20} \times \frac{5}{5} = \frac{25}{100} = 25\%$$

Measurement and Geometry

Side Lengths of Shapes When Given the Area or Perimeter

The **perimeter** of a polygon is the distance around the outside of the two-dimensional figure or the sum of the lengths of all the sides. Perimeter is a one-dimensional measurement and is therefore expressed in linear units such as centimeters (*cm*), feet (*ft*), and miles (*mi*). The perimeter (*P*) of a figure can be calculated by adding together each of the sides.

Properties of certain polygons allow that the perimeter may be obtained by using formulas. A regular polygon is one in which all sides have equal length and all interior angles have equal measures, such as a square and an equilateral triangle. To find the perimeter of a regular polygon, the length of one side is multiplied by the number of sides.

A rectangle consists of two sides called the length (*l*), which have equal measures, and two sides called the width (*w*), which have equal measures. Therefore, the perimeter (*P*) of a rectangle can be expressed as $P = l + l + w + w$. This can be simplified to produce the following formula to find the perimeter of a rectangle: $P = 2l + 2w$ or $P = 2(l + w)$.

The perimeter of a square is measured by adding together all of the sides. Since a square has four equal sides, its perimeter can be calculated by multiplying the length of one side by 4. Thus, the formula is $P = 4 \times s$, where s equals one side. For example, the following square has side lengths of 5 meters:

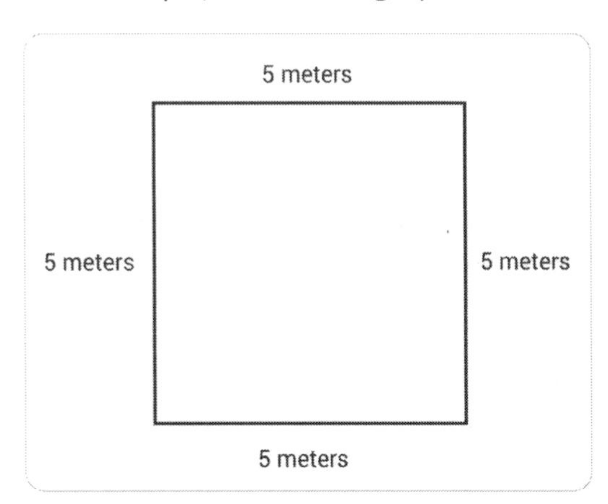

The perimeter is 20 meters because 4 times 5 is 20.

Like a square, a rectangle's perimeter is measured by adding together all of the sides. But as the sides are unequal, the formula is different. A rectangle has equal values for its lengths (long sides) and equal values for its widths (short sides), so the perimeter formula for a rectangle is:

$$P = l + l + w + w = 2l + 2w$$

l equals length
w equals width

Consider the following problem:

The total perimeter of a rectangular garden is 36m. If the length of each side is 12m, what is the width?

The formula for the perimeter of a rectangle is $P = 2L + 2W$, where P is the perimeter, L is the length, and W is the width. The first step is to substitute all of the data into the formula:

$$36 = 2(12) + 2W$$

Simplify by multiplying 2×12:

$$36 = 24 + 2W$$

Simplify this further by subtracting 24 on each side, which gives:

$$36 - 24 = 24 - 24 + 2W$$

$$12 = 2W$$

Divide by 2:

$$6 = W$$

The width is 6m. Remember to test this answer by substituting this value into the original formula:

$$36 = 2(12) + 2(6)$$

A triangle's perimeter is measured by adding together the three sides, so the formula is $P = a + b + c$, where $a, b,$ and c are the values of the three sides. The area is calculated by multiplying the length of the base times the height times ½, so the formula is:

$$A = \frac{1}{2} \times b \times h = \frac{bh}{2}$$

The base is the bottom of the triangle, and the height is the distance from the base to the peak. If a problem asks to calculate the area of a triangle, it will provide the base and height.

A circle's perimeter—also known as its circumference—is measured by multiplying the diameter (the straight line measured from one end to the direct opposite end of the circle) by π, so the formula is $\pi \times d$. This is sometimes expressed by the formula $C = 2 \times \pi \times r$, where r is the radius of the circle. These formulas are equivalent, as the radius equals half of the diameter.

Missing side lengths can be determined using subtraction. For example, if you are told that a triangle has a perimeter of 34 inches and that one side is 12 inches, another side is 16 inches, and the third side is unknown, you can calculate the length of that unknown side by setting up the following subtraction problem:

$$34\ inches = 12\ inches + 16\ inches + x$$

$$34\ inches = 28\ inches + x$$

$$6\ inches = x$$

Therefore, the missing side length is 6 inches.

Area and Perimeter of Two-Dimensional Shapes

As mentioned, the **perimeter** of a polygon is the distance around the outside of the two-dimensional figure. Perimeter is a one-dimensional measurement and is therefore expressed in linear units such as centimeters (*cm*), feet (*ft*), and miles (*mi*). The perimeter (*P*) of a figure can be calculated by adding together each of the sides.

Properties of certain polygons allow that the perimeter may be obtained by using formulas. A rectangle consists of two sides called the length (*l*), which have equal measures, and two sides called the width (*w*), which have equal measures. Therefore, the perimeter (*P*) of a rectangle can be expressed as:

$$P = l + l + w + w$$

This can be simplified to produce the following formula to find the perimeter of a rectangle:

$$P = 2l + 2w \text{ or } P = 2(l + w)$$

A regular polygon is one in which all sides have equal length and all interior angles have equal measures, such as a square and an equilateral triangle. To find the perimeter of a regular polygon, the length of one side is multiplied by the number of sides. For example, to find the perimeter of an equilateral triangle with a side of length of 4 feet, 4 feet is multiplied by 3 (number of sides of a triangle). The perimeter of a regular octagon (8 sides) with a side of length of $\frac{1}{2}$ cm is:

$$\frac{1}{2}cm \times 8 = 4cm$$

The **area** of a polygon is the number of square units needed to cover the interior region of the figure. Area is a two-dimensional measurement. Therefore, area is expressed in square units, such as square centimeters (cm^2), square feet (ft^2), or square miles (mi^2). Regarding the area of a rectangle with sides of length *x* and *y*, the area is given by xy. For a triangle with a base of length *b* and a height of length *h*, the area is $\frac{1}{2}bh$. To find the area (*A*) of a parallelogram, the length of the base (*b*) is multiplied by the length of the height (*h*) $\rightarrow A = b \times h$. Similar to triangles, the height of the parallelogram is measured from one base to the other at a 90° angle (or perpendicular).

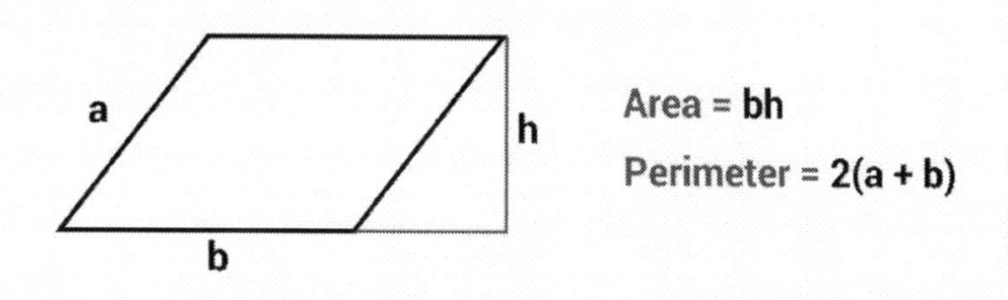

The area of a trapezoid can be calculated using the formula: $A = \frac{1}{2} \times h(b_1 + b_2)$, where *h* is the height and b_1 and b_2 are the parallel bases of the trapezoid.

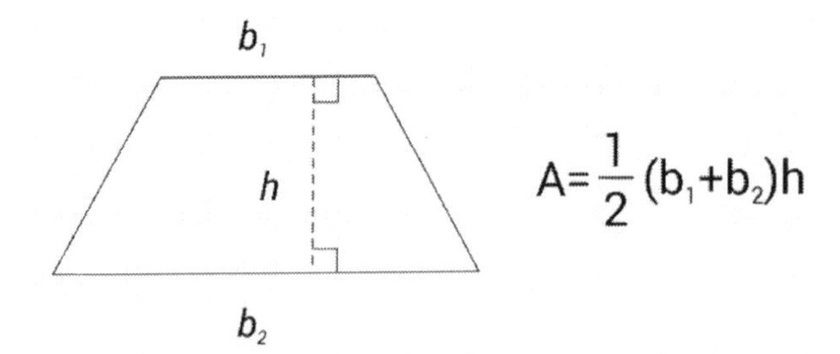

The area of a regular polygon can be determined by using its perimeter and the length of the **apothem**. The apothem is a line from the center of the regular polygon to any of its sides at a right angle. (Note that the perimeter of a regular polygon can be determined given the length of only one side.) The formula for the area (*A*) of a regular polygon is $A = \frac{1}{2} \times a \times P$, where *a* is the length of the apothem and *P* is the perimeter of the figure. Consider the following regular pentagon:

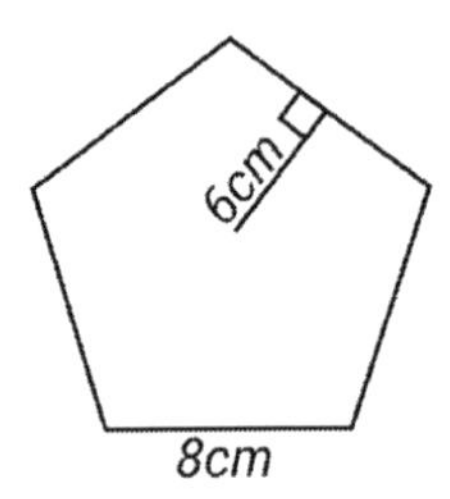

To find the area, the perimeter (*P*) is calculated first: $8cm \times 5 \rightarrow P = 40cm$. Then the perimeter and the apothem are used to find the area (*A*):

$$A = \frac{1}{2} \times a \times P$$

$$A = \frac{1}{2} \times (6cm) \times (40cm)$$

$$A = 120cm^2$$

Note that the unit is:

$$cm^2 \rightarrow cm \times cm = cm^2$$

The area of irregular polygons is found by decomposing, or breaking apart, the figure into smaller shapes. When the area of the smaller shapes is determined, the area of the smaller shapes will produce the area of the original figure when added together. Consider the example below:

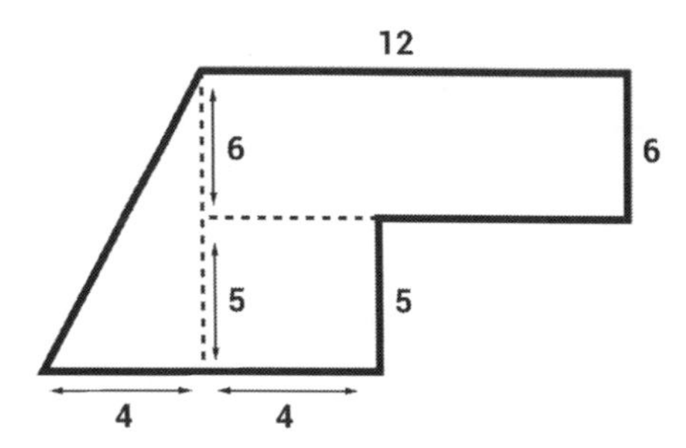

The irregular polygon is decomposed into two rectangles and a triangle. The area of the large rectangles ($A = l \times w \rightarrow A = 12 \times 6$) is 72 square units. The area of the small rectangle is 20 square units:

$$A = 4 \times 5$$

The area of the triangle:

$$A = \frac{1}{2} \times b \times h$$

$$A = \frac{1}{2} \times 4 \times 11$$

22 square units

The sum of the areas of these figures produces the total area of the original polygon:

$$A = 72 + 20 + 22$$

A = 114 square units

The perimeter (*P*) of the figure below is calculated by: $P = 9m + 5m + 4m + 6m + 8m \rightarrow P = 32\ m.$

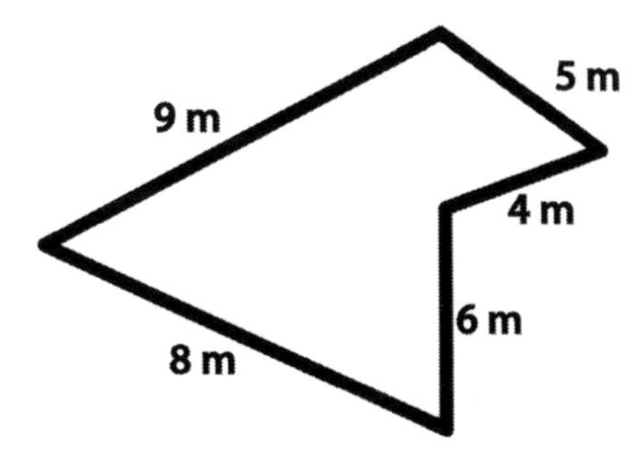

Area, Circumference, Radius, and Diameter of a Circle

A **circle** can be defined as the set of all points that are the same distance (known as the **radius**, *r*) from a single point (known as the **center** of the circle). The center has coordinates (h, k), and any point on the circle can be labelled with coordinates (x, y).

The **circumference** of a circle is the distance traveled by following the edge of the circle for one complete revolution, and the length of the circumference is given by $2\pi r$, where *r* is the radius of the circle. The formula for circumference is $C = 2\pi r$.

The area of a circle is calculated through the formula $A = \pi \times r^2$. The test will indicate either to leave the answer with π attached or to calculate to the nearest decimal place, which means multiplying by 3.14 for π.

Given two points on the circumference of a circle, the path along the circle between those points is called an **arc** of the circle. For example, the arc between *B* and *C* is denoted by a thinner line:

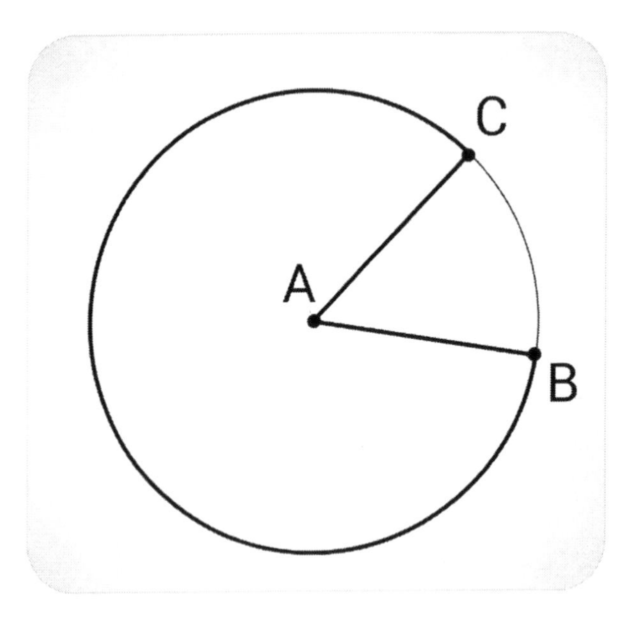

The length of the path along an arc is called the **arc length**. If the circle has radius *r*, then the arc length is given by multiplying the measure of the angle in radians by the radius of the circle.

Pythagorean Theorem

The Pythagorean theorem is an important result in geometry. It states that for right triangles, the sum of the squares of the two shorter sides will be equal to the square of the longest side (also called the **hypotenuse**). The longest side will always be the side opposite to the 90° angle. If this side is called *c*, and the other two sides are *a* and *b*, then the Pythagorean theorem states that $c^2 = a^2 + b^2$. Since lengths are always positive, this also can be written as:

$$c = \sqrt{a^2 + b^2}$$

A diagram to show the parts of a triangle using the Pythagorean theorem is below.

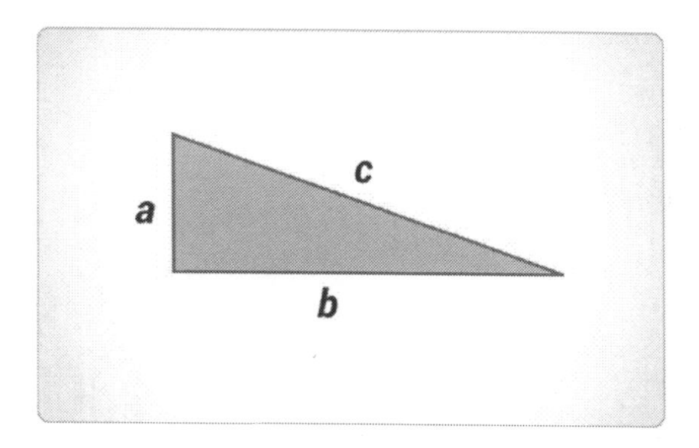

As an example of the theorem, suppose that Shirley has a rectangular field that is 5 feet wide and 12 feet long, and she wants to split it in half using a fence that goes from one corner to the opposite corner. How long will this fence need to be? To figure this out, note that this makes the field into two right triangles, whose hypotenuse will be the fence dividing it in half. Therefore, the fence length will be given by $\sqrt{5^2 + 12^2} = \sqrt{169} = 13$ feet long.

Volume and Surface Area of Three-Dimensional Shapes

Geometry in three dimensions is similar to geometry in two dimensions. The main new feature is that three points now define a unique **plane** that passes through each of them. Three-dimensional objects can be made by putting together two-dimensional figures in different surfaces. Below, some of the possible three-dimensional figures will be provided, along with formulas for their volumes and surface areas.

Volume is the measurement of how much space an object occupies, like how much space is in the cube. Volume questions will ask how much of something is needed to completely fill the object. The most common surface area and volume questions deal with spheres, cubes, and rectangular prisms.

Surface area of a three-dimensional figure refers to the number of square units needed to cover the entire surface of the figure. This concept is similar to using wrapping paper to completely cover the outside of a box. For example, if a triangular pyramid has a surface area of 17 square inches (written $17in^2$), it will take 17 squares, each with sides one inch in length, to cover the entire surface of the pyramid. Surface area is also measured in square units.

A **rectangular prism** is a box whose sides are all rectangles meeting at 90° angles. Such a box has three dimensions: length, width, and height. If the length is *x*, the width is *y*, and the height is *z*, then the volume is given by $V = xyz$.

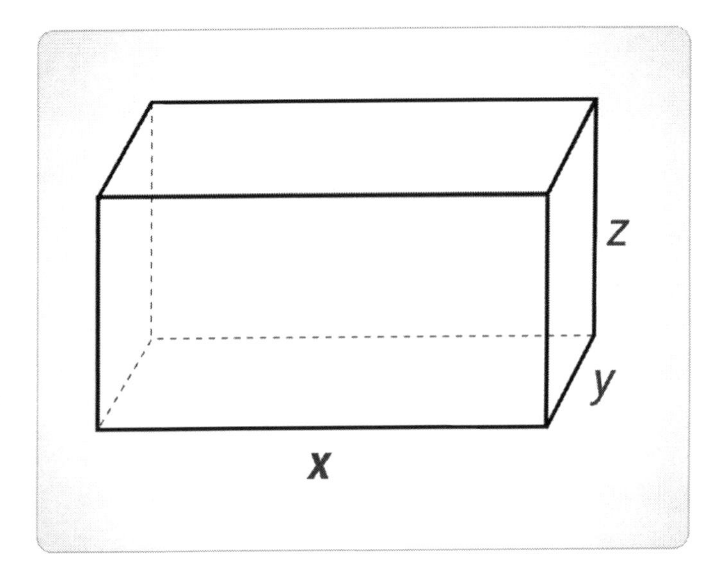

The **surface area** will be given by computing the surface area of each rectangle and adding them together. There is a total of six rectangles. Two of them have sides of length *x* and *y*, two have sides of length *y* and *z*, and two have sides of length *x* and *z*. Therefore, the total surface area will be given by:

$$SA = 2xy + 2yz + 2xz$$

A **cube** is a special type of rectangular solid in which its length, width, and height are the same. If this length is s, then the formula for the volume of a cube is $V = s \times s \times s$. The surface area of a cube is $SA = 6s^2$.

A **rectangular pyramid** is a figure with a rectangular base and four triangular sides that meet at a single vertex. If the rectangle has sides of length *x* and *y*, then the volume will be given by $V = \frac{1}{3}xyh$.

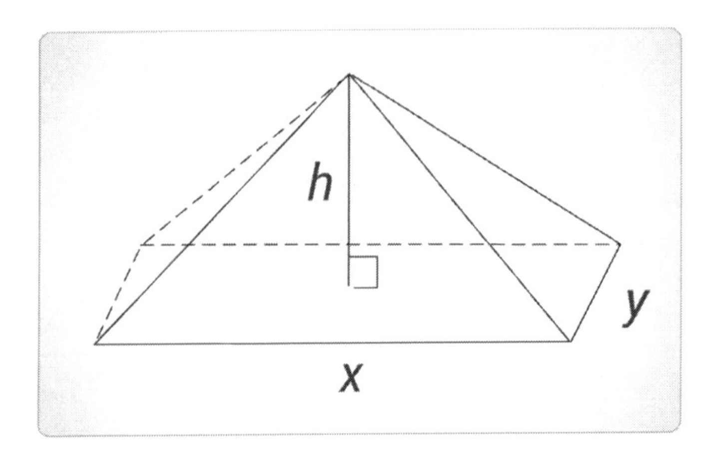

Many three-dimensional figures (solid figures) can be represented by nets consisting of rectangles and triangles. The surface area of such solids can be determined by adding the areas of each of its faces and bases. Finding the surface area using this method requires calculating the areas of rectangles and triangles. To find the area (*A*) of a rectangle, the length (*l*) is multiplied by the width $(w) \rightarrow A = l \times w$. The area of a rectangle with a length of 8cm and a width of 4cm is calculated:

$$A = (8cm) \times (4cm) \rightarrow A = 32cm^2$$

To calculate the area (*A*) of a triangle, the product of $\frac{1}{2}$, the base (*b*), and the height (*h*) is found:

$$A = \frac{1}{2} \times b \times h$$

Note that the height of a triangle is measured from the base to the vertex opposite of it forming a right angle with the base. The area of a triangle with a base of 11cm and a height of 6cm is calculated:

$$A = \frac{1}{2} \times (11cm) \times (6cm)$$

$$A = 33cm^2$$

Consider the following triangular prism, which is represented by a net consisting of two triangles and three rectangles.

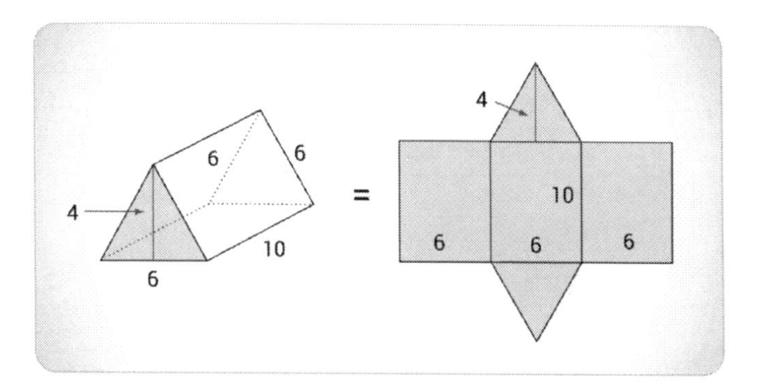

The surface area of the prism can be determined by adding the areas of each of its faces and bases. The surface area (*SA*) = area of triangle + area of triangle + area of rectangle + area of rectangle + area of rectangle.

$$SA = \left(\frac{1}{2} \times b \times h\right) + \left(\frac{1}{2} \times b \times h\right) + (l \times w) + (l \times w) + (l \times w)$$

$$SA = \left(\frac{1}{2} \times 6 \times 4\right) + \left(\frac{1}{2} \times 6 \times 4\right) + (6 \times 10) + (6 \times 10) + (6 \times 10)$$

$$SA = (12) + (12) + (60) + (60) + (60)$$

$$SA = 204 \textit{ square units}$$

A **sphere** is a set of points all of which are equidistant from some central point. It is like a circle, but in three dimensions. The volume of a sphere of radius *r* is given by:

$$V = \frac{4}{3}\pi r^3$$

The surface area is given by $A = 4\pi r^2$.

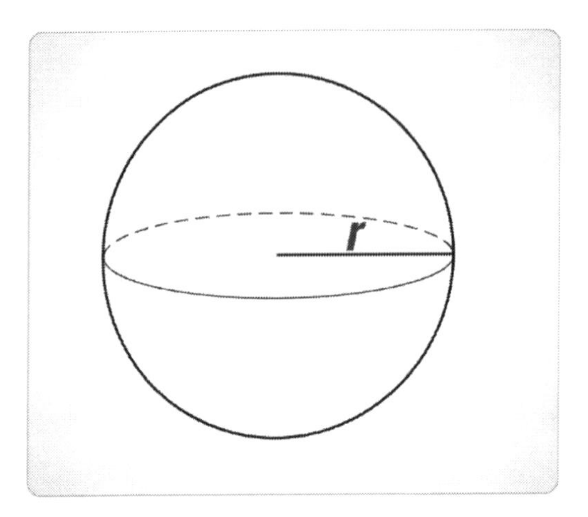

The volume of a **cylinder** is then found by adding a third dimension onto the circle. Volume of a cylinder is calculated by multiplying the area of the base (which is a circle) by the height of the cylinder. Doing so results in the equation $V = \pi r^2 h$. The volume of a **cone** is $\frac{1}{3}$ of the volume of a cylinder. Therefore, the formula for the volume of a **cone** is:

$$\frac{1}{3}\pi r^2 h$$

Solving Three-Dimensional Problems

Three-dimensional objects can be simplified into related two-dimensional shapes to solve problems. This simplification can make problem-solving a much easier experience. An isometric representation of a three-dimensional object can be completed so that important properties (e.g., shape, relationships of faces and surfaces) are noted. Edges and vertices can be translated into two-dimensional objects as well.

For example, below is a three-dimensional object that's been partitioned into two-dimensional representations of its faces:

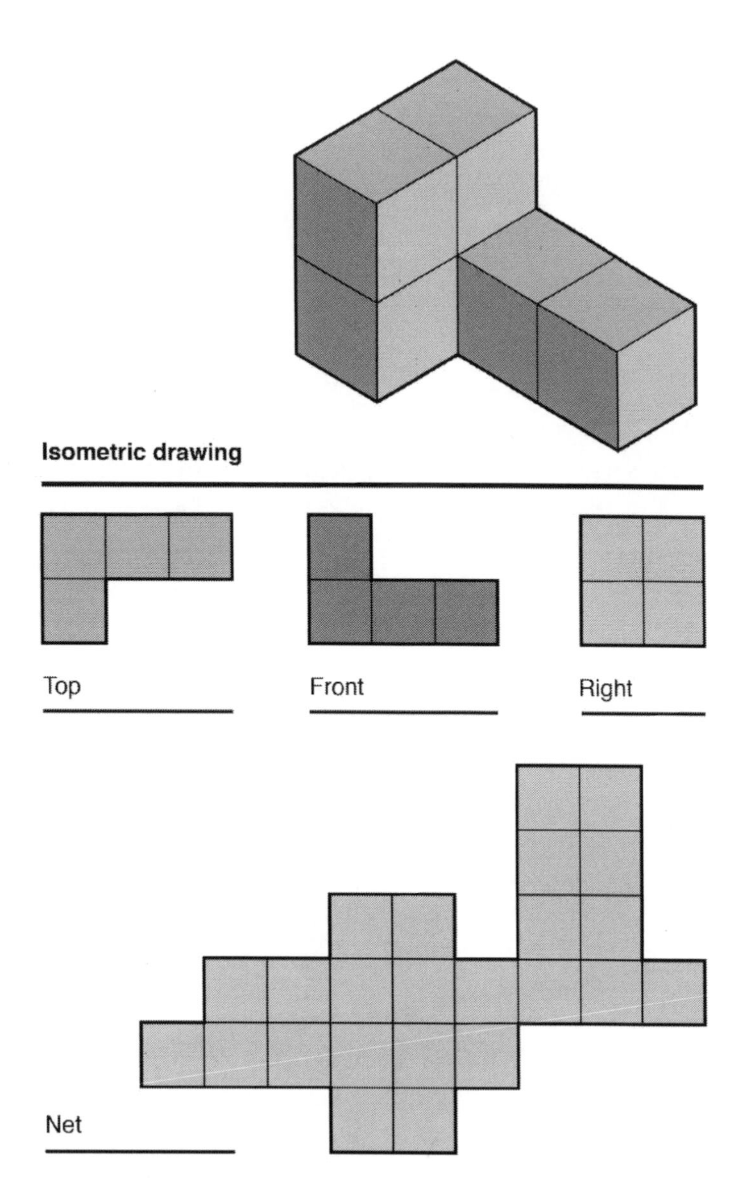

The net represents the sum of the three different faces. Depending on the problem, using a smaller portion of the given shape may be helpful, by simplifying the steps necessary to solve.

Many objects in the real world consist of three-dimensional shapes such as prisms, cylinders, and spheres. Surface area problems involve quantifying the outside area of such a three-dimensional object, and volume problems involve quantifying how much space the object takes up. Surface area of a prism is the sum of the areas, which is simplified into $SA = 2A + Bh$, where A is the area of the base, B is the perimeter of the base, and h is the height of the prism. The volume of the same prism is $V = Ah$. The surface area of a cylinder is equal to the sum of the areas of each end and the side, which is:

$$SA = 2\pi rh + 2\pi r^2$$

and its volume is:

$$V = \pi r^2 h$$

Finally, the surface area of a sphere is $SA = 4\pi r^2$, and its volume is $V = \frac{4}{3}\pi r^3$.

An example when one of these formulas should be used would be when calculating how much paint is needed for the outside of a house. In this scenario, surface area must be used. The sum of all individual areas of each side of the house must be found. Also, when calculating how much water a cylindrical tank can hold, a volume formula is used. Therefore, the amount of water that a cylindrical tank that is 8 feet tall with a radius of 3 feet is:

$$\pi \times 3^2 \times 8 = 226.1 \text{ cubic feet}$$

The formula used to calculate the volume of a cone is $\frac{1}{3}\pi r^2 h$. Essentially, the area of the base of the cone is multiplied by the cone's height. In a real-life example where the radius of a cone is 2 meters and the height of a cone is 5 meters, the volume of the cone is calculated by utilizing the formula $\frac{1}{3}\pi 2^2 \times 5$. After substituting 3.14 for π, the volume is $20.9\ m^3$.

Data Analysis, Statistics, and Probability

Graphical Data Including Graphs, Tables, and More

A set of data can be visually displayed in various forms allowing for quick identification of characteristics of the set. **Histograms**, such as the one shown below, display the number of data points (vertical axis) that fall into given intervals (horizontal axis) across the range of the set. The histogram below displays the heights of black cherry trees in a certain city park. Each rectangle represents the number of trees with heights between a given five-point span. For example, the furthest bar to the right indicates that two trees are between 85 and 90 feet. Histograms can describe the center, spread, shape, and any unusual characteristics of a data set.

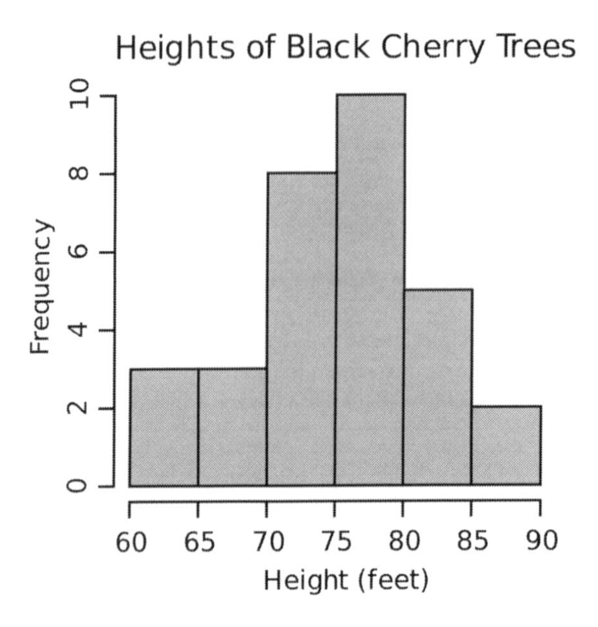

A **box plot**, also called a **box-and-whisker plot**, divides the data points into four groups and displays the five-number summary for the set as well as any outliers. The five-number summary consists of:

- The lower extreme: the lowest value that is not an outlier
- The higher extreme: the highest value that is not an outlier
- The median of the set: also referred to as the second quartile or Q_2
- The first quartile or Q_1: the median of values below Q_2
- The third quartile or Q_3: the median of values above Q_2

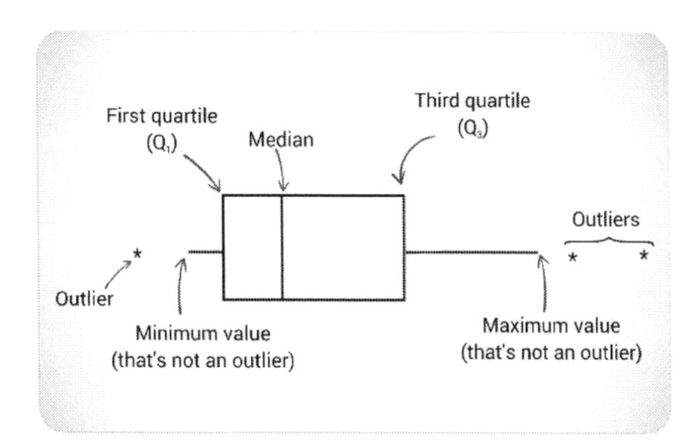

Suppose the box plot displays IQ scores for 12th grade students at a given school. The five-number summary of the data consists of: lower extreme (67); upper extreme (127); Q_2or median (100); Q_1 (91); Q_3 (108); and outliers (135 and 140). Although all data points are not known from the plot, the points are divided into four quartiles each, including 25% of the data points. Therefore, 25% of students scored between 67 and 91, 25% scored between 91 and 100, 25% scored between 100 and 108, and 25% scored between 108 and 127. These percentages include the normal values for the set and exclude the outliers. This information is useful when comparing a given score with the rest of the scores in the set.

A **scatter plot** is a mathematical diagram that visually displays the relationship or connection between two variables. The independent variable is placed on the *x*-axis, or horizontal axis, and the dependent variable is placed on the *y*-axis, or vertical axis. When visually examining the points on the graph, if the points model a linear relationship, or a **line of best fit** can be drawn through the points with the points relatively close on either side, then a correlation exists. If the line of best fit has a positive slope (rises from left to right), then the variables have a positive correlation. If the line of best fit has a negative slope (falls from left to right), then the variables have a negative correlation. If a line of best fit cannot be drawn, then no correlation exists. A positive or negative correlation can be categorized as strong or weak, depending on how closely the points are graphed around the line of best-fit.

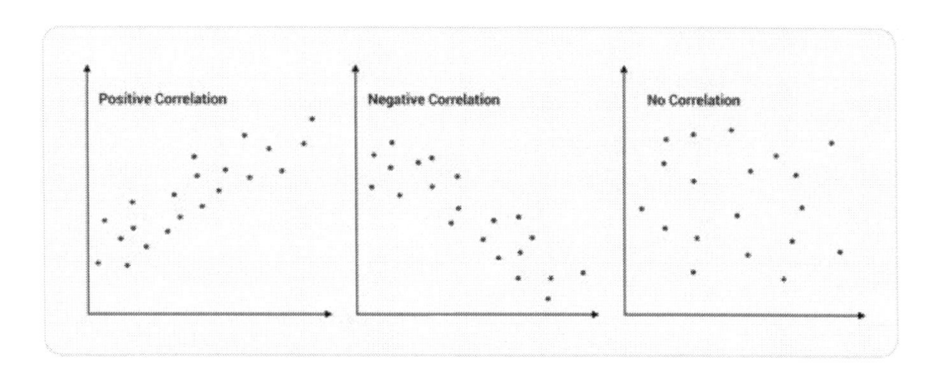

Like a scatter plot, a **line graph** compares variables that change continuously, typically over time. Paired data values (ordered pairs) are plotted on a coordinate grid with the x- and y-axis representing the variables. A line is drawn from each point to the next, going from left to right. The line graph below displays cell phone use for given years (two variables) for men, women, and both sexes (three data sets).

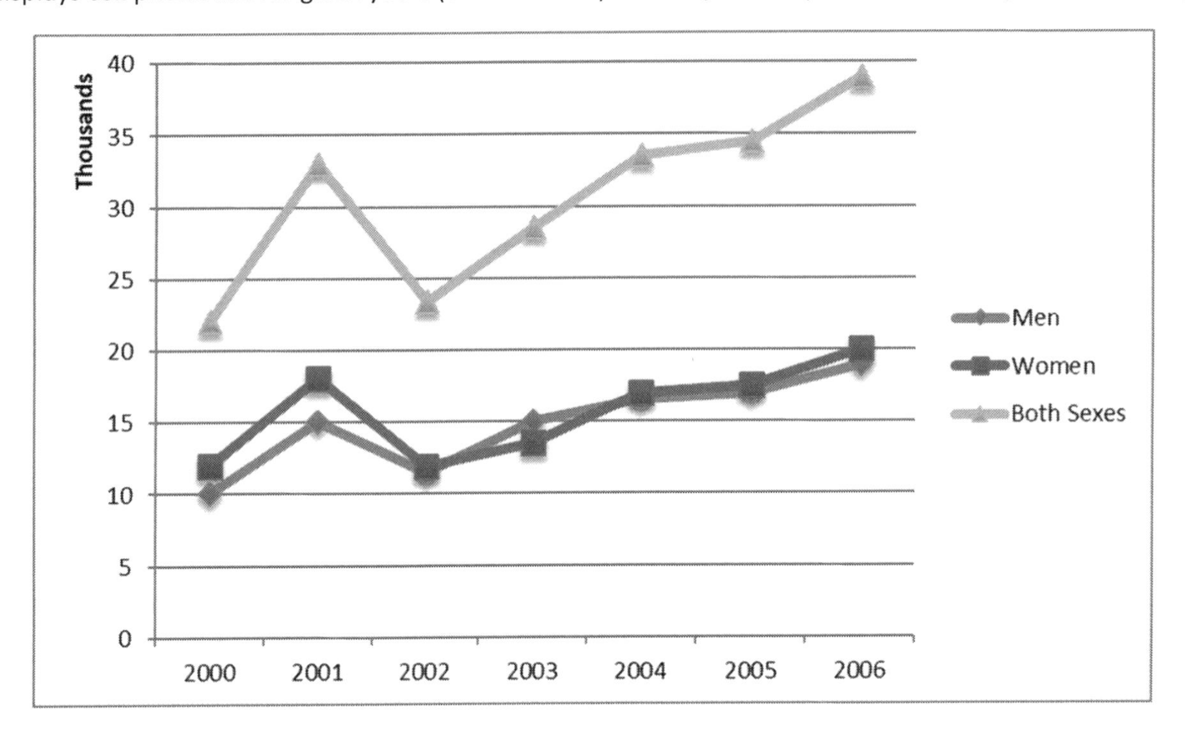

A **line plot**, also called **dot plot**, displays the frequency of data (numerical values) on a number line. To construct a line plot, a number line is used that includes all unique data values. It is marked with x's or dots above the value the number of times that the value occurs in the data set.

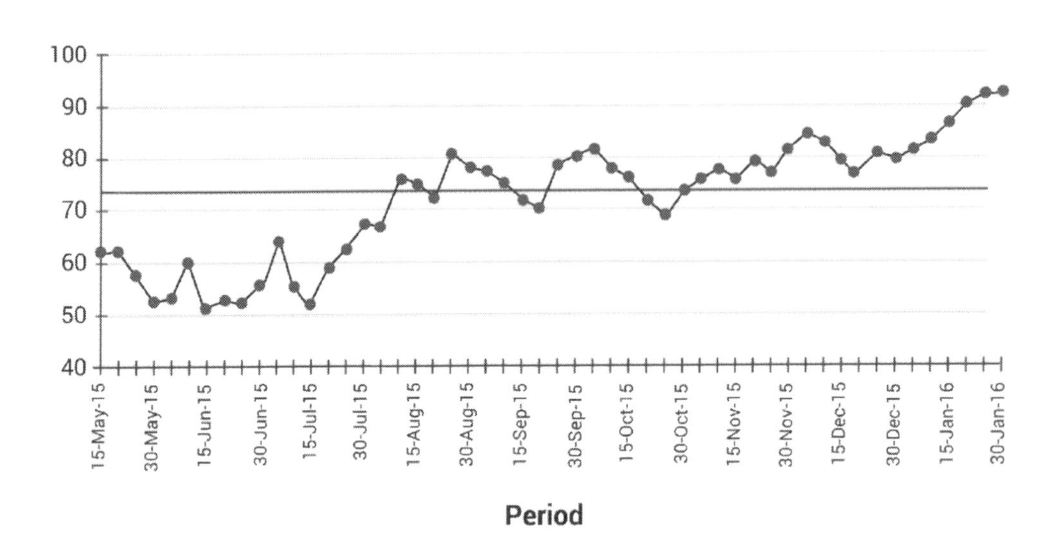

A **bar graph** is a diagram in which the quantity of items within a specific classification is represented by the height of a rectangle. Each type of classification is represented by a rectangle of equal width.

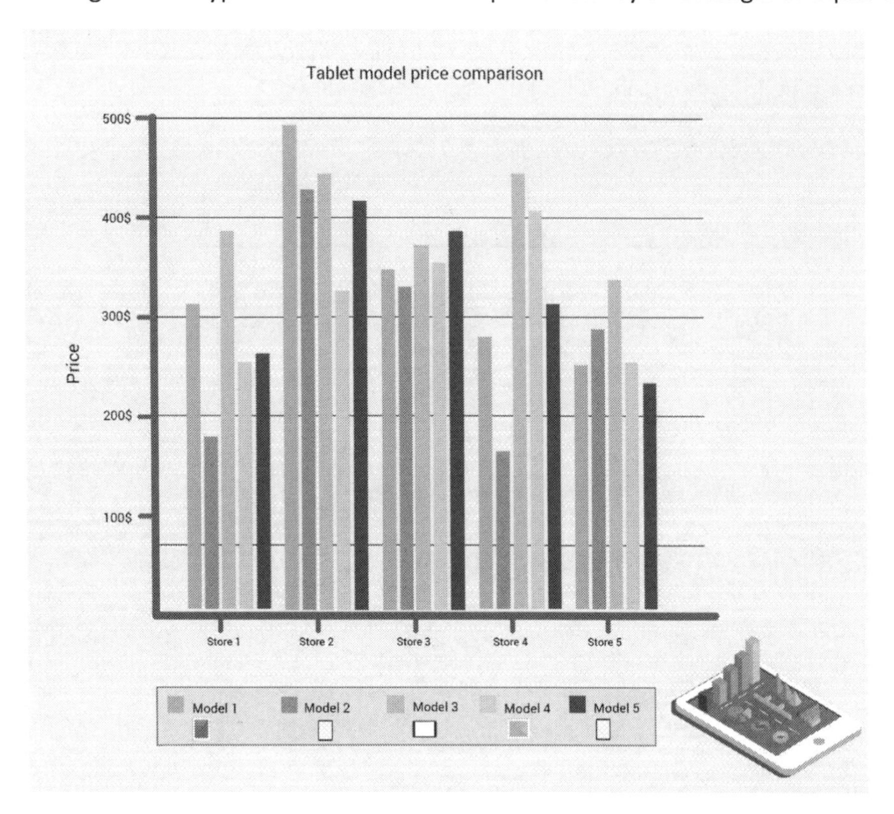

A **circle graph**, also called a **pie chart**, shows categorical data with each category representing a percentage of the whole data set. To make a circle graph, the percent of the data set for each category must be determined. To do so, the frequency of the category is divided by the total number of data points and converted to a percent. For example, if 80 people were asked what their favorite sport is and 20 responded basketball, basketball makes up 25% of the data ($\frac{20}{80}$ =.25=25%). Each category in a data set is represented by a slice of the circle proportionate to its percentage of the whole.

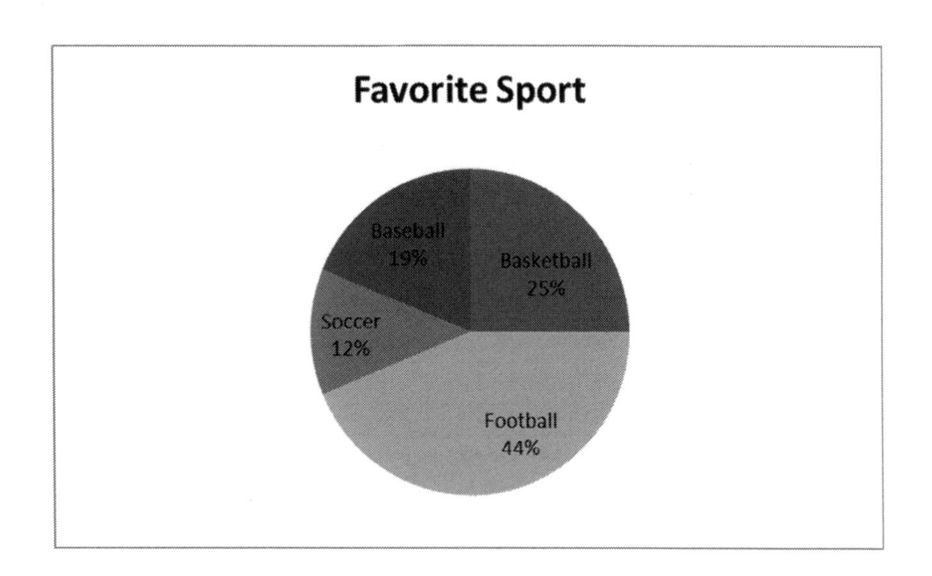

A **stem-and-leaf plot** is a method of displaying sets of data by organizing numbers by their stems (usually the tens digit) and different leaf values (usually the ones digit).

For example, to organize a number of movie critic's ratings, as listed below, a stem-and-leaf plot could be utilized to display the information in a more condensed manner.

Movie critic scores: 47, 52, 56, 59, 61, 64, 66, 68, 68, 70, 73, 75, 79, 81, 83, 85, 86, 88, 88, 89, 90, 90, 91, 93, 94, 96, 96, 99.

	Movie Ratings
4	7
5	2 6 9
6	1 4 6 8 8
7	0 3 5 9
8	1 3 5 6 8 8 9
9	0 0 1 3 4 6 6 9
Key	6 | 1 represents 61

Looking at this stem and leaf plot, it is easy to ascertain key features of the data set. For example, what is the range of the data in the stem-and-leaf plot?

Using this method, it is easier to visualize the distribution of the scores and answer the question pertaining to the range of scores, which is:

$$99 - 47 = 52$$

A **tally chart** is a diagram in which tally marks are utilized to represent data. Tally marks are a means of showing a quantity of objects within a specific classification. Here is an example of a tally chart:

Number of days with rain	Number of weeks
0	II
1	~~IIII~~
2	~~IIII~~
3	~~IIII~~
4	~~IIII~~ ~~IIII~~ ~~IIII~~ IIII
5	~~IIII~~ I
6	~~IIII~~ I
7	IIII

Data is often recorded using fractions, such as half a mile, and understanding fractions is critical because of their popular use in real-world applications. Also, it is extremely important to label values with their units when using data. For example, regarding length, the number 2 is meaningless unless it is attached to a unit. Writing 2 cm shows that the number refers to the length of an object.

Mean, Median, Mode, and Range

Suppose that X is a set of data points $(x_1, x_2, x_3, \dots x_n)$, and some description of the general properties of this data need to be found.

The first property that can be defined for this set of data is the **mean**. To find the mean, add up all the data points, then divide by the total number of data points. This can be expressed using **summation notation** as:

$$\bar{X} = \frac{x_1 + x_2 + x_3 + \cdots + x_n}{n} = \frac{1}{n}\sum_{i=1}^{n} x_i$$

For example, suppose that in a class of 10 students, the scores on a test were 50, 60, 65, 65, 75, 80, 85, 85, 90, 100. Therefore, the average test score will be:

$$\frac{1}{10}(50 + 60 + 65 + 65 + 75 + 80 + 85 + 85 + 90 + 100) = 75.5$$

The mean is a useful number if the distribution of data is normal (more on this later), which roughly means that the frequency of different outcomes has a single peak and is roughly equally distributed on both sides of that peak. However, it is less useful in some cases where the data might be split or where there are some outliers. **Outliers** are data points that are far from the rest of the data. For example, suppose there are 90 employees and 10 executives at a company. The executives make $1000 per hour, and the employees make $10 per hour. Therefore, the average pay rate will be:

$$\frac{1000 \times 10 + 10 \times 90}{100} = \$109 \textit{ per hour}$$

In this case, this average is not very descriptive.

Another useful measurement is the **median**. In a data set *X* consisting of data points $x_1, x_2, x_3, \dots x_n$, the median is the point in the middle. The middle refers to the point where half the data comes before it and half comes after, when the data is recorded in numerical order. If *n* is odd, then the median is:

$$x_{\frac{n+1}{2}}$$

If *n* is even, it is defined as $\frac{1}{2}\left(x_{\frac{n}{2}} + x_{\frac{n}{2}+1}\right)$, the mean of the two data points closest to the middle of the data points. In the previous example of test scores, the two middle points are 75 and 80. Since there is no single point, the average of these two scores needs to be found. The average is:

$$\frac{75+80}{2} = 77.5$$

The median is generally a good value to use if there are a few outliers in the data. It prevents those outliers from affecting the "middle" value as much as when using the mean.

Since an outlier is a data point that is far from most of the other data points in a data set, this means an outlier also is any point that is far from the median of the data set. The outliers can have a substantial effect on the mean of a data set, but usually do not change the median or mode, or do not change them by a large quantity. For example, consider the data set (3, 5, 6, 6, 6, 8). This has a median of 6 and a mode of 6, with a mean of $\frac{34}{6} \approx 5.67$. Now, suppose a new data point of 1000 is added so that the data set is now (3, 5, 6, 6, 6, 8, 1000). This does not change the median or mode, which are both still 6. However, the average is now $\frac{1034}{7}$, which is approximately 147.7. In this case, the median and mode will be better descriptions for most of the data points.

The reason for outliers in a given data set is a complicated problem. It is sometimes the result of an error by the experimenter, but often they are perfectly valid data points that must be taken into consideration.

One additional measure to define for *X* is the **mode**. This is the data point that appears more frequently. If two or more data points all tie for the most frequent appearance, then each of them is considered a mode. In the case of the test scores, where the numbers were 50, 60, 65, 65, 75, 80, 85, 85, 90, 100, there are two modes: 65 and 85.

The **first quartile** of a set of data *X* refers to the largest value from the first ¼ of the data points. In practice, there are sometimes slightly different definitions that can be used, such as the median of the first half of the data points (excluding the median itself if there are an odd number of data points). The term also has a slightly different use: when it is said that a data point lies in the first quartile, it means it is less than or equal to the median of the first half of the data points. Conversely, if it lies *at* the first quartile, then it is equal to the first quartile.

When it is said that a data point lies in the **second quartile**, it means it is between the first quartile and the median.

The **third quartile** refers to data that lies between ½ and ¾ of the way through the data set. Again, there are various methods for defining this precisely, but the simplest way is to include all of the data that lie between the median and the median of the top half of the data.

Data that lies in the **fourth quartile** refers to all of the data above the third quartile.

Percentiles may be defined in a similar manner to quartiles. Generally, this is defined in the following manner:

If a data point lies *in* the n-th percentile, this means it lies in the range of the first *n*% of the data.

If a data point lies *at* the *n*-th percentile, then it means that *n*% of the data lies below this data point.

Given a data set *X* consisting of data points $(x_1, x_2, x_3, \dots x_n)$, the **variance of X** is defined to be:

$$\frac{\sum_{i=1}^{n}(x_i - \bar{X})^2}{n}$$

This means that the variance of *X* is the average of the squares of the differences between each data point and the mean of *X*. In the formula, $\bar{X}$ is the mean of the values in the data set, and x_i represents each individual value in the data set. The sigma notation indicates that the sum should be found with n being the number of values to add together. $i = 1$ means that the values should begin with the first value.

Given a data set *X* consisting of data points $(x_1, x_2, x_3, \dots x_n)$, the **standard deviation of X** is defined to be

$$s_x = \sqrt{\frac{\sum_{i=1}^{n}(x_i - \bar{X})^2}{n}}$$

In other words, the standard deviation is the square root of the variance.

Both the variance and the standard deviation are measures of how much the data tend to be spread out. When the standard deviation is low, the data points are mostly clustered around the mean. When the standard deviation is high, this generally indicates that the data are quite spread out, or else that there are a few substantial outliers.

As a simple example, compute the standard deviation for the data set (1, 3, 3, 5). First, compute the mean, which will be:

$$\frac{1+3+3+5}{4} = \frac{12}{4} = 3$$

Now, find the variance of *X* with the formula:

$$\sum_{i=1}^{4}(x_i - \bar{X})^2 = (1-3)^2 + (3-3)^2 + (3-3)^2 + (5-3)^2 = -2^2 + 0^2 + 0^2 + 2^2 = 8$$

Therefore, the variance is $\frac{8}{4} = 2$. Taking the square root, the standard deviation will be $\sqrt{2}$.

Note that the standard deviation only depends upon the mean, not upon the median or mode(s). Generally, if there are multiple modes that are far apart from one another, the standard deviation will be high. A high standard deviation does not always mean there are multiple modes, however.

Describing a Set of Data

A set of data can be described in terms of its center, spread, shape and any unusual features. The center of a data set can be measured by its mean, median, or mode. The spread of a data set refers to how far the data points are from the center (mean or median). The spread can be measured by the range or by the quartiles and interquartile range. A data set with data points clustered around the center will have a small spread. A data set covering a wide range will have a large spread.

When a data set is displayed as a **histogram** or frequency distribution plot, the shape indicates if a sample is normally distributed, symmetrical, or has measures of skewness or kurtosis. When graphed, a data set with a **normal distribution** will resemble a bell curve.

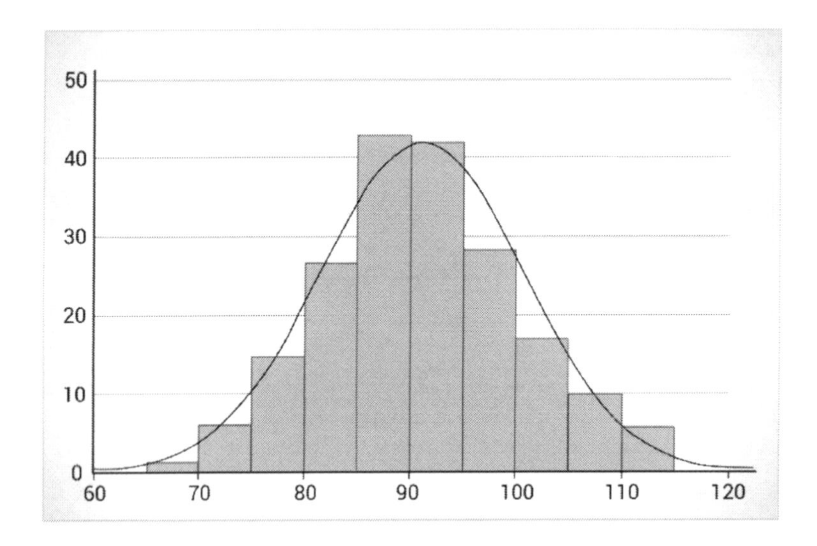

If the data set is symmetrical, each half of the graph when divided at the center is a mirror image of the other. If the graph has fewer data points to the right, the data is **skewed right**. If it has fewer data points to the left, the data is **skewed left**.

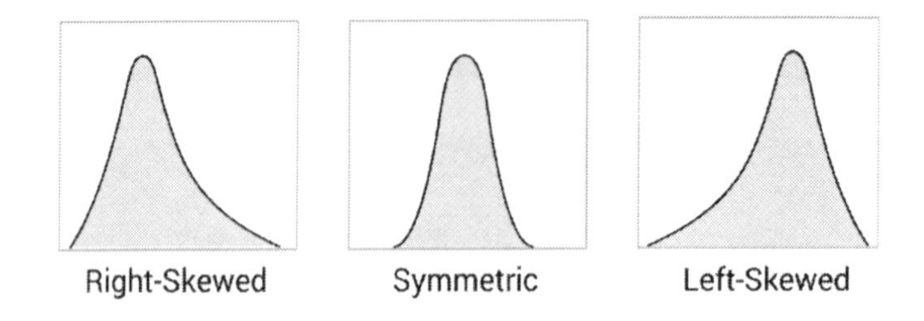

Kurtosis is a measure of whether the data is heavy-tailed with a high number of outliers, or light-tailed with a low number of outliers.

A description of a data set should include any unusual features such as gaps or outliers. A **gap** is a span within the range of the data set containing no data points. An **outlier** is a data point with a value either extremely large or extremely small when compared to the other values in the set.

Counting Techniques

The **addition rule** for probabilities states that the probability of *A* or *B* happening is:

$$P(A \cup B) = P(A) + P(B) - P(A \cap B)$$

Note that the subtraction of $P(A \cap B)$ must be performed, or else it would result in double counting any outcomes that lie in both *A* and in *B*. For example, suppose that a 20-sided die is being rolled. Fred bets that the outcome will be greater than 10, while Helen bets that it will be greater than 4 but less than 15. What is the probability that at least one of them is correct?

We apply the rule:

$$P(A \cup B) = P(A) + P(B) - P(A \cap B)$$

where *A* is that outcome *x* is in the range $x > 10$, and *B* is that outcome *x* is in the range $4 < x < 15$.

$$P(A) = 10 \times \frac{1}{20} = \frac{1}{2}$$

$$P(B) = 10 \times \frac{1}{20} = \frac{1}{2}$$

$P(A \cap B)$ can be computed by noting that $A \cap B$ means the outcome *x* is in the range $10 < x < 15$, so

$$P(A \cap B) = 4 \times \frac{1}{20} = \frac{1}{5}$$

Therefore:

$$P(A \cup B) = P(A) + P(B) - P(A \cap B)$$

$$\frac{1}{2} + \frac{1}{2} - \frac{1}{5} = \frac{4}{5}$$

Note that in this particular example, we could also have directly reasoned about the set of possible outcomes $A \cup B$, by noting that this would mean that *x* must be in the range $5 \leq x$. However, this is not always the case, depending on the given information.

The **multiplication rule** for probabilities states the probability of *A* and *B* both happening is:

$$P(A \cap B) = P(A)P(B|A)$$

As an example, suppose that when Jamie wears black pants, there is a ½ probability that she wears a black shirt as well, and that she wears black pants ¾ of the time. What is the probability that she is wearing both a black shirt and black pants?

To figure this, use the above formula, where *A* will be "Jamie is wearing black pants," while *B* will be "Jamie is wearing a black shirt." It is known that $P(A)$ is ¾. It is also known that $P(B|A) = \frac{1}{2}$. Multiplying the two, the probability that she is wearing both black pants and a black shirt is:

$$P(A)P(B|A) = \frac{3}{4} \times \frac{1}{2} = \frac{3}{8}$$

Probability of an Event

Given a set of possible outcomes *X*, a **probability distribution** of *X* is a function that assigns a probability to each possible outcome. If the outcomes are $(x_1, x_2, x_3, \dots x_n)$, and the probability distribution is *p*, then the following rules are applied.

- $0 \leq p(x_i) \leq 1$, for any i.
- $\sum_{i=1}^{n} p(x_i) = 1$.

In other words, the probability of a given outcome must be between zero and 1, while the total probability must be 1.

If $p(x_i)$ is constant, then this is called a **uniform probability distribution**, and $p(x_i) = \frac{1}{n}$. For example, on a six-sided die, the probability of each of the six outcomes will be $\frac{1}{6}$.

If seeking the probability of an outcome occurring in some specific range *A* of possible outcomes, written $P(A)$, add up the probabilities for each outcome in that range. For example, consider a six-sided die, and figure the probability of getting a 3 or lower when it is rolled. The possible rolls are 1, 2, 3, 4, 5, and 6. So, to get a 3 or lower, a roll of 1, 2, or 3 must be completed. The probabilities of each of these is $\frac{1}{6}$, so add these to get:

$$p(1) + p(2) + p(3) = \frac{1}{6} + \frac{1}{6} + \frac{1}{6} = \frac{1}{2}$$

An outcome occasionally lies within some range of possibilities *B*, and the probability that the outcomes also lie within some set of possibilities *A* needs to be figured. This is called a **conditional probability**. It is written as $P(A|B)$, which is read "the probability of *A* given *B*." The general formula for computing conditional probabilities is:

$$P(A|B) = \frac{P(A \cap B)}{P(B)}$$

However, when dealing with uniform probability distributions, simplify this a bit. Write $|A|$ to indicate the number of outcomes in *A*. Then, for uniform probability distributions, write:

$$P(A|B) = \frac{|A \cap B|}{|B|}$$

(recall that $A \cap B$ means "*A* intersect *B*," and consists of all of the outcomes that lie in both *A* and *B*)

This means that all possible outcomes do not need to be known. To see why this formula works, suppose that the set of outcomes *X* is $(x_1, x_2, x_3, \dots x_n)$, so that $|X| = n$. Then, for a uniform probability distribution:

$$P(A) = \frac{|A|}{n}$$

However, this means:

$$(A|B) = \frac{P(A \cap B)}{P(B)} = \frac{\frac{|A \cap B|}{n}}{\frac{|B|}{n}} = \frac{|A \cap B|}{|B|}$$

since the *n*'s cancel out.

For example, suppose a die is rolled and it is known that it will land between 1 and 4. However, how many sides the die has is unknown. Figure the probability that the die is rolled higher than 2. To figure this, $P(3)$ or $P(4)$ does not need to be determined, or any of the other probabilities, since it is known that a fair die has a uniform probability distribution. Therefore, apply the formula $\frac{|A \cap B|}{|B|}$. So, in this case *B* is (1, 2, 3, 4) and $A \cap B$ is (3, 4). Therefore:

$$\frac{|A \cap B|}{|B|} = \frac{2}{4} = \frac{1}{2}$$

Conditional probability is an important concept because, in many situations, the likelihood of one outcome can differ radically depending on how something else comes out. The probability of passing a test given that one has studied all of the material is generally much higher than the probability of passing a test given that one has not studied at all. The probability of a person having heart trouble is much lower if that person exercises regularly. The probability that a college student will graduate is higher when his or her SAT scores are higher, and so on. For this reason, there are many people who are interested in conditional probabilities.

Note that in some practical situations, changing the order of the conditional probabilities can make the outcome very different. For example, the probability that a person with heart trouble has exercised regularly is quite different than the probability that a person who exercises regularly will have heart trouble. The probability of a person receiving a military-only award, given that he or she is or was a soldier, is generally not very high, but the probability that a person being or having been a soldier, given that he or she received a military-only award, is 1.

However, in some cases, the outcomes do not influence one another this way. If the probability of *A* is the same regardless of whether *B* is given; that is, if $P(A|B) = P(A)$, then *A* and *B* are considered **independent**. In this case:

$$P(A|B) = \frac{P(A \cap B)}{P(B)} = P(A)$$

so $P(A \cap B) = P(A)P(B)$. In fact, if $P(A \cap B) = P(A)P(B)$, it can be determined that $P(A|B) = P(A)$ and $P(A|B) = P(B)$ by working backward. Therefore, *B* is also independent of *A*.

An example of something being independent can be seen in rolling dice. In this case, consider a red die and a green die. It is expected that when the dice are rolled, the outcome of the green die should not depend in any way on the outcome of the red die. Or, to take another example, if the same die is rolled repeatedly, then the next number rolled should not depend on which numbers have been rolled previously. Similarly, if a coin is flipped, then the next flip's outcome does not depend on the outcomes of previous flips.

This can sometimes be counter-intuitive, since when rolling a die or flipping a coin, there can be a streak of surprising results. If, however, it is known that the die or coin is fair, then these results are just the result of the fact that over long periods of time, it is very likely that some unlikely streaks of outcomes will occur. Therefore, avoid making the mistake of thinking that when considering a series of independent outcomes, a particular outcome is "due to happen" simply because a surprising series of outcomes has already been seen.

There is a second type of common mistake that people tend to make when reasoning about statistical outcomes: the idea that when something of low probability happens, this is surprising. It would be surprising that something with low probability happened after just one attempt. However, with so much happening all at once, it is easy to see at least something happen in a way that seems to have a very low probability. In fact, a lottery is a good example. The odds of winning a lottery are very small, but the odds that somebody wins the lottery each week are actually fairly high. Therefore, no one should be surprised when some low probability things happen.

A **simple event** consists of only one outcome. The most popular simple event is flipping a coin, which results in either heads or tails. A **compound event** results in more than one outcome and consists of more than one simple event. An example of a compound event is flipping a coin while tossing a die. The result is either heads or tails on the coin and a number from one to six on the die. The probability of a simple event is calculated by dividing the number of possible outcomes by the total number of outcomes. Therefore, the probability of obtaining heads on a coin is $\frac{1}{2}$, and the probability of rolling a 6 on a die is $\frac{1}{6}$. The probability of compound events is calculated using the basic idea of the probability of simple events. If the two events are independent, the probability of one outcome is equal to the product of the probabilities of each simple event. For example, the probability of obtaining heads on a coin and rolling a 6 is equal to $\frac{1}{2} \times \frac{1}{6} = \frac{1}{12}$. The probability of either A or B occurring is equal to the sum of the probabilities minus the probability that both A and B will occur. Therefore, the probability of obtaining either heads on a coin or rolling a 6 on a die is:

$$\frac{1}{2} + \frac{1}{6} - \frac{1}{12} = \frac{7}{12}$$

The two events aren't mutually exclusive because they can happen at the same time. If two events are mutually exclusive, and the probability of both events occurring at the same time is zero, the probability of event A or B occurring equals the sum of both probabilities. An example of calculating the probability of two mutually exclusive events is determining the probability of pulling a king or a queen from a deck of cards. The two events cannot occur at the same time.

Algebra, Functions, and Patterns

Adding, Subtracting, Multiplying, and Factoring Linear Expressions

Algebraic expressions look similar to equations, but they do not include the equal sign. Algebraic expressions are comprised of numbers, variables, and mathematical operations. Some examples of algebraic expressions are:

$$8x + 7y - 12z, 3a^2, and\ 5x^3 - 4y^4$$

Algebraic expressions consist of variables, numbers, and operations. A term of an expression is any combination of numbers and/or variables, and terms are separated by addition and subtraction. For example, the expression $5x^2 - 3xy + 4 - 2$ consists of 4 terms: $5x^2$, -3*xy*, 4*y*, and -2. Note that each term includes its given sign (+ or −). The variable part of a term is a letter that represents an unknown quantity. The coefficient of a term is the number by which the variable is multiplied. For the term 4*y*, the variable is *y,* and the coefficient is 4. Terms are identified by the power (or exponent) of its variable.

A number without a variable is referred to as a constant. If the variable is to the first power (x^1 or simply *x*), it is referred to as a linear term. A term with a variable to the second power (x^2) is quadratic and a term to the third power (x^3) is cubic. Consider the expression $x^3 + 3x - 1$. The constant is -1. The linear term is 3*x*. There is no quadratic term. The cubic term is x^3.

An algebraic expression can also be classified by how many terms exist in the expression. Any like terms should be combined before classifying. A monomial is an expression consisting of only one term. Examples of monomials are: 17, 2*x*, and $-5ab^2$. A binomial is an expression consisting of two terms separated by addition or subtraction. Examples include $2x - 4$ and $-3y^2 + 2y$. A trinomial consists of 3 terms. For example, $5x^2 - 2x + 1$ is a trinomial.

Algebraic expressions and equations can represent real-life situations and model the behavior of different variables. For example, $2x + 5$ could represent the cost to play games at an arcade. In this case, 5 represents the price of admission to the arcade and 2 represents the cost of each game played. To calculate the total cost, use the number of games played for x, multiply it by 2, and add 5.

Adding and Subtracting Linear Algebraic Expressions

An algebraic expression is simplified by combining like terms. A term is a number, variable, or product of a number, and variables separated by addition and subtraction. For the algebraic expression $3x^2 - 4x + 5 - 5x^2 + x - 3$, the terms are $3x^2$, -4*x*, 5, -5x^2, *x*, and -3. Like terms have the same variables raised to the same powers (exponents). The like terms for the previous example are $3x^2$ and -5x^2, -4x and x, 5 and -3. To combine like terms, the coefficients (numerical factors of the term including sign) are added, and the variables and their powers are kept the same. Note that if a coefficient is not written, it is an implied coefficient of 1 ($x = 1x$). The previous example will simplify to $-2x^2 - 3x + 2$.

When adding or subtracting algebraic expressions, each expression is written in parenthesis. The negative sign is distributed when necessary, and like terms are combined. Consider the following:

$$\text{add } 2a + 5b - 2 \text{ to } a - 2b + 8c - 4$$

The sum is set as follows:

$$(a - 2b + 8c - 4) + (2a + 5b - 2)$$

In front of each set of parentheses is an implied positive one, which, when distributed, does not change any of the terms. Therefore, the parentheses are dropped and like terms are combined:

$$a - 2b + 8c - 4 + 2a + 5b - 2$$

$$3a + 3b + 8c - 6$$

Consider the following problem:

Subtract $2a + 5b - 2$ from $a - 2b + 8c - 4$

The difference is set as follows:

$$(a - 2b + 8c - 4) - (2a + 5b - 2)$$

The implied one in front of the first set of parentheses will not change those four terms. However, distributing the implied -1 in front of the second set of parentheses will change the sign of each of those three terms:

$$a - 2b + 8c - 4 - 2a - 5b + 2$$

Combining like terms yields the simplified expression: $-a - 7b + 8c - 2$.

Distributive Property

The **distributive property** states that multiplying a sum (or difference) by a number produces the same result as multiplying each value in the sum (or difference) by the number and adding (or subtracting) the products. Using mathematical symbols, the distributive property states $a(b + c) = ab + ac$. The expression $4(3 + 2)$ is simplified using the order of operations. Simplifying inside the parenthesis first produces 4×5, which equals 20. The expression $4(3 + 2)$ can also be simplified using the distributive property:

$$4(3 + 2)$$

$$4 \times 3 + 4 \times 2$$

$$12 + 8$$

$$20$$

Consider the following example: $4(3x - 2)$. The expression cannot be simplified inside the parenthesis because 3*x* and -2 are not like terms, and therefore cannot be combined. However, the expression can be simplified by using the distributive property and multiplying each term inside of the parenthesis by the term outside of the parenthesis: $12x - 8$. The resulting equivalent expression contains no like terms, so it cannot be further simplified.

Consider the expression:

$$(3x + 2y + 1) - (5x - 3) + 2(3y + 4)$$

Again, there are no like terms, but the distributive property is used to simplify the expression. Note there is an implied one in front of the first set of parentheses and an implied -1 in front of the second set of parentheses. Distributing the one, -1, and 2 produces:

$$1(3x) + 1(2y) + 1(1) - 1(5x) - 1(-3) + 2(3y) + 2(4)$$

$$3x + 2y + 1 - 5x + 3 + 6y + 8$$

This expression contains like terms that are combined to produce the simplified expression:

$$-2x + 8y + 12$$

Algebraic expressions are tested to be equivalent by choosing values for the variables and evaluating both expressions (see 2.A.4). For example, $4(3x - 2)$ and $12x - 8$ are tested by substituting 3 for the variable *x* and calculating to determine if equivalent values result.

Evaluating Algebraic Expressions

To evaluate the expression, the given values for the variables are substituted (or replaced) and the expression is simplified using the order of operations. Parenthesis should be used when substituting. Consider the following: Evaluate $a - 2b + ab$ for $a = 3$ and $b = -1$. To evaluate, any variable *a* is replaced with 3 and any variable *b* with -1, producing:

$$(3) - 2(-1) + (3)(-1)$$

Next, the order of operations is used to calculate the value of the expression, which is 2.

Here's another example:

$$\text{Evaluate } a - 2b + ab \; for \; a = 3 \text{ and } b = -1$$

To evaluate an expression, the given values should be substituted for the variables and simplified using the order of operations. In this case:

$$(3) - 2(-1) + (3)(-1)$$

Parentheses are used when substituting.

Given an algebraic expression, students may be asked to simplify the expression. For example:

$$\text{Simplify } 5x^2 - 10x + 2 - 8x^2 + x - 1.$$

Simplifying algebraic expressions requires combining like terms. A term is a number, variable, or product of a number and variables separated by addition and subtraction. The terms in the above expressions are: $5x^2, -10x, 2, -8x^2, x,$ and -1. Like terms have the same variables raised to the same powers (exponents). To combine like terms, the coefficients (numerical factor of the term including sign) are added, while the variables and their powers are kept the same. The example above simplifies to:

$$-3x^2 - 9x + 1$$

Let's try two more.

Evaluate $\frac{1}{2}x^2 - 3, x = 4$.

The first step is to substitute in 4 for *x* in the expression:

$$\frac{1}{2}(4)^2 - 3$$

Then, the order of operations is used to simplify.

The exponent comes first, $\frac{1}{2}(16) - 3$, then the multiplication 8 – 3, and then, after subtraction, the solution is 5.

Evaluate $4|5 - x| + 2y, x = 4, y = -3$.

The first step is to substitute 4 in for *x* and -3 in for *y* in the expression:

$$4|5 - 4| + 2(-3)$$

Then, the absolute value expression is simplified, which is:

$$|5 - 4| = |1| = 1$$

The expression is $4(1) + 2(-3)$ which can be simplified using the order of operations.

First is the multiplication, $4 + (-6)$; then addition yields an answer of -2.

Creating Algebraic Expressions

A linear expression is a statement about an unknown quantity expressed in mathematical symbols. The statement "five times a number added to forty" can be expressed as $5x + 40$. A linear equation is a statement in which two expressions (at least one containing a variable) are equal to each other. The statement "five times a number added to forty is equal to ten" can be expressed as $5x + 40 = 10$.

Real world scenarios can also be expressed mathematically. Suppose a job pays its employees $300 per week and $40 for each sale made. The weekly pay is represented by the expression $40x + 300$ where *x* is the number of sales made during the week.

Consider the following scenario: Bob had $20 and Tom had $4. After selling 4 ice cream cones to Bob, Tom has as much money as Bob. The cost of an ice cream cone is an unknown quantity and can be represented by a variable (*x*). The amount of money Bob has after his purchase is four times the cost of an ice cream cone subtracted from his original $\$20 \rightarrow 20 - 4x$. The amount of money Tom has after his sale is four times the cost of an ice cream cone added to his original $\$4 \rightarrow 4x + 4$. After the sale, the amounts of money that Bob and Tom have are equal $\rightarrow 20 - 4x = 4x + 4$.

When expressing a verbal or written statement mathematically, it is key to understand words or phrases that can be represented with symbols. The following are examples:

Symbol	Phrase
$+$	added to, increased by, sum of, more than
$-$	decreased by, difference between, less than, take away
x	multiplied by, 3 (4, 5 . . .) times as large, product of
$\div$	divided by, quotient of, half (third, etc.) of
$=$	is, the same as, results in, as much as
$x, t, n, etc.$	a number, unknown quantity, value of

Adding, Subtracting, Multiplying, Dividing, and Factoring Polynomials

An expression of the form ax^n, where *n* is a non-negative integer, is called a **monomial** because it contains one term. A sum of monomials is called a **polynomial**. For example, $-4x^3 + x$ is a polynomial, while $5x^7$ is a monomial. A function equal to a polynomial is called a **polynomial function**.

The monomials in a polynomial are also called the **terms** of the polynomial.

The constants that precede the variables are called **coefficients**.

The highest value of the exponent of x in a polynomial is called the **degree** of the polynomial. So, $-4x^3 + x$ has a degree of 3, while $-2x^5 + x^3 + 4x + 1$ has a degree of 5. When multiplying polynomials, the degree of the result will be the sum of the degrees of the two polynomials being multiplied.

Addition and subtraction operations can be performed on polynomials with like terms. **Like terms** refers to terms that have the same variable and exponent. The two following polynomials can be added together by collecting like terms:

$$(x^2 + 3x - 4) + (4x^2 - 7x + 8)$$

The x^2 terms can be added as $x^2 + 4x^2 = 5x^2$. The x terms can be added as $3x + -7x = -4x$, and the constants can be added as $-4 + 8 = 4$. The following expression is the result of the addition:

$$5x^2 - 4x + 4$$

Let's try another:

$$(-2x^5 + x^3 + 4x + 1) + (-4x^3 + x)$$

$$-2x^5 + (1 - 4)x^3 + (4 + 1)x + 1$$

$$-2x^5 - 3x^3 + 5x + 1$$

Likewise, subtraction of polynomials is performed by subtracting coefficients of like powers of *x*. So,

$$(-2x^5 + x^3 + 4x + 1) - (-4x^3 + x)$$

$$-2x^5 + (1 + 4)x^3 + (4 - 1)x + 1$$

$$-2x^5 + 5x^3 + 3x + 1$$

To multiply two polynomials, multiply each term of the first polynomial by each term of the second polynomial and add the results. For example:

$$(4x^2 + x)(-x^3 + x)$$

$$4x^2(-x^3) + 4x^2(x) + x(-x^3) + x(x)$$

$$-4x^5 + 4x^3 - x^4 + x^2$$

In the case where each polynomial has two terms, like in this example, some students find it helpful to remember this as multiplying the First terms, then the Outer terms, then the Inner terms, and finally the Last terms, with the mnemonic FOIL. For longer polynomials, the multiplication process is the same, but there will be, of course, more terms, and there is no common mnemonic to remember each combination.

Factors for polynomials are similar to factors for integers—they are numbers, variables, or polynomials that, when multiplied together, give a product equal to the polynomial in question. One polynomial is a factor of a second polynomial if the second polynomial can be obtained from the first by multiplying by a third polynomial.

$6x^6 + 13x^4 + 6x^2$ can be obtained by multiplying together $(3x^4 + 2x^2)(2x^2 + 3)$. This means $2x^2 + 3$ and $3x^4 + 2x^2$ are factors of:

$$6x^6 + 13x^4 + 6x^2$$

In general, finding the factors of a polynomial can be tricky. However, there are a few types of polynomials that can be factored in a straightforward way.

If a certain monomial divides each term of a polynomial, it can be factored out:

$$x^2 + 2xy + y^2 = (x + y)^2$$

$$x^2 - 2xy + y^2 = (x - y)^2$$

$$x^2 - y^2 = (x + y)(x - y)$$

$$x^3 + y^3 = (x + y)(x^2 - xy + y^2)$$

$$x^3 - y^3 = (x - y)(x^2 + xy + y^2)$$

$$x^3 + 3x^2y + 3xy^2 + y^3 = (x + y)^3$$

$$x^3 - 3x^2y + 3xy^2 - y^3 = (x - y)^3$$

These rules can be used in many combinations with one another. For example, the expression $3x^3 - 24$ factors to:

$$3(x^3 - 8) = 3(x - 2)(x^2 + 2x + 4)$$

When factoring polynomials, a good strategy is to multiply the factors to check the result.

Let's try another example:

$$4x^3 + 16x^2 = 4x^2(x + 4).$$

$$x^2 + 2xy + y^2 = (x + y)^2 \text{ or } x^2 - 2xy + y^2 = (x - y)^2$$

$$x^2 - y^2 = (x + y)(x - y)$$

$$x^3 + y^3 = (x + y)(x^2 - xy + y^2)$$

$$x^3 - y^3 = (x - y)(x^2 + xy + y^2)$$

$$x^3 + 3x^2y + 3xy^2 + y^3 = (x + y)^3 \text{ and } x^3 - 3x^2y + 3xy^2 - y^3 = (x - y)^3$$

It sometimes can be necessary to rewrite the polynomial in some clever way before applying the above rules. Consider the problem of factoring $x^4 - 1$. This does not immediately look like any of the cases for which there are rules. However, it's possible to think of this polynomial as $x^4 - 1 = (x^2)^2 - (1^2)^2$, and now apply the third rule in the above list to simplify this:

$$(x^2)^2 - (1^2)^2$$

$$(x^2 + 1^2)(x^2 - 1^2)$$

$$(x^2 + 1)(x^2 - 1)$$

Creating Polynomials from Written Descriptions

Polynomials that represent mathematical or real-world problems can also be created from written descriptions, much like algebraic expressions. For example, polynomials might be created when working with formulas. Formulas are mathematical expressions that define the value of one quantity, given the value of one or more different quantities. Formulas look like equations because they contain variables, numbers, operators, and an equal sign. All formulas are equations but not all equations are formulas. A formula must have more than one variable. For example, $2x + 7 = y$ is an equation and a formula (it relates the unknown quantities *x* and *y*). However, $2x + 7 = 3$ is an equation but not a formula (it only expresses the value of the unknown quantity *x*).

Formulas are typically written with one variable alone (or isolated) on one side of the equal sign. This variable can be thought of as the **subject** in that the formula is stating the value of the subject in terms of the relationship between the other variables. Consider the distance formula: $distance = rate \times time$ or $d = rt$. The value of the subject variable *d* (distance) is the product of the variable *r* and *t* (rate and time). Given the rate and time, the distance traveled can easily be determined by substituting the values into the formula and evaluating.

The formula $P = 2l + 2w$ expresses how to calculate the perimeter of a rectangle (*P*) given its length (*l*) and width (*w*). To find the perimeter of a rectangle with a length of 3ft and a width of 2ft, these values are substituted into the formula for *l* and *w*: $P = 2(3ft) + 2(2ft)$. Following the order of operations, the perimeter is determined to be 10ft. When working with formulas such as these, including units is an important step.

Given a formula expressed in terms of one variable, the formula can be manipulated to express the relationship in terms of any other variable. In other words, the formula can be rearranged to change which variable is the *subject*. To solve for a variable of interest by manipulating a formula, the equation may be solved as if all other variables were numbers. The same steps for solving are followed, leaving operations in terms of the variables instead of calculating numerical values. For the formula $P = 2l + 2w$, the perimeter is the subject expressed in terms of the length and width. To write a formula to calculate the width of a rectangle, given its length and perimeter, the previous formula relating the three variables is solved for the variable *w*. If *P* and *l* were numerical values, this is a two-step linear equation solved by subtraction and division. To solve the equation $P = 2l + 2w$ for *w*, 2*l* is first subtracted from both sides:

$$P - 2l = 2w$$

Then both sides are divided by 2:

$$\frac{P - 2l}{2} = w$$

Test questions may involve creating a polynomial based on a formula. For example, using the perimeter of a rectangle formula, a problem may ask for the perimeter of a rectangle with a length of $2x + 12$ and a width of $x + 1$. Using the formula $P = 2l + 2w$, the perimeter would then be:

$$P = 2(2x + 12) + 2(x + 1)$$

This equals:

$$4x + 24 + 2x + 2 = 6x + 26$$

The area of the same rectangle, which uses the formula $A = l \times w$, would be:

$$A = (2x + 12)(x + 1)$$

$$2x^2 + 2x + 12x + 12$$

$$2x^2 + 14x + 12$$

Adding, Subtracting, Multiplying, Dividing Rational Expressions

A fraction, or ratio, wherein each part is a polynomial, defines **rational expressions**. Some examples include $\frac{2x+6}{x}$, $\frac{1}{x^2-4x+8}$, and $\frac{z^2}{x+5}$. Exponents on the variables are restricted to whole numbers, which means roots and negative exponents are not included in rational expressions.

Rational expressions can be transformed by factoring. For example, the expression $\frac{x^2-5x+6}{(x-3)}$ can be rewritten by factoring the numerator to obtain:

$$\frac{(x - 3)(x - 2)}{(x - 3)}$$

Therefore, the common binomial $(x - 3)$ can cancel so that the simplified expression is:

$$\frac{(x - 2)}{1} = (x - 2)$$

Additionally, other rational expressions can be rewritten to take on different forms. Some may be factorable in themselves, while others can be transformed through arithmetic operations. Rational expressions are closed under addition, subtraction, multiplication, and division by a nonzero expression. **Closed** means that if any one of these operations is performed on a rational expression, the result will still be a rational expression. The set of all real numbers is another example of a set closed under all four operations.

Adding and subtracting rational expressions is based on the same concepts as adding and subtracting simple fractions. For both concepts, the denominators must be the same for the operation to take place. For example, here are two rational expressions:

$$\frac{x^3 - 4}{(x-3)} + \frac{x+8}{(x-3)}$$

Since the denominators are both $(x - 3)$, the numerators can be combined by collecting like terms to form:

$$\frac{x^3 + x + 4}{(x-3)}$$

If the denominators are different, they need to be made common (the same) by using the **least common denominator (LCD)**. Each denominator needs to be factored, and the LCD contains each factor that appears in any one denominator the greatest number of times it appears in any denominator. The original expressions need to be multiplied by a form of 1 such as 5/5 or x-2/x-2, which will turn each denominator into the LCD. This process is like adding fractions with unlike denominators. It is also important when working with rational expressions to define what value of the variable makes the denominator zero. For this particular value, the expression is undefined.

Multiplication of rational expressions is performed like multiplication of fractions. The numerators are multiplied; then, the denominators are multiplied. The final fraction is then simplified. The expressions are simplified by factoring and cancelling out common terms. In the following example, the numerator of the second expression can be factored first to simplify the expression before multiplying:

$$\frac{x^2}{(x-4)} \times \frac{x^2 - x - 12}{2}$$

$$\frac{x^2}{(x-4)} \times \frac{(x-4)(x+3)}{2}$$

The $(x - 4)$ on the top and bottom cancel out:

$$\frac{x^2}{1} \times \frac{(x+3)}{2}$$

Then multiplication is performed, resulting in:

$$\frac{x^3 + 3x^2}{2}$$

Dividing rational expressions is similar to the division of fractions, where division turns into multiplying by a reciprocal. Thus, the following expression can be rewritten as a multiplication problem:

$$\frac{x^2 - 3x + 7}{x - 4} \div \frac{x^2 - 5x + 3}{x - 4}$$

$$\frac{x^2 - 3x + 7}{x - 4} \times \frac{x - 4}{x^2 - 5x + 3}$$

The $x - 4$ cancels out, leaving:

$$\frac{x^2 - 3x + 7}{x^2 - 5x + 3}$$

The final answers should always be completely simplified. If a function is composed of a rational expression, the zeros of the graph can be found from setting the polynomial in the numerator as equal to zero and solving. The values that make the denominator equal to zero will either exist on the graph as a **hole** or a **vertical asymptote**.

A **complex fraction** is a fraction in which the numerator and denominator are themselves fractions, of the form:

$$\frac{(\frac{a}{b})}{(\frac{c}{d})}$$

These can be simplified by following the usual rules for the order of operations, or by remembering that dividing one fraction by another is the same as multiplying by the reciprocal of the divisor. This means that any complex fraction can be rewritten using the following form:

$$\frac{(\frac{a}{b})}{(\frac{c}{d})} = \frac{a}{b} \times \frac{d}{c}$$

The following problem is an example of solving a complex fraction:

$$\frac{(\frac{5}{4})}{(\frac{3}{8})} = \frac{5}{4} \times \frac{8}{3} = \frac{40}{12} = \frac{10}{3}$$

Writing an Expression from a Written Description

When expressing a verbal or written statement mathematically, it is vital to understand words or phrases that can be represented with symbols. The following are examples:

Symbol	Phrase
+	Added to; increased by; sum of; more than
−	Decreased by; difference between; less than; take away
×	Multiplied by; 3(4,5...) times as large; product of
÷	Divided by; quotient of; half (third, etc.) of
=	Is; the same as; results in; as much as; equal to
x, t, n, etc.	A number; unknown quantity; value of; variable

Addition and subtraction are **inverse operations**. Adding a number and then subtracting the same number will cancel each other out, resulting in the original number, and vice versa. For example, $8 + 7 - 7 = 8$ and $137 - 100 + 100 = 137$. Similarly, multiplication and division are inverse operations. Therefore, multiplying by a number and then dividing by the same number results in the original number, and vice versa. For example, $8 \times 2 \div 2 = 8$ and $12 \div 4 \times 4 = 12$. Inverse operations are used to work backwards to solve problems. In the case that 7 and a number add to 18, the inverse operation of subtraction is used to find the unknown value ($18 - 7 = 11$). If a school's entire 4th grade was divided evenly into 3 classes each with 22 students, the inverse operation of multiplication is used to determine the total students in the grade ($22 \times 3 = 66$). Additional scenarios involving inverse operations are included in the tables below.

Recall that a rational expression is a fraction where the numerator and denominator are both polynomials. Some examples of rational expressions include the following: $\frac{4x^3y^5}{3z^4}$, $\frac{4x^3+3x}{x^2}$, and $\frac{x^2+7x+10}{x+2}$. Since these refer to expressions and not equations, they can be simplified but not solved. Using the rules in the previous Exponents and Roots sections, some rational expressions with monomials can be simplified. Other rational expressions such as the last example,

$$\frac{x^2 + 7x + 10}{x + 2}$$

take more steps to be simplified. First, the polynomial on top can be factored from $x^2 + 7x + 10$ into $(x + 5)(x + 2)$. Then the common factors can be canceled, and the expression can be simplified to $(x + 5)$.

Consider this problem as an example of using rational expressions. Reggie wants to lay sod in his rectangular backyard. The length of the yard is given by the expression $4x + 2$ and the width is unknown. The area of the yard is $20x + 10$. Reggie needs to find the width of the yard. Knowing that the area of a rectangle is length multiplied by width, an expression can be written to find the width: $\frac{20x+10}{4x+2}$, area divided by length. Simplifying this expression by factoring out 10 on the top and 2 on the bottom leads to this expression:

$$\frac{10(2x + 1)}{2(2x + 1)}$$

By cancelling out the $2x+1$, that results in $\frac{10}{2}=5$. The width of the yard is found to be 5 by simplifying a rational expression.

Using Linear Equations to Solve Real-World Problems

Linear relationships describe the way two quantities change with respect to each other. The relationship is defined as linear because a line is produced if all the sets of corresponding values are graphed on a coordinate grid. When expressing the linear relationship as an equation, the equation is often written in the form $y=mx+b$ (slope-intercept form) where *m* and *b* are numerical values and *x* and *y* are variables (for example, $y=5x+10$). Given a linear equation and the value of either variable (*x* or *y*), the value of the other variable can be determined.

Imagine the following problem: The sum of a number and 5 is equal to -8 times the number.

To find this unknown number, a simple equation can be written to represent the problem. Key words such as difference, equal, and times are used to form the following equation with one variable: $n+5=-8n$. When solving for n, opposite operations are used. First, n is subtracted from $-8n$ across the equals sign, resulting in $5=-9n$. Then, -9 is divided on both sides, leaving $n=-\frac{5}{9}$. This solution can be graphed on the number line with a dot as shown below:

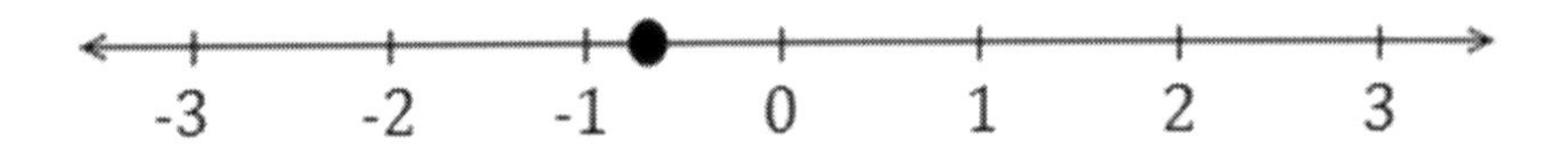

Suppose a teacher is grading a test containing 20 questions with 5 points given for each correct answer, adding a curve of 10 points to each test. This linear relationship can be expressed as the equation $y=5x+10$ where *x* represents the number of correct answers and *y* represents the test score. To determine the score of a test with a given number of correct answers, the number of correct answers is substituted into the equation for *x* and evaluated. For example, for 10 correct answers, 10 is substituted for *x*: $y=5(10)+10 \rightarrow y=60$. Therefore, 10 correct answers will result in a score of 60. The number of correct answers needed to obtain a certain score can also be determined. To determine the number of correct answers needed to score a 90, 90 is substituted for *y* in the equation (*y* represents the test score) and solved: $90=5x+10 \rightarrow 80=5x \rightarrow 16=x$. Therefore, 16 correct answers are needed to score a 90.

Linear relationships may be represented by a table of 2 corresponding values. Certain tables may determine the relationship between the values and predict other corresponding sets. Consider the table below, which displays the money in a checking account that charges a monthly fee:

Month	0	1	2	3	4
Balance	$210	$195	$180	$165	$150

An examination of the values reveals that the account loses $15 every month (the month increases by one and the balance decreases by 15). This information can be used to predict future values. To determine what the value will be in month 6, the pattern can be continued, and it can be concluded that the balance will be $120. To determine which month the balance will be $0, $210 is divided by $15 (since the balance decreases $15 every month), resulting in month 14.

Solving a System of Two Linear Equations

A **system of equations** is a group of equations that have the same variables or unknowns. These equations can be linear, but they are not always so. Finding a solution to a system of equations means finding the values of the variables that satisfy each equation. For a linear system of two equations and two variables, there could be a single solution, no solution, or infinitely many solutions.

A single solution occurs when there is one value for x and y that satisfies the system. This would be shown on the graph where the lines cross at exactly one point. When there is no solution, the lines are parallel and do not ever cross. With infinitely many solutions, the equations may look different, but they are the same line. One equation will be a multiple of the other, and on the graph, they lie on top of each other.

The process of elimination can be used to solve a system of equations. For example, the following equations make up a system:

$$x + 3y = 10 \text{ } and \text{ } 2x - 5y = 9$$

Immediately adding these equations does not eliminate a variable, but it is possible to change the first equation by multiplying the whole equation by -2. This changes the first equation to

$$-2x - 6y = -20$$

The equations can be then added to obtain $-11y = -11$. Solving for y yields $y = 1$. To find the rest of the solution, 1 can be substituted in for y in either original equation to find the value of $x = 7$. The solution to the system is (7, 1) because it makes both equations true, and it is the point in which the lines intersect. If the system is **dependent**—having infinitely many solutions—then both variables will cancel out when the elimination method is used, resulting in an equation that is true for many values of *x* and *y*. Since the system is dependent, both equations can be simplified to the same equation or line.

A system can also be solved using **substitution.** This involves solving one equation for a variable and then plugging that solved equation into the other equation in the system. This equation can be solved for one variable, which can then be plugged in to either original equation and solved for the other variable. For example, $x - y = -2$ and $3x + 2y = 9$ can be solved using substitution. The first equation can be solved for x, where $x = -2 + y$. Then it can be plugged into the other equation:

$$3(-2 + y) + 2y = 9$$

Solving for y yields:

$$-6 + 3y + 2y = 9$$

That shows that $y = 3$. If $y = 3$, then $x = 1$.

This solution can be checked by plugging in these values for the variables in each equation to see if it makes a true statement.

Finally, a solution to a system of equations can be found graphically. The solution to a linear system is the point or points where the lines cross. The values of x and y represent the coordinates (x, y) where the lines intersect. Using the same system of equation as above, they can be solved for y to put them in slope-intercept form, $y = mx + b$. These equations become $y = x + 2$ and $y = -\frac{3}{2}x + 4.5$. The slope is the coefficient of x, and the y-intercept is the constant value.

This system with the solution is shown below:

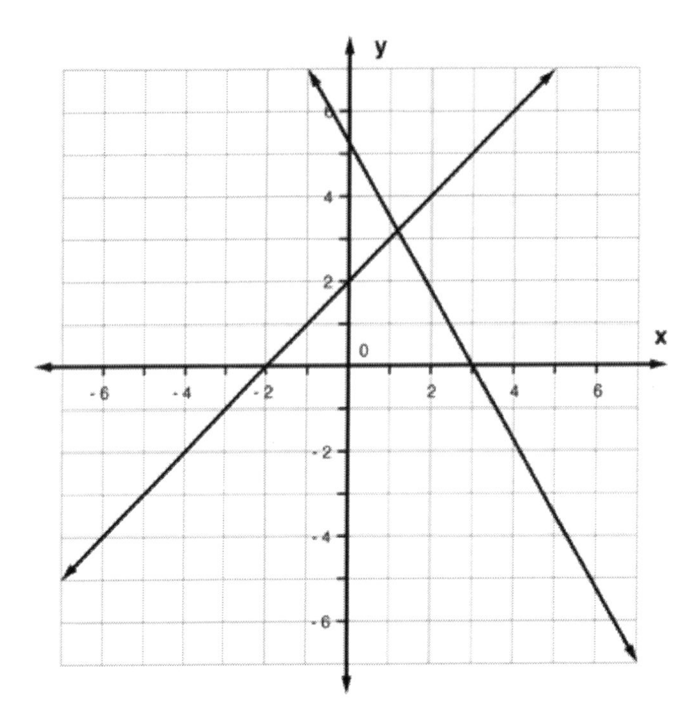

If the lines intersect, the point of intersection is the solution to the system. Every point on a line represents an ordered pair that makes its equation true. The ordered pair represented by this point of intersection lies on both lines and therefore makes both equations true. This ordered pair should be checked by substituting its values into both of the original equations of the system. Note that given a system of equations and an ordered pair, the ordered pair can be determined to be a solution or not by checking it in both equations.

If, when graphed, the lines representing the equations of a system do not intersect, then the two lines are parallel to each other or they are the same exact line. Parallel lines extend in the same direction without ever meeting. A system consisting of parallel lines has no solution. If the equations for a system represent the same exact line, then every point on the line is a solution to the system. In this case, there would be an infinite number of solutions. A system consisting of intersecting lines is referred to as independent; a system consisting of parallel lines is referred to as inconsistent; and a system consisting of coinciding lines is referred to as dependent.

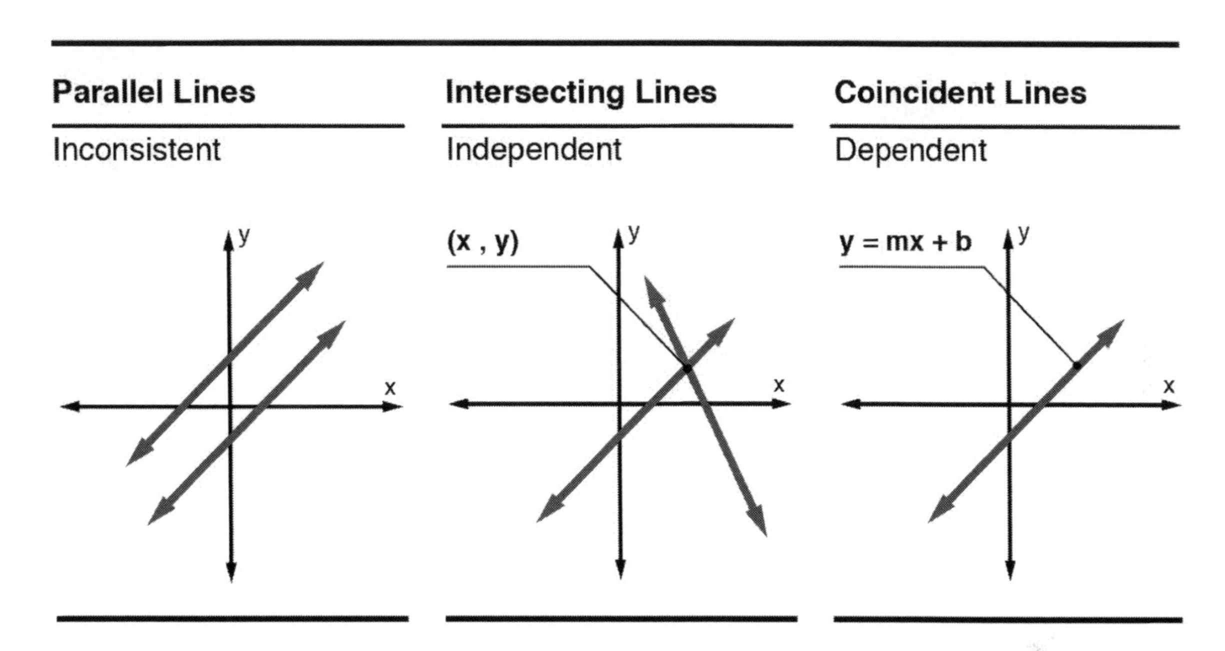

Matrices can also be used to solve systems of linear equations. Specifically, for systems, the coefficients of the linear equations in standard form are the entries in the matrix. Using the same system of linear equations as above, $x - y = -2$ and $3x + 2y = 9$, the matrix to represent the system is:

$$\begin{bmatrix} 1 & -1 \\ 3 & 2 \end{bmatrix} \begin{bmatrix} x \\ y \end{bmatrix} = \begin{bmatrix} -2 \\ 9 \end{bmatrix}$$

To solve this system using matrices, the inverse matrix must be found. For a general 2×2 matrix:

$$\begin{bmatrix} a & b \\ c & d \end{bmatrix}$$

The inverse matrix is found by the expression:

$$\frac{1}{ad - bc} \begin{bmatrix} d & -b \\ -c & a \end{bmatrix}$$

The inverse matrix for the system given above is:

$$\frac{1}{2 - -3} \begin{bmatrix} 2 & 1 \\ -3 & 1 \end{bmatrix} = \frac{1}{5} \begin{bmatrix} 2 & 1 \\ -3 & 1 \end{bmatrix}$$

The next step in solving is to multiply this identity matrix by the system matrix above. This is given by the following equation:

$$\frac{1}{5}\begin{bmatrix} 2 & 1 \\ -3 & 1 \end{bmatrix}\begin{bmatrix} 1 & -1 \\ 3 & 2 \end{bmatrix}\begin{bmatrix} x \\ y \end{bmatrix} = \begin{bmatrix} 2 & 1 \\ -3 & 1 \end{bmatrix}\begin{bmatrix} -2 \\ 9 \end{bmatrix}\frac{1}{5}$$

which simplifies to

$$\frac{1}{5}\begin{bmatrix} 5 & 0 \\ 0 & 5 \end{bmatrix}\begin{bmatrix} x \\ y \end{bmatrix} = \frac{1}{5}\begin{bmatrix} 5 \\ 15 \end{bmatrix}$$

Solving for the solution matrix, the answer is:

$$\begin{bmatrix} 1 & 0 \\ 0 & 1 \end{bmatrix}\begin{bmatrix} x \\ y \end{bmatrix} = \begin{bmatrix} 1 \\ 3 \end{bmatrix}$$

Since the first matrix is the identity matrix, the solution is $x = 1$ and $y = 3$.

Finding solutions to systems of equations is essentially finding what values of the variables make both equations true. It is finding the input value that yields the same output value in both equations. For functions $g(x)$ and $f(x)$, the equation $g(x) = f(x)$ means the output values are being set equal to each other. Solving for the value of x means finding the x-coordinate that gives the same output in both functions. For example, $f(x) = x + 2$ and $g(x) = -3x + 10$ is a system of equations. Setting $f(x) = g(x)$ yields the equation $x + 2 = -3x + 10$. Solving for x, gives the x-coordinate $x = 2$ where the two lines cross. This value can also be found by using a table or a graph. On a table, both equations can be given the same inputs, and the outputs can be recorded to find the point(s) where the lines cross. Any method of solving finds the same solution, but some methods are more appropriate for some systems of equations than others.

Solving Inequalities and Graphing the Answer on a Number Line

Linear inequalities and linear equations are both comparisons of two algebraic expressions. However, unlike equations in which the expressions are equal to each other, linear inequalities compare expressions that are unequal. Linear equations typically have one value for the variable that makes the statement true. Linear inequalities generally have an infinite number of values that make the statement true.

Linear inequalities are a concise mathematical way to express the relationship between unequal values. More specifically, they describe in what way the values are unequal. A value could be greater than ($>$); less than ($<$); greater than or equal to ($\geq$); or less than or equal to ($\leq$) another value. The statement "five times a number added to forty is more than sixty-five" can be expressed as $5x + 40 > 65$. Common words and phrases that express inequalities are:

Symbol	Phrase
$<$	is under, is below, smaller than, beneath
$>$	is above, is over, bigger than, exceeds
$\leq$	no more than, at most, maximum
$\geq$	no less than, at least, minimum

If a problem were to say, "The sum of a number and 5 is greater than -8 times the number," then an inequality would be used instead of an equation. Using key words again, *greater than* is represented by the symbol >. The inequality $n + 5 > -8n$ can be solved using the same techniques, resulting in $n < -\frac{5}{9}$. The only time solving an inequality differs from solving an equation is when a negative number is either multiplied by or divided by each side of the inequality. The sign must be switched in this case. For this example, the graph of the solution changes to the following graph because the solution represents all real numbers less than $-\frac{5}{9}$. Not included in this solution is $-\frac{5}{9}$ because it is a *less than* symbol, not *equal to*.

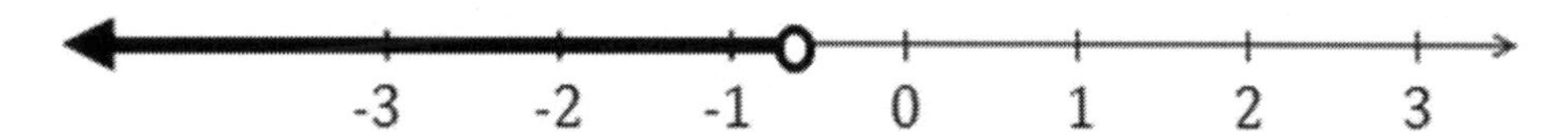

When solving a linear inequality, the solution is the set of all numbers that makes the statement true. The inequality $x + 2 \geq 6$ has a solution set of 4 and every number greater than 4 (4.0001, 5, 12, 107, etc.). Adding 2 to 4 or any number greater than 4 would result in a value that is greater than or equal to 6. Therefore, $x \geq 4$ would be the solution set.

Solution sets for linear inequalities often will be displayed using a number line. If a value is included in the set ($\geq$ or $\leq$), there is a shaded dot placed on that value and an arrow extending in the direction of the solutions. For a variable $>$ or $\geq$ a number, the arrow would point right on the number line (the direction where the numbers increase); and if a variable is $<$ or $\leq$ a number, the arrow would point left (where the numbers decrease). If the value is not included in the set ($>$ or $<$), an open circle on that value would be used with an arrow in the appropriate direction.

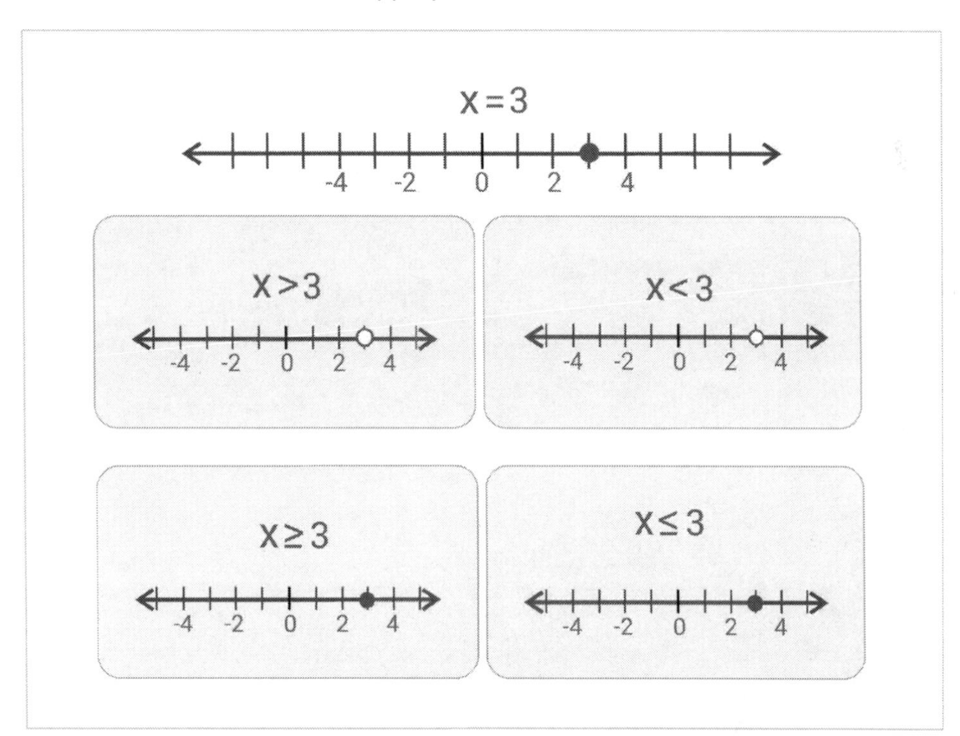

Students may be asked to write a linear inequality given a graph of its solution set. To do so, they should identify whether the value is included (shaded dot or open circle) and the direction in which the arrow is pointing.

In order to algebraically solve a linear inequality, the same steps should be followed as in solving a linear equation. The inequality symbol stays the same for all operations EXCEPT when dividing by a negative number. If dividing by a negative number while solving an inequality, the relationship reverses (the sign flips). Dividing by a positive does not change the relationship, so the sign stays the same. In other words, $>$ switches to $<$ and vice versa. An example is shown below.

Solve $-2(x + 4) \leq 22$ for the value of x.

First, distribute -2 to the binomial by multiplying:

$$-2x - 8 \leq 22$$

Next, add 8 to both sides to isolate the variable:

$$-2x \leq 30$$

Divide both sides by -2 to solve for x:

$$x \geq -15$$

With a single equation in two variables, the solutions are limited only by the situation the equation represents. When two equations or inequalities are used, more constraints are added. For example, in a system of linear equations, there is often—although not always—only one answer. The point of intersection of two lines is the solution. For a system of inequalities, there are infinitely many answers.

The intersection of two solution sets gives the solution set of the system of inequalities. In the following graph, the darker shaded region is where two inequalities overlap. Any set of x and y found in that region satisfies both inequalities. The line with the positive slope is solid, meaning the values on that line are included in the solution.

The line with the negative slope is dotted, so the coordinates on that line are not included.

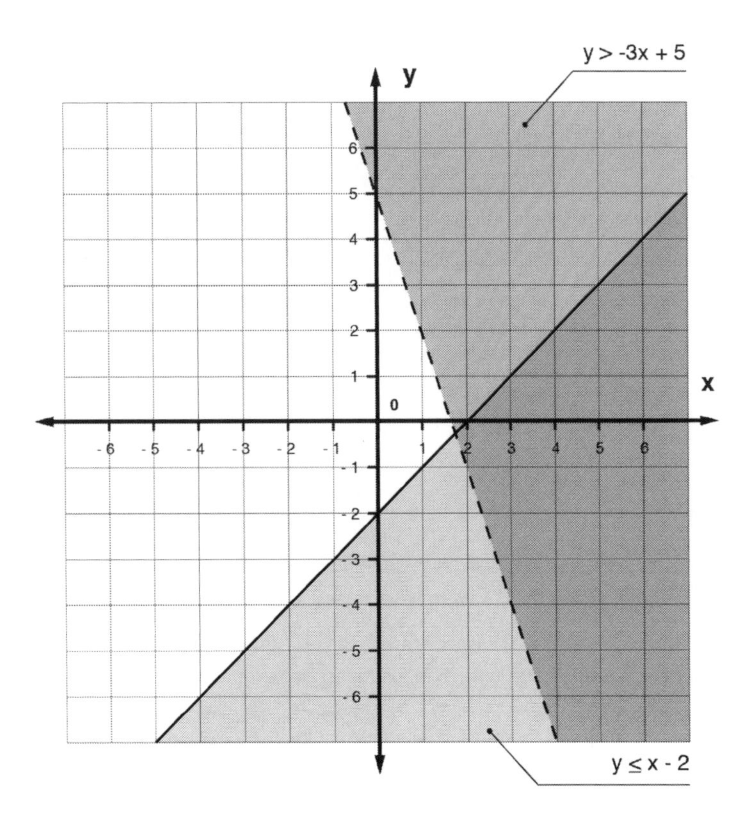

Quadratic Equations with One Variable

A **quadratic equation** can be written in the form $y = ax^2 + bx + c$. The u-shaped graph of a quadratic equation is called a **parabola**. The graph can either open up or open down (upside down u). The graph is symmetric about a vertical line, called the **axis of symmetry**. Corresponding points on the parabola are directly across from each other (same *y*-value) and are the same distance from the axis of symmetry (on either side). The axis of symmetry intersects the parabola at its **vertex**. The *y*-value of the vertex represents the minimum or maximum value of the function. If the graph opens up, the value of *a* in its equation is positive and the vertex represents the minimum of the function. If the graph opens down, the value of *a* in its equation is negative and the vertex represents the maximum of the function.

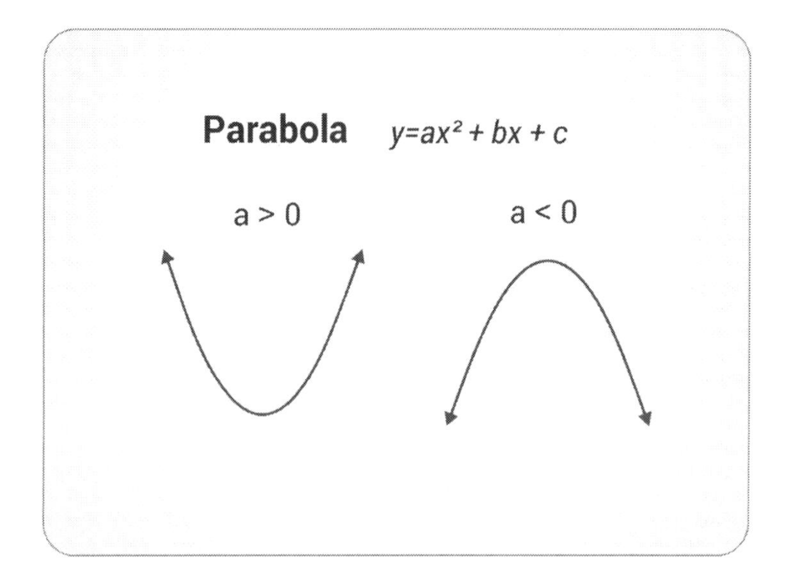

A quadratic equation can be written in the standard form: $y = ax^2 + bx + c$. It can be represented by a u-shaped graph called a parabola. For a quadratic equation where the value of *a* is positive, as the inputs increase, the outputs increase until a certain value (maximum of the function) is reached. As inputs increase past the value that corresponds with the maximum output, the relationship reverses and the outputs decrease. For a quadratic equation where *a* is negative, as the inputs increase, the outputs (1) decrease, (2) reach a maximum, and (3) then increase.

Consider a ball thrown straight up into the air. As time passes, the height of the ball increases until it reaches its maximum height. After reaching the maximum height, as time increases, the height of the ball decreases (it is falling toward the ground). This relationship can be expressed as a quadratic equation where time is the input (*x*), and the height of the ball is the output (*y*).

Equations with one variable (linear equations) can be solved using the addition principle and multiplication principle. If $a = b$, then $a + c = b + c$, and $ac = bc$. Given the equation $2x - 3 = 5x + 7$, the first step is to combine the variable terms and the constant terms. Using the principles, expressions can be added and subtracted onto and off both sides of the equals sign, so the equation turns into $-10 = 3x$. Dividing by 3 on both sides through the multiplication principle with $c = \frac{1}{3}$ results in the final answer of $x = \frac{-10}{3}$.

However, this same process cannot be used to solve nonlinear equations, including quadratic equations. Quadratic equations have a higher degree than linear ones (2 versus 1) and are not solved by simply using opposite operations. When an equation has a degree of 2, completing the square is an option. For example, the quadratic equation $x^2 - 6x + 2 = 0$ can be rewritten by completing the square. The goal of completing the square is to get the equation into the form $(x - p)^2 = q$. Using the example, the constant term 2 first needs to be moved over to the opposite side by subtracting. Then, the square can be completed by adding 9 to both sides, which is the square of half of the coefficient of the middle term $-6x$.

The current equation is $x^2 - 6x + 9 = 7$. The left side can be factored into a square of a binomial, resulting in $(x - 3)^2 = 7$. To solve for x, the square root of both sides should be taken, resulting in:

$$(x - 3) = \pm\sqrt{7} \; and \; x = 3 \pm \sqrt{7}$$

Other ways of solving quadratic equations include graphing, factoring, and using the quadratic formula. The equation $y = x^2 - 4x + 3$ can be graphed on the coordinate plane, and the solutions can be observed where it crosses the x-axis. The graph will be a parabola that opens up with two solutions at 1 and 3.

If quadratic equations take the form $ax^2 - b = 0$, then the equation can be solved by adding *b* to both sides and dividing by *a* to get:

$$x^2 = \frac{b}{a} \text{ or } x = \pm\sqrt{\frac{b}{a}}$$

Note that this is actually two separate solutions, unless *b* happens to be 0.

If a quadratic equation has no constant—so that it takes the form $ax^2 + bx = 0$—then the *x* can be factored out to get $x(ax + b) = 0$. Then, the solutions are $x = 0$, together with the solutions to $ax + b = 0$. Both factors x and $(ax + b)$ can be set equal to zero to solve for x because one of those values must be zero for their product to equal zero. For an equation $ab = 0$ to be true, either $a = 0$, or $b = 0$.

A given quadratic equation $x^2 + bx + c$ can be factored into $(x + A)(x + B)$, where $A + B = b$, and $AB = c$. Finding the values of A and B can take time, but such a pair of numbers can be found by guessing and checking. Looking at the positive and negative factors for *c* offers a good starting point.

For example, in $x^2 - 5x + 6$, the factors of 6 are 1, 2, and 3. Now,$(-2)(-3) = 6$, and $-2 - 3 = -5$. In general, however, this may not work, in which case another approach may need to be used.

A quadratic equation of the form $x^2 + 2xb + b^2 = 0$ can be factored into $(x + b)^2 = 0$. Similarly, $x^2 - 2xy + y^2 = 0$ factors into $(x - y)^2 = 0$.

The first method of completing the square can be used in finding the second method, the quadratic formula. It can be used to solve any quadratic equation. This formula may be the longest method for solving quadratic equations and is commonly used as a last resort after other methods are ruled out.

It can be helpful in memorizing the formula to see where it comes from, so here are the steps involved.

The most general form for a quadratic equation is:

$$ax^2 + bx + c = 0$$

First, dividing both sides by *a* leaves us with:

$$x^2 + \frac{b}{a}x + \frac{c}{a} = 0$$

To complete the square on the left-hand side, c/a can be subtracted on both sides to get:

$$x^2 + \frac{b}{a}x = -\frac{c}{a}$$

$(\frac{b}{2a})^2$ is then added to both sides.

This gives:

$$x^2 + \frac{b}{a}x + (\frac{b}{2a})^2 = (\frac{b}{2a})^2 - \frac{c}{a}$$

The left can now be factored and the right-hand side simplified to give:

$$(x + \frac{b}{2a})^2 = \frac{b^2 - 4ac}{4a}$$

Taking the square roots gives:

$$x + \frac{b}{2a} = \pm\frac{\sqrt{b^2 - 4ac}}{2a}$$

Solving for x yields the quadratic formula:

$$x = \frac{-b \pm \sqrt{b^2 - 4ac}}{2a}$$

It isn't necessary to remember how to get this formula but memorizing the formula itself is the goal.

If an equation involves taking a root, then the first step is to move the root to one side of the equation and everything else to the other side. That way, both sides can be raised to the index of the radical in order to remove it, and solving the equation can continue.

Locating Points and Graphing Equations

The coordinate plane, sometimes referred to as the Cartesian plane, is a two-dimensional surface consisting of a horizontal and a vertical number line. The horizontal number line is referred to as the *x*-axis, and the vertical number line is referred to as the *y*-axis. The *x*-axis and *y*-axis intersect (or cross) at a point called the origin. At the origin, the value of the *x*-axis is zero and the value of the *y*-axis is zero. The coordinate plane identifies the exact location of a point that is plotted on the two-dimensional surface. Like a map, the location of all points on the plane are in relation to the origin. Along the *x*-axis (horizontal line), numbers to the right of the origin are positive and increasing in value (1, 2, 3, . . .) and to the left of the origin numbers are negative and decreasing in value (-1, -2, -3, . . .). Along the *y*-axis (vertical line), numbers above the origin are positive and increasing in value and numbers below the origin are negative and decreasing in value.

The *x*- and *y*-axis divide the coordinate plane into four sections. These sections are referred to as quadrant one, quadrant two, quadrant three, and quadrant four, and are often written with Roman numerals I, II, III, and IV.

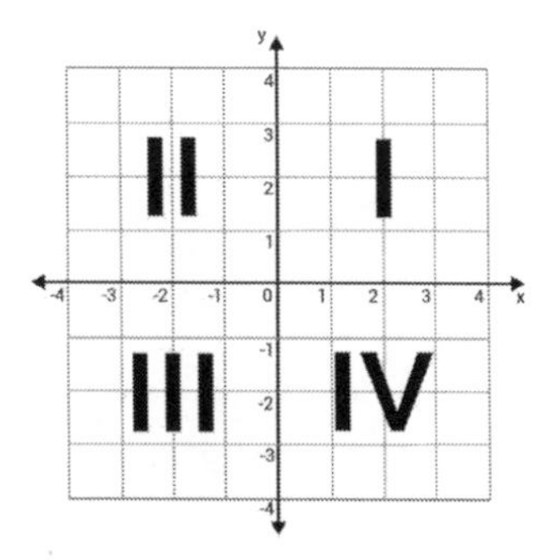

The upper right section is Quadrant I and consists of points with positive *x*-values and positive *y*-values. The upper left section is Quadrant II and consists of points with negative *x*-values and positive *y*-values. The bottom left section is Quadrant III and consists of points with negative *x*-values and negative *y*-values. The bottom right section is Quadrant IV and consists of points with positive *x*-values and negative *y*-values.

Graphing in the Coordinate Plane

The coordinate plane consists of the intersection of two number lines. A number line is a straight line in which numbers are marked at evenly spaced intervals with tick marks. The horizontal number line in the coordinate plane represents the *x*-axis, and the vertical number line represents the *y*-axis. The coordinate plane represents a representation of real-world space, and any point within the plane can be defined by a set of **coordinates** (x, y). The coordinates consist of two numbers, x and y, which represent a position on each number line. The coordinates can also be referred to as an **ordered pair,** and (0, 0) is the ordered pair known as the **vertex**, or the origin, the point in which the axes intersect. Positive x-coordinates go to the right of the vertex, and positive y-coordinates go up. Negative x-coordinates go left, and negative y-coordinates go down.

Here is an example of the coordinate plane with a point plotted:

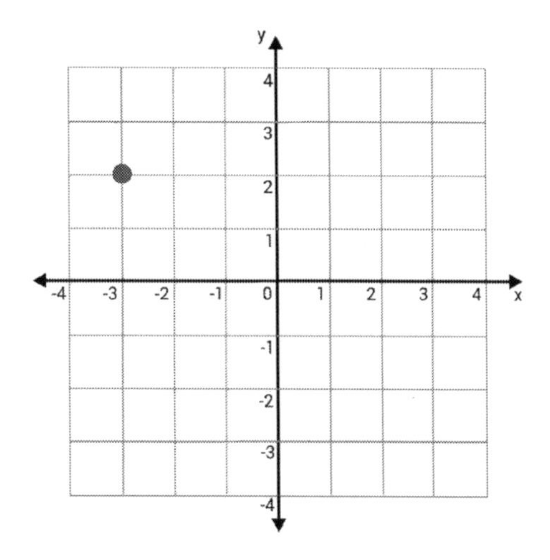

In order to plot a point on the coordinate plane, each coordinate must be considered individually. The value of x represents how many units away from the vertex the point lies on the x-axis. The value of y represents the number of units away from the vertex that the point lies on the y-axis.

For example, given the ordered pair (5, 4), the x-coordinate, 5, is the distance from the origin along the x-axis, and the y-coordinate, 4, is the distance from the origin along the y-axis. This is determined by counting 5 units to the right from (0, 0) along the x-axis and then counting 4 units up from that point, to reach the point where $x = 5$ and $y = 4$. In order to graph the single point, the point should be marked there with a dot and labeled as (5, 4). Every point on the plane has its own ordered pair.

Graphing on the Coordinate Plane Using Mathematical Problems, Tables, and Patterns

Data can be recorded using a coordinate plane. Graphs are utilized frequently in real-world applications and can be seen in many facets of everyday life. A relationship can exist between the *x*- and *y*-coordinates that are plotted on a graph, and those values can represent a set of data that can be listed in a table. Going back and forth between the table and the graph is an important concept and defining the relationship between the variables is the key that links the data to a real-life application.

For example, temperature increases during a summer day. The *x*-coordinate can be used to represent hours in the day, and the *y*-coordinate can be used to represent the temperature in degrees. The graph would show the temperature at each hour of the day. Time is almost always plotted on the *x*-axis, and utilizing different units on each axis, if necessary, is important. Labeling the axes with units is also important.

Within the first quadrant of the coordinate plane, both the x and y values are positive. Most real-world problems can be plotted in this quadrant because most real-world quantities, such as time and distance, are positive. Consider the following table of values:

X	Y
1	2
2	4
3	6
4	8

Each row gives a coordinate pair. For example, the first row gives the coordinates (1,2). Each *x*-value tells you how far to move from the origin, the point (0,0), to the right, and each *y*-value tells you how far to move up from the origin.

Here is the graph of the points listed above in the table in addition to the origin:

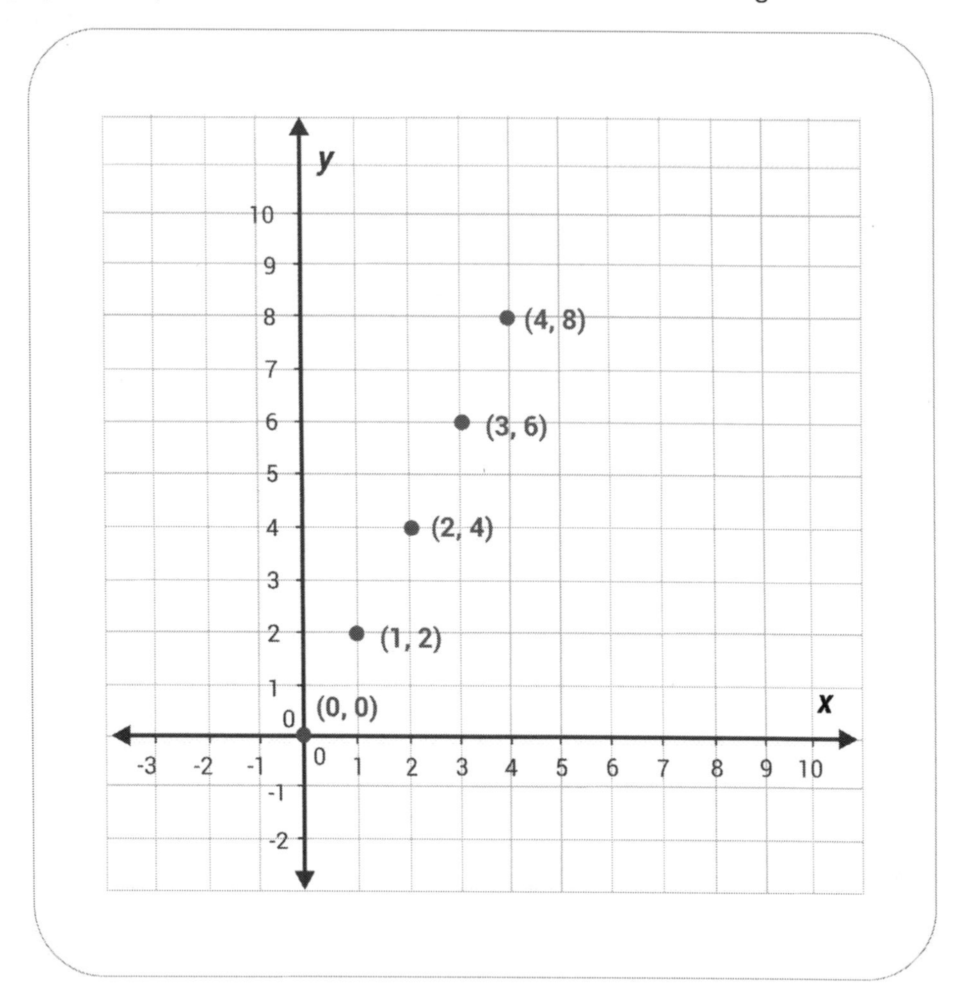

Notice that each *y*-value is found by doubling the *x*-value that forms the other portion of its coordinate pair.

Determining the Slope of a Line from a Graph, Equation, or Table

Rate of change for any line calculates the steepness of the line over a given interval. Rate of change is also known as the **slope** or rise/run. The slope of a linear function is given by the change in *y* divided by the change in *x*. So, the formula looks like this:

$$slope = \frac{y_2 - y_1}{x_2 - x_1}$$

In the graph below, two points are plotted. The first has the coordinates of (0,1) and the second point is (2,3). Remember that the x coordinate is always placed first in coordinate pairs. Work from left to right when identifying coordinates. Thus, the point on the left is point 1 (0,1) and the point on the right is point 2 (2,3).

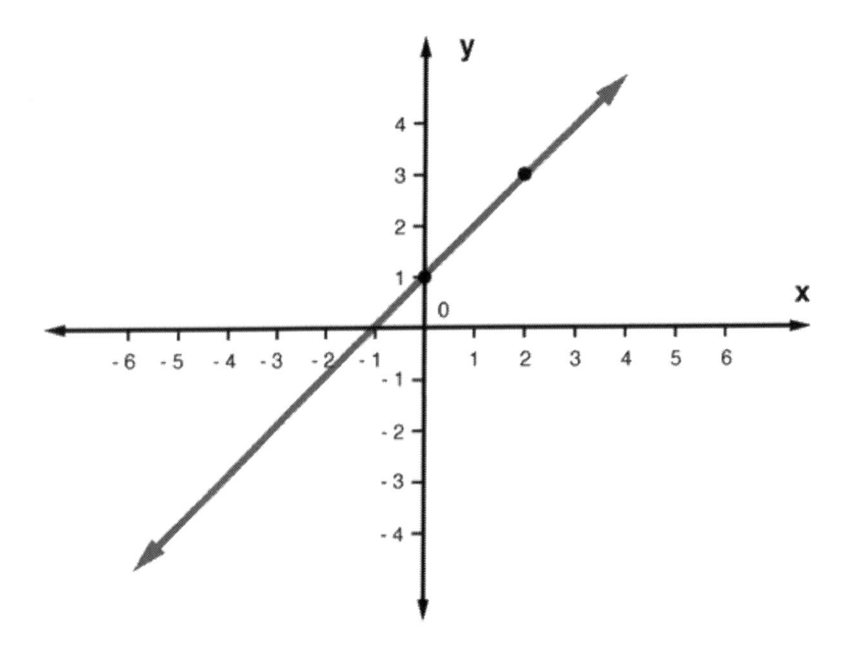

Now we need to just plug those numbers into the equation:

$$slope = \frac{3-1}{2-0}$$

$$slope = \frac{2}{2}$$

$$slope = 1$$

This means that for every increase of 1 for x, y also increased by 1. You can see this in the line. When x equalled 0, y equalled 1, and when x was increased to 1, y equalled 2.

Slope can be thought of as determining the rise over run:

$$slope = \frac{rise}{run}$$

The rise being the change vertically on the y axis and the run being the change horizontally on the x axis.

Proportional Relationships for Equations and Graphs

The rate of change for a linear function is constant and can be determined based on a few representations. One method is to place the equation in slope-intercept form: $y = mx + b$. Thus, m is the slope, and b is the y-intercept. In the graph below, the equation is $y = x + 1$, where the slope is 1 and the y-intercept is 1. For every vertical change of 1 unit, there is a horizontal change of 1 unit.

The x-intercept is -1, which is the point where the line crosses the x-axis:

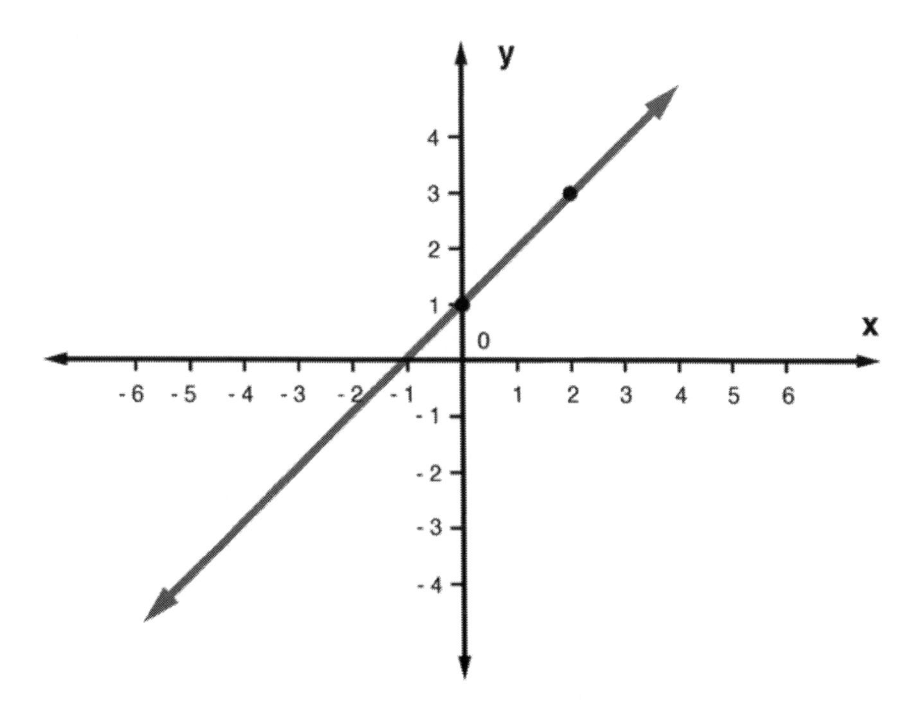

Let's look at an example of a proportional, or linear relationship, seen in the real world.

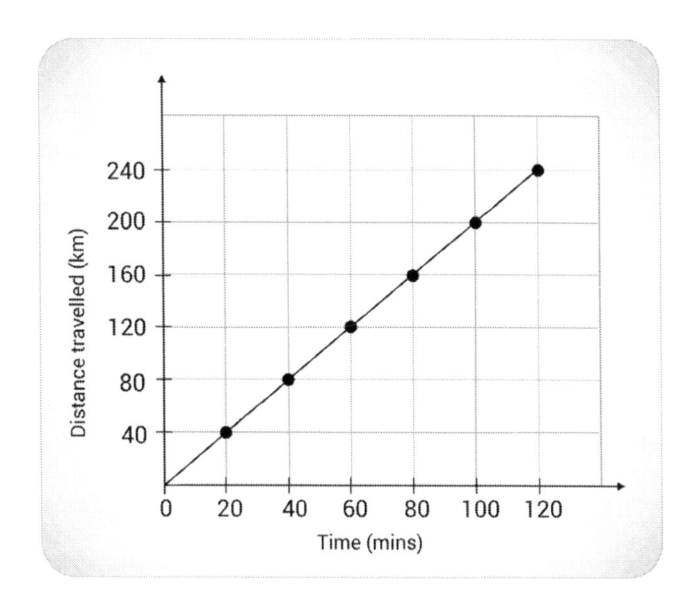

The graph above represents the relationship between distance traveled and time. To find the distance traveled in 80 minutes, the mark for 80 minutes is located at the bottom of the graph. By following this mark directly up on the graph, the corresponding point for 80 minutes is directly across from the 160-kilometer mark. This information indicates that the distance travelled in 80 minutes is 160 kilometers. To predict information not displayed on the graph, the way in which the variables change with respect to one another is determined. In this case, distance increases by 40 kilometers as time increases by 20 minutes. This information can be used to continue the data in the graph or convert the values to a table.

Let's try another example. Jim owns a car wash and charges $40 per car. The rent for the facility is $350 per month. An equation can be written to relate the number of cars Jim cleans to the money he makes per month. Let x represent the number of cars and y represent the profit Jim makes each month from the car wash. The equation $y = 40x - 350$ can be used to show Jim's profit or loss. Since this equation has two variables, the coordinate plane can be used to show the relationship and predict profit or loss for Jim. The following graph shows that Jim must wash at least nine cars to pay the rent, where $x = 9$. Anything nine cars and above yield a profit shown in the value on the y-axis.

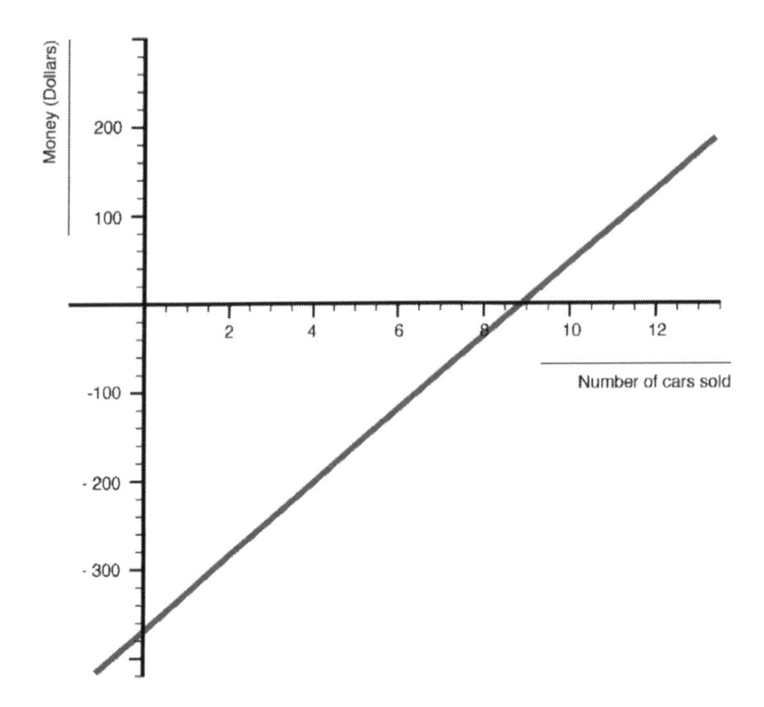

Formulas with two variables are equations used to represent a specific relationship. For example, the formula $d = rt$ represents the relationship between distance, rate, and time. If Bob travels at a rate of 35 miles per hour on his road trip from Westminster to Seneca, the formula $d = 35t$ can be used to represent his distance traveled in a specific length of time. Formulas can also be used to show different roles of the variables, transformed without any given numbers. Solving for r, the formula becomes $\frac{d}{t} = r$. The t is moved over by division so that **rate** is a function of distance and time.

Features of Graphs and Tables for Linear and Nonlinear Relationships

As mentioned, linear relationships describe the way two quantities change with respect to each other. The relationship is defined as linear because a line is produced if all the sets of corresponding values are graphed on a coordinate grid. When expressing the linear relationship as an equation, the equation is often written in the form $y = mx + b$ (**slope-intercept form**) where *m* and *b* are numerical values and *x* and *y* are variables (for example, $y = 5x + 10$). The slope is the coefficient of x, and the y-intercept is the constant value. The slope of the line containing the same two points is $m = \frac{y_2 - y_1}{x_2 - x_1}$ and is also equal to rise/run. Given a linear equation and the value of either variable (*x* or *y*), the value of the other variable can be determined.

With polynomial functions such as quadratics, the x-intercepts represent zeros of the function. Finding the **zeros of polynomial functions** is the same process as finding the solutions of polynomial equations.

These are the points at which the graph of the function crosses the x-axis. In the following quadratic equation, factoring the binomial leads to finding the zeros of the function: $x^2 - 5x + 6 = y$. This equations factors into $(x - 3)(x - 2) = y$, where 2 and 3 are found to be the zeros of the function when y is set equal to zero. The zeros of any function are the x-values where the graph of the function on the coordinate plane crosses the x-axis, which is the same as an x-intercept.

Selecting an Equation that Best Represents a Graph

Three common functions used to model different relationships between quantities are linear, quadratic, and exponential functions. **Linear functions** are the simplest of the three, and the independent variable x has an exponent of 1. Written in the most common form, $y = mx + b$, the coefficient of x tells how fast the function grows at a constant rate, and the b-value tells the starting point. A **quadratic function** has an exponent of 2 on the independent variable x. Standard form for this type of function is $y = ax^2 + bx + c$, and the graph is a parabola. These type functions grow at a changing rate. **An exponential function** has an independent variable in the exponent $y = ab^x$. The graph of these types of functions is described as **growth** or **decay**, based on whether the **base**, b, is greater than or less than 1. These functions are different from quadratic functions because the base stays constant. A common base is base *e*.

The following three functions model a linear, quadratic, and exponential function respectively: $y = 2x$, $y = x^2$, and $y = 2^x$. Their graphs are shown below. The first graph, modeling the linear function, shows that the growth is constant over each interval. With a horizontal change of 1, the vertical change is 2. It models constant positive growth. The second graph shows the quadratic function, which is a curve that is symmetric across the y-axis. The growth is not constant, but the change is mirrored over the axis. The last graph models the exponential function, where the horizontal change of 1 yields a vertical change that increases more and more with each iteration of horizontal change. The exponential graph gets very close to the x-axis, but never touches it, meaning there is an asymptote there. The y-value can never be zero because the base of 2 can never be raised to an input value that yields an output of zero.

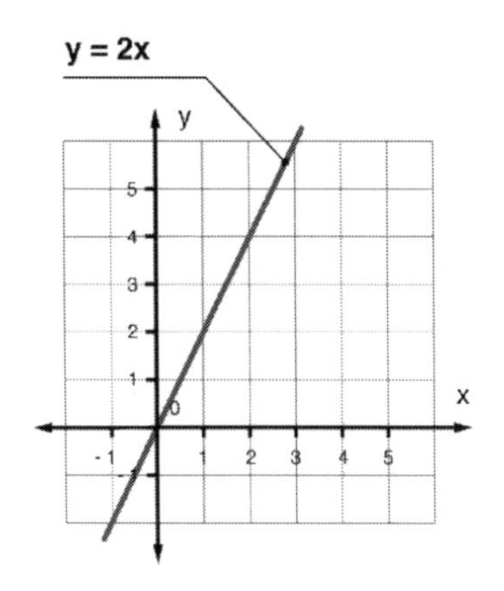

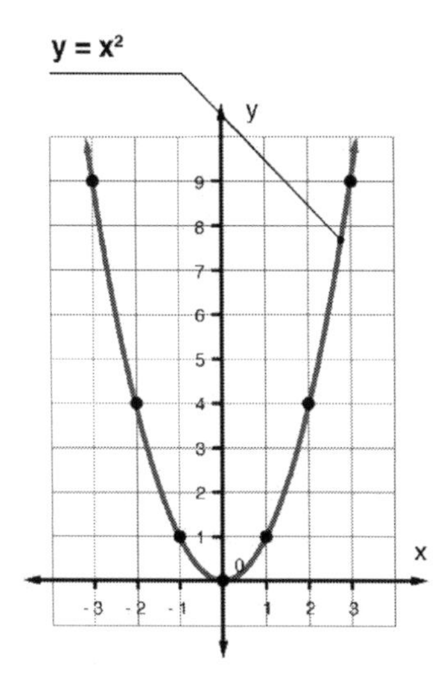

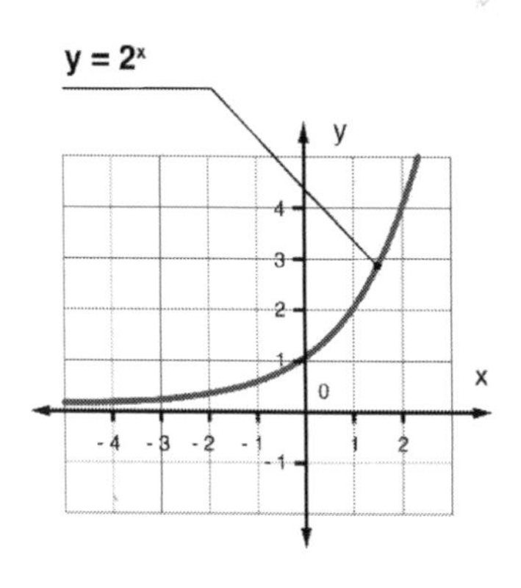

Determining the Graphical Properties and Sketch a Graph Given an Equation

Graphing a Linear Function

The process for graphing a line depends on the form in which its equation is written: slope-intercept form or standard form.

When an equation is written in slope-intercept form, $y = mx + b$, *m* represents the slope of the line and *b* represents the *y*-intercept. The *y*-intercept is the value of *y* when $x = 0$ and the point at which the graph of the line crosses the *y*-axis. The slope is the rate of change between the variables, expressed as a fraction. The fraction expresses the change in *y* compared to the change in *x*. If the slope is an integer, it should be written as a fraction with a denominator of 1. For example, 5 would be written as $\frac{5}{1}$.

To graph a line given an equation in slope-intercept form, the *y*-intercept should first be plotted. For example, to graph $y = -\frac{2}{3}x + 7$, the *y*-intercept of 7 would be plotted on the *y*-axis (vertical axis) at the point (0, 7). Next, the slope would be used to determine a second point for the line. Note that all that is necessary to graph a line is two points on that line. The slope will indicate how to get from one point on the line to another. The slope expresses vertical change (*y*) compared to horizontal change (*x*) and therefore is sometimes referred to as $\frac{rise}{run}$. The numerator indicates the change in the *y* value (move up for positive integers and move down for negative integers), and the denominator indicates the change in the *x* value. For the previous example, using the slope of $-\frac{2}{3}$, from the first point at the *y*-intercept, the second point should be found by counting down 2 and to the right 3. This point would be located at (3, 5).

When an equation is written in standard form, $Ax + By = C$, it is easy to identify the *x*- and *y*-intercepts for the graph of the line. Just as the *y*-intercept is the point at which the line intercepts the *y*-axis, the *x*-intercept is the point at which the line intercepts the *x*-axis. At the *y*-intercept, $x = 0$; and at the *x*-intercept, $y = 0$. Given an equation in standard form, $x = 0$ should be used to find the *y*-intercept. Likewise, $y = 0$ should be used to find the *x*-intercept. For example, to graph $3x + 2y = 6$, 0 for *y* results in $3x + 2(0) = 6$. Solving for *y* yields $x = 2$; therefore, an ordered pair for the line is (2, 0). Substituting 0 for *x* results in $3(0) + 2y = 6$. Solving for *y* yields $y = 3$; therefore, an ordered pair for the line is (0, 3). The two ordered pairs (the *x*- and *y*-intercepts) can be plotted and a straight line through them can be constructed.

T - chart

x	y
0	3
2	0

Intercepts

x - intercept : (2,0)

y - intercept : (0,3)

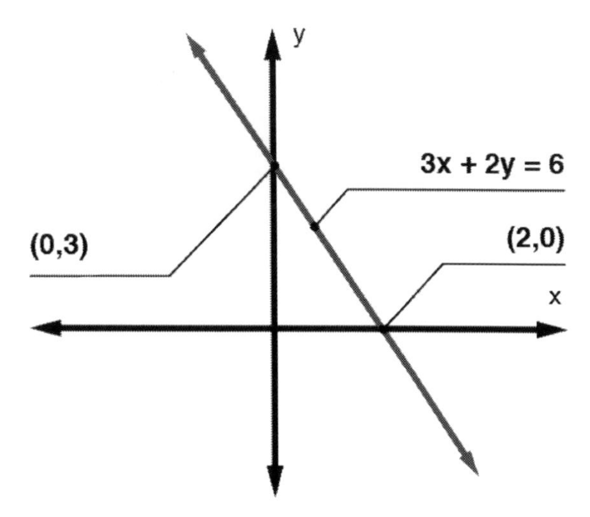

Graphing a Quadratic Function

The standard form of a quadratic function is $y = ax^2 + bx + c$. The graph of a quadratic function is a u-shaped (or upside-down u) curve, called a parabola, which is symmetric about a vertical line (axis of symmetry). To graph a parabola, its vertex (high or low point for the curve) and at least two points on each side of the axis of symmetry need to be determined.

Given a quadratic function in standard form, the axis of symmetry for its graph is the line $x = -\frac{b}{2a}$. The vertex for the parabola has an x-coordinate of $-\frac{b}{2a}$. To find the y-coordinate for the vertex, the calculated x-coordinate needs to be substituted. To complete the graph, two different x-values need to be selected and substituted into the quadratic function to obtain the corresponding y-values. This will give two points on the parabola. These two points and the axis of symmetry are used to determine the two points corresponding to these. The corresponding points are the same distance from the axis of symmetry (on the other side) and contain the same y-coordinate. Plotting the vertex and four other points on the parabola allows for constructing the curve.

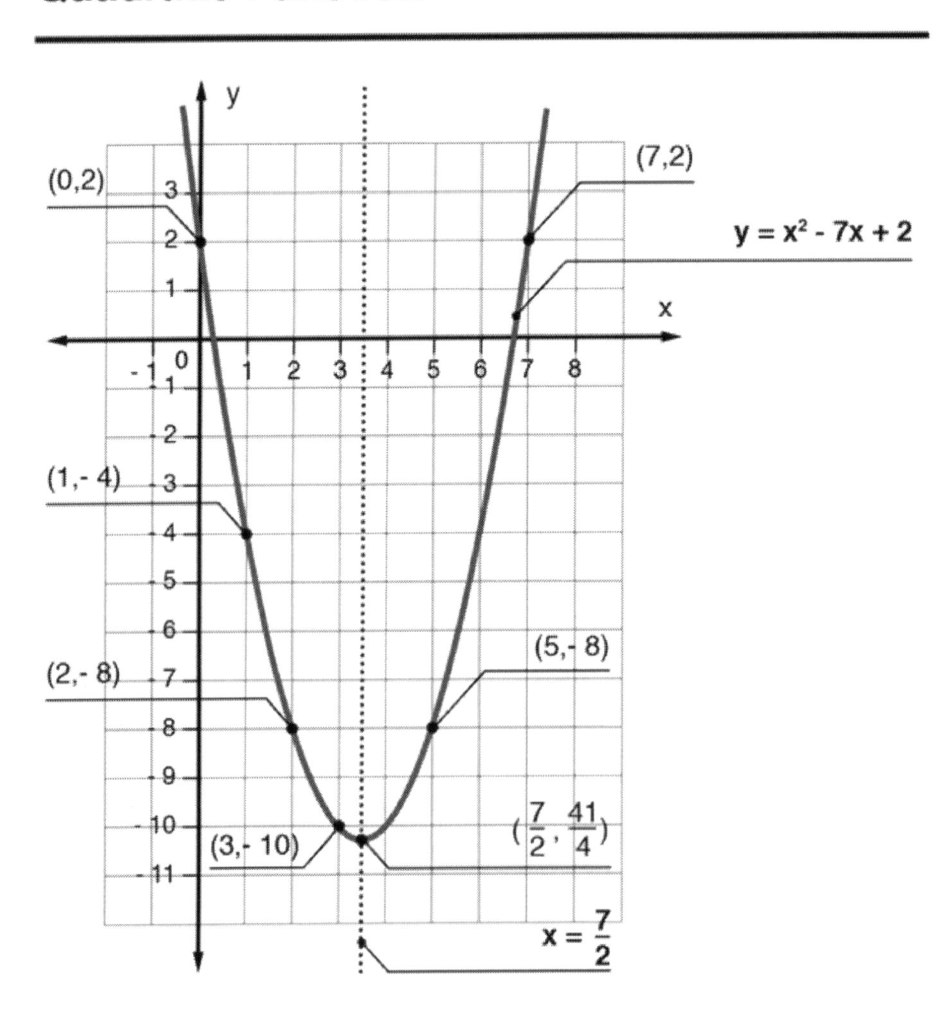

Graphing an Exponential Function

Exponential functions have a general form of $y = a \times b^x$. The graph of an exponential function is a curve that slopes upward or downward from left to right. The graph approaches a line, called an asymptote, as x or y increases or decreases. To graph the curve for an exponential function, x-values are selected and then substituted into the function to obtain the corresponding y-values. A general rule of thumb is to select three negative values, zero, and three positive values. Plotting the seven points on the graph for an exponential function should allow for constructing a smooth curve through them.

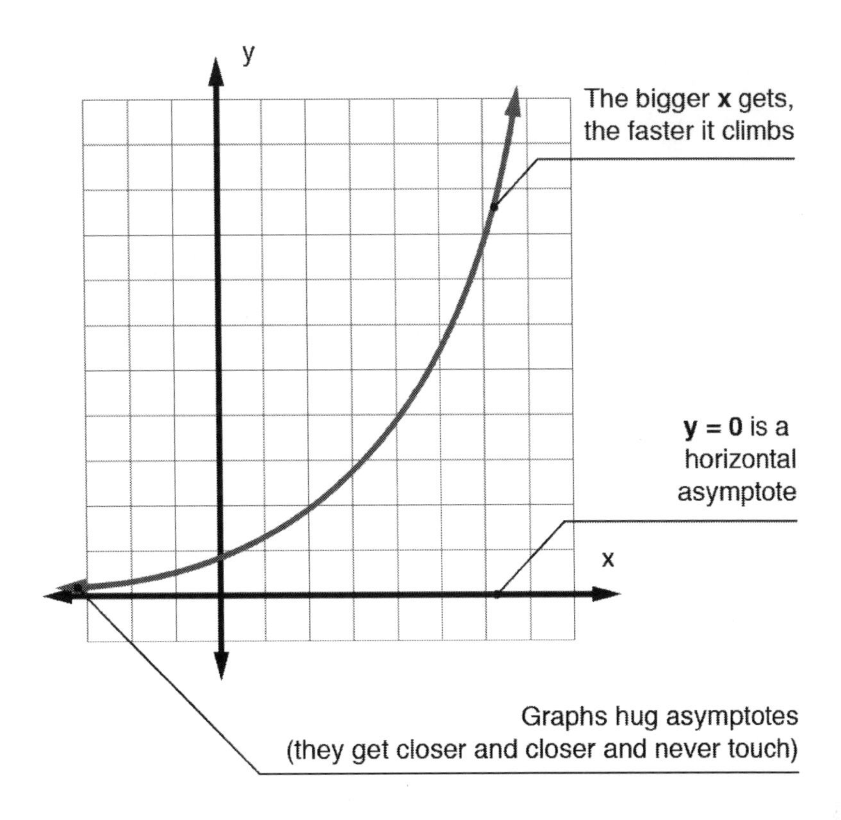

Equation of a Line from the Slope and a Point on a Line

Linear equations are written in slope-intercept form, $y = mx + b$, where *m* represents the slope of the line and *b* represents the *y*-intercept. The slope is the rate of change between the variables, usually expressed as a whole number or fraction. The *y*-intercept is the value of *y* when *x* = 0 (the point where the line intercepts the *y*-axis on a graph). Given the slope and *y*-intercept of a line, the values are substituted for *m* and *b* into the equation. A line with a slope of ½ and *y*-intercept of -2 would have an equation $y = \frac{1}{2}x - 2$.

The point-slope form of a line, $y - y_1 = m(x - x_1)$, is used to write an equation when given an ordered pair (point on the equation's graph) for the function and its rate of change (slope of the line). The values for the slope, *m*, and the point (x_1, y_1) are substituted into the point-slope form to obtain the equation of the line. A line with a slope of 3 and an ordered pair (4, -2) would have an equation:

$$y - (-2) = 3(x - 4)$$

If a question specifies that the equation be written in slope-intercept form, the equation should be manipulated to isolate *y*:

Solve: $y - (-2) = 3(x - 4)$

Distribute: $y + 2 = 3x - 12$

Subtract 2 from both sides: $y = 3x - 14$

Equation of a Line from Two Points

Given two ordered pairs for a function, (x_1, y_1) and (x_2, y_2), it is possible to determine the rate of change between the variables (slope of the line). To calculate the slope of the line, m, the values for the ordered pairs should be substituted into the formula: $m = \frac{y_2 - y_1}{x_2 - x_1}$. The expression is substituted to obtain a whole number or fraction for the slope. Once the slope is calculated, the slope and either of the ordered pairs should be substituted into the point-slope form to obtain the equation of the line.

Using Slope of a Line

Two lines are parallel if they have the same slope and different intercept. Two lines are **perpendicular** if the product of their slope equals -1. Parallel lines never intersect unless they are the same line, and perpendicular lines intersect at a right angle. If two lines aren't parallel, they must intersect at one point. If lines do cross, they're labeled as **intersecting lines** because they "intersect" at one point. If they intersect at more than one point, they're the same line. Determining equations of lines based on properties of parallel and perpendicular lines appears in word problems. To find an equation of a line, both the slope and a point the line goes through are necessary. Therefore, if an equation of a line is needed that's parallel to a given line and runs through a specified point, the slope of the given line and the point are plugged into the point-slope form of an equation of a line. Secondly, if an equation of a line is needed that's perpendicular to a given line running through a specified point, the negative reciprocal of the slope of the given line and the point are plugged into the **point-slope form**. Also, if the point of intersection of two lines is known, that point will be used to solve the set of equations. Therefore, to solve a system of equations, the point of intersection must be found. If a set of two equations with two unknown variables has no solution, the lines are parallel.

The **Parallel Postulate** states that if two parallel lines are cut by a transversal, then the corresponding angles are equal. Here is a picture that highlights this postulate:

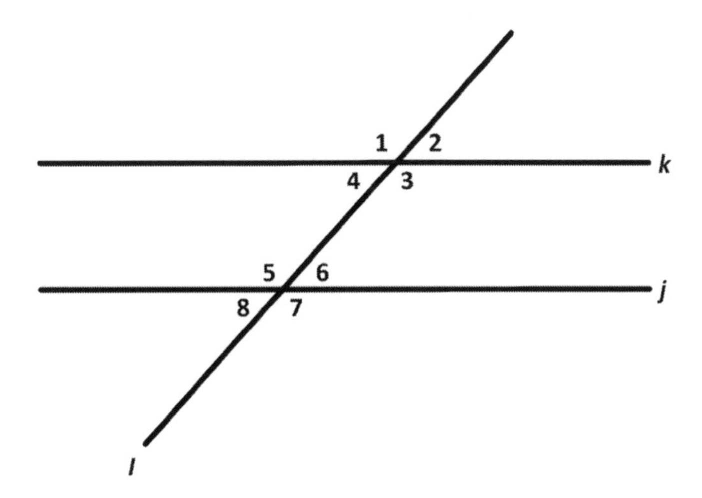

Because lines *k* and *i* are parallel, when cut by transversal *l,* angles 1 and 5 are equal, angles 2 and 6 are equal, angles 4 and 8 are equal, and angles 3 and 7 are equal. Note that angles 1 and 2, 3 and 4, 5 and 6, and 7 and 8 add up to 180 degrees.

This statement is equivalent to the **Alternate Interior Angle Theorem**, which states that when two parallel lines are cut by a transversal, the resultant interior angles are congruent. In the picture above, angles 3 and 5 are congruent, and angles 4 and 6 are congruent.

The Parallel Postulate or the Alternate Interior Angle Theorem can be used to find the missing angles in the following picture:

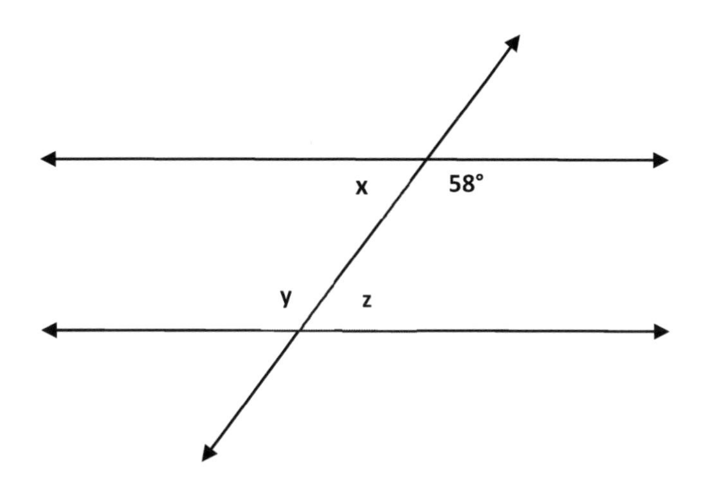

Assuming that the lines are parallel, angle x is found to be 122 degrees. Angle x and the 58-degree angle add up to 180 degrees. The Alternate Interior Angle Theorem states that angle y is equal to 58 degrees. Also, angles y and z add up to 180 degrees, so angle z is 122 degrees. Note that angles x and z are also alternate interior angles, so their equivalence can be used to find angle z as well.

An equivalent statement to the Parallel Postulate is that the sum of all angles in a triangle is 180 degrees. Therefore, given any triangle, if two angles are known, the third can be found accordingly.

Functions Shown in Different Ways

First, it's important to understand the definition of a **relation**. Given two variables, *x* and *y*, which stand for unknown numbers, a relation between *x* and *y* is an object that splits all of the pairs (*x*, *y*) into those for which the relation is true and those for which it is false. For example, consider the relation of $x^2 = y^2$. This relationship is true for the pair (1, 1) and for the pair (-2, 2), but false for (2, 3). Another example of a relation is $x \leq y$. This is true whenever *x* is less than or equal to *y*.

A **function** is a special kind of relation where, for each value of *x*, there is only a single value of *y* that satisfies the relation. So, $x^2 = y^2$ is *not* a function because in this case, if *x* is 1, *y* can be either 1 or -1: the pair (1, 1) and (1, -1) both satisfy the relation. More generally, for this relation, any pair of the form $(a, \pm a)$ will satisfy it. On the other hand, consider the following relation: $y = x^2 + 1$. This is a function because for each value of *x*, there is a unique value of *y* that satisfies the relation. Notice, however, there are multiple values of *x* that give us the same value of *y*. This is perfectly acceptable for a function. Therefore, *y* is a function of *x*.

To determine if a relation is a function, check to see if every *x* value has a unique corresponding *y* value.

A function can be viewed as an object that has *x* as its input and outputs a unique *y*-value. It is sometimes convenient to express this using **function notation**, where the function itself is given a name, often *f*. To emphasize that *f* takes *x* as its input, the function is written as $f(x)$. In the above example, the equation could be rewritten as $f(x) = x^2 + 1$. To write the value that a function yields for some specific value of *x*, that value is put in place of *x* in the function notation. For example, $f(3)$ means the value that the function outputs when the input value is 3. If $f(x) = x^2 + 1$, then:

$$f(3) = 3^2 + 1 = 10$$

Another example of a function would be $f(x) = 4x + 4$, read "*f* of *x* is equal to four times *x* plus four." In this example, the input would be *x* and the output would be f(*x*). Ordered pairs would be represented as (*x*, f(*x*)). To find the output for an input value of 3, 3 would be substituted for *x* into the function as follows: $f(3) = 4(3) + 4$, resulting in $f(3) = 16$. Therefore, the ordered pair $(3, f(3)) = (3, 16)$. Note f(*x*) is a function of *x* denoted by *f*. Functions of *x* could be named *g*(*x*), read "*g* of *x*"; *p*(*x*), read "*p* of *x*"; etc.

As an example, the following function is in function notation:

$$f(x) = 3x - 4$$

The $f(x)$ represents the output value for an input of x. If $x = 2$, the equation becomes:

$$f(2) = 3(2) - 4 = 6 - 4 = 2$$

The input of 2 yields an output of 2, forming the ordered pair $(2, 2)$. The following set of ordered pairs corresponds to the given function: $(2, 2), (0, -4), (-2, -10)$. The set of all possible inputs of a function is its **domain**, and all possible outputs is called the **range**. By definition, each member of the domain is paired with only one member of the range.

Functions can also be defined recursively. In this form, they are not defined explicitly in terms of variables. Instead, they are defined using previously evaluated function outputs, starting with either $f(0)$ or $f(1)$. An example of a recursively defined function is:

$$f(1) = 2, f(n) = 2f(n-1) + 2n, n > 1$$

The domain of this function is the set of all integers.

A function can also be viewed as a table of pairs (*x*, *y*), which lists the value for *y* for each possible value of *x*.

Functions in Tables and Graphs

The domain and range of a function can be found by observing a table. The table below shows the input values $x = -1$ to $x = 1$ for the function $f(x) = x^2 - 3$. The range, or output, for these inputs results in a minimum of -3. On each side of $x = 0$, the numbers increase, showing that the range is all real numbers greater than or equal to -3.

x (domain/input)	y (range/output)
-2	1
-1	-2
0	-3
-1	-2
2	1

Determining the Domain and Range from a Given Graph of a Function

The domain and range of a function can also be found visually by its plot on the coordinate plane. In the function $f(x) = x^2 - 3$, for example, the domain is all real numbers because the parabola can stretch infinitely far left and right with no restrictions. This means that any input value from the real number system will yield an output in the real number system. For the range, the inequality $y \geq -3$ would be used to describe the possible output values because the parabola has a minimum at $y = -3$. This means there will not be any real output values less than -3 because -3 is the lowest value the function reaches on the y-axis.

Determining the Domain and Range of a Given Function

The set of all possible values for *x* in $f(x)$ is called the domain of the function, and the set of all possible outputs is called the range of the function. Note that usually the domain is assumed to be all real numbers, except those for which the expression for $f(x)$ is not defined, unless the problem specifies otherwise. An example of how a function might not be defined is in the case of $f(x) = \frac{1}{x+1}$, which is not defined when $x = -1$ (which would require dividing by zero). Therefore, in this case the domain would be all real numbers except $x = -1$.

Interpreting Domain and Range in Real-World Settings

A function can be built from the information given in a situation. For example, the relationship between the money paid for a gym membership and the number of months that someone has been a member can be described through a function. If the one-time membership fee is $40 and the monthly fee is $30, then the function can be written $f(x) = 30x + 40$. The x-value represents the number of months the person has been part of the gym, while the output is the total money paid for the membership. The table below shows this relationship. It is a representation of the function because the initial cost is $40, and the cost increases each month by $30.

x (months)	f(x) (money paid to gym)
0	40
1	70
2	100
3	130

In this situation, the domain of the function is real numbers greater than or equal to zero because it represents the number of months that a membership is held. We aren't told if the gym prorates memberships for partial months (if you join 10 days into a month, for example). If not, the domain would only be whole numbers plus zero, since there's a meaningful data point of $40, as a fee for joining. The range is real numbers greater than or equal to 40, because the range represents the total cost of the gym membership. Because there is a one-time fee of $40, the cost of carrying a membership will never be less than $40, so this is the minimum value.

When working through any word problem, the domain and range of the function should be considered in terms of the real-world context that the function models. For example, considering the above function for the cost of a gym membership, it would be nonsensical to include negative numbers in either the domain or range because there can't be negative months that someone holds a membership and similarly, the gym isn't going to pay a person for months prior to becoming a member. Therefore, while the function to model the situation (defined as $f(x) = 30x + 40$) theoretically could result in a true mathematical statement if negative values of are inputted, this would not make sense in the real-world context for which the function applies. Therefore, defining the domain as whole numbers and the range as all real numbers greater than or equal to 40 is important.

Evaluating Functions

To evaluate functions, plug in the given value everywhere the variable appears in the expression for the function. For example, find $f(-2)$ where:

$$f(x) = 2x^2 - \frac{4}{x}$$

To complete the problem, plug in -2 in the following way:

$$f(-2) = 2(-2)^2 - \frac{4}{-2}$$

$$2 \times 4 + 2$$

$$8 + 2 = 10$$

Practice Questions

1. Which of the following numbers has the greatest value?
   a. 1.4378
   b. 1.07548
   c. 1.43592
   d. 0.89409
   e. 0.94739

2. The value of 6×12 is the same as:
   a. $2 \times 4 \times 4 \times 2$
   b. $7 \times 4 \times 3$
   c. $6 \times 6 \times 3$
   d. $3 \times 3 \times 4 \times 2$
   e. $5 \times 9 \times 8$

3. This chart indicates how many sales of CDs, vinyl records, and MP3 downloads occurred over the last year. Approximately what percentage of the total sales was from CDs?

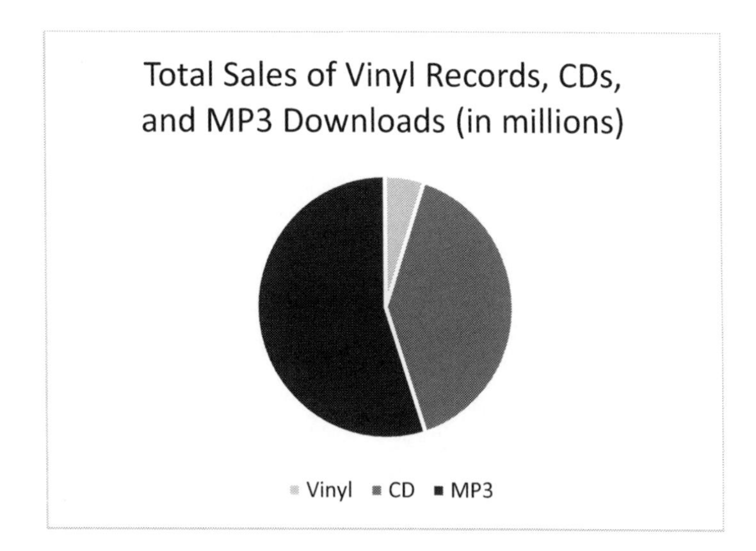

   a. 55%
   b. 25%
   c. 40%
   d. 5%
   e. 15%

4. Alan currently weighs 200 pounds, but he wants to lose weight to get down to 175 pounds. What is this difference in kilograms? (1 pound is approximately equal to 0.45 kilograms.)
   a. 9 kg
   b. 18.55 kg
   c. 78.75 kg
   d. 90 kg
   e. 11.25 kg

5. Johnny earns $2334.50 from his job each month. He pays $1437 for monthly expenses. Johnny is planning a vacation in 3 months' time that he estimates will cost $1750 total. How much will Johnny have left over from three months' of saving once he pays for his vacation?

a. $948.50
b. $584.50
c. $852.50
d. $942.50
e. $742.50

Answer Explanations

1. A: Compare each numeral after the decimal point to figure out which overall number is greatest. In answers *A* (1.43785) and *C* (1.43592), both have the same tenths (4) and hundredths (3). However, the thousandths is greater in answer *A* (7), so *A* has the greatest value overall.

2. D: By grouping the four numbers in the answer into factors of the two numbers of the question (6 and 12), it can be determined that:

$$(3 \times 2) \times (4 \times 3) = 6 \times 12$$

Alternatively, each of the answer choices could be prime factored or multiplied out and compared to the original value.

6×12 has a value of 72 and a prime factorization of $2^3 \times 3^2$.

The answer choices respectively have values of 64, 84, 108, 72, and 360, so Choice *D* is correct.

3. C: The sum total percentage of a pie chart must equal 100%. Since the CD sales take up less than half of the chart and more than a quarter (25%), it can be determined to be 40% overall. This can also be measured with a protractor. The angle of a circle is 360°. Since 25% of 360° would be 90° and 50% would be 180°, the angle percentage of CD sales falls in between; therefore, it would be Choice *C*.

4. B: Using the conversion rate, multiply the projected weight loss of 25 lb. by 0.45 $\frac{kg}{lb}$ to get the amount in kilograms (11.25 kg).

5. D: First, subtract $1437 from $2334.50 to find Johnny's monthly savings; this equals $897.50. Then, multiply this amount by 3 to find out how much he will have in three months before he pays for his vacation; this equals $2692.50. Finally, subtract the cost of the vacation ($1750) from this amount to find how much Johnny will have left: $942.50.

Reading

Comprehension and Analysis

Putting Events in Order

One of the most crucial skills for conquering the GED Reading Test is the ability to recognize the sequences of events for each passage and place them in the correct order. Every passage has a plot, whether it is from a short story, a manual, a newspaper article or editorial, or a history text. And each plot has a logical order, which is also known as a sequence. Some of the most straightforward sequences can be found in technology directions, science experiments, instructional materials, and recipes. These forms of writing list actions that must occur in a proper sequence in order to get sufficient results. Other forms of writing, however, use style and ideas in ways that completely change the sequence of events. Poetry, for instance, may introduce repetitions that make the events seem cyclical. Postmodern writers are famous for experimenting with different concepts of place and time, creating "cut scenes" that distort straightforward sequences and abruptly transport the audience to different contexts or times. Even everyday newspaper articles, editorials, and historical sources may experiment with different sequential forms for stylistic effect.

Most questions that call for test takers to apply their sequential knowledge use key words such as **sequence**, **sequence of events**, or **sequential order** to cue the test taker into the task at hand. In social studies or history passages, the test questions might employ key words such as **chronology** or **chronological order** to cue the test taker. In some cases, sequence can be found through comprehension techniques. These literal passages number the sequences, or they use key words such as *firstly*, *secondly*, *finally*, *next*, or *then*. The sequences of these stories can be found by rereading the passage and charting these numbers or key words. In most cases, however, readers have to correctly order events through inferential and evaluative reading techniques; they have to place events in a logical order without explicit cues.

Understanding Main Ideas and Details

It is very important to know the difference between the topic and the main idea of the text. Even though these two are similar because they both present the central point of a text, they have distinctive differences. A **topic** is the subject of the text; it can usually be described in a one- to two-word phrase and appears in the simplest form. On the other hand, the **main idea** is more detailed and provides the author's central point of the text. It can be expressed through a complete sentence and is often found in the beginning, middle, or end of a paragraph. In most nonfiction books, the first sentence of the passage usually (but not always) states the main idea.

Review the passage below to explore the topic versus the main idea:

> Cheetahs are one of the fastest mammals on the land, reaching up to 70 miles an hour over short distances. Even though cheetahs can run as fast as 70 miles an hour, they usually only have to run half that speed to catch up with their choice of prey. Cheetahs cannot maintain a fast pace over long periods of time because their bodies will overheat. After a chase, cheetahs need to rest for approximately 30 minutes prior to eating or returning to any other activity.

In the example above, the topic of the passage is "Cheetahs" simply because that is the subject of the text. The main idea of the text is "Cheetahs are one of the fastest mammals on the land but can only maintain a fast pace for shorter distances." While it covers the topic, it is more detailed and refers to the text in its entirety. The text continues to provide additional details called **supporting details**, which will be discussed in the next section.

How Details Develop the Main Idea

Supporting details help readers better develop and understand the main idea. Supporting details answer questions like *who, what, where, when, why,* and *how*. Different types of supporting details include examples, facts and statistics, anecdotes, and sensory details.

Persuasive and informative texts often use supporting details. In persuasive texts, authors attempt to make readers agree with their points of view, and supporting details are often used as "selling points." If authors make a statement, they need to support the statement with evidence in order to adequately persuade readers. Informative texts use supporting details such as examples and facts to inform readers. Review the previous "Cheetahs" passage to find examples of supporting details.

> Cheetahs are one of the fastest mammals on the land, reaching up to 70 miles an hour over short distances. Even though cheetahs can run as fast as 70 miles an hour, they usually only have to run half that speed to catch up with their choice of prey. Cheetahs cannot maintain a fast pace over long periods of time because their bodies will overheat. After a chase, cheetahs need to rest for approximately 30 minutes prior to eating or returning to any other activity.

In the example, supporting details include:

- Cheetahs reach up to 70 miles per hour over short distances.
- They usually only have to run half that speed to catch up with their prey.
- Cheetahs will overheat if they exert a high speed over longer distances.
- Cheetahs need to rest for 30 minutes after a chase.

Look at the diagram below (applying the cheetah example) to help determine the hierarchy of topic, main idea, and supporting details.

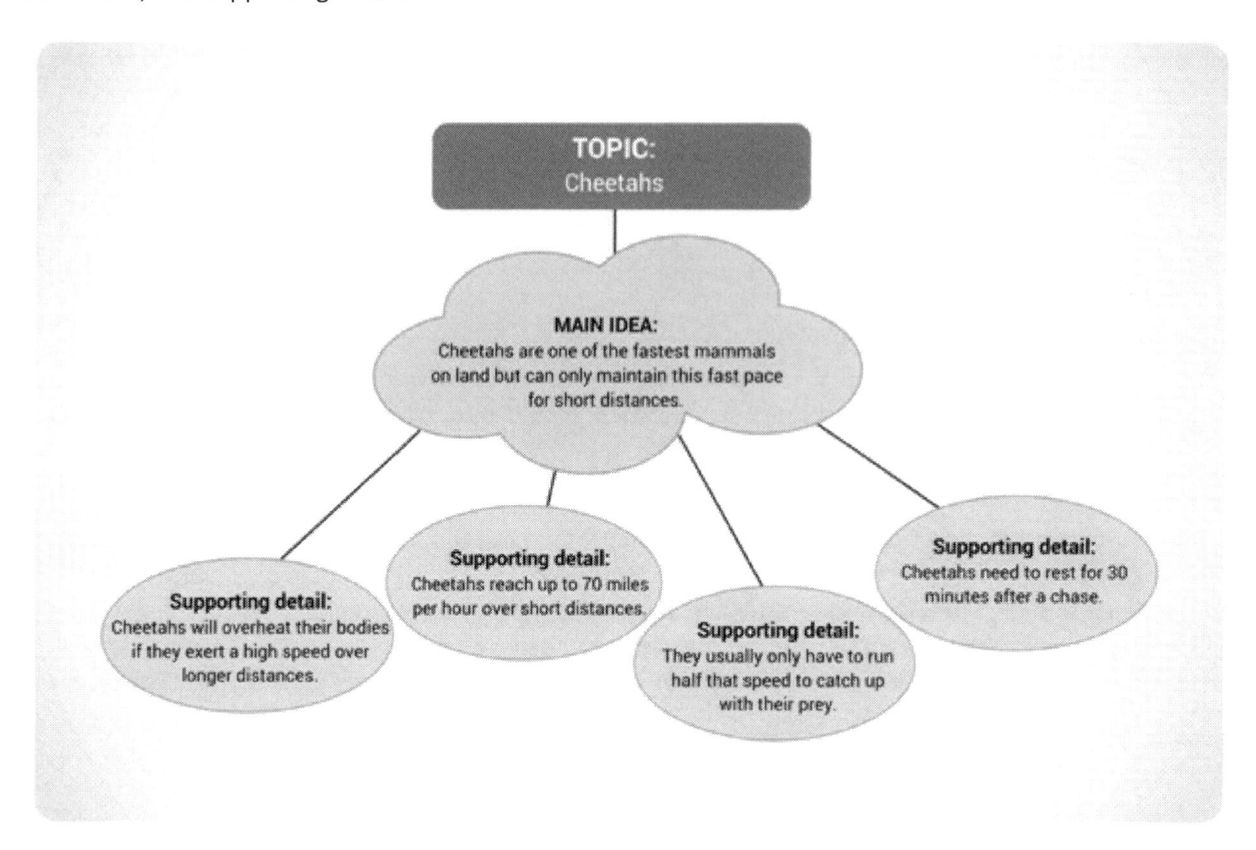

Summarizing Information from a Passage

Summarizing is an effective way to draw a conclusion from a passage. A summary is a shortened version of the original text, written by the reader in his/her own words. Focusing on the main points of the original text and including only the relevant details can help readers reach a conclusion. It's important to retain the original meaning of the passage.

Like summarizing, **paraphrasing** can also help a reader fully understand different parts of a text. Paraphrasing calls for the reader to take a small part of the passage and list or describe its main points. Paraphrasing is more than rewording the original passage, though. It should be written in the reader's own words, while still retaining the meaning of the original source. This will indicate an understanding of the original source, yet still help the reader expand on his/her interpretation.

Readers should pay attention to the **sequence**, or the order in which details are laid out in the text, as this can be important to understanding its meaning as a whole. Writers will often use transitional words to help the reader understand the order of events and to stay on track. Words like *next, then, after*, and *finally* show that the order of events is important to the author. In some cases, the author omits these transitional words, and the sequence is implied. Authors may even purposely present the information out of order to make an impact or have an effect on the reader. An example might be when a narrative writer uses **flashback** to reveal information.

How Authors Develop Theme

Authors employ a variety of techniques to present a theme. They may compare or contrast characters, events, places, ideas, or historical or invented settings to speak thematically. They may use analogies, metaphors, similes, allusions, or other literary devices to convey the theme. An author's use of diction, syntax, and tone can also help convey the theme. Authors will often develop themes through the development of characters, use of the setting, repetition of ideas, use of symbols, and through contrasting value systems. Authors of both fiction and nonfiction genres will use a variety of these techniques to develop one or more themes.

Regardless of the literary genre, there are commonalities in how authors, playwrights, and poets develop themes or central ideas.

Authors often do research, the results of which contributes to theme. In prose fiction and drama, this research may include real historical information about the setting the author has chosen or include elements that make fictional characters, settings, and plots seem realistic to the reader. In nonfiction, research is critical since the information contained within this literature must be accurate and, moreover, accurately represented.

In fiction, authors present a narrative conflict that will contribute to the overall theme. In fiction, this conflict may involve the storyline itself and some trouble within characters that needs resolution. In nonfiction, this conflict may be an explanation or commentary on factual people and events.

Authors will sometimes use character motivation to convey theme, such as in the example from *Hamlet* regarding revenge. In fiction, the characters an author creates will think, speak, and act in ways that effectively convey the theme to readers. In nonfiction, the characters are factual, as in a biography, but authors pay particular attention to presenting those motivations to make them clear to readers.

Authors also use literary devices as a means of conveying theme. For example, the use of moon symbolism in Mary Shelley's *Frankenstein* is significant, as its phases can be compared to the phases that the Creature undergoes as he struggles with his identity.

The selected point of view can also contribute to a work's theme. The use of first-person point of view in a fiction or non-fiction work engages the reader's response differently than third person point of view. The central idea or theme from a first-person narrative may differ from a third-person limited text.

In literary nonfiction, authors usually identify the purpose of their writing, which differs from fiction, where the general purpose is to entertain. The purpose of nonfiction is usually to inform, persuade, or entertain the audience. The stated purpose of a non-fiction text will drive how the central message or theme, if applicable, is presented.

Authors identify an audience for their writing, which is critical in shaping the theme of the work. For example, the audience for J.K. Rowling's *Harry Potter* series would be different than the audience for a biography of George Washington. The audience an author chooses to address is closely tied to the purpose of the work. The choice of an audience also drives the choice of language and level of diction an author uses. Ultimately, the intended audience determines the level to which that subject matter is presented and the complexity of the theme.

Analyzing Relationships within Passages

Inferences are useful in gaining a deeper understanding of how people, events, and ideas are connected in a passage. Readers can use the same strategies used with general inferences and analyzing texts—paying attention to details and using them to make reasonable guesses about the text—to read between the lines and get a more complete picture of how (and why) characters are thinking, feeling, and acting. Read the following passage from O. Henry's story "The Gift of the Magi":

> One dollar and eighty-seven cents. That was all. And sixty cents of it was in pennies. Pennies saved one and two at a time by bulldozing the grocer and the vegetable man and the butcher until one's cheeks burned with the silent imputation of parsimony that such close dealing implied. Three times Della counted it. One dollar and eighty-seven cents. And the next day would be Christmas. There was clearly nothing to do but flop down on the shabby little couch and howl. So Della did it.

This paragraph introduces the reader to the character Della. Even though the author doesn't include a direct description of Della, the reader can already form a general impression of her personality and emotions. One detail that should stick out to the reader is repetition: "one dollar and eighty-seven cents." This amount is repeated twice in the paragraph, along with other descriptions of money: "sixty cents of it was in pennies," "pennies saved one and two at a time." The story's preoccupation with money parallels how Della herself is constantly thinking about her finances: "three times Della counted" her meager savings. Already the reader can guess that Della is having money problems. Next, think about her emotions. The paragraph describes haggling over groceries "until one's cheeks burned"—another way to describe blushing. People tend to blush when they are embarrassed or ashamed, so readers can infer that Della is ashamed by her financial situation. This inference is also supported when she flops down and howls on her "shabby little couch." Clearly, she's in distress. Without saying, "Della has no money and is embarrassed to be poor," O. Henry is able to communicate the same impression to readers through his careful inclusion of details.

A character's **motive** is their reason for acting a certain way. Usually, characters are motivated by something that they want. In the passage above, why is Della upset about not having enough money? There's an important detail at the end of the paragraph: "the next day would be Christmas." Why is money especially important around Christmas? Christmas is a holiday when people exchange gifts. If Della is struggling with money, she's probably also struggling to buy gifts. A shrewd reader should be able to guess that Della's motivation is wanting to buy a gift for someone—but she's currently unable to afford it, leading to feelings of shame and frustration.

In order to understand characters in a text, readers should keep the following questions in mind:

- What words does the author use to describe the character? Are these words related to any specific emotions or personality traits (for example, characteristics like rude, friendly, unapproachable, or innocent)?
- What does the character say? Does their dialogue seem to be straightforward, or are they hiding some thoughts or emotions?
- What actions can be observed from this character? How do their actions reflect their feelings?
- What does the character want? What do they do to get it?

Extending Your Understanding to New Situations

Combining Information from Different Sources

Synthesizing, or combining, ideas and information from different sources is a skill that helps test takers pass the GED and also thrive in the workforce. The theories and concepts offered in different passages cannot just haphazardly be tossed together. Every test taker has to come up with their own recipe for success when it comes to synthesizing separate sources.

One way for test takers to think about synthesizing sources is to imagine their written responses as empty homes that need to be decorated. They can then imagine the words, concepts, and theories in the different sources as their desired décor. At times, two different sources combine to create perfectly matched décor—the words, concepts, and theories blend ceaselessly upon the walls of the test taker's literary home, creating a balance. At other times, two different sources clash, forcing test takers to sort and separate the ideas into different rooms (for example, different paragraphs or sentences). At still other times, the two sources are incomplete, so test takers need to combine materials with their own interests and statements. If sources contradict one another, it is best to highlight these contradictions. A test taker should take note of the contradictions and use their best judgment in choosing which source is more aligned with their own theories. At times, the test taker may even disagree with information in both articles. It is perfectly acceptable to make the audience aware of all contradictions and disagreements.

Writers, like interior designers, must hone their craft through experience. The best way to begin synthesizing sources is to *practice*. There are four practical ways test takers can start practicing synthesis. Firstly, they need to learn how to properly identify and cite captivating quotations. Secondly, they need to learn how to summarize ideas succinctly in their own words. Thirdly, they need to create unique sentences that are part quotation and part summary. And, lastly, they need to ensure that all of the above is backed by sound grammar, syntax, and organization. The best way to ensure quality is to read other high-quality works and enlist a group of friends or colleagues to edit.

Drawing Conclusions

Determining conclusions requires being an active reader, as a reader must make a prediction and analyze facts to identify a conclusion. A reader should identify key words in a passage to determine the logical conclusion or determination that flows from the information presented. Consider the passage below:

> Lindsay, covered in flour, moved around the kitchen frantically. Her mom yelled from another room, "Lindsay, we're going to be late!

You can conclude that Lindsay's next steps are to finish baking, clean herself up, and head off somewhere with her baked goods. Notice that the conclusion cannot be verified factually. Many conclusions are not spelled out specifically in the text, thus they have to be identified and drawn out by the reader.

Transferring Information to New Situations

A natural extension of being able to make an inference from a given set of information is also being able to apply that information to a new context. This is especially useful in non-fiction or informative writing. Considering the facts and details presented in the text, readers should consider how the same information might be relevant in a different situation. The following is an example of applying an inferential conclusion to a different context:

> Often, individuals behave differently in large groups than they do as individuals. One example of this is the psychological phenomenon known as the bystander effect. According to the bystander effect, the more people who witness an accident or crime occur, the less likely each individual bystander is to respond or offer assistance to the victim. A classic example of this is the murder of Kitty Genovese in New York City in the 1960s. Although there were over thirty witnesses to her killing by a stabber, none of them intervened to help Kitty or contact the police.

Considering the phenomenon of the bystander effect, what would probably happen if somebody tripped on the stairs in a crowded subway station?

a. Everybody would stop to help the person who tripped
b. Bystanders would point and laugh at the person who tripped
c. Someone would call the police after walking away from the station
d. Few if any bystanders would offer assistance to the person who tripped

This question asks readers to apply the information they learned from the passage, which is an informative paragraph about the bystander effect. According to the passage, this is a concept in psychology that describes the way people in groups respond to an accident—the more people are present, the less likely any one person is to intervene. While the passage illustrates this effect with the example of a woman's murder, the question asks readers to apply it to a different context—in this case, someone falling down the stairs in front of many subway passengers. Although this specific situation is not discussed in the passage, readers should be able to apply the general concepts described in the paragraph. The definition of the bystander effect includes any instance of an accident or crime in front of a large group of people. The question asks about a situation that falls within the same definition, so the general concept should still hold true: in the midst of a large crowd, few individuals are likely to actually respond to an accident. In this case, Choice *D* is the best response.

Making Generalizations Based on Evidence

One way to make generalizations is to look for main topics. When doing so, pay particular attention to any titles, headlines, or opening statements made by the author. Topic sentences or repetitive ideas can be clues in gleaning inferred ideas. For example, if a passage contains the phrase *DNA testing, while some consider it infallible, is an inherently flawed technique,* the test taker can infer the rest of the passage will contain information that points to DNA testing's infallibility.

The test taker may be asked to make a generalization based on prior knowledge but may also be asked to make predictions based on new ideas. For example, the test taker may have no prior knowledge of DNA other than its genetic property to replicate. However, if the reader is given passages on the flaws of DNA testing with enough factual evidence, the test taker may arrive at the inferred conclusion or generalization that the author does not support the infallibility of DNA testing in all identification cases.

When making generalizations, it is important to remember that the critical thinking process involved must be fluid and open to change. While a reader may infer an idea from a main topic, general statement, or other clues, they must be open to receiving new information within a particular passage. New ideas presented by an author may require the test taker to alter a generalization. Similarly, when asked questions that require making an inference, it's important to read the entire test passage and all of the answer options. Often, a test taker will need to refine a generalization based on new ideas that may be presented within the text itself.

Predictions

Some texts use suspense and foreshadowing to captivate readers. For example, an intriguing aspect of murder mysteries is that the reader is never sure of the culprit until the author reveals the individual's identity. Authors often build suspense and add depth and meaning to a work by leaving clues to provide hints or predict future events in the story; this is called foreshadowing. While some instances of foreshadowing are subtle, others are quite obvious.

Inferences

Another way to read actively is to identify examples of inference within text. Making an inference requires the reader to read between the lines and look for what is implied rather than what is explicitly stated. That is, using information that is known from the text, the reader is able to make a logical assumption about information that is not explicitly stated but is probably true.

Authors employ literary devices such as tone, characterization, and theme to engage the audience by showing details of the story instead of merely telling them. For example, if an author said *Bob is selfish*, there's little left to infer. If the author said, *Bob cheated on his test, ignored his mom's calls, and parked illegally*, the reader can infer that Bob is selfish. Authors also make implications through character dialogue, thoughts, effects on others, actions, and looks. Like in life, readers must assemble all the clues to form a complete picture.

Read the following passage:

> "Hey, do you want to meet my new puppy?" Jonathan asked.
>
> "Oh, I'm sorry but please don't—" Jacinta began to protest, but before she could finish, Jonathan had already opened the passenger side door of his car and a perfect white ball of fur came bouncing towards Jacinta.
>
> "Isn't he the cutest?" beamed Jonathan.
>
> "Yes—achoo!—he's pretty—aaaachooo!!—adora—aaa—aaaachoo!" Jacinta managed to say in between sneezes. "But if you don't mind, I—I—achoo!—need to go inside."

Which of the following can be inferred from Jacinta's reaction to the puppy?

a. she hates animals
b. she is allergic to dogs
c. she prefers cats to dogs
d. she is angry at Jonathan

An inference requires the reader to consider the information presented and then form their own idea about what is probably true. Based on the details in the passage, what is the best answer to the question? Important details to pay attention to include the tone of Jacinta's dialogue, which is overall polite and apologetic, as well as her reaction itself, which is a long string of sneezes. Answer choices (a) and (d) both express strong emotions ("hates" and "angry") that are not evident in Jacinta's speech or actions. Answer choice (c) mentions cats, but there is nothing in the passage to indicate Jacinta's feelings about cats. Answer choice (b), "she is allergic to dogs," is the most logical choice—based on the fact that she began sneezing as soon as a fluffy dog approached her, it makes sense to guess that Jacinta might be allergic to dogs. So even though Jacinta never directly states, "Sorry, I'm allergic to dogs!" using the clues in the passage, it is still reasonable to guess that this is true.

Making inferences is crucial for readers of literature, because literary texts often avoid presenting complete and direct information to readers about characters' thoughts or feelings, or they present this information in an unclear way, leaving it up to the reader to interpret clues given in the text. In order to make inferences while reading, readers should ask themselves:

- What details are being presented in the text?
- Is there any important information that seems to be missing?
- Based on the information that the author *does* include, what else is probably true?
- Is this inference reasonable based on what is already known?

Using Main Ideas to Draw Conclusions

Determining conclusions requires being an active reader, as a reader must make a prediction and analyze facts to identify a conclusion. There are a few ways to determine a logical conclusion, but careful reading is the most important. It's helpful to read a passage a few times, noting details that seem important to the text. A reader should also identify key words in a passage to determine the logical conclusion or determination that flows from the information presented.

Textual evidence helps readers draw a conclusion about a passage. **Textual evidence** refers to information—facts and examples that support the main point; it will likely come from outside sources and can be in the form of quoted or paraphrased material. In order to draw a conclusion from evidence, it's important to examine the credibility and validity of that evidence as well as how (and if) it relates to the main idea.

If an author presents a differing opinion or a **counterargument** in order to refute it, the reader should consider how and why the information is being presented. It is meant to strengthen the original argument and shouldn't be confused with the author's intended conclusion, but it should also be considered in the reader's final evaluation.

Sometimes, authors explicitly state the conclusion they want readers to understand. Alternatively, a conclusion may not be directly stated. In that case, readers must rely on the implications to form a logical conclusion:

> On the way to the bus stop, Michael realized his homework wasn't in his backpack. He ran back to the house to get it and made it back to the bus just in time.

In this example, though it's never explicitly stated, it can be inferred that Michael is a student on his way to school in the morning. When forming a conclusion from implied information, it's important to read the text carefully to find several pieces of evidence in the text to support the conclusion.

Describing the Steps of an Argument

Strong arguments tend to follow a fairly defined format. In the introduction, background information regarding the problem is shared, the implications of the issue are stated, and the author's thesis or claims are given. Supporting evidence is then presented in the body paragraphs, along with the counterargument, which then gets refuted with specific evidence. Lastly, in the conclusion, the author summarizes the points and claims again.

Evidence Used to Support a Claim or Conclusion

Premises are the evidence or facts supporting why a **conclusion** is logical and valid. Take the following argument for example:

> Julie is a Canadian track athlete. She is the star of the number one collegiate team in the country. Her times are consistently at the top of national rankings. Julie is extremely likely to represent Canada at the upcoming Olympics.

In this example, the conclusion is that she will likely be on the Canadian Olympic team. The author supports this conclusion with two premises. First, Julie is the star of an elite track team. Second, she runs some of the best times of the country. This is the *why* behind the conclusion. The following builds off this basic argument:

> Julie is a Canadian track athlete. She's the star of the number one collegiate team in the country. Her times are consistently at the top of national rankings. Julie is extremely likely to represent Canada at the upcoming Olympics. Julie will continue to develop after the Olympic trials. She will be a frontrunner for the gold. Julie is likely to become a world-famous track star.

These additions to the argument make the conclusion different. Now, the conclusion is that Julie is likely to become a world-famous track star. The previous conclusion, Julie will likely be on the Olympic team, functions as a **sub-conclusion** in this argument. Like conclusions, premises must adequately support sub-conclusions. However, sub-conclusions function like premises, since sub-conclusions also support the overall conclusion.

Determining Whether Evidence is Relevant and Sufficient

A **hasty generalization** involves an argument relying on insufficient statistical data or inaccurately generalizing. One common generalization occurs when a group of individuals under observation have some quality or attribute that is asserted to be universal or true for a much larger number of people than actually documented. Here's an example of a hasty generalization:

> A man smokes a lot of cigarettes, but so did his grandfather. The grandfather smoked nearly two packs per day since his World War II service until he died at ninety years of age. Continuing to smoke cigarettes will clearly not impact the grandson's long-term health.

This argument is a hasty generalization because it assumes that one person's addiction and lack of consequences will naturally be reflected in a different individual. There is no reasonable justification for such extrapolation. It is common knowledge that any smoking is detrimental to everyone's health. The fact that the man's grandfather smoked two packs per day and lived a long life has no logical connection with the grandson engaging in similar behavior. The hasty generalization doesn't take into account other reasons behind the grandfather's longevity. Nor does the author offer evidence that might support the idea that the man would share a similar lifetime if he smokes. It might be different if the author stated that the man's family shares some genetic trait rendering them immune to the effects of tar and chemicals on the lungs.

Determining Whether a Statement Is or Is Not Supported

The basic tenet of reading comprehension is the ability to read and understand text. One way to understand text is to look for information that supports the author's main idea, topic, or position statement. This information may be factual, or it may be based on the author's opinion.

In order to identify factual information within one or more text passages, begin by looking for statements of fact. Factual statements can be either true or false. Identifying factual statements as opposed to opinion statements is important in demonstrating full command of evidence in reading. For example, the statement *The temperature outside was unbearably hot* may seem like a fact; however, it's not. While anyone can point to a temperature gauge as factual evidence, the statement itself reflects only an opinion. Some people may find the temperature unbearably hot. Others may find it comfortably warm. Thus, the sentence, *The temperature outside was unbearably hot,* reflects the opinion of the author who found it unbearable. If the text passage followed up the sentence with atmospheric conditions indicating heat indices above 140 degrees Fahrenheit, then the reader knows there is factual information that supports the author's assertion of *unbearably hot*.

In looking for information that can be proven or disproven, it's helpful to scan for dates, numbers, timelines, equations, statistics, and other similar data within any given text passage. These types of indicators will point to proven particulars. For example, the statement, *The temperature outside was unbearably hot on that summer day, July 10, 1913,* most likely indicates factual information, even if the reader is unaware that this is the hottest day on record in the United States. Be careful when reading biased words from an author. Biased words indicate opinion, as opposed to fact. See the list of biased words below and keep in mind that it's not an inclusive list:

- Good/bad
- Great/greatest
- Better/best/worst
- Amazing
- Terrible/bad/awful
- Beautiful/handsome/ugly
- More/most
- Exciting/dull/boring
- Favorite
- Very
- Probably/should/seem/possibly

Remember, most of what is written is actually opinion or carefully worded information that seems like fact when it isn't. To say, *duplicating DNA results is not cost-effective* sounds like it could be a scientific fact, but it isn't. Factual information can be verified through independent sources.

The simplest type of test question may provide a text passage, then ask the test taker to distinguish the correct factual supporting statement that best answers the corresponding question on the test. However, be aware that most questions may ask the test taker to read more than one text passage and identify which answer best supports an author's topic. While the ability to identify factual information is critical, these types of questions require the test taker to identify chunks of details, and then relate them to one another.

Assessing Whether an Argument is Valid

Although different from conditions and if/then statements, **reasonableness** is another important foundational concept. Evaluating an argument for reasonableness and validity entails evaluating the evidence presented by the author to justify their conclusions. Everything contained in the argument should be considered, but remember to ignore outside biases, judgments, and knowledge. For the purposes of this test, the test taker is a one-person jury at a criminal trial using a standard of reasonableness under the circumstances presented by the argument.

These arguments are encountered on a daily basis through social media, entertainment, and cable news. An example is:

> Although many believe it to be a natural occurrence, some believe that the red tide that occurs in Florida each year may actually be a result of human sewage and agricultural runoff. However, it is arguable that both natural and human factors contribute to this annual phenomenon. On one hand, the red tide has been occurring every year since the time of explorers like Cabeza de Vaca in the 1500's. On the other hand, the red tide seems to be getting worse each year, and scientists from the Florida Fish & Wildlife Conservation say the bacteria found inside the tide feed off of nutrients found in fertilizer runoff.

The author's conclusion is that both natural phenomena and human activity contribute to the red tide that happens annually in Florida. The author backs this information up by historical data to prove the natural occurrence of the red tide, and then again with scientific data to back up the human contribution to the red tide. Both of these statements are examples of the premises in the argument. Evaluating the strength of the logical connection between the premises and conclusion is how reasonableness is determined. Another example is:

> The local railroad is a disaster. Tickets are exorbitantly priced, bathrooms leak, and the floor is sticky.

The author is clearly unhappy with the railroad service. They cite three examples of why they believe the railroad to be a disaster. An argument more familiar to everyday life is:

> Alexandra said the movie she just saw was amazing. We should go see it tonight.

Although not immediately apparent, this is an argument. The author is making the argument that they should go see the movie. This conclusion is based on the premise that Alexandra said the movie was amazing. There's an inferred note that Alexandra is knowledgeable on the subject, and she's credible enough to prompt her friends to go see the movie. This seems like a reasonable argument. A less reasonable argument is:

> Alexandra is a film student, and she's written the perfect romantic comedy script. We should put our life savings toward its production as an investment in our future.

The author's conclusion is that they should invest their life savings into the production of a movie, and it is justified by referencing Alexandra's credibility and current work. However, the premises are entirely too weak to support the conclusion. Alexandra is only a film *student*, and the script is seemingly her first work. This is not enough evidence to justify investing one's life savings in the film's success.

Assumptions in an Argument

Think of assumptions as unwritten premises. Although they never explicitly appear in the argument, the author is relying on it to defend the argument, just like a premise. Assumptions are the most important part of an argument that will never appear in an argument.

An argument in the abstract is: The author concludes Z based on W and X premises. But the W and X premises actually depend on the unmentioned assumption of Y. Therefore, the author is really saying that X, W, and Y make Z correct, but Y is assumed.

People assume all of the time. Assumptions and inferences allow the human mind to process the constant flow of information. Many assumptions underlie even the most basic arguments. However, in the world of Legal Reasoning arguments, assumptions must be avoided. An argument must be fully presented to be valid; relying on an assumption is considered weak. The test requires that test takers identify these underlying assumptions. One example is:

> Peyton Manning is the most over-rated quarterback of all time. He lost more big games than anyone else. Plus, he allegedly assaulted his female trainer in college. Peyton clearly shouldn't make the Hall of Fame.

The author certainly relies on a lot of assumptions. A few assumptions are:

- Peyton Manning plays quarterback.
- He is considered to be a great quarterback by at least some people.
- He played in many big games.
- Allegations and past settlements without any admission of guilt from over a decade ago can be relied upon as evidence against Hall of Fame acceptance.
- The Hall of Fame voters factor in off-the-field incidents, even if true.
- The best players should make the Hall of Fame.
- Losing big games negates, at least in part, the achievement of making it to those big games
- Peyton Manning is retired, and people will vote on whether he makes the Hall of Fame at some point in the future.

The author is relying on all of these assumptions. Some are clearly more important to his argument than others. In fact, disproving a necessary assumption can destroy a premise and possibly an entire conclusion. For example, what if the Hall of Fame did not factor in any of the off-the-field incidents? Then the alleged assault no longer factors into the argument. Even worse, what if making the big games actually was more important than losing those games in the eyes of the Hall of Fame voters? Then the whole conclusion falls apart and is no longer justified if that premise is disproven.

Assumption questions test this exact point by asking the test taker to identify which assumption the argument relies upon. If the author is making numerous assumptions, then the most important *one* assumption must be chosen.

If the author truly relies on an assumption, then the argument will completely fall apart if the assumption isn't true. **Negating** a necessary assumption will *always* make the argument fall apart. This is a universal rule of logic and should be the first thing done in testing answer choices.

Here are some ways that underlying assumptions will appear as questions:

- Which of the following is a hidden assumption that the author makes to advance his argument?
- Which assumption, if true, would support the argument's conclusion (make it more logical)?
- The strength of the argument depends on which of the following?
- Upon which of the following assumptions does the author rely?
- Which assumption does the argument presuppose?

An example is:

> Frank Underwood is a terrible president. The man is a typical spend, spend, spend liberal. His employment program would exponentially increase the annual deficit and pile on the national debt. Not to mention, Underwood is also on the verge of starting a war with Russia.

Upon which of the following assumptions does the author's argument most rely?

a. Frank Underwood is a terrible president.
b. The United States cannot afford Frank Underwood's policy plans without spending more than the country raises in revenue.
c. No spend, spend, spend liberal has ever succeeded as president.
d. Starting a war with Russia is beneficial to the United States.

Use the negation rule to find the correct answer in the choices below.

Choice *A* is not an assumption—it is the author's conclusion. This type of restatement will never be the correct answer, but test it anyway. After negating the choice, what remains is: *Frank Underwood is a fantastic president*. Does this make the argument fall apart? No, it just becomes the new conclusion. The argument is certainly worse since it does not seem reasonable for someone to praise a president for being a spend, spend, spend liberal or raising the national debt; however, the argument still makes *logical* sense. Eliminate this choice.

Choice *B* is certainly an assumption. It underlies the premises that the country cannot afford Underwood's economic plans. When reversed to: *The United States can afford Frank Underwood's policy plans without spending more than the country raises in revenue,* this destroys the argument. If the United States can afford his plans, then the annual deficit and national debt won't increase; therefore, Underwood being a terrible president would only be based on the final premise. The argument is much weaker without the two sentences involving the financials. Keep it as a benchmark while working through the remaining choices.

Choice *C* is irrelevant. The author is not necessarily claiming that all loose-pocket liberals make for bad presidents. His argument specifically pertains to Underwood. Negate it— *Some spend, spend, spend liberals have succeeded as president.* This does not destroy the argument. Some other candidate could have succeeded as president. However, the author is pointing out that those policies would be disastrous considering the rising budget and debt. The author is not making an appeal to historical precedent. Although not a terrible choice, it is certainly weaker than Choice *B*. Eliminate this choice.

Choice *D* is definitely not an assumption made by the author. The author is assuming that a war with Russia is disastrous. Negate it anyway—*Starting a war with Russia is not beneficial for the United States.* This does not destroy the argument; it makes it stronger. Eliminate this choice.

Analyzing Two Arguments and Evaluating the Types of Evidence Used to Support Each Claim

Arguments use evidence and reasoning to support a position or prove a point. Claims are typically controversial and may be faced with some degree of contention. Thus, authors support claims with evidence. Two arguments might present different types of evidence that readers will need to evaluate for merit, worthiness, accuracy, relevance, and impact. Evidence can take on many forms such as numbers (statistics, measurements, numerical data, etc.), expert opinions or quotes, testimonies, anecdotal evidence or stories from individuals, and textual evidence, such as that obtained from documents like diaries, newspapers, and laws.

Data, Graphs, or Pictures as Evidence

Some writing in the test contains **infographics** such as charts, tables, or graphs. In these cases, interpret the information presented and determine how well it supports the claims made in the text. For example, if the writer makes a case that seat belts save more lives than other automobile safety measures, they might want to include a graph (like the one below) showing the number of lives saved by seat belts versus those saved by air bags.

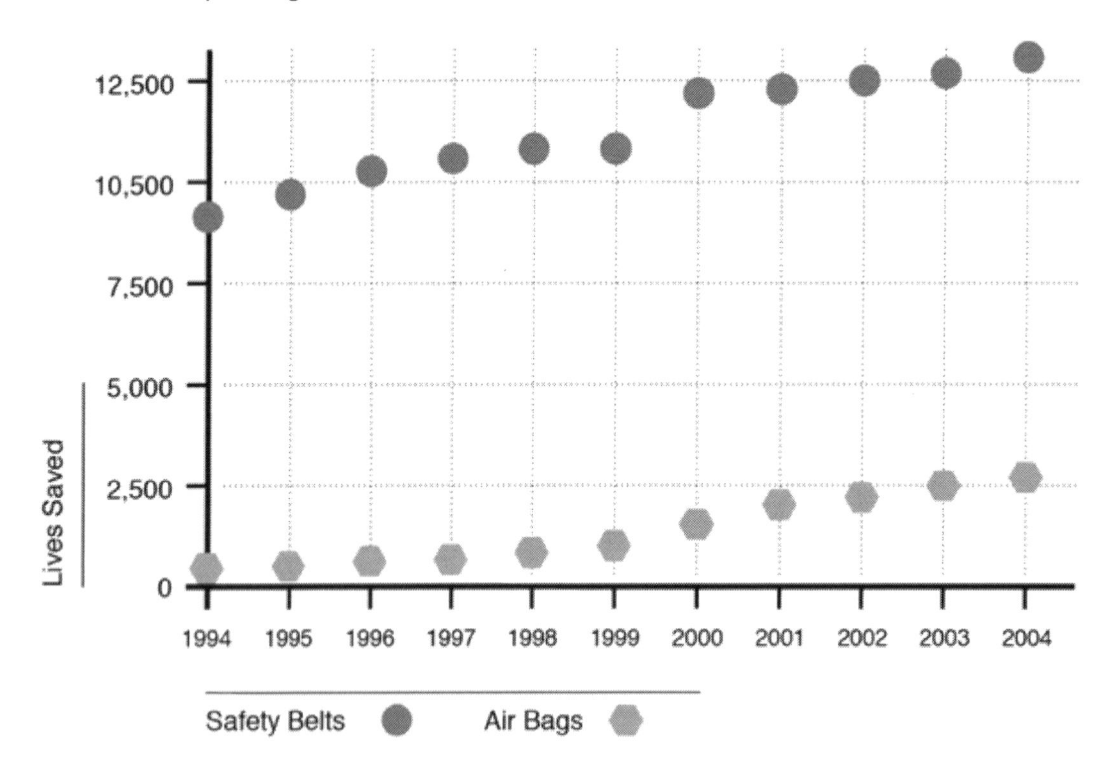

Based on data from the National Highway Traffic Safety Administration

If the graph clearly shows a higher number of lives are saved by seat belts, then it's effective. However, if the graph shows air bags save more lives than seat belts, then it doesn't support the writer's case.

Finally, graphs should be easy to understand. Their information should immediately be clear to the reader at a glance. Here are some basic things to keep in mind when interpreting infographics:

- In a **bar graph**, higher bars represent larger numbers. Lower bars represent smaller numbers.
- **Line graphs** are the same, but often show trends over time. A line that consistently ascends from left to right shows a steady increase over time. A line that consistently descends from left to right shows a steady decrease over time. If the line bounces up and down, this represents instability or inconsistency in the trend. When interpreting a line graph, determine the point the writer is trying to make, and then see if the graph supports that point.
- **Pie charts** are used to show proportions or percentages of a whole but are less effective in showing change over time.
- **Tables** present information in numerical form, not as graphics. When interpreting a table, make sure to look for patterns in the numbers.

There can also be timelines, illustrations, or maps on the test. When interpreting these, keep in mind the writer's intentions and determine whether or not the graphic supports the case.

Author's Point of View and Purpose

When it comes to an author's writing, readers should always identify a **position** or **stance**. No matter how objective a text may seem, readers should assume the author has preconceived beliefs. One can reduce the likelihood of accepting an invalid argument by looking for multiple articles on the topic, including those with varying opinions. If several opinions point in the same direction and are backed by reputable peer-reviewed sources, it's more likely that the author has a valid argument. Positions that run contrary to widely held beliefs and existing data should invite scrutiny. There are exceptions to the rule, so readers should be careful consumers of information.

While themes, symbols, and motifs are buried deep within the text and can sometimes be difficult to infer, an author's **purpose** is usually obvious from the beginning. There are four purposes of writing: to inform, to persuade, to describe, and to entertain. **Informative** writing presents facts in an accessible way. **Persuasive** writing appeals to emotions and logic to inspire the reader to adopt a specific stance. Readers should be wary of this type of writing, as it can mask a lack of objectivity with powerful emotion. **Descriptive** writing is designed to paint a picture in the reader's mind, while texts that **entertain** are often narratives designed to engage and delight the reader.

The various writing styles are usually blended, with one purpose dominating the rest. A persuasive text, for example, might begin with a humorous tale to make readers more receptive to the persuasive message, or a recipe in a cookbook designed to inform might be preceded by an entertaining anecdote that makes the recipes more appealing.

Author's Position and Response to Different Viewpoints

If an author presents a differing opinion or a counterargument in order to refute it, the reader should consider how and why the information is being presented. It is meant to strengthen the original argument and shouldn't be confused with the author's intended conclusion, but it should also be considered in the reader's final evaluation.

Authors can also use bias if they ignore the opposing viewpoint or present their side in an unbalanced way. A strong argument considers the opposition and finds a way to refute it. Critical readers should look for an unfair or one-sided presentation of the argument and be skeptical, as a bias may be present. Even if this bias is unintentional, if it exists in the writing, the reader should be wary of the validity of the argument. Readers should also look for the use of stereotypes, which refer to specific groups. Stereotypes are often negative connotations about a person or place and should always be avoided. When a critical reader finds stereotypes in a piece of writing, they should be critical of the argument, and consider the validity of anything the author presents. Stereotypes reveal a flaw in the writer's thinking and may suggest a lack of knowledge or understanding about the subject.

Inferring the Author's Purpose in the Passage

In nonfiction writing, authors employ argumentative techniques to present their opinion to readers in the most convincing way. First of all, persuasive writing usually includes at least one type of appeal: an appeal to logic (**logos**), emotion (**pathos**), or credibility and trustworthiness (**ethos**). When a writer appeals to logic, they are asking readers to agree with them based on research, evidence, and an established line of reasoning. An author's argument might also appeal to readers' emotions, perhaps by including personal stories and anecdotes (a short narrative of a specific event). A final type of appeal—appeal to authority—asks the reader to agree with the author's argument on the basis of their expertise or credentials. Consider three different approaches to arguing the same opinion:

Logic (Logos)

Below is an example of an appeal to logic. The author uses evidence to disprove the logic of the school's rule (the rule was supposed to reduce discipline problems; the number of problems has not been reduced; therefore, the rule is not working) and he or she calls for its repeal.

> Our school should abolish its current ban on campus cell phone use. The ban was adopted last year as an attempt to reduce class disruptions and help students focus more on their lessons. However, since the rule was enacted, there has been no change in the number of disciplinary problems in class. Therefore, the rule is ineffective and should be done away with.

Emotion (Pathos)

An author's argument might also appeal to readers' emotions, perhaps by including personal stories and anecdotes. The next example presents an appeal to emotion. By sharing the personal anecdote of one student and speaking about emotional topics like family relationships, the author invokes the reader's empathy in asking them to reconsider the school rule.

> Our school should abolish its current ban on campus cell phone use. If students aren't able to use their phones during the school day, many of them feel isolated from their loved ones. For example, last semester, one student's grandmother had a heart attack in the morning. However, because he couldn't use his cell phone, the student didn't know about his grandmother's condition until the end of the day—when she had already passed away and it was too late to say goodbye. By preventing students from contacting their friends and family, our school is placing undue stress and anxiety on students.

Credibility (Ethos)

Finally, an appeal to authority includes a statement from a relevant expert. In this case, the author uses a doctor in the field of education to support the argument. All three examples begin from the same opinion—the school's phone ban needs to change—but rely on different argumentative styles to persuade the reader.

> Our school should abolish its current ban on campus cell phone use. According to Dr. Bartholomew Everett, a leading educational expert, "Research studies show that cell phone usage has no real impact on student attentiveness. Rather, phones provide a valuable technological resource for learning. Schools need to learn how to integrate this new technology into their curriculum." Rather than banning phones altogether, our school should follow the advice of experts and allow students to use phones as part of their learning.

Rhetorical Questions

Another commonly used argumentative technique is asking **rhetorical questions**, questions that do not actually require an answer but that push the reader to consider the topic further.

> I wholly disagree with the proposal to ban restaurants from serving foods with high sugar and sodium contents. Do we really want to live in a world where the government can control what we eat? I prefer to make my own food choices.

Here, the author's rhetorical question prompts readers to put themselves in a hypothetical situation and imagine how they would feel about it.

How Words Affect Tone

Tone refers to the writer's attitude toward the subject matter. For example, the tone conveys how the writer feels about the topic he or she is writing about. A lot of nonfiction writing has a neutral tone, which is an important tone for the writer to take. A neutral tone demonstrates that the writer is presenting a topic impartially and letting the information speak for itself. On the other hand, nonfiction writing can be just as effective and appropriate if the tone isn't neutral. For instance, consider this example:

> Seat belts save more lives than any other automobile safety feature. Many studies show that airbags save lives as well; however, not all cars have airbags. For instance, some older cars don't. Furthermore, air bags aren't entirely reliable. For example, studies show that in 15% of accidents, airbags don't deploy as designed; but, on the other hand, seat belt malfunctions are extremely rare. The number of highway fatalities has plummeted since laws requiring seat belt usage were enacted.

In this passage, the writer mostly chooses to retain a neutral tone when presenting information. If the writer would instead include their own personal experience of losing a friend or family member in a car accident, the tone would change dramatically. The tone would no longer be neutral and would show that the writer has a personal stake in the content, allowing them to interpret the information in a different way. When analyzing tone, consider what the writer is trying to achieve in the text and how they *create* the tone using style.

An author's choice of words—also referred to as **diction**—helps to convey his or her meaning in a particular way. Through diction, an author can convey a particular tone—e.g., a humorous tone, a serious tone—in order to support the thesis in a meaningful way to the reader.

Connotation and Denotation

Connotation is when an author chooses words or phrases that invoke ideas or feelings other than their literal meaning. An example of the use of connotation is the word *cheap*, which suggests something is poor in value or negatively describes a person as reluctant to spend money. When something or someone is described this way, the reader is more inclined to have a particular image or feeling about it or him/her. Thus, connotation can be a very effective language tool in creating emotion and swaying opinion. However, connotations are sometimes hard to pin down because varying emotions can be associated with a word. Generally, though, connotative meanings tend to be fairly consistent within a specific cultural group.

Denotation refers to words or phrases that mean exactly what they say. It is helpful when a writer wants to present hard facts or vocabulary terms with which readers may be unfamiliar. Some examples of denotation are the words *inexpensive* and *frugal. Inexpensive* refers to the cost of something, not its value, and *frugal* indicates that a person is conscientiously watching his or her spending. These terms do not elicit the same emotions that *cheap* does.

Authors sometimes choose to use both, but what they choose and when they use it is what critical readers need to differentiate. One method isn't inherently better than the other; however, one may create a better effect, depending upon an author's intent. If, for example, an author's purpose is to inform, to instruct, and to familiarize readers with a difficult subject, his or her use of connotation may be helpful. However, it may also undermine credibility and confuse readers. An author who wants to create a credible, scholarly effect in his or her text would most likely use denotation, which emphasizes literal, factual meaning and examples.

How Figurative Language Affects the Meaning of Words

It's important to be able to recognize and interpret **figurative,** or non-literal, language. Literal statements rely directly on the denotations of words and express exactly what's happening in reality. Figurative language uses non-literal expressions to present information in a creative way. Consider the following sentences:

a. His pillow was very soft, and he fell asleep quickly.

b. His pillow was a fluffy cloud and he floated away on it to the dream world.

Sentence *A* is literal, employing only the real meanings of each word. Sentence *B* is figurative. It employs a metaphor by stating that his pillow was a cloud. Of course, he isn't actually sleeping on a cloud, but the reader can draw on images of clouds as light, soft, fluffy, and relaxing to get a sense of how the character felt as he fell asleep. Also, in sentence *B*, the pillow becomes a vehicle that transports him to a magical dream world. The character isn't literally floating through the air—he's simply falling asleep! But by utilizing figurative language, the author creates a scene of peace, comfort, and relaxation that conveys stronger emotions and more creative imagery than the purely literal sentence. While there are countless types of figurative language, there are a few common ones that any reader should recognize.

Simile and **metaphor** are comparisons between two things, but their formats differ slightly. A simile says that two things are *similar* and makes a comparison using "like" or "as"—*A* is like *B*, or *A* is as [some characteristic] as *B*—whereas a metaphor states that two things are exactly the same—*A* is *B*. In both cases, simile and metaphor invite the reader to think more deeply about the characteristics of the two subjects and consider where they overlap. An example of metaphor can be found in the above sentence about the sleeper ("His pillow was a fluffy cloud"). For an example of simile, look at the first line of Robert Burns' famous poem:

> My love is like a red, red rose

This is comparison using "like," and the two things being compared are love and a rose. Some characteristics of a rose are that it's fragrant, beautiful, blossoming, colorful, vibrant—by comparing his love to a rose, Burns asks the reader to apply these qualities to his love. In this way, he implies that his love is also fresh, blossoming, and brilliant.

Similes can also compare things that appear dissimilar. Here's a song lyric from Florence and the Machine:

> Happiness hit her like a bullet in the back

"Happiness" has a very positive connotation, but getting "a bullet in the back" seems violent and aggressive, not at all related to happiness. By using an unexpected comparison, the writer forces readers to think more deeply about the comparison and ask themselves how could getting shot be similar to feeling happy. "A bullet in the back" is something that she doesn't see coming; it's sudden and forceful; and presumably, it has a strong impact on her life. So, in this way, the author seems to be saying that unexpected happiness made a sudden and powerful change in her life.

Another common form of figurative language is **personification,** when a non-human object is given human characteristics. William Blake uses personification here:

> . . . the stars threw down their spears,
>
> And watered heaven with their tears

He imagines the stars as combatants in a heavenly battle, giving them both action (throwing down their spears) and emotion (the sadness and disappointment of their tears). Personification helps to add emotion or develop relationships between characters and non-human objects. In fact, most people use personification in their everyday lives:

> My alarm clock betrayed me! It didn't go off this morning!
>
> The last piece of chocolate cake was staring at me from the refrigerator.

Next is **hyperbole,** a type of figurative language that uses extreme exaggeration. Sentences like, "I love you to the moon and back," or "I will love you for a million years," are examples of hyperbole. They aren't literally true—unfortunately, people cannot jump to outer space or live for a million years—but they're creative expressions that communicate the depth of feeling of the author.

Another way that writers add deeper meaning to their work is through **allusions.** An allusion is a reference to something from history, literature, or another cultural source. When the text is from a different culture or a time period, readers may not be familiar with every allusion. However, allusions tend to be well-known because the author wants the reader to make a connection between what's happening in the text and what's being referenced.

> I can't believe my best friend told our professor that I was skipping class to finish my final project! What a Judas!

This sentence contains a Biblical allusion to Judas, a friend and follower of Jesus who betrayed Jesus to the Romans. In this case, the allusion to Judas is used to give a deeper impression of betrayal and disloyalty from a trusted friend. Commonly used allusions in Western texts may come from the Bible, Greek or Roman mythology, or well-known literature such as Shakespeare. By familiarizing themselves with these touchstones of history and culture, readers can be more prepared to recognize allusions.

How Figurative Language Influences the Author's Purpose

A **rhetorical strategy**—also referred to as a **rhetorical mode**—is the structural way an author chooses to present his/her argument. Though the terms noted below are similar to the organizational structures noted earlier, these strategies do not imply that the entire text follows the approach. For example, a cause-and-effect organizational structure is solely that, nothing more. A persuasive text may use cause and effect as a strategy to convey a singular point. Thus, an argument may include several of the strategies as the author strives to convince his or her audience to take action or accept a different point of view. It is important that readers are able to identify an author's thesis and position on the topic in order to be able to identify the careful construction through which the author speaks to the reader.

The following are some of the more common rhetorical strategies:

- **Cause and effect**—establishing a logical correlation or causation between two ideas
- **Classification/division**—the grouping of similar items together or division of something into parts
- **Comparison/contrast**—the distinguishing of similarities/differences to expand on an idea
- **Definition**—used to clarify abstract ideas, unfamiliar concepts, or to distinguish one idea from another
- **Description**—use of vivid imagery, active verbs, and clear adjectives to explain ideas
- **Exemplification**—the use of examples to explain an idea
- **Narration**—anecdotes or personal experience to present or expand on a concept
- **Problem/Solution**—presentation of a problem or problems, followed by proposed solution(s)

How Rhetorical Language Conveys Meaning, Emotion, or Persuades Readers

A **rhetorical device** is the phrasing and presentation of an idea that reinforces and emphasizes a point in an argument. A rhetorical device is often quite memorable. One of the more famous uses of a rhetorical device is in John F. Kennedy's 1961 inaugural address: "Ask not what your country can do for you, ask what you can do for your country." The contrast of ideas presented in the phrasing is an example of the rhetorical device of antimetabole. Some other common examples are provided below, but test takers should be aware that this is not a complete list.

Device	Definition	Example
Alliteration	Repeating the same beginning sound or letter in a phrase for emphasis	The busy baby babbled.
Allusion	A reference to a famous person, event, or significant literary text as a form of significant comparison	"We are apt to shut our eyes against a painful truth, and listen to the song of that siren till she transforms us into beasts." Patrick Henry
Anaphora	The repetition of the same words at the beginning of successive words, phrases, or clauses, designed to emphasize an idea	"We shall not flag or fail. We shall go on to the end. We shall fight in France, we shall fight on the seas and oceans, we shall fight with growing confidence ... we shall fight in the fields and in the streets, we shall fight in the hills. We shall never surrender." Winston Churchill
Antithesis	A part of speech where a contrast of ideas is expressed by a pair of words that are opposite of each other.	"That's one small step for man, one giant leap for mankind." Neil Armstrong
Foreshadowing	Giving an indication that something is going to happen later in the story	I wasn't aware at the time, but I would come to regret those words.
Hyperbole	Using exaggeration not meant to be taken literally	The girl weighed less than a feather.
Idiom	Using words with predictable meanings to create a phrase with a different meaning	The world is your oyster.
Imagery	Appealing to the senses by using descriptive language	The sky was painted with red and pink and streaked with orange.
Metaphor	Compares two things as if they are the same	He was a giant teddy bear.
Onomatopoeia	Using words that imitate sound	The tire went off with a bang and a crunch.
Parallelism	A syntactical similarity in a structure or series of structures used for impact of an idea, making it memorable	"A penny saved is a penny earned." Ben Franklin
Personification	Attributing human characteristics to an object or an animal	The house glowered menacingly with a dark smile.

Device	Definition	Example
Rhetorical question	A question posed that is not answered by the writer though there is a desired response, most often designed to emphasize a point	"Can anyone look at our reduced standing in the world today and say, 'Let's have four more years of this?'" Ronald Reagan
Simile	Compares two things using "like" or "as"	Her hair was like gold.
Symbolism	Using symbols to represent ideas and provide a different meaning	The ring represented the bond between us.
Understatement	A statement meant to portray a situation as less important than it actually is to create an ironic effect	"The war in the Pacific has not necessarily developed in Japan's favor." Emperor Hirohito, surrendering Japan in World War II

Sarcasm

Depending on the tone of voice or the words used, sarcasm can be expressed in many different ways. **Sarcasm** is defined as a bitter or ambiguous declaration that intends to cut or taunt. Most of the ways we use sarcasm is saying something and not really meaning it. In a way, sarcasm is a contradiction that is understood by both the speaker and the listener to convey the opposite meaning. For example, let's say Bobby is struggling to learn how to play the trumpet. His sister, Gloria, walks in and tells him: "What a great trumpet player you've become!" This is a sort of verbal irony known as sarcasm. Gloria is speaking a contradiction, but Bobby and Gloria both know the truth behind what she's saying: that Bobby is not a good trumpet player. Sarcasm can also be accompanied by nonverbal language, such as a smirk or a head tilt. Remember that sarcasm is not always clear to the listener; sometimes sarcasm can be expressed by the speaker but lost on the listener.

Irony

Irony is a device that authors use when pitting two contrasting items or ideas against each other in order to create an effect. It's frequently used when an author wants to employ humor or convey a sarcastic tone. Additionally, it's often used in fictional works to build tension between characters, or between a particular character and the reader. An author may use **verbal irony** (sarcasm), **situational irony** (where actions or events have the opposite effect than what's expected), and **dramatic irony** (where the reader knows something a character does not). Examples of irony include:

- Dramatic Irony: An author describing the presence of a hidden killer in a murder mystery, unbeknownst to the characters but known to the reader.
- Situational Irony: An author relating the tale of a fire captain who loses her home in a five-alarm conflagration.
- Verbal Irony: This is where an author or character says one thing but means another. For example, telling a police officer "Thanks a lot" after receiving a ticket.

Understatement

Making an **understatement** means making a statement that gives the illusion of something being smaller than it actually is. Understatement is used, in some instances, as a humorous rhetorical device. Let's say that there are two friends. One of the friends, Kate, meets the other friend's, Jasmine's, boyfriend. Jasmine's boyfriend, in Kate's opinion, is attractive, funny, and intelligent. After Kate meets her friend's boyfriend, Kate says to Jasmine, "You could do worse." Kate and Jasmine both know from Kate's tone that this means Kate is being ironic—Jasmine could do much, much worse, because her boyfriend is considered a "good catch." The understatement was a rhetorical device used by Kate to let Jasmine know she approves.

Comparing Different Ways of Presenting Ideas

Evaluating Two Different Texts

Every passage has its own unique scope, purpose, and emphasis, or what it covers, why it is written, and what its specific focus is centered upon. Additionally, each passage is written with a particular audience in mind, and each passage affects its audience differently. The scope, purpose, and emphasis of each passage can be found by comparing the parts of the piece with the whole framework of the piece. Word choices, grammatical choices, and syntactical choices can help the reader figure out the scope, purpose, and emphasis. These choices are embedded in the words and sentences of the passage (the parts). They help show the intentions and goals of the author (the "whole"). For example, if an author uses strong language like *enrage*, *ignite*, *infuriate*, and *antagonize*, then they may be cueing the reader in to their own rage, or they may be trying to incite anger in others. Likewise, if an author continually uses short, simple sentences, he or she might be trying to incite excitement or nervousness. These different choices and styles affect the overall message, or purpose. Sometimes the subject matter or audience is discussed explicitly, but often, on GED tests, test takers have to break a passage down, also known as decoding the passage. In this way, test takers can find the passage's target audience and intentions. Meanwhile, the impact of the article can be personal or historical, depending upon the passage—it can either speak to the test taker personally or capture a historical era.

When two passages are analyzed in juxtaposition—or side-by-side—it can help the audience have a clearer picture of the scope, purpose, emphasis, audience, and impact. Evaluating and comparing passages side-by-side helps shed light on similarities and differences that are helpful for test takers. The key is to figure out both the parts and the "wholes" of each passage. Compare the word choices, grammatical choices, and syntactical choices of each passage, and then compare the big picture of each passage. As a result, test takers will have a stronger basis for understanding the intricate details and broader frameworks of all passages they encounter.

Evaluating Two Different Passages

Every passage also has its own view, tone, style, organization, purpose, or impact. It is extremely important to compare the parts to the "wholes" of each passage. Additionally, these parts and "wholes" are better understood through **intertextual analysis** (for example, comparing the texts as if they were side by side). The viewpoint of the text can be found through a close analysis of the author's biases, or personal opinions or perceptions. All biases are embedded in the word choices, grammatical choices, and syntactical choices of each passage. For example, if an author continually uses negative words like *dislike*, *hate*, *despise*, *detrimental*, or *loathe*, they are trying to illustrate their own hatred of something or convey a character's hatred of something. These negative terms inevitably affect the view and tone of a passage. Comparing terminologies and biases can help a test taker better understand the similarities and differences of two or more passages. Similarly, the purposes of each can be better highlighted with a closer examination of word choices, grammatical choices, and syntactical choices.

Organization, on the other hand, is more easily understood by studying the passage on its own. Organizational differences on the page are likely to jump out at the test taker. A poetry passage, for instance, is traditionally organized differently than a prose passage. Test takers can see the differences in structures: one uses paragraphs, while the other uses stanzas. If organizational differences cannot be deduced through visual analysis, test takers should try to take a closer look at the sequence of events or the content of each paragraph. Organization, nevertheless, is not something that is separate from view, tone, purpose, style, and impact. It can skew or connect a viewpoint, shift or solidify tone, reinforce or undermine purpose, express or conceal a particular style, or establish or disestablish the impact of passage. Organization is the backbone of every form of written expression—it unifies all the parts and sets the parameters for the wholes. Thus, organization should be analyzed strategically when comparing passages.

Literary Text

Poetry

The genre of **poetry** refers to literary works that focus on the expression of feelings and ideas through the use of structure and linguistic rhythm to create a desired effect.

Different poetic structures and devices are used to create the various major forms of poetry. Some of the most common forms are discussed in the following chart.

Type	Poetic Structure	Example
Ballad	A poem or song passed down orally which tells a story and in English tradition usually uses an ABAB or ABCB rhyme scheme	William Butler Yeats' "The Ballad of Father O'Hart"
Epic	A long poem from ancient oral tradition which narrates the story of a legendary or heroic protagonist	Homer's The Odyssey Virgil's The Aeneid
Haiku	A Japanese poem of three unrhymed lines with five, seven, and five syllables (in English) with nature as a common subject matter	Matsuo Bashō An old silent pond... A frog jumps into the pond, splash! Silence again.
Limerick	A five-line poem written in an AABBA rhyme scheme, with a witty focus	From Edward Lear's Book of Nonsense— "There was a Young Person of Smyrna Whose grandmother threatened to burn her..."
Ode	A formal lyric poem that addresses and praises a person, place, thing, or idea	Edna St. Vincent Millay's "Ode to Silence"
Sonnet	A fourteen-line poem written in iambic pentameter	Shakespeare's Sonnets 18 and 130

Understanding Poetic Devices and Structure

Poetic Devices

Rhyme is the poet's use of corresponding word sounds in order to create an effect. Most rhyme occurs at the ends of a poem's lines, which is how readers arrive at the **rhyme scheme**. Each line that has a corresponding rhyming sound is assigned a letter—A, B, C, and so on. When using a rhyme scheme, poets will often follow lettered patterns. Robert Frost's *"The Road Not Taken"* uses the ABAAB rhyme scheme:

Two roads diverged in a yellow wood,	A
And sorry I could not travel both	B
And be one traveler, long I stood	A
And looked down one as far as I could	A
To where it bent in the undergrowth;	B

Another important poetic device is **rhythm**—metered patterns within poetry verses. When a poet develops rhythm through **meter**, he or she is using a combination of stressed and unstressed syllables to create a sound effect for the reader.

Rhythm is created by the use of poetic feet—individual rhythmic units made up of the combination of stressed and unstressed syllables. A line of poetry is made up of one or more poetic feet. There are five standard types in English poetry, as depicted in the chart below.

Foot Type	Rhythm	Pattern
Iamb	buh Buh	Unstressed/stressed
Trochee	Buh buh	Stressed/unstressed
Spondee	Buh Buh	Stressed/stressed
Anapest	buh buh Buh	Unstressed/unstressed/stressed
Dactyl	Buh buh buh	Stressed/unstressed/unstressed

Structure

Poetry is most easily recognized by its structure, which varies greatly. For example, a structure may be strict in the number of lines it uses. It may use rhyming patterns or may not rhyme at all. There are three main types of poetic structures:

- *Verse*—poetry with a consistent meter and rhyme scheme
- *Blank verse*—poetry with consistent meter but an inconsistent rhyme scheme
- *Free verse*—poetry with inconsistent meter or rhyme

Verse poetry is most often developed in the form of **stanzas**—groups of word lines. Stanzas can also be considered *verses*. The structure is usually formulaic and adheres to the protocols for the form. For example, the English sonnet form uses a structure of fourteen lines and a variety of different rhyming patterns. The English ode typically uses three ten-line stanzas and has a particular rhyming pattern.

Poets choose poetic structure based on the effect they want to create. Some structures—such as the ballad and haiku—developed out of cultural influences and common artistic practice in history, but in more modern poetry, authors choose their structure to best fit their intended effect.

History of Poetry's Development

Ancient Times

Poetry has been in existence for thousands of years and even predates literacy. Early poems were passed on through oral tradition and were sung or recited as a way to remember a culture's history. The ancient Greek poets were the first to write down their poetry in the seventh to fourth century BC.

Greek poetry took three forms: epic, lyric, and dramatic. It used meter, and Greeks were the first to introduce iambic pentameter. Some of the greatest poets of this time included Homer, Aeschylus, and Euripides. Homer's *The Iliad* and *The Odyssey,* which focused on Greek mythology, are two of the most famous epic poems from this time period. Lyric poets of ancient Greece included Alcaeus, Sappho, and Pindar. Sappho was a female poet from the island of Lesbos. Most of her poetry exists in fragments, though her most complete poem, "Ode to Aphrodite," is influenced by and makes references to Homer's *Iliad.* The dramatic poetry of ancient Greece was written by Aeschylus and Sophocles, among others. It was divided into the same categories seen today in modern drama of tragedy and comedy.

When Greece was conquered by the Romans, their works were borrowed and adapted and eventually became the basis for modern literature. Romans wrote in Latin, and the Greek influence is clear in their poetry. Some of the oldest Roman literature is actually a translation of Greek works. Greco-Roman poet Andronicus first translated Homer's *The Odyssey* from Greek into Latin for a Roman audience. Though this was done for educational purposes at first, it became the basis for more Roman works and helped to develop Roman literature. Ovid was one of the most famous Roman poets. He is best known for *Metamorphoses,* a narrative poem consisting of fifteen texts. It spans the history of the world in myths from creation to the time of Julius Caesar and is one of the most influential texts in poetry to date. Another of Rome's great poets, Virgil, is best known for his *Aeneid.* This epic poem, written between 29 BC and 19 BC, tells the story of mythological Trojan hero Aeneas and his journey to Italy.

Middle Ages to Seventeenth Century

Poems in the Middle Ages were influenced by the historical events of the time, including religious movements. They were often religious in nature and typically written in Latin, as this was the predominant language of the Roman Catholic Church. Geoffrey Chaucer, a famous medieval writer, experimented with using the vernacular, or common, language of the people, writing works such as *The Parliament of Birds* and *The Canterbury Tales* in English. *Beowulf,* written by an anonymous Anglo-Saxon poet, is probably the most well-known poem to come out of the medieval period. It is an epic written in Old English, the vernacular language of the time. While its origins are unknown, this is the first time *Beowulf* was taken from oral tradition to written format.

The Renaissance brought with it one of the most prolific times in poetry's history. This cultural movement saw creative advancements in literature, art, and music, lasting from the fifteenth to the seventeenth century. Some of the most famous poetry to come from this time period includes William Shakespeare's sonnets. During the Enlightenment period that followed, there was a return to the style of the ancient Greeks, with a concentration on the epic poem. Alexander Pope was the most famous poet of this time, known for his satirical works and the use of the heroic couplet. The end of the eighteenth century brought about the birth of Romanticism and such famous British poets as William Blake, Lord Byron, William Wordsworth, and John Keats. Many of these poets got their inspiration from the natural world, which was in contrast to the religious themes in the poetry that came before it.

Nineteenth Century

Romanticism carried over into the nineteenth century, when there was also a rise in American poetry. These early American poets included Walt Whitman, Robert Frost, and Henry Wadsworth Longfellow. Whitman was known as the father of free verse. Most of these poets were known as transcendentalists, focusing on themes of spirituality, nature, and utopian values. Whitman's *Leaves of Grass*, published in 1855, was his idea of an American epic. The poetry in *Leaves of Grass* does not follow any rhythmic or metric patterns. Its themes of sensual pleasures and the natural world were novel and controversial at this time, but it has become one of the most influential works of American poetry.

Poetry of the nineteenth century also saw more working-class themes and authors from working-class backgrounds. This poetry had its roots in politics, rather than religion, nature, or romantic themes. This was also an important time for women in poetry, with the works of Emily Dickinson, Emily Bronte, and Elizabeth Browning gaining popularity. Dickinson's lyric poems used short lines, unconventional spelling and punctuation, and often lacked titles. While she wrote hundreds of poems in her lifetime, because of its unconventional style, most of Dickinson's work was published after her death. Her "Because I Could Not Stop for Death," published posthumously in 1890, exemplified her unique and influential style of meter and rhyme. Modernist poetry also got its start in the late nineteenth century. This format is marked by a movement from the personal to the world around the individual. It was born in the nineteenth century but flourished in the early twentieth century.

Twentieth Century

Modernist poetry continued into the twentieth century, with poets such as T. S. Eliot, Ezra Pound, and Gertrude Stein. Modernist poetry is characterized by the use of allusion and fragmented language. Some smaller movements that grew out of modernism include free verse, Dadaism, and surrealism. The twentieth century also saw a rise in African American poets with the Harlem Renaissance spanning the 1920s and 1930s. The publishing industry sought out African American writers whose poetry focused on realistic portrayals of their lives. One of the most important writers to come out of this movement was Langston Hughes, whose poetry honestly depicted his life and struggles as a black man living in America.

Another important poetry movement of the twentieth century was the Beat movement. Beat poetry, born out of New York and San Francisco, was characterized by anti-conformist themes and influenced by ideas of sexual freedom and drug use. Some of the most famous Beat poets include Allen Ginsberg and Lawrence Ferlinghetti, whose works challenged the social norms of the time. The second half of the twentieth century saw the advent of confessional poetry with the works of Sylvia Plath. Her poetry was deeply personal and rooted in natural themes in a lyrical style. Near the end of the nineteenth century, the New Formalism movement brought back the meter and rhyme of more traditional poetry. Poets such as Charles Martin, Brad Leithauser, and Molly Peacock adopted this New Formalist style.

Twenty-First Century

Contemporary poetry takes many forms and abides by no strict rules. Poets of the twenty-first century write about anything, from technology to love to civil rights. Their poetry tends to be more realistic in nature, and liberties are taken with form and structure. Poets of the twenty-first century often inject humor into their poetry and do not take form quite as seriously as those that came before them. Sherman Alexie, for example, writes about the plight of Native Americans but does so using irony and dark humor. Rita Dove, a poet laureate to both the state of Virginia and the Library of Congress, is another important twenty-first-century poet. Her work often has historical aspects and uses themes from other art forms, such as music and dance. Dove's *Sonata Mullatica*, published in 2009, is a collection of poetry about the life of George Bridgetower, a biracial musician, and his friendship with

Beethoven. Poetry in the twenty-first century has been affected by the digital age, and many poems are now exclusively written and published in electronic format. The Internet has created a new forum for poetry that has helped it to continue to develop into the modern age.

Drama

Drama is a type of fiction that is based on a script that is meant to be performed. Works of drama are called **plays**. Plays are intended to be performed on a stage by actors in front of an audience. Like other works of fiction, plays contain characters, plot, setting, theme, symbolism, and imagery. The main difference is that plays are sectioned into acts and scenes rather than chapters or stanzas. Drama is one of the oldest forms of literature, and it has evolved from the first Greek tragedies, such as *Antigone* and *Prometheus Bound*, into what is performed on modern stages today.

Like prose fiction, drama has several genres. The following are the most common ones:

- Comedy: a humorous play designed to amuse and entertain, often with an emphasis on the common person's experience, generally resolved in a positive way—e.g., Richard Sheridan's *School for Scandal*, Shakespeare's *Taming of the Shrew*, Neil Simon's *The Odd Couple*
- History: a play based on recorded history where the fate of a nation or kingdom is at the core of the conflict—e.g., Christopher Marlowe's *Edward II*, Shakespeare's *King Richard III*, Arthur Miller's *The Crucible*
- Tragedy: a serious play that often involves the downfall of the protagonist. In modern tragedies, the protagonist is not necessarily in a position of power or authority—e.g., Jean Racine's *Phèdre*, Arthur Miller's *Death of a Salesman*, John Steinbeck's *Of Mice and Men*
- Melodrama: a play that emphasizes heightened emotion and sensationalism, generally with stereotypical characters in exaggerated or realistic situations and with moral polarization—e.g., Jean-Jacques Rousseau's *Pygmalion*
- Tragicomedy: a play that has elements of both tragedy—a character experiencing a tragic loss—and comedy—the resolution is often positive with no clear distinctive mood for either—e.g., Shakespeare's *The Merchant of Venice*, Anton Chekhov's *The Cherry Orchard*

Structural Elements

The text of the play is called a **script**, and it is made up of both stage directions and dialogue. Stage directions are sections of the play that set the scene. Set directions might include information on what the scenery should look like and where the actors should stand. Plays also contain **dialogue**, which refers to the actual words the actors should speak. The difference in a play and other literary forms is in its construction and that it is intended to be performed. Plays are typically made up of acts. A playwright might use acts to indicate a change in time, setting, or mood. Acts can also be divided into scenes. A scene change may be used to indicate a change in the action, to introduce new characters, or to indicate a change in setting at the same time. In a play, there may be a protagonist, or central character, and an antagonist, who opposes the protagonist. A play, like fictional prose, often uses the following plot structure known as dramatic structure or Freytag's pyramid: exposition, rising action, climax, falling action, and denouement.

- *Exposition*—The first part of the play that introduces background information about setting, characters, plot, backstories, etc.

- *Rising action*—A series of events that build up to the main event of the story
- *Climax*—The main event or turning point of the play, when things turn around for a protagonist in a comedy or start to go bad for the protagonist in a tragedy
- *Falling action*—The part when the plot slows down and starts moving toward a conclusion, often the logical consequence of the climax
- *Denouement*—The ending of the play when conflicts are resolved

A longer play may also contain subplots, which are secondary or in contrast to the main plot. A play can have one or more themes, depending on its length.

As an example, the following dramatic structure is used in Shakespeare's *Romeo and Juliet*:

Exposition: The setting of Verona, Italy; protagonists Romeo and Juliet are introduced; and the feud between the Capulets and the Montagues is revealed.

Rising action: Romeo and Juliet meet and fall in love but cannot be together because of the feud.

Climax: Juliet's cousin Tybalt kills Mercutio, igniting the feud. Romeo kills Tybalt and is banished. Romeo and Juliet secretly marry.

Falling action: Juliet fakes her death to avoid an arranged marriage and be with Romeo; Romeo plans his own suicide when he learns she has (seemingly) died.

Denouement: Romeo commits suicide at Juliet's tomb. She wakes to find him dead and also commits suicide. When the families learn they were secretly married, they resolve to end their long-standing feud.

History of Drama's Development

Ancient Times

The ancient Greeks are widely accepted as the inventors of drama. The word *drama* comes from the Greek word meaning action. The earliest plays were religious in nature, focusing on the Greek gods. Greek drama included comedy, which was satirical and made light of the foils of men in power. The Greek tragedies were a bit more involved, including themes of love and loss. They typically involved the downfall of the protagonist, an otherwise good person, because of a fatal flaw. For example, in Sophocles' *Antigone*, the title character's tragic flaw is her loyalty to the gods.

Greek drama also included a chorus, masked performers who represent the voice of society. They spoke in unison and offered commentary on the dramatic action of the play. The chorus also sang, danced, and recited poetry during the play. Important playwrights of this time included Aeschylus, Sophocles, and Euripides. These early Greek playwrights greatly influenced the writers who came after them, with Aeschylus being called the father of modern drama. Aeschylus' *Oresteia* is likely the first example of a play in trilogy format. Other important works of this time included Sophocles' *Antigone* and *Oedipus Rex*, Euripides' *Medea*, and Aeschylus' *Prometheus Bound.*

Middle Ages

In the Middle Ages, drama continued to be influenced by religion. Three types of drama began to emerge in the medieval period: mystery play, miracle play, and morality play. The **mystery play** focused on biblical stories. These plays contained multiple acts and were performed by religious figures such as priests or monks. The **miracle play**, also with a religious theme, focused on the life of a saint. The **morality play** was meant to teach the audience a lesson based on the rules of the church. Two kinds of stages were invented for medieval drama: the fixed stage and the movable stage. While there were some comic elements to medieval plays, their religious nature made them mostly serious in tone. Secular plays were less popular at this time but did exist, particularly in France, where the farce was typically performed by professional actors in public forums.

Renaissance

The Renaissance was a prolific time for drama. This time period gave birth to the Elizabethan drama, popularized by playwrights such as Ben Jonson, Christopher Marlowe, and William Shakespeare. Shakespeare wrote both drama and comedy, and his plays such as *Romeo and Juliet, Hamlet, Macbeth,* and many more are some of the most recognized and acclaimed plays of all time. He also popularized a new type of drama, the romantic play, which did not fit in either of the previous categories. Elizabethan drama saw a break from the religious themes of the plays that came before it and a shift in focus from God to people. Like in the Greek tragedies before them, the tragedies of this period were marked by a protagonist with a central flaw, which ultimately brings him to his downfall.

An important development during this period was the establishment of permanent theaters. These large theaters were profitable and gave playwrights a designated place to showcase their plays. Having designated theaters allowed the creation of theater companies made up of common men, and young boys often played the roles of women. Women were not allowed to act in plays until after 1660, as it was not deemed a suitable profession for them. Queen Elizabeth I loved drama and was a patron of Shakespeare. Her interest in and support for the theater helped it to flourish during her reign.

Seventeenth and Eighteenth Centuries

The Elizabethan playwrights continued to develop plays in the seventeenth century, but the Puritanical government shut down theaters for a time. When King Charles II came into power in 1660, the theater ban was lifted. Theaters once again flourished after the English Restoration. Women were now able to perform in these dramas, bringing life to the intended female roles, with Margaret Hughes credited as the first female actress in English theater. This time period also saw the first recognized female playwright, Aphra Behn. Her two-part play, *The Rover*, was written in 1677. New types of drama that were developed at this time included heroic drama and Restoration comedy, which made use of immoral themes. The eighteenth century saw the fall of Restoration comedy and the rise of musical comedies and themes much more geared to musical entertainment than serious drama. John Gay's *The Beggar's Opera*, for example, was written to the tune of popular music of the time.

Nineteenth Century to the Present

In the nineteenth century, drama was influenced by the Victorian era. **Closet drama**, a type of dramatic play that is meant to be read rather than performed, became more popular. As for the stage, melodrama became very popular at this time. Melodrama used music to enhance the more dramatic scenes of plays. Shorter musical acts were also included in nineteenth-century productions and often interspersed between acts of plays. Toward the end of the nineteenth century, modernist plays such as Henrik Ibsen's *A Doll's House,* written in 1879, tackled such issues as the emancipation of women. Russian playwright Anton Chekhov also wrote modernist plays at this time. His works were unique in that the most meaningful parts of the play were not in the words but in the set direction for the actors. In the early twentieth century, playwrights such as T. S. Eliot and American playwrights such as Arthur Miller and Tennessee Williams saw their plays not only produced for the stage, but also the screen. The advent of television and film created a new format and a wider audience for these dramatic plays. Miller's *Death of a Salesman* and *The Crucible* were made into television and motion picture films, respectively. Williams' *A Streetcar Named Desire* was made into a major motion picture that went on to win four Academy Awards. More recent contemporary playwrights such as David Mamet often write both stage plays and screenplays for films. Mamet has won the Pulitzer Prize for his dramatic plays *Speed-the-Plow* and *Glengarry Glen Ross* and earned Oscar nominations for his screenplays.

Prose Fiction

Elements of Fiction

Literary Elements of Fiction

There is no one, final definition of what literary elements are. They can be considered features or characteristics of fiction, but they are really more of a way that readers can unpack a text for the purpose of analysis and understanding the meaning. The elements contribute to a reader's literary interpretation of a passage as to how they function to convey the central message of a work. The most common literary elements used for analysis are the following:

- The **theme** is the central message of a fictional work, whether that work is structured as prose, drama, or poetry. It is the heart of what an author is trying to say to readers through the writing, and theme is largely conveyed through literary elements and techniques. Poetic elements overlap these elements and will be addressed separately.

- The **plot** is what happens in the story. Plots may be singular, containing one problem, or they may be very complex, with many sub-plots. All plots have exposition, a conflict, a climax, and a resolution. The *conflict* drives the plot and is something that the reader expects to be resolved. The plot carries those events along until there is a resolution to the conflict.

- **Characters** are the story's figures that assume primary, secondary, or minor roles. **Central** or major characters are those integral to the story—the plot cannot be resolved without them. A central character can be a **protagonist** or hero. There may be more than one protagonist, and he/she doesn't always have to possess good characteristics. A character can also be an **antagonist**—the force against a protagonist.

  Character development is when the author takes the time to create dynamic characters that add uniqueness and depth to the story. *Dynamic* characters are characters that change over the course of the plot time. **Stock** characters are those that appear across genres and embrace stereotypes—e.g., the cowboy of the Wild West or the blonde bombshell in a detective novel. A **flat** character is one that does not present a lot of complexity or depth, while a **rounded** character does. Sometimes, the **narrator** of a story or the speaker in a poem can be a character—e.g., Nick Carraway in Fitzgerald's *The Great Gatsby* or the speaker in Browning's "My Last Duchess." The narrator might also function as a character in prose, though not be part of the story—e.g., Dicken's narrator of *A Christmas Carol*.

- The **setting** is the time, place, or set of surroundings in which the story occurs. It includes time or time span, place(s), climates, geography—man-made or natural—or cultural environments. Emily Dickenson's poem "Because I could not stop for Death" has a simple setting—the narrator's symbolic ride with Death through town towards the local graveyard. Conversely, Leo Tolstoy's War and Peace encompasses numerous settings within settings in the areas affected by the Napoleonic Wars, spanning 1805 to 1812.

- The **point of view** is the position the narrator takes when telling the story in prose. If a narrator is incorporated in a drama, the point of view may vary; in poetry, point of view refers to the position the speaker in a poem takes.
    - The **first-person** point of view is when the writer uses the word "I" in the text. Poetry often uses first person, e.g., William Wordsworth's "I Wandered Lonely as a Cloud." Two examples of prose written in first person are Suzanne Collins's *The Hunger Games* and Anthony Burgess's *A Clockwork Orange*.
    - The **second person** point of view is when the writer uses the pronoun "you." It is not widely used in prose fiction, but as a technique, it has been used by writers such as William Faulkner in *Absalom, Absalom* and Albert Camus in *The Fall*. It is more common in poetry—e.g., Pablo Neruda's "If You Forget Me."
    - **Third person** point of view is when the writer utilizes pronouns such as him, her, or them. It may be the most utilized point of view in prose as it provides flexibility to an author and is the one with which readers are most familiar. There are two main types of third person used in fiction:
        - *Third person omniscient*—narrator is all-knowing, relating the story by conveying and interpreting thoughts/feelings of all characters
        - *Third person limited*—narrator relates the story through the perspective of one character's thoughts/feelings, usually the main character

Genres

Fiction written in prose can be further broken down into **fiction genres**—types of fiction. Some of the more common genres of fiction are as follows:

- **Classical fiction**: a work of fiction considered timeless in its message or theme, remaining noteworthy and meaningful over decades or centuries—e.g., Charlotte Brontë's *Jane Eyre*, Mark Twain's *Adventures of Huckleberry Finn*
- **Fables**: short fiction that generally features animals, fantastic creatures, or other forces within nature that assume human-like characters and has a moral lesson for the reader—e.g., *Aesop's Fables*
- **Fairy tales**: children's stories with magical characters in imaginary, enchanted lands, usually depicting a struggle between good and evil, a sub-genre of folklore—e.g., Hans Christian Anderson's *The Little Mermaid, Cinderella* by the Brothers Grimm
- **Fantasy**: fiction with magic or supernatural elements that cannot occur in the real world, sometimes involving medieval elements in language, usually includes some form of sorcery or witchcraft and sometimes set on a different world—e.g., J.R.R. Tolkien's *The Hobbit*, J.K. Rowling's *Harry Potter and the Sorcerer's Stone*, George R.R. Martin's *A Game of Thrones*
- **Folklore**: types of fiction passed down from oral tradition, stories indigenous to a particular region or culture, with a local flavor in tone, designed to help humans cope with their condition in life and validate cultural traditions, beliefs, and customs—e.g., William Laughead's *Paul Bunyan and The Blue Ox*, the Buddhist story of "The Banyan Deer"

- **Mythology**: closely related to folklore but more widespread, features mystical, otherworldly characters and addresses the basic question of why and how humans exist, relies heavily on allegory and features gods or heroes captured in some sort of struggle—e.g., Greek myths, Arthurian legends

- **Science fiction**: fiction that uses the principle of extrapolation—loosely defined as a form of prediction—to imagine future realities and problems of the human experience—e.g., Robert Heinlein's *Stranger in a Strange Land*, Ayn Rand's *Anthem*, Isaac Asimov's *I, Robot*, Philip K. Dick's *Do Androids Dream of Electric Sheep?*

- **Short stories**: short works of prose fiction with fully-developed themes and characters, focused on mood, generally developed with a single plot, with a short period of time for settings—e.g., Edgar Allan Poe's "Fall of the House of Usher," Shirley Jackson's "The Lottery," Isaac Bashevis Singer's "Gimpel the Fool"

Identifying Literary Contexts

Understanding that works of literature emerged either because of a particular context—or perhaps despite a context—is key to analyzing them effectively.

Historical Context

The **historical context** of a piece of literature can refer to the time period, setting, or conditions of living at the time it was written as well as the context of the work. For example, Hawthorne's *The Scarlet Letter* was published in 1850, though the setting of the story is 1642-1649. Historically, then, when Hawthorne wrote his novel, the United States found itself at odds as the beginnings of a potential Civil War were in view. Thus, the historical context is potentially significant as it pertains to the ideas of traditions and values, which Hawthorne addresses in his story of Hester Prynne in the era of Puritanism.

Cultural Context

The **cultural context** of a piece of literature refers to cultural factors, such as the beliefs, religions, and customs that surround and are in a work of literature. The Puritan's beliefs, religion, and customs in Hawthorne's novel would be significant as they are at the core of the plot—the reason Hester wears the A and why Arthur kills himself. The customs of people in the Antebellum Period, though not quite as restrictive, were still somewhat similar. This would impact how the audience of the time received the novel.

Literary Context

Literary context refers to the consideration of the genre, potentially at the time the work was written. In 1850, Realism and Romanticism were the driving forces in literature in the U.S., with depictions of life as it was at the time in which the work was written or the time it was written *about* as well as some works celebrating the beauty of nature. Thus, an audience in Hawthorne's time would have been well satisfied with the elements of both offered in the text. They would have been looking for details about everyday things and people (Realism), but they also would appreciate his approach to description of nature and the focus on the individual (American Romanticism). The contexts would be significant as they would pertain to evaluating the work against those criteria.

Here are some questions to use when considering context:

- When was the text written?
- What was society like at the time the text was written, or what was it like, given the work's identified time period?
- Who or what influenced the writer?
- What political or social influences might there have been?
- What influences may there have been in the genre that may have affected the writer?

Additionally, test takers should familiarize themselves with literary periods such as Old and Middle English, American Colonial, American Renaissance, American Naturalistic, and British and American Modernist and Post-Modernist movements. Most students of literature will have had extensive exposure to these literary periods in history, and while it is not necessary to recognize every major literary work on sight and associate that work to its corresponding movement or cultural context, the test taker should be familiar enough with the historical and cultural significance of each test passage in order to be able to address test questions correctly.

The following brief description of some literary contexts and their associated literary examples follows. It is not an all-inclusive list. The test taker should read each description, then follow up with independent study to clarify each movement, its context, its most familiar authors, and their works.

Metaphysical Poetry

Metaphysical poetry is the descriptor applied to 17th century poets whose poetry emphasized the lyrical quality of their work. These works contain highly creative poetic conceits or metaphoric comparisons between two highly dissimilar things or ideas. **Metaphysical poetry** is characterized by highly prosaic language and complicated, often layered, metaphor.

Poems such as John Donne's "The Flea," Andrew Marvell's "To His Coy Mistress," George Herbert's "The Collar," Henry Vaughan's "The World," and Richard Crashaw's "A Song" are associated with this type of poetry.

British Romanticism

British Romanticism was a cultural and literary movement within Europe that developed at the end of the 18th century and extended into the 19th century. It occurred partly in response to aristocratic, political, and social norms and partly in response to the Industrial Revolution of the day. Characterized by intense emotion, major literary works of **British Romanticism** embrace the idea of aestheticism and the beauty of nature. Literary works exalted folk customs and historical art and encouraged spontaneity of artistic endeavor. The movement embraced the heroic ideal and the concept that heroes would raise the quality of society.

Authors who are classified as British Romantics include Samuel Taylor Coleridge, John Keats, George Byron, Mary Shelley, Percy Bysshe Shelley, and William Blake. Well-known works include Samuel Taylor Coleridge's "Kubla Khan," John Keats's "Ode on a Grecian Urn," George Byron's "Childe Harold's Pilgrimage," Mary Shelley's *Frankenstein*, Percy Bysshe Shelley's "Ode to the West Wind," and William Blake's "The Tyger."

American Romanticism

American Romanticism occurred within the American literary scene beginning early in the 19th century. While many aspects were similar to British Romanticism, it is further characterized as having gothic aspects and the idea that individualism was to be encouraged. **American Romanticism** also embraced the concept of the *noble savage*—the idea that indigenous culture uncorrupted by civilization is better than advanced society.

Well-known authors and works include Nathanial Hawthorne's *The House of the Seven Gables*, Edgar Allan Poe's "The Raven" and "The Cask of Amontillado," Emily Dickinson's "I Felt a Funeral in My Brain" and James Fenimore Cooper's *The Last of the Mohicans.*

Transcendentalism

Transcendentalism was a movement that applied to a way of thinking that developed within the United States, specifically New England, around 1836. While this way of thinking originally employed philosophical aspects, **transcendentalism** spread to all forms of art, literature, and even to the ways people chose to live. It was born out of a reaction to traditional rationalism and purported concepts such as a higher divinity, feminism, humanitarianism, and communal living. Transcendentalism valued intuition, self-reliance, and the idea that human nature was inherently good.

Well-known authors include Ralph Waldo Emerson, Henry David Thoreau, Louisa May Alcott, and Ellen Sturgis Hooper. Works include Ralph Waldo Emerson's "Self-Reliance" and "Uriel," Henry David Thoreau's *Walden* and *Civil Disobedience*, Louisa May Alcott's *Little Women*, and Ellen Sturgis Hooper's "I Slept, and Dreamed that Life was Beauty."

The Harlem Renaissance

The Harlem Renaissance is the descriptor given to the cultural, artistic, and social boom that developed in Harlem, New York, at the beginning of the 20th century, spanning the 1920s and 1930s. Originally termed *The New Negro Movement*, it emphasized African American urban cultural expression and migration across the United States. It had strong roots in African American Christianity, discourse, and intellectualism. The **Harlem Renaissance** heavily influenced the development of music and fashion as well. Its singular characteristic was to embrace Pan-American culturalisms; however, strong themes of the slavery experience and African American folk traditions also emerged. A hallmark of the Harlem Renaissance was that it laid the foundation for the future Civil Rights Movement in the United States.

Well-known authors and works include Zora Neale Hurston's *Their Eyes Were Watching God*, Richard Wright's *Native Son*, Langston Hughes' "I, Too," and James Weldon Johnson's "God's Trombones: Seven Negro Sermons in Verse" and *The Book of American Negro Poetry.*

History of the Novel's Development

Eighteenth Century

The novel as it is known today first appeared in the eighteenth century. The word **novel** comes from the Italian word *novella,* which means new, referring to a new type of writing. Before this time, prose existed, but it wasn't realistic. Religion's hold over literature started to wane at this time, and authors began to write about the world around them. The novel was popularized by the use of realistic characters, set in real geographical locations and engaged in real-life situations. The earliest novels included *Robinson Crusoe* and *Moll Flanders* by Daniel Defoe. These works used the common man and woman as characters, unlike the prose that came before them, which used plots that centered on heroes, legends, and gods. Other popular novelists of this time included Jonathan Swift, Henry Fielding, and Samuel Richardson. Swift's *Gulliver's Travels* is considered an early version of the fantasy novel. The novel also flourished at this time with the growth of the printing industry. Books were readily available, and for the first time, the middle class was able to afford them. Many of the readers of these realistic stories were women. Writers took this into account and created characters that represented a wider range of people, including women and the middle class. For example, in the late eighteenth century, Jane Austen's books were populated with female characters, and Defoe's books centered on the common man. Another development that aided the popularity of novels in the eighteenth century was the creation of libraries. The ability to borrow books made literature much more widely available to lower classes, and novels flourished in this environment.

Nineteenth Century

The trend in nineteenth-century novels was again realism. Characters and settings were realistic, and they encountered realistic situations. Some of the most important and critically acclaimed novelists come out of this era, including Jane Austen, Charles Dickens, and the Bronte sisters. Austen's novels such as *Pride and Prejudice* and *Sense and Sensibility* were extremely popular and dealt with the issue of women and their dependence on marriage in the nineteenth century for social status and economic standing. Dickens' work focused on the people of London in the Victorian age. He used realistic characters and injected humor into his works, such as *Great Expectations* and *Oliver Twist.*

In America, writers like Nathaniel Hawthorne and Mark Twain were crafting the great American novel. Twain's *Adventures of Huckleberry Finn* and Hawthorne's *The Scarlet Letter* are revered as some of the best novels of all time, each making social commentary on the American experience. Hawthorne's work in particular featured the people of Puritan New England. His work often made social commentary on his anti-Puritan views. Hawthorne was considered part of a dark Romantic movement, focusing on themes of intense emotion such as horror, apprehension, and awe. Edgar Allan Poe was also part of the Romantic movement, and his novels included these same themes, particularly horror.

Twentieth Century

The novel continued its popularity in the twentieth century. Novelists such as George Orwell, James Joyce, F. Scott Fitzgerald, Ernest Hemingway, John Steinbeck, Edith Wharton, Toni Morrison, and Franz Kafka developed novels that are still read in high school classrooms all over the world. Varying subtypes of the novel developed at this time as well. Because of the great number of novels being written, they could be classified into categories such as romance, science fiction, mystery, fantasy, and historical fiction. The science fiction novel was even further popularized by authors such as George Orwell and H. G. Wells. Readers loved this new type of novel, and Wells' *War of the Worlds* was famously read on the radio in 1938, sparking fear in listeners that it was an actual news report of an invasion by Martians.

The first part of the century focused on modernism in literature, which centered more on the decline in civilization. Authors such as D. H. Lawrence and James Joyce wrote modernist novels, with Joyce's *Ulysses* being a prime example. Virginia Woolf, along with her husband, Leonard Woolf, established the Hogarth Press, which printed her own books as well as works by T. S. Eliot. Woolf's works, such as *Mrs. Dalloway* and *To the Lighthouse,* were experimental in form, and she is considered the foremost lyrical novelist of her time. Other authors were developing their own type of novels with writers like Agatha Christie popularizing the mystery novel in the 1930s.

The world was profoundly changed by World War I and World War II, and this can be seen in the novels of the second half of the twentieth century. Postwar novelists were from many different backgrounds as a reflection of the changes in the world after the war. They were the children of immigrants in America, and authors such as Truman Capote and Tennessee Williams are homosexual or bisexual. Many novelists wrote about World War II and its impact, such as Norman Mailer's critically acclaimed *The Naked and the Dead*, published in 1948. Mailer used journalistic style in his account of the lives of a platoon of soldiers stationed in the Pacific during World War II. Joseph Heller's *Catch 22,* written after his own experience as a bombardier in the US Air Force, took a more satirical approach to the war and is considered an example of dark surrealism. Postwar novels were also influenced by the Beat movement, with Jack Kerouac's *On the Road* being one of the prime examples of this movement.

Twenty-First Century

The twenty-first century saw a blurring of the lines of fiction. Novelists tended to cross genres, combining the elements of romance, fantasy, science fiction, and mystery into one book. The Young Adult category was created and gained much popularity with adult readers. An example of this is the Harry Potter series by J. K. Rowling. These novels cannot be confined to a category, as they contain elements of mystery, fantasy, and romance. They were also intended for young adults but were widely read by both children and adults alike. Another important development was the advancement of the graphic novel, which combines art along with the narrative. More than a comic book, the first graphic novel was published in the late twentieth century, but it became more widely published in the twenty-first century. Graphic novels are written in many genres, but supernatural themes are widely used. The so-called death of print also has an effect on the novel as e-books and e-readers become available.

The Short Story

The short story is a work of prose fiction that is typically much shorter than a novel. It is meant to be read in one sitting. Because of its length, the short story focuses on just one plot, with a single main character, central theme, and minimal secondary characters. There is no set length of a short story, though they are typically between one thousand and twenty thousand words. The idea of a short story dates back to the time of oral traditions such as fables, parables, and fairy tales. The father of the modern American short story is said to be Edgar Allan Poe. His works are widely used as an example of the short story. Other important short story writers include Ernest Hemingway, James Joyce, and Joyce Carol Oates.

The structure of the short story is similar to that of a novel, but it differs in that a novel may contain multiple themes, plots, or characters, and a short story typically contains just one of each. The short story focuses on a single main character who faces a challenge. The plot of the short story is the arrangement of the events. Many short story authors use a chronological order of events, telling the story from beginning to end in the order that it happened. Other authors break from this chronological order using devices such as flashback. Telling the events out of order can have an effect on the writing, creating suspense or allowing the reader to have knowledge of a later event throughout the story. An example of flashback would be revealing the ending at the beginning of the story. The effect might be to show that the journey is more important than the destination. Short stories usually contain a singular central theme. The theme is the overall idea that an author wants to convey. It can also be considered a lesson in some short stories. Setting is where and when a short story takes place. Setting can refer to time of day, time of year, or time in history. Other elements, such as the author's point of view, symbolism, and imagery, are important elements to the structure of a short story.

History of the Short Story's Development

Middle Ages

Like the poetry of the Middle Ages, fiction was also influenced by religion. While the short story wasn't actually called this, it did exist in some forms. At this time, it was more like a tale than the formal short story format of today. These tales were a reflection of the norms of the cultures that created them. The most notable form was written by author Geoffrey Chaucer. His *Canterbury Tales* is a collection of short narrative works about a group of Pilgrims on a journey from London to Canterbury Cathedral. This work is well loved for its vivid characters and is widely accepted as a predecessor to the modern short story. *The Canterbury Tales* is also significant because it was one of the first major works of literature written in the English language. After this time, works of short fiction saw a decline in the seventeenth and eighteenth centuries in favor of other types of literature.

Nineteenth Century

The modern short story as it's known today was created in the nineteenth century. Short story writers emerged in both Europe and America. Perhaps the most famous of these writers was Edgar Allan Poe. Poe's tales of horror such as "The Telltale Heart" and "The Cask of Amontillado" included all the elements of the modern short story, such as a single main character or narrator, a limited setting, and a singular central theme. Another subgenre that emerged at this time was southern gothic writing, which centered on the American South. The characters in these stories were deeply flawed and even disturbed. Rural communities were often the setting in the works of authors such as William Faulkner and Flannery O'Connor. O'Connor's "Good Country People" is an example of a southern gothic story, set in rural Georgia, populated by the common people of the South. O'Connor used this setting to make social commentary on the lack of vision or knowledge of these people.

Other important short story writers from this time included Nathaniel Hawthorne, Mark Twain, and Guy de Maupassant. Modernism was an important literary movement in the late nineteenth century, and the short story was influenced by it. Examples of the modernist movement can be seen in Anton Chekhov's "Gusev," the story of discharged soldiers dying of consumption. The late nineteenth century also included more women authors, and works by Charlotte Perkins Gilman and Kate Chopin were published. Kate Chopin wrote important short stories about women's issues, such as independence and their reliance on marriage for status. Chopin's most popular works include "Paul's Case," "Desiree's Baby," and "The Story of an Hour."

Twentieth Century

The short story continued to gain popularity in the twentieth century. Many writers of poems, plays, and novels also began to produce short stories. Writers such as Ernest Hemingway, William Faulkner, D. H. Lawrence, Kate Chopin, and James Joyce published short stories in addition to their other works at this time. Short story writers found a forum for their work in magazine format. This was the perfect way to deliver these short narratives to readers. Writers often found they could finance their larger projects such as novels and plays by selling short stories to magazines. Herman Melville said that he hated writing stories and only did so to make money.

The rise of the film industry took its toll on the short story and decreased the need for these short works. In turn, the short story also began to evolve into a new format. Writers in the nineteenth century were more concerned with plot and its resolution. In the second half of the twentieth century, short story authors began to experiment with form and often centered their stories on motifs rather than plot. Writers like Raymond Carver revived the short story a bit through this experimentation with form, and though the short story still exists today, it has never again reached the heights it did in the early part of the twentieth century.

Nonfiction Text

Nonfiction Prose

Nonfiction works are best characterized by their subject matter, which must be factual and real, describing true life experiences. There are several common types of literary non-fiction.

Biography

A biography is a work written about a real person (historical or currently living). It involves factual accounts of the person's life, often in a re-telling of those events based on available, researched factual information. The re-telling and dialogue, especially if related within quotes, must be accurate and reflect reliable sources. A **biography** reflects the time and place in which the person lived, with the goal of creating an understanding of the person and his/her human experience. Examples of well-known biographies include *The Life of Samuel Johnson* by James Boswell and *Steve Jobs* by Walter Isaacson.

Autobiography

An autobiography is a factual account of a person's life written by that person. It may contain some or all of the same elements as a biography, but the author is the subject matter. An **autobiography** will be told in first person narrative. Examples of well-known autobiographies in literature include *Night* by Elie Wiesel and *Margaret Thatcher: The Autobiography* by Margaret Thatcher.

Memoir

A memoir is a historical account of a person's life and experiences written by one who has personal, intimate knowledge of the information. The line between memoir, autobiography, and biography is often muddled, but generally speaking, a memoir covers a specific timeline of events as opposed to the other forms of nonfiction. A **memoir** is less all-encompassing. It is also less formal in tone and tends to focus on the emotional aspect of the presented timeline of events. Some examples of memoirs in literature include *Angela's Ashes* by Frank McCourt and *All Creatures Great and Small* by James Herriot.

Journalism

Some forms of **journalism** can fall into the category of literary non-fiction—e.g., travel writing, nature writing, sports writing, the interview, and sometimes, the essay. Some examples include Elizabeth Kolbert's "The Lost World, in the Annals of Extinction series for *The New Yorker* and Gary Smith's "Ali and His Entourage" for ***Sports Illustrated.***

The Essay

An essay is a short piece of nonfiction writing. It typically uses the opinion of the author on a single subject. Essay authors can write about virtually any subject, as there are no rules for the content of an essay. It is often used to convey a point about a subject but can also be written simply for pleasure. **Essays** can make use of different writing modes, such as argument, persuasion, causal analysis, critique, or observation. They can be formal or informal, serious in tone, satirical, or even humorous. Organization is up to the author and does not follow any strict rules. Virginia Woolf wrote a series of essays, most notably about the struggle for survival in her essay "Death of a Moth." George Orwell's essay "Shooting an Elephant" includes his graphic account of shooting an elephant as social commentary on anti-colonialism.

The Speech

A speech is a formal address meant to be spoken aloud to an audience. The purpose of a speech can vary, but it is typically to persuade, argue, inform, or inspire the audience. **Speeches** can be political in nature but don't have to be. Some of the most famous speeches include Martin Luther King's "I Have a Dream" speech, Abraham Lincoln's Gettysburg Address, and John F. Kennedy's inaugural address. Some important components of a speech are the style, substance, and impact of the words. While the writing of a speech is important, the delivery is important as well.

Speeches can be delivered in four methods: **impromptu**, with little to no preparation; **extemporaneous**, which involves preparation but is not read directly from cards; **manuscript**, which is read directly from a script or teleprompter; and **memorized**, which is a written speech delivered from the speaker's memory. The speech delivery type depends on the situation and reaction a speaker wants from the audience. Extemporaneous speaking is the most common delivery method, and many great speakers have used this format because it appears more natural to the audience. It is widely known that Martin Luther King's "I Have a Dream" speech was delivered extemporaneously, as he largely improvised the second half of the speech, including the portion where he states, "I have a dream." George W. Bush's Bullhorn speech, delivered at ground zero on September 14, 2001, in the wake of the September 11 terror attacks, is an example of an impromptu speech. Though he had not planned to speak at the event and was even advised against it, Bush took a bullhorn and spoke to the crowd, aiming to lift the spirits of a broken nation, delivering a memorable impromptu speech.

Visual and Performing Arts Reviews

Visual and performing arts includes film, theater, poetry performance, photography, painting, sculpting, public speech, dance, and music. Anything performed for a live audience is in the realm of this genre. A passage in the GED may be a critical review of one of these performances. Critical reviews go deeper than simply summary; they analyze the work with thoughtfulness to each of its components and how they relate to its contribution to the artform. For example, a reviewer might raise the question of how a photographer's use of light lends to the meaning of the subject in the photo. In visual and performing arts, meaning goes beyond subject; there are usually multiple components that enhance a piece's meaning, such as a dancer's technique within a solo. In order to analyze successfully, reviewers and test takers must be aware of critical theory and historical contexts. This will aid test takers in understanding how the reviewers go beyond summary of an artwork.

Critical Theory

A **critical theory** can be considered a methodology for understanding art. It asks, "What is art?" and offers readers a working set of principles to understand common themes, ideas, and intent. Classifications of critical theory are often referred to as *schools of thought*. These schools are based on subdivisions in historical perspective and in philosophical thinking across analysts and critics.

Romanticism/Aestheticism

Romanticism/Aestheticism spanned the 19th century and developed in response to the idea that enlightenment and reason were the source of all truth and authority in philosophy. **Romanticism** and **Aestheticism** embraced the tenet that *aesthetics*—all that is beautiful and natural—in art and literature should be considered the highest-held principle, overriding all others. Popular authors include Oscar Wilde, Edgar Allan Poe, Mary Shelley, and John Keats.

Marxism

Marxism as a literary theory developed in the early twentieth century after the Russian October Revolution of 1917. It loosely embraced the idea that social realism was the highest form of art and that the social classes' struggle for progress was the most important concept art could emphasize. Examples of authors include Simone de Beauvoir and Bertolt Brecht.

Structuralism

Structuralism included all aspects of philosophy, linguistics, anthropology, and literary theory. Beginning in the early 1900s, this school of thought focused on ideas surrounding how human culture is understood within its larger structures and how those structures influence people's thoughts and actions. Specifically, structuralism examines how literature is interconnected through structure. It examines common elements in the stories and the myths that contribute to literature as a whole. Popular theorists and writers include Claude Levi-Strauss, Umberto Eco, and Roland Barthes.

Post-Structuralism and Deconstruction

Post-Structuralism and deconstruction developed out of structuralism in the twentieth century. It expanded on the idea of overall structure in literature, but both theories argue varying analytical concepts of how that structure should be examined and utilized. For example, while structuralism acknowledges oppositional relationships in literature—e.g., male/female, beginning/end, rational/emotional—**post-structuralism** and **deconstruction** began de-emphasizing the idea that one idea is always dominant over another. Both also assert that studying text also means studying the knowledge that produced the text. Popular theorists and writers include Roland Barthes and Michel Foucault.

New Criticism

New Criticism dominated American culture in the mid-twentieth century. It purports that close, critical reading was necessary to understanding artwork, especially poetry. Popular theory also focused on the inherent beauty of artwork itself. **New Criticism** rejected the previous critical focus of how history, use of language, and the author's experience influence art, asserting those ideas as being too loosely interpretive in examining art. As a movement, it tended to separate art from historical context and an author's intent. It embraced the idea that formal study of structure and text should not be separated. Theorists of note include Stephen Greenblatt and Jonathan Goldberg.

Workplace and Community Documents

Following a Given Set of Directions

When you read a comic or magazine, it's not necessary to understand everything. However, other more technical readings, such as directions for setting up a coffeemaker or a new phone app, require more attention to detail. Read each step all the way through, and don't skip ahead. While you may think that you know all or some of the steps, it's important to read directions in the manner that the writer intended. This aids in comprehension and ensures that you catch all relevant information. Take your time and reread sentences and passages if necessary. Look up unfamiliar words or concepts, and jot comments, called annotations, in the margins.

Identifying Specific Information from a Printed Communication

While expository in nature, memorandums (memos) are designed to convey basic information in a specific and concise message. Memos have a heading, which includes the information **to**, **from**, **date**, and **subject**, and a body, which is either in paragraph form or bullet points that detail what was in the subject line.

Though e-mails often replace memos in the modern workplace, printed memos still have a place. For example, if a supervisor wants to relate information, such as a company-wide policy change, to a large group, posting a memo in a staff lounge or other heavily traveled area is an efficient way to do so.

Posted announcements are useful to convey information to a large group of people. Announcements, however, take on a more informal tone than a memo. Common announcement topics include items for sale, services offered, lost pets, or business openings. Since posted announcements are found in public places, like grocery or hardware stores, they include contact information, purpose, meeting times, and prices, as well as pictures, graphics, and colors to attract the reader's eye.

Classified advertisements are another useful medium to convey information to large groups. Consider using classified advertisements when you want to buy and sell items or look for services. Classified ads are found in newspapers, or online through **Craigslist**, **eBay**, or similar websites and blogs. While newspapers rely on ads to help fund their publications and often provide only local exposure, online sites provide a statewide or even global platform, thus shipping costs are an important consideration when looking at the cost of the item.

Regardless of the medium, all advertisements offer basic information, such as the item in question, a description, picture, cost, and the seller's contact information. It may also note a willingness to negotiate on the price or offer an option to trade in lieu of a sale. As websites like **Craigslist** and **Buy/Sell/Trade** increase in popularity, more localities offer "safe zones," where purchases and trades are conducted in supervised environments.

Identifying Information from a Graphic Representation of Information

Texts may have graphic representations to help illustrate and visually support assertions made. For example, graphics can be used to express samples or segments of a population or demonstrate growth or decay. Three of the most popular graphic formats include line graphs, bar graphs, and pie charts.

Line graphs rely on a horizontal X axis and a vertical Y axis to establish baseline values. Dots are plotted where the horizontal and vertical axes intersect, and those dots are connected with lines. Compared to bar graphs or pie charts, line graphs are more useful for looking at the past and present and predicting future outcomes. For instance, a potential investor would look for stocks that demonstrated steady growth over many decades when examining the stock market. Note that severe spikes up and down indicate instability, while line graphs that display a slow but steady increase may indicate good returns.

Bar graphs are usually displayed on a vertical Y axis. The bars themselves can be two- or three-dimensional, depending on the designer's tastes. Unlike a line graph, which shows the fluctuation of only one variable, the X axis on a bar graph is excellent for making comparisons, because it shows differences between several variables. For instance, if a consumer wanted to buy a new tablet, she could narrow the selection down to a few choices by using a bar graph to plot the prices side by side. The tallest bar would be the most expensive tablet, the shortest bar would be the cheapest.

A pie chart is divided into wedges that represent a numerical piece of the whole. Pie charts are useful for demonstrating how different categories add up to 100 percent. However, pie charts are not useful in comparing dissimilar items. High schools tend to use pie charts to track where students end up after graduation. Each wedge, for instance, might be labeled **vocational school**, **two-year college**, **four-year college**, **workforce**, or **unemployed**. By calculating the size of each wedge, schools can offer classes in the same ratios as where students will end up after high school. Pie charts are also useful for tracking finances. Items such as car payments, insurance, rent, credit cards, and entertainment would each get their own wedge proportional to the amount spent in a given time period. If one wedge is inordinately bigger than the rest, or if a wedge is expendable, it might be time to create a new financial strategy.

Identifying Scale Readings

Most measuring instruments have scales to allow someone to determine precise measurement. Many of these instruments are becoming digitized, such as their screens' output measurements; for example, weighing scales and tire-pressure gauges often have digital screens. However, it's still important to know how to read scales. Many rulers have scales for inches on one side and scales for centimeters on the other side. On the inches' side, the longest black lines indicate the inch marks, and the slightly shorter lines indicate the half-inch marks. Progressively shorter black lines indicate the quarter-inch, eighth-inch, and sometimes even sixteenth-inch marks.

Using Legends and Map Keys

Legends and map keys are placed on maps to identify what the symbols on the map represent. Generally, map symbols stand for things like railroads, national or state highways, and public parks. Legends and maps keys can generally be found in the bottom right corner of a map. They are necessary to avoid the needless repetition of the same information because of the large amounts of information condensed onto a map. In addition, there may be a compass rose that shows the directions of north, south, east, and west. Most maps are oriented such that the top of the map is north.

Maps also have scales, which are a type of legend or key that show relative distances between fixed points. If you were on a highway and nearly out of gas, a map's scale would help you determine if you could make it to the next town before running out of fuel.

Evaluating Product Information to Determine the Most Economical Buy

When evaluating product information, be on the lookout for bolded and italicized words and numbers, which indicate the information is especially important. Also be on the lookout for repeated or similar information, which indicates importance. If you're trying to find the best deal, it might be useful to do a side-by-side comparison. Using software, like Microsoft Excel, can help you organize and compare costs systematically.

In addition, being a savvy shopper in today's market means not only having a decent grasp of math, but also understanding how retailers use established techniques to encourage consumers to spend more. Look for units of measurement—pounds, ounces, liters, grams, etc.—then divide the amount by the cost. By comparing this way, you may find products on sale cost more than ones that are not.

You should also take into consideration any tax or shipping costs. Obviously, the more an item costs, the more tax or shipping tends to cost too. Most brick-and-mortar establishments, like Target, Walmart, and Sears, are required to charge tax based on location, but some internet sites, like Amazon, Overstock, and eBay, will offer no tax or free shipping as an incentive. Comparisons between local stores and internet sites can aid in finding the best deal.

Practice Questions

Questions 1–5 are based on the following passage:

"Mademoiselle Eugénie is pretty—I think I remember that to be her name."

"Very pretty, or rather, very beautiful," replied Albert, "but of that style of beauty which I don't appreciate; I am an ungrateful fellow."

"Really," said Monte Cristo, lowering his voice, "you don't appear to me to be very enthusiastic on the subject of this marriage."

"Mademoiselle Danglars is too rich for me," replied Morcerf, "and that frightens me."

"Bah," exclaimed Monte Cristo, "that's a fine reason to give. Are you not rich yourself?"

"My father's income is about 50,000 francs per annum; and he will give me, perhaps, ten or twelve thousand when I marry."

"That, perhaps, might not be considered a large sum, in Paris especially," said the count; "but everything doesn't depend on wealth, and it's a fine thing to have a good name, and to occupy a high station in society. Your name is celebrated, your position magnificent; and then the Comte de Morcerf is a soldier, and it's pleasing to see the integrity of a Bayard united to the poverty of a Duguesclin; disinterestedness is the brightest ray in which a noble sword can shine. As for me, I consider the union with Mademoiselle Danglars a most suitable one; she will enrich you, and you will ennoble her."

Albert shook his head and looked thoughtful. "There is still something else," said he.

"I confess," observed Monte Cristo, "that I have some difficulty in comprehending your objection to a young lady who is both rich and beautiful."

"Oh," said Morcerf, "this repugnance, if repugnance it may be called, isn't all on my side."

"Whence can it arise, then? for you told me your father desired the marriage."

"It's my mother who dissents; she has a clear and penetrating judgment and doesn't smile on the proposed union. I cannot account for it, but she seems to entertain some prejudice against the Danglars."

"Ah," said the count, in a somewhat forced tone, "that may be easily explained; the Comtesse de Morcerf, who is aristocracy and refinement itself, doesn't relish the idea of being allied by your marriage with one of ignoble birth; that is natural enough."

Excerpt from The Count of Monte Cristo, by Alexandre Dumas, 1844

1. The meaning of the word "repugnance" is closest to:
    a. Strong resemblance
    b. Strong dislike
    c. Extreme shyness
    d. Extreme dissimilarity
    e. Intense suffering

2. What can be inferred about Albert's family?
    a. Their finances are uncertain.
    b. Albert is the only son in his family.
    c. Their name is more respected than the Danglars'.
    d. Albert's mother and father both agree on their decisions.
    e. No one wants Albert to be married at all.

3. What is Albert's attitude towards his impending marriage?
    a. Pragmatic
    b. Romantic
    c. Indifferent
    d. Apprehensive
    e. Animosity

4. What is the best description of the Count's relationship with Albert?
    a. He's like a strict parent, criticizing Albert's choices.
    b. He's like a wise uncle, giving practical advice to Albert.
    c. He's like a close friend, supporting all of Albert's opinions.
    d. He's like a suspicious investigator, asking many probing questions.
    e. He's like a stranger, listening politely but offering no advice.

5. Which sentence is true of Albert's mother?
    a. She belongs to a noble family.
    b. She often makes poor choices.
    c. She is primarily occupied with money.
    d. She is unconcerned about her son's future.
    e. She is supportive of the upcoming marriage.

Answer Explanations

1. B: Strong dislike. This vocabulary question can be answered using context clues. Based on the rest of the conversation, the reader can gather that Albert isn't looking forward to his marriage. As the Count notes that "you don't appear to me to be very enthusiastic on the subject of this marriage," and also remarks on Albert's "objection to a young lady who is both rich and beautiful," readers can guess Albert's feelings. The answer choice that most closely matches "objection" and "not . . . very enthusiastic" is *B*, "strong dislike."

2. C: Their name is more respected than the Danglars'. This inference question can be answered by eliminating incorrect answers. Choice *A* is tempting, considering that Albert mentions money as a concern in his marriage. However, although he may not be as rich as his fiancée, his father still has a stable income of 50,000 francs a year. Choice *B* isn't mentioned at all in the passage, so it's impossible to make an inference. Choices *D* and *E* are both false because Albert's father arranged his marriage, but his mother doesn't approve of it. Evidence for Choice *C* can be found in the Count's comparison of Albert and Eugénie: "she will enrich you, and you will ennoble her." In other words, the Danglars are wealthier, but the Morcerf family has a nobler background.

3. D: Apprehensive. There are many clues in the passage that indicate Albert's attitude towards his marriage—far from enthusiastic, he has many reservations. This question requires test takers to understand the vocabulary in the answer choices. "Pragmatic" is closest in meaning to "realistic," and "indifferent" means "uninterested." The word "animosity" is a bit strong, meaning extreme anger. The only word related to feeling worried, uncertain, or unfavorable about the future is "apprehensive."

4. B: He is like a wise uncle, giving practical advice to Albert. Choice *A* is incorrect because the Count's tone is friendly and conversational. Choice *C* is also incorrect because the Count questions why Albert doesn't want to marry a young, beautiful, and rich girl. While the Count asks many questions, he isn't particularly "probing" or "suspicious"—instead, he's asking to find out more about Albert's situation and then give him advice about marriage. The two men are not strangers, as the count does offer Albert some advice on his marriage.

5. A: She belongs to a noble family. Though Albert's mother doesn't appear in the scene, there's more than enough information to answer this question. More than once is his family's noble background mentioned (not to mention that Albert's mother is the Comtess de Morcerf, a noble title). The other answer choices can be eliminated—she is deeply concerned about her son's future; money isn't her highest priority because otherwise she would favor a marriage with the wealthy Danglars; she is not supportive of her son's marriage; and Albert describes her "clear and penetrating judgment," meaning she makes good decisions.

Writing

Organization

How a Section Fits into a Passage and Helps Develop the Ideas

Being able to determine what is most important while reading is critical to synthesis. It is the difference between being able to tell what is necessary to full comprehension and that which is interesting but not necessary.

When determining the importance of an author's ideas, consider the following:

- Ask how critical an author's particular idea, assertion, or concept is to the overall message.
- Ask "is this an interesting fact or is this information essential to understanding the author's main idea?"
- Make a simple chart. On one side, list all of the important, essential points an author makes and on the other, list all of the interesting yet non-critical ideas.
- Highlight, circle, or underline any dates or data in non-fiction passages. Pay attention to headings, captions, and any graphs or diagrams.
- When reading a fictional passage, delineate important information such as theme, character, setting, conflict (what the problem is), and resolution (how the problem is fixed). Most often, these are the most important aspects contained in fictional text.
- If a non-fiction passage is instructional in nature, take physical note of any steps in the order of their importance as presented by the author. Look for words such as *first*, *next*, *then*, and *last*.

Determining the importance of an author's ideas is critical to synthesis in that it requires the test taker to parse out any unnecessary information and demonstrate they have the ability to make sound determination on what is important to the author, and what is merely a supporting or less critical detail.

Analyzing How a Text is Organized

Depending on what the author is attempting to accomplish, certain formats or text structures work better than others. For example, a sequence structure might work for narration but not for identifying similarities and differences between concepts. Similarly, a comparison-contrast structure is not useful for narration. It's the author's job to put the right information in the correct format.

Readers should be familiar with the five main literary structures:

Sequence Structure

Sequence structure (sometimes referred to as the order structure) is when the order of events proceeds in a predictable order. In many cases, this means the text goes through the plot elements: exposition, rising action, climax, falling action, and resolution. Readers are introduced to characters, setting, and conflict in the **exposition**. In the **rising action**, there's an increase in tension and suspense. The **climax** is the height of tension and the point of no return. **Tension** decreases during the falling action. In the

resolution, any conflicts presented in the exposition are resolved, and the story concludes. An informative text that is structured sequentially will often go in order from one step to the next.

Problem-Solution

In the **problem-solution structure**, authors identify a potential problem and suggest a solution. This form of writing is usually divided into two paragraphs and can be found in informational texts. For example, cell phone, cable, and satellite providers use this structure in manuals to help customers troubleshoot or identify problems with services or products.

Comparison-Contrast

When authors want to discuss similarities and differences between separate concepts, they arrange thoughts in a **comparison-contrast paragraph structure**. **Venn diagrams** are an effective graphic organizer for comparison-contrast structures because they feature two overlapping circles that can be used to organize similarities and differences. A comparison-contrast essay organizes one paragraph based on similarities and another based on differences. A comparison-contrast essay can also be arranged with the similarities and differences of individual traits addressed within individual paragraphs. Words such as *however*, *but*, and *nevertheless* help signal a contrast in ideas.

Descriptive

Descriptive writing is designed to appeal to your senses. Much like an artist who constructs a painting, good descriptive writing builds an image in the reader's mind by appealing to the five senses: *sight, hearing, taste, touch,* and *smell.* However, overly descriptive writing can become tedious; likewise, sparse descriptions can make settings and characters seem flat. Good authors strike a balance by applying descriptions only to facts that are integral to the passage.

Cause and Effect

Passages that use the **cause-and-effect structure** are simply asking *why* by demonstrating some type of connection between ideas. Words such as *if*, *since*, *because*, *then*, or *consequently* indicate a relationship. By switching the order of a complex sentence, the writer can rearrange the emphasis on different clauses. Saying, *If Sheryl is late, we'll miss the dance*, is different from saying *We'll miss the dance if Sheryl is late*. One emphasizes Sheryl's tardiness while the other emphasizes missing the dance. Paragraphs can also be arranged in a cause-and-effect format. Since the format—before and after—is sequential, it is useful when authors wish to discuss the impact of choices. Researchers often apply this paragraph structure to the scientific method.

Understanding the Meaning and Purpose of Transition Words

The writer should act as a guide, showing the reader how all the sentences fit together. Consider this example:

> Seat belts save more lives than any other automobile safety feature. Many studies show that airbags save lives as well. Not all cars have airbags. Many older cars don't. Air bags aren't entirely reliable. Studies show that in 15% of accidents, airbags don't deploy as designed. Seat belt malfunctions are extremely rare.

There's nothing wrong with any of these sentences individually, but together they're disjointed and difficult to follow. The best way for the writer to communicate information is through the use of transition words.

Here are examples of transition words and phrases that tie sentences together, enabling a more natural flow:

- To show causality: as a result, therefore, and consequently
- To compare and contrast: *however*, *but*, and *on the other hand*
- To introduce examples: *for instance*, *namely*, and *including*
- To show order of importance: *foremost*, *primarily*, *secondly*, and *lastly*

Note: This is not a complete list of transitions. There are many more that can be used; however, most fit into these or similar categories. The point is that the words should clearly show the relationship between sentences, supporting information, and the main idea.

Here is an update to the previous example using transition words. These changes make it easier to read and bring clarity to the writer's points:

> Seat belts save more lives than any other automobile safety feature. Many studies show that airbags save lives as well; however, not all cars have airbags. For instance, some older cars don't. Furthermore, air bags aren't entirely reliable. For example, studies show that in 15% of accidents, airbags don't deploy as designed; but, on the other hand, seat belt malfunctions are extremely rare.

Also, be prepared to analyze whether the writer is using the best transition word or phrase for the situation. Take this sentence for example: "As a result, seat belt malfunctions are extremely rare." This sentence doesn't make sense in the context above because the writer is trying to show the contrast between seat belts and airbags, not the causality.

How the Passage Organization Supports the Author's Ideas

Even if the writer includes plenty of information to support their point, the writing is only coherent when the information is in a logical order. **Logical sequencing** is really just common sense, but it's an important writing technique. First, the writer should introduce the main idea, whether for a paragraph, a section, or the entire piece. Second, they should present evidence to support the main idea by using transitional language. This shows the reader how the information relates to the main idea and the sentences around it. The writer should then take time to interpret the information, making sure necessary connections are obvious to the reader. Finally, the writer can summarize the information in a closing section.

Note: Though most writing follows this pattern, it isn't a set rule. Sometimes writers change the order for effect. For example, the writer can begin with a surprising piece of supporting information to grab the reader's attention, and then transition to the main idea. Thus, if a passage doesn't follow the logical order, don't immediately assume it's wrong. However, most writing usually settles into a logical sequence after a nontraditional beginning.

Introductions and Conclusions

Examining the writer's strategies for introductions and conclusions puts the reader in the right mindset to interpret the rest of the text. Look for methods the writer might use for **introductions** such as:

- Stating the main point immediately, followed by outlining how the rest of the piece supports this claim.
- Establishing important, smaller pieces of the main idea first, and then grouping these points into a case for the main idea.
- Opening with a quotation, anecdote, question, seeming paradox, or other piece of interesting information, and then using it to lead to the main point.
- Whatever method the writer chooses, the introduction should make their intention clear, establish their voice as a credible one, and encourage a person to continue reading.

Conclusions tend to follow a similar pattern. In them, the writer restates their main idea a final time, often after summarizing the smaller pieces of that idea. If the introduction uses a quote or anecdote to grab the reader's attention, the conclusion often makes reference to it again. Whatever way the writer chooses to arrange the conclusion, the final restatement of the main idea should be clear and simple for the reader to interpret. Finally, conclusions shouldn't introduce any new information.

Sentence Structure

Eliminating Dangling or Misplaced Modifiers

Modifiers are words or phrases (often adjectives or nouns) that add detail to, explain, or limit the meaning of other parts of a sentence. Look at the following example:

A big pine tree is in the yard.

In the sentence, the words *big* (an adjective) and *pine* (a noun) modify *tree* (the head noun).

All related parts of a sentence must be placed together correctly. **Misplaced** and **dangling modifiers** are common writing mistakes. In fact, they're so common that many people are accustomed to seeing them and can decipher an incorrect sentence without much difficulty. On the test, expect to be asked to identify and correct this kind of error.

Misplaced Modifiers

Since modifiers refer to something else in the sentence (*big* and *pine* refer to *tree* in the example above), they need to be placed close to what they modify. If a modifier is so far away that the reader isn't sure what it's describing, it becomes a **misplaced modifier**. For example:

Seat belts almost saved 5,000 lives in 2009.

It's likely that the writer means that the total number of lives saved by seat belts in 2009 is close to 5,000. However, due to the misplaced modifier (*almost*), the sentence actually says there are 5,000 instances when seat belts *almost saved lives*. In this case, the position of the modifier is actually the difference between life and death (at least in the meaning of the sentence).

A clearer way to write the sentence is:

Seat belts saved almost 5,000 lives in 2009.

Now that the modifier is close to the 5,000 lives it references, the sentence's meaning is clearer.

Another common example of a misplaced modifier occurs when the writer uses the modifier to begin a sentence. For example:

Having saved 5,000 lives in 2009, Senator Wilson praised the seat belt legislation.

It seems unlikely that Senator Wilson saved 5,000 lives on her own, but that's what the writer is saying in this sentence. To correct this error, the writer should move the modifier closer to the intended object it modifies. Here are two possible solutions:

Having saved 5,000 lives in 2009, the seat belt legislation was praised by Senator Wilson.

Senator Wilson praised the seat belt legislation, which saved 5,000 lives in 2009.

When choosing a solution for a misplaced modifier, look for an option that places the modifier close to the object or idea it describes.

Dangling Modifiers

A modifier must have a target word or phrase that it's modifying. Without this, it's a **dangling modifier**. Dangling modifiers are usually found at the beginning of sentences:

After passing the new law, there is sure to be an improvement in highway safety.

This sentence doesn't say anything about who is passing the law. Therefore, "After passing the new law" is a dangling modifier because it doesn't modify anything in the sentence. To correct this type of error, determine what the writer intended the modifier to point to:

After passing the new law, legislators are sure to see an improvement in highway safety.

"After passing the new law" now points to *legislators*, which makes the sentence clearer and eliminates the dangling modifier.

Editing Sentences for Parallel Structure and Correct Use of Conjunctions

Parallel Structure

Parallel structure occurs when phrases or clauses within a sentence contain the same structure. Parallelism increases readability and comprehensibility because it is easy to tell which sentence elements are paired with each other in meaning.

Jennifer enjoys cooking, knitting, and to spend time with her cat.

This sentence is not parallel because the items in the list appear in two different forms. Some are **gerunds**, which is the verb + ing: *cooking, knitting*. The other item uses the **infinitive** form, which is to + verb: *to spend*. To create parallelism, all items in the list may reflect the same form:

Jennifer enjoys cooking, knitting, and spending time with her cat.

All of the items in the list are now in gerund forms, so this sentence exhibits parallel structure. Here's another example:

The company is looking for employees who are responsible and with a lot of experience.

Again, the items that are listed in this sentence are not parallel. "Responsible" is an adjective, yet "with a lot of experience" is a prepositional phrase. The sentence elements do not utilize parallel parts of speech.

The company is looking for employees who are responsible and experienced.

"Responsible" and "experienced" are both adjectives, so this sentence now has parallel structure.

Conjunctions

Conjunctions join words, phrases, clauses, or sentences together, indicating the type of connection between these elements.

I like pizza, *and* I enjoy spaghetti.

I like to play baseball, *but* I'm allergic to mitts.

Some conjunctions are **coordinating**, meaning they give equal emphasis to two main clauses. Coordinating conjunctions are short, simple words that can be remembered using the mnemonic FANBOYS: for, and, nor, but, or, yet, so. Other conjunctions are subordinating. **Subordinating conjunctions** introduce dependent clauses and include words such as *because*, *since*, *before*, *after*, *if*, and *while*.

Conjunctions can also be classified as follows:

- **Cumulative conjunctions** add one statement to another.
  - Examples: and, both, also, as well as, not only
  - I.e., The juice is sweet *and* sour.
- **Adversative conjunctions** are used to contrast two clauses.
  - Examples: but, while, still, yet, nevertheless
  - I.e., She was tired, *but* she was happy.
- **Alternative conjunctions** express two alternatives.
  - Examples: or, either, neither, nor, else, otherwise
  - I.e., He must eat, *or* he will die.

Editing for Subject-Verb and Pronoun-Antecedent Agreement

Subject-Verb Agreement

The subject of a sentence and its verb must agree. The cornerstone rule of subject-verb agreement is that subject and verb must agree in number. Whether the subject is singular or plural, the verb must follow suit.

Incorrect: The houses is new.

Correct: The houses are new.

Also Correct: The house is new.

In other words, a singular subject requires a singular verb; a plural subject requires a plural verb.

The words or phrases that come between the subject and verb do not alter this rule.

Incorrect: The houses built of brick is new.

Correct: The houses built of brick are new.

Incorrect: The houses with the sturdy porches is new.

Correct: The houses with the sturdy porches are new.

The subject will always follow the verb when a sentence begins with *here* or *there.* Identify these with care.

Incorrect: Here *is* the *houses* with sturdy porches.

Correct: Here *are* the *houses* with sturdy porches.

The subject in the sentences above is not *here*, it is *houses*. Remember, *here* and *there* are never subjects. Be careful that contractions such as *here's* or *there're* do not cause confusion!

Two subjects joined by *and* require a plural verb form, except when the two combine to make one thing:

Incorrect: Garrett and Jonathan is over there.

Correct: Garrett and Jonathan are over there.

Incorrect: Spaghetti and meatballs are a delicious meal!

Correct: Spaghetti and meatballs is a delicious meal!

In the example above, *spaghetti and meatballs* is a compound noun. However, *Garrett and Jonathan* is not a compound noun.

Two singular subjects joined by *or, either/or,* or *neither/nor* call for a singular verb form.

Incorrect: Butter or syrup are acceptable.

Correct: Butter or syrup is acceptable.

Plural subjects joined by *or*, *either/or*, or *neither/nor* are, indeed, plural.

The chairs or the boxes are being moved next.

If one subject is singular and the other is plural, the verb should agree with the closest noun.

Correct: The chair or the boxes are being moved next.

Correct: The chairs or the box is being moved next.

Some plurals of money, distance, and time call for a singular verb.

Incorrect: Three dollars *are* enough to buy that.

Correct: Three dollars *is* enough to buy that.

For words declaring degrees of quantity such as *many of, some of,* or *most of,* let the noun that follows *of* be the guide:

Incorrect: Many of the books is in the shelf.

Correct: Many of the books are in the shelf.

Incorrect: Most of the pie *are* on the table.

Correct: Most of the pie *is* on the table.

For indefinite pronouns like anybody or everybody, use singular verbs.

Everybody *is* going to the store.

However, the pronouns *few, many, several, all, some,* and *both* have their own rules and use plural forms.

Some *are* ready.

Some nouns like *crowd* and *congress* are called **collective nouns** and they require a singular verb form.

Congress *is* in session.

The news *is* over.

Books and movie titles, though, including plural nouns such as *Great Expectations*, also require a singular verb. Remember that only the subject affects the verb. While writing tricky subject-verb arrangements, say them aloud. Listen to them. Once the rules have been learned, one's ear will become sensitive to them, making it easier to pick out what's right and what's wrong.

Pronoun-Antecedent Agreement

An **antecedent** is the noun to which a pronoun refers; it needs to be written or spoken before the pronoun is used. For many pronouns, antecedents are imperative for clarity. In particular, a lot of the personal, possessive, and demonstrative pronouns need antecedents. Otherwise, it would be unclear who or what someone is referring to when they use a pronoun like *he* or *this*.

Pronoun reference means that the pronoun should refer clearly to one, clear, unmistakable noun (the antecedent).

Pronoun-antecedent agreement refers to the need for the antecedent and the corresponding pronoun to agree in gender, person, and number. Here are some examples:

> The *kidneys* (plural antecedent) are part of the urinary system. *They* (plural pronoun) serve several roles.
>
> The kidneys are part of the *urinary system* (singular antecedent). *It* (singular pronoun) is also known as the renal system.

Eliminating Wordiness or Awkward Sentence Structure

A great Facebook posts or Twitter story is witty, to the point, and even moving. Good writing is like a good social media post—it needs to be seamless, succinct, and sound in its organization. Alternatively, there are also social media rants so jumbled that they do not make sense or are so endless that readers lose interest. The most captivating social media entries are the ones that meet a high standard of organization. Likewise, the most captivating essays follow these same standards.

Wordiness and awkward sentence structure can happen as a result of many factors. Firstly, they can result from poor grammar or run-on sentences. In order to avoid this, test takers should try using punctuation with fidelity and breaking up independent and dependent clauses into simpler, bite-size nuggets of knowledge. Secondly, wordiness and awkward sentence structure can stem from the overuse of adjectives and adverbs. Test takers should try to limit adverbs and adjectives to ensure clarity. Lastly, wordiness and awkward sentence structure can be the product of flawed organization. Not only should sentences be succinct, but paragraphs and pages should also be succinct—they should use space efficiently and effectively. Test takers should try conveying a message using the fewest words possible.

Below are examples of ways to rectify wordiness and awkward sentences in writing.

WORDINESS:

BEFORE: Science is an important subject of study, and it is important for all students to learn because it focuses on the way the world works and it has important subfields like biology, physics, and chemistry.	AFTER: Science—which is composed of important subfields like biology, physics, and chemistry—helps students understand the way the world works.
BEFORE: History is about important people, places, events, movements, and eras in the past it is a really, really interesting field with lots of different lenses of study such as economic history, political history, and cultural history to name a few types of history.	AFTER: History can be studied through many lenses: economics, politics, and culture. However, all types of history focus on interesting people, places, events, movements, and eras in the past.

AWKWARDNESS:

BEFORE: The administrative expertise of George Washington's presidential administration was known for its powerfully powerful administrators.	AFTER: George Washington's presidential administration was known for its powerful leadership and expertise.
BEFORE: I want to study history, working hard, and becoming a historian.	AFTER: I want to study history, work hard, and become a historian.

Eliminating Run-On Sentences and Sentence Fragments

A **sentence fragment** is a failed attempt to create a complete sentence because it's missing a required noun or verb. Fragments don't function properly because there isn't enough information to understand the writer's intended meaning. For example:

> Seat belt use corresponds to a lower rate of hospital visits, reducing strain on an already overburdened healthcare system. Insurance claims as well.

Look at the last sentence: *Insurance claims as well*. What does this mean? This is a fragment because it has a noun but no verb, and it leaves the reader guessing what the writer means about insurance claims. Many readers can probably infer what the writer means, but this distracts them from the flow of the writer's argument. Choosing a suitable replacement for a sentence fragment may be one of the questions on the test. The fragment is probably related to the surrounding content, so look at the overall point the writer is trying to make and choose the answer that best fits that idea.

Remember that sometimes a fragment can *look* like a complete sentence or have all the nouns and verbs it needs to make sense. Consider the following two examples:

> Seat belt use corresponds to a lower rate of hospital visits.

> Although seat belt use corresponds to a lower rate of hospital visits.

Both examples above have nouns and verbs, but only the first sentence is correct. The second sentence is a fragment, even though it's actually longer. The key is the writer's use of the word *although*. Starting a sentence with *although* turns that part into a *subordinate clause* (more on that next). Keep in mind that one doesn't have to remember that it's called a subordinate clause on the test. Just be able to recognize that the words form an incomplete thought and identify the problem as a sentence fragment.

A **run-on sentence** is, in some ways, the opposite of a fragment. It contains two or more sentences that have been improperly forced together into one. An example of a run-on sentence looks something like this:

> Seat belt use corresponds to a lower rate of hospital visits it also leads to fewer insurance claims.

Here, there are two separate ideas in one sentence. It's difficult for the reader to follow the writer's thinking because there is no transition from one idea to the next. On the test, choose the best way to correct the run-on sentence.

Here are two possibilities for the sentence above:

> Seat belt use corresponds to a lower rate of hospital visits. It also leads to fewer insurance claims.

> Seat belt use corresponds to a lower rate of hospital visits, but it also leads to fewer insurance claims.

Both solutions are grammatically correct, so which one is the best choice? That depends on the point that the writer is trying to make. Always read the surrounding text to determine what the writer wants to demonstrate, and choose the option that best supports that thought.

Usage

Correcting Errors with Frequently Confused Words

There are a handful of words in the English language that writers often confuse with other words because they sound similar or identical. Errors involving these words are hard to spot because they *sound* right even when they're wrong. Also, because these mistakes are so pervasive, many people think they're correct. Here are a few examples that may be encountered on the test:

They're vs. Their vs. There

This set of words is probably the all-time winner of misuse. The word *they're* is a contraction of "they are." Remember that contractions combine two words, using an apostrophe to replace any eliminated letters. If a question asks whether the writer is using the word *they're* correctly, change the word to "they are" and reread the sentence. Look at the following example:

> Legislators can be proud of they're work on this issue.

This sentence *sounds* correct, but replace the contraction *they're* with "they are" to see what happens:

> Legislators can be proud of they are work on this issue.

The result doesn't make sense, which shows that it's an incorrect use of the word *they're*. Did the writer mean to use the word *their* instead? The word *their* indicates possession because it shows that something *belongs* to something else. Now put the word *their* into the sentence:

> Legislators can be proud of their work on this issue.

To check the answer, find the word that comes right after the word *their* (which in this case is *work*). Pose this question: whose *work* is it? If the question can be answered in the sentence, then the word signifies possession. In the sentence above, it's the legislators' work. Therefore, the writer is using the word *their* correctly.

If the words *they're* and *their* don't make sense in the sentence, then the correct word is almost always *there*. The word *there* can be used in many different ways, so it's easy to remember to use it when *they're* and *their* don't work. Now test these methods with the following sentences:

> Their going to have a hard time passing these laws.
>
> Enforcement officials will have there hands full.
>
> They're are many issues to consider when discussing car safety.

In the first sentence, asking the question "Whose going is it?" doesn't make sense. Thus, the word *their* is incorrect. However, when replaced with the conjunction *they're* (or *they are*), the sentence works. Thus, the correct word for the first sentence should be *they're*.

In the second sentence, ask this question: "Whose hands are full?" The answer (*enforcement officials*) is correct in the sentence. Therefore, the word *their* should replace *there* in this sentence.

In the third sentence, changing the word *they're* to "they are" ("They are are many issues") doesn't make sense. Ask this question: "Whose are is it?" This makes even less sense, since neither of the words *they're* or *their* makes sense. Therefore, the correct word must be *there*.

Who's vs. Whose

Who's is a contraction of "who is" while the word *whose* indicates possession. Look at the following sentence:

> Who's job is it to protect America's drivers?

The easiest way to check for correct usage is to replace the word *who's* with "who is" and see if the sentence makes sense:

> Who is job is it to protect America's drivers?

By changing the contraction to "Who is" the sentence no longer makes sense. Therefore, the correct word must be *whose*.

Your vs. You're

The word *your* indicates possession, while *you're* is a contraction for "you are." Look at the following example:

> Your going to have to write your congressman if you want to see action.

Again, the easiest way to check correct usage is to replace the word *Your* with "You are" and see if the sentence still makes sense.

> You are going to have to write your congressman if you want to see action.

By replacing Your with "You are," the sentence still makes sense. Thus, in this case, the writer should have used "You're."

Its vs. It's

Its is a word that indicates possession, while the word *it's* is a contraction of "it is." Once again, the easiest way to check for correct usage is to replace the word with "it is" and see if the sentence makes sense. Look at the following sentence:

> It's going to take a lot of work to pass this law.

Replacing *it's* with "it is" results in this: "It is going to take a lot of work to pass this law." This makes sense, so the contraction (*it's*) is correct. Now look at another example:

> The car company will have to redesign it's vehicles.

Replacing *it's* with "it is" results in this: "The car company will have to redesign it is vehicles." This sentence doesn't make sense, so the contraction (*it's*) is incorrect.

Than vs. Then

Than is used in sentences that involve comparisons, while *then* is used to indicate an order of events. Consider the following sentence:

> Japan has more traffic fatalities than the U.S.

The use of the word *than* is correct because it compares Japan to the U.S. Now look at another example:

> Laws must be passed, and then we'll see a change in behavior.

Here the use of the word *then* is correct because one thing happens after the other.

Affect vs. Effect

Affect is a verb that means to change something, while *effect* is a noun that indicates such a change. Look at the following sentence:

> There are thousands of people affected by the new law.

This sentence is correct because *affected* is a verb that tells what's happening. Now look at this sentence:

> The law will have a dramatic effect.

This sentence is also correct because *effect* is a noun and the thing that happens.

Note that a noun version of *affect* is occasionally used. It means "emotion" or "desire," usually in a psychological sense.

Two vs. Too vs. To

Two is the number (2). *Too* refers to an amount of something, or it can mean *also*. *To* is used for everything else. Look at the following sentence:

> Two senators still haven't signed the bill.

This is correct because there are *two* (2) senators. Here's another example:

> There are too many questions about this issue.

In this sentence, the word *too* refers to an amount ("too many questions"). Now here's another example:

> Senator Wilson is supporting this legislation, too.

In this sentence, the word *also* can be substituted for the word *too*, so it's also correct. Finally, one last example:

> I look forward to signing this bill into law.

In this sentence, the tests for *two* and *too* don't work. Thus, the word *to* fits the bill!

Other Common Writing Confusions

In addition to all of the above, there are other words that writers often misuse. This doesn't happen because the words sound alike, but because the writer is not aware of the proper way to use them.

Correcting Subject-Verb Agreement Errors

In English, verbs must agree with the subject. The form of a verb may change depending on whether the subject is singular or plural, or whether it is first, second, or third person. For example, the verb *to be* has various forms:

> I am a student.
>
> You are a student.
>
> She is a student.
>
> We are students.
>
> They are students.

Errors occur when a verb does not agree with its subject. Sometimes, the error is readily apparent:

> We is hungry.

Is is not the appropriate form of *to be* when used with the third person plural *we*.

> We are hungry.

This sentence now has correct subject-verb agreement.

However, some cases are trickier, particularly when the subject consists of a lengthy noun phrase with many modifiers:

> Students who are hoping to accompany the anthropology department on its annual summer trip to Ecuador needs to sign up by March 31st.

The verb in this sentence is *needs*. However, its subject is not the noun adjacent to it—Ecuador. The subject is the noun at the beginning of the sentence—students. Because *students* is plural, *needs* is the incorrect verb form.

Students who are hoping to accompany the anthropology department on its annual summer trip to Ecuador *need* to sign up by March 31st.

This sentence now uses correct agreement between *students* and *need*.

Another case to be aware of is a **collective noun**. A collective noun refers to a group of many things or people but can be singular in itself—e.g., *family, committee, army, pair team, council, jury*. Whether or not a collective noun uses a singular or plural verb depends on how the noun is being used. If the noun refers to the group performing a collective action as one unit, it should use a singular verb conjugation:

The family is moving to a new neighborhood.

The whole family is moving together in unison, so the singular verb form *is* is appropriate here.

The committee has made its decision.

The verb *has* and the possessive pronoun *its* both reflect the word *committee* as a singular noun in the sentence above; however, when a collective noun refers to the group as individuals, it can take a plural verb:

The newlywed pair spend every moment together.

This sentence emphasizes the love between two people in a pair, so it can use the plural verb *spend*.

The council are all newly elected members.

The sentence refers to the council in terms of its individual members and uses the plural verb *are*.

Overall, though, American English is more likely to pair a collective noun with a singular verb, while British English is more likely to pair a collective noun with a plural verb.

Which of the following sentences is correct?

A large crowd of protesters was on hand.

A large crowd of protesters were on hand.

Many people would say the second sentence is correct, but they'd be wrong. However, they probably wouldn't be alone. Most people just look at two words: *protesters were*. Together they make sense. They sound right. The problem is that the verb *were* doesn't refer to the word *protesters*. Here, the word *protesters* is part of a prepositional phrase that clarifies the actual subject of the sentence (*crowd*).

Take the phrase "of protesters" away and re-examine the sentences:

A large crowd was on hand.

A large crowd were on hand.

Without the prepositional phrase to separate the subject and verb, the answer is obvious. The first sentence is correct. On the test, look for confusing prepositional phrases when answering questions about subject-verb agreement. Take the phrase away, and then recheck the sentence.

Correcting Pronoun Errors

Pronoun Person

Pronoun person refers to the narrative voice the writer uses in a piece of writing. A great deal of nonfiction is written in third person, which uses pronouns like *he, she, it,* and *they* to convey meaning. Occasionally a writer uses first person (*I, me, we*, etc.) or second person (*you*). Any choice of pronoun person can be appropriate for a particular situation, but the writer must remain consistent and logical.

Test questions may cover examining samples that should stay in a single pronoun person, be it first, second, or third. Look out for shifts between words like *you* and *I* or *he* and *they.*

Pronoun Clarity

Pronouns always refer back to a noun. However, as the writer composes longer, more complicated sentences, the reader may be unsure which noun the pronoun should replace. For example:

> An amendment was made to the bill, but now it has been voted down.

Was the amendment voted down or the entire bill? It's impossible to tell from this sentence. To correct this error, the writer needs to restate the appropriate noun rather than using a pronoun:

> An amendment was made to the bill, but now the bill has been voted down.

Pronouns in Combination

Writers often make mistakes when choosing pronouns to use in combination with other nouns. The most common mistakes are found in sentences like this:

> Please join Senator Wilson and I at the event tomorrow.

Notice anything wrong? Though many people think the sentence sounds perfectly fine, the use of the pronoun *I* is actually incorrect. To double-check this, take the other person out of the sentence:

> Please join I at the event tomorrow.

Now the sentence is obviously incorrect, as it should read, "Please join *me* at the event tomorrow." Thus, the first sentence should replace *I* with *me*:

> Please join Senator Wilson and me at the event tomorrow.

For many people, this sounds wrong because they're used to hearing and saying it incorrectly. Take extra care when answering this kind of question and follow the double-checking procedure.

Eliminating Non-Standard English Words or Phrases

Non-standard English words and phrases, such as slang, should be eliminated, as it not only reduces the professionalism and formality of a text, but it also opens the door for confusion. Slang tends to evolve quickly, and it is less universally understood than standard English. Therefore, unless working on a narrative fiction piece that purposely includes non-standard English as part of the dialogue, writers should make every effort to eliminate this type of language from their writing.

Mechanics

Transition Words

Transitions are the glue that helps put ideas together seamlessly, within sentences and paragraphs, between them, and (in longer documents) even between sections. Transitions may be single words, sentences, or whole paragraphs (as in the prior example). Transitions help readers to digest and understand what to feel about what has gone on and clue readers in on what is going on, what will be, and how they might react to all these factors. Transitions are like good clues left at a crime scene.

Recall this list of some common transition words and phrases:

- To show causality: as a result, therefore, and consequently
- To compare and contrast: *however, but*, and *on the other hand*
- To introduce examples: *for instance, namely*, and *including*
- To show order of importance: *foremost, primarily, secondly*, and *lastly*

Capitalization, Punctuation, and Apostrophes

Correct Capitalization

Here's a non-exhaustive list of things that should be capitalized.

- The first word of every sentence
- The first word of every line of poetry
- The first letter of proper nouns (World War II)
- Holidays (Valentine's Day)
- The days of the week and months of the year (Tuesday, March)
- The first word, last word, and all major words in the titles of books, movies, songs, and other creative works (In the novel, *To Kill a Mockingbird*, note that *a* is lowercase since it's not a major word, but *to* is capitalized since it's the first word of the title.)
- Titles when preceding a proper noun (President Roberto Gonzales, Aunt Judy)

When simply using a word such as president or secretary, though, the word is not capitalized.

> Officers of the new business must include a *president* and *treasurer*.

Seasons—spring, fall, etc.—are not capitalized.

North, *south*, *east*, and *west* are capitalized when referring to regions but are not when being used for directions. In general, if it's preceded by *the* it should be capitalized.

> I'm from the South.
>
> I drove south.

Using Apostrophes with Possessive Nouns Correctly

Possessives

In grammar, **possessive nouns** show ownership, which was seen in previous examples like *mine, yours,* and *theirs*.

Singular nouns are generally made possessive with an apostrophe and an *s* (*'s*).

> My *uncle's* new car is silver.
>
> The *dog's* bowl is empty.
>
> *James's* ties are becoming outdated.

Plural nouns ending in *s* are generally made possessive by just adding an apostrophe ('):

> The pistachio nuts' saltiness is added during roasting. (The saltiness of pistachio nuts is added during roasting.)
>
> The students' achievement tests are difficult. (The achievement tests of the students are difficult.)

If the plural noun does not end in an *s* such as *women,* then it is made possessive by adding an **apostrophe** *s* (*'s*)—*women's.*

Indefinite possessive pronouns such as *nobody* or *someone* become possessive by adding an *apostrophe s— nobody's* or *someone's.*

Using Correct Punctuation

Ellipses

An **ellipsis** (...) consists of three handy little dots that can speak volumes on behalf of irrelevant material. Writers use them in place of words, lines, phrases, list content, or paragraphs that might just as easily have been omitted from a passage of writing. This can be done to save space or to focus only on the specifically relevant material.

> Exercise is good for some unexpected reasons. Watkins writes, "Exercise has many benefits such as...reducing cancer risk."

In the example above, the ellipsis takes the place of the other benefits of exercise that are more expected.

The ellipsis may also be used to show a pause in sentence flow.

> "I'm wondering...how this could happen," Dylan said in a soft voice.

Commas

A **comma** (,) is the punctuation mark that signifies a pause—breath—between parts of a sentence. It denotes a break of flow. As with so many aspects of writing structure, authors will benefit by reading their writing aloud or mouthing the words. This can be particularly helpful if one is uncertain about whether the comma is needed.

In a complex sentence—one that contains a **subordinate** (**dependent**) clause or clauses—the use of a comma is dictated by where the subordinate clause is located. If the subordinate clause is located before the main clause, a comma is needed between the two clauses.

> Because I don't have enough money, I will not order steak.

Generally, if the subordinate clause is placed after the main clause, no punctuation is needed.

I did well on my exam because I studied two hours the night before.

Notice how the last clause is dependent because it requires the earlier independent clauses to make sense.

Use a comma on both sides of an interrupting phrase.

I will pay for the ice cream, chocolate and vanilla, and then will eat it all myself.

The words forming the phrase in italics are nonessential (extra) information. To determine if a phrase is nonessential, try reading the sentence without the phrase and see if it's still coherent.

A comma is not necessary in this next sentence because no interruption—nonessential or extra information—has occurred. Read sentences aloud when uncertain.

I will pay for the chocolate and vanilla ice cream and then eat it all myself.

If the nonessential phrase comes at the beginning of a sentence, a comma should only go at the end of the phrase. If the phrase comes at the end of a sentence, a comma should only go at the beginning of the phrase.

Other types of interruptions include the following:

- Interjections: Oh no, I am not going.
- Abbreviations: Barry Potter, M.D., specializes in heart disorders.
- Direct addresses: Yes, Claudia, I am tired and going to bed.
- Parenthetical phrases: His wife, lovely as she was, was not helpful.
- Transitional phrases: Also, it is not possible.

The second comma in the following sentence is called an Oxford comma.

I will pay for ice cream, syrup, and pop.

It is a comma used after the second-to-last item in a series of three or more items. It comes before the word *or* or *and*. Not everyone uses the Oxford comma; it is optional, but many believe it is needed. The comma functions as a tool to reduce confusion in writing. So, if omitting the Oxford comma would cause confusion, then it's best to include it.

Commas are used in math to mark the place of thousands in numerals, breaking them up so they are easier to read. Other uses for commas are in dates (*March 19, 2016*), letter greetings (*Dear Sally,*), and in between cities and states (*Louisville, KY*).

Semicolons

The **semicolon** (;) might be described as a heavy-handed comma. Take a look at these two examples:

I will pay for the ice cream, but I will not pay for the steak.

I will pay for the ice cream; I will not pay for the steak.

What's the difference? The first example has a comma and a conjunction separating the two independent clauses. The second example does not have a conjunction, but there are two independent clauses in the sentence, so something more than a comma is required. In this case, a semicolon is used.

Two independent clauses can only be joined in a sentence by either a comma and conjunction or a semicolon. If one of those tools is not used, the sentence will be a run-on. Remember that while the clauses are independent, they need to be closely related in order to be contained in one sentence.

Another use for the semicolon is to separate items in a list when the items themselves require commas.

> The family lived in Phoenix, Arizona; Oklahoma City, Oklahoma; and Raleigh, North Carolina.

Colons

Colons (:) have many miscellaneous functions. Colons can be used to proceed further information or a list. In these cases, a colon should only follow an independent clause.

> Humans take in sensory information through five basic senses: sight, hearing, smell, touch, and taste.

The meal includes the following components:

- Caesar salad
- Spaghetti
- Garlic bread
- Cake

The family got what they needed: a reliable vehicle.

While a comma is more common, a colon can also precede a formal quotation.

> He said to the crowd: "Let's begin!"

The colon is used after the greeting in a formal letter.

> Dear Sir:
> To Whom It May Concern:

In the writing of time, the colon separates the minutes from the hour (*4:45 p.m.*). The colon can also be used to indicate a ratio between two numbers (*50:1*).

Hyphens

The **hyphen** (-) is a little hash mark that can be used to join words to show that they are linked.

Hyphenate two words that work together as a single adjective (a compound adjective).

> honey-covered biscuits

Some words always require hyphens, even if not serving as an adjective.

> merry-go-round

Hyphens always go after certain prefixes like *anti-* & *all-*.

Hyphens should also be used when the absence of the hyphen would cause a strange vowel combination (*semi-engineer*) or confusion. For example, *re-collect* should be used to describe something being gathered twice rather than being written as *recollect*, which means to remember.

Parentheses and Dashes

Parentheses are half-round brackets that look like this: *( )*. They set off a word, phrase, or sentence that is an afterthought, explanation, or side note relevant to the surrounding text but not essential. A pair of commas is often used to set off this sort of information, but parentheses are generally used for information that would not fit well within a sentence or that the writer deems not important enough to be structurally part of the sentence.

> The picture of the heart (see above) shows the major parts you should memorize.
>
> Mount Everest is one of three mountains in the world that are over 28,000 feet high (K2 and Kanchenjunga are the other two).

See how the sentences above are complete without the parenthetical statements? In the first example, *see above* would not have fit well within the flow of the sentence. The second parenthetical statement could have been a separate sentence, but the writer deemed the information not pertinent to the topic.

The **dash** (—) is a mark longer than a hyphen used as a punctuation mark in sentences and to set apart a relevant thought. Even after plucking out the line separated by the dash marks, the sentence will be intact and make sense.

> Looking out the airplane window at the landmarks—Lake Clarke, Thompson Community College, and the bridge—she couldn't help but feel excited to be home.

The dashes use is similar to that of parentheses or a pair of commas. So, what's the difference? Many believe that using dashes makes the clause within them stand out while using parentheses is subtler. It's advised to not use dashes when commas could be used instead.

Quotation Marks

Here are some instances where **quotation marks** should be used:

- Dialogue for characters in narratives. When characters speak, the first word should always be capitalized, and the punctuation goes inside the quotes. For example:

  > Janie said, "The tree fell on my car during the hurricane."

- Around titles of songs, short stories, essays, and chapters in books
- To emphasize a certain word
- To refer to a word as the word itself

Apostrophes

The apostrophe (') is a versatile punctuation mark. It has a few different functions:

- Quotes: Apostrophes are used when a second quote is needed within a quote.

  > In my letter to my friend, I wrote, "The girl had to get a new purse, and guess what Mary did? She said, 'I'd like to go with you to the store.' I knew Mary would buy it for her."

- Contractions: Another use for an apostrophe in the quote above is a contraction. *I'd is* used for *I would.*

- Possession: An apostrophe followed by the letter *s* shows possession (*Mary's* purse). If the possessive word is plural, the apostrophe generally just follows the word.

Essay

Brainstorming

One of the most important steps in writing an essay is prewriting. Before drafting an essay, it's helpful to think about the topic for a moment or two, in order to gain a more solid understanding of the task. Then, spending about five minutes jotting down the immediate ideas that could work for the essay is recommended. It is a way to get some words on the page and offer a reference for ideas when drafting. Scratch paper is provided for writers to use any prewriting techniques such as webbing, free writing, or listing. The goal is to get ideas out of the mind and onto the page.

Considering Opposing Viewpoints

In the planning stage, it's important to consider all aspects of the topic, including different viewpoints on the subject. There are more than two ways to look at a topic, and a strong argument considers those opposing viewpoints. Considering opposing viewpoints can help writers present a fair, balanced, and informed essay that shows consideration for all readers. This approach can also strengthen an argument by recognizing and potentially refuting opposing viewpoint(s).

Drawing from personal experience may help to support ideas. For example, if the goal for writing is a personal narrative, then the story should come from the writer's own life. Many writers find it helpful to draw from personal experience, even in an essay that is not strictly narrative. Personal anecdotes or short stories can help to illustrate a point in other types of essays as well.

Moving from Brainstorming to Planning

Once the ideas are on the page, it's time to turn them into a solid plan for the essay. The best ideas from the brainstorming results can then be developed into a more formal outline. An outline typically has one main point (the thesis) and at least three sub-points that support the main point. Here's an example:

Main Idea

- Point #1
- Point #2
- Point #3

Of course, there will be details under each point, but this approach is the best for dealing with timed writing.

Staying on Track

Basing the essay on the outline aids in both organization and coherence. The goal is to ensure that there is enough time to develop each sub-point in the essay, roughly spending an equal amount of time on each idea. Keeping an eye on the time will help. If there are fifteen minutes left to draft the essay, then it makes sense to spend about 5 minutes on each of the ideas. Staying on task is critical to success, and timing out the parts of the essay can help writers avoid feeling overwhelmed.

Parts of the Essay

The introduction has to do a few important things:

- Establish the topic of the essay in original wording (i.e., not just repeating the prompt)
- Clarify the significance/importance of the topic or purpose for writing (not too many details, a brief overview)
- Offer a thesis statement that identifies the writer's own viewpoint on the topic (typically one-two brief sentences as a clear, concise explanation of the main point on the topic)

Body paragraphs reflect the ideas developed in the outline. Three-four points is probably sufficient for a short essay, and they should include the following:

- A topic sentence that identifies the sub-point (e.g., a reason why, a way how, a cause or effect)
- A detailed explanation of the point, explaining why the writer thinks this point is valid
- Illustrative examples, such as personal examples or real-world examples, that support and validate the point (i.e., "prove" the point)
- A concluding sentence that connects the examples, reasoning, and analysis to the point being made

The conclusion, or final paragraph, should be brief and should reiterate the focus, clarifying why the discussion is significant or important. It is important to avoid adding specific details or new ideas to this paragraph. The purpose of the conclusion is to sum up what has been said to bring the discussion to a close.

Don't Panic!

Writing an essay can be overwhelming, and performance panic is a natural response. The outline serves as a basis for the writing and helps writers keep focused. Getting stuck can also happen, and it's helpful to remember that brainstorming can be done at any time during the writing process. Following the steps of the writing process is the best defense against writer's block.

Timed essays can be particularly stressful, but assessors are trained to recognize the necessary planning and thinking for these timed efforts. Using the plan above and sticking to it helps with time management. Timing each part of the process helps writers stay on track. Sometimes writers try to cover too much in their essays. If time seems to be running out, this is an opportunity to determine whether all of the ideas in the outline are necessary. Three body paragraphs are sufficient, and more than that is probably too much to cover in a short essay.

More isn't always better in writing. A strong essay will be clear and concise. It will avoid unnecessary or repetitive details. It is better to have a concise, five-paragraph essay that makes a clear point, than a ten-paragraph essay that doesn't. The goal is to write one to two pages of quality writing. Paragraphs should also reflect balance; if the introduction goes to the bottom of the first page, the writing may be going off-track or be repetitive. It's best to fall into the one-two page range, but a complete, well-developed essay is the ultimate goal.

Practice Questions

Questions 1–5 are based on the following passage:

(1) Seeing a lasting social change for African American people Fred Hampton desired to see through nonviolent means and community recognition. (2) As a result, he became an African American activist during the American Civil Rights Movement and led the Chicago chapter of the Black Panther Party.

Hampton's Education

(3) Born and raised in Maywood of Chicago, Illinois in 1948 was Hampton. (4) He was gifted academically and a natural athlete, he became a stellar baseball player in high school. (5) After graduating from Proviso East High School in 1966, he later went on to study law at Triton Junior College.

While studying at Triton, Hampton joined and became a leader of the National Association for the Advancement of Colored People (NAACP). The NAACP gained more than 500 members resulting from his membership. Hampton worked relentlessly to acquire recreational facilities in the neighborhood and improve the educational resources provided to the impoverished black community of Maywood.

1. Which of the following would be the best choice for this sentence (reproduced below)?

   (31) Seeing a lasting social change for African American people Fred Hampton desired to see through nonviolent means and community recognition.

   a. NO CHANGE
   b. Desiring to see a lasting social change for African American people, Fred Hampton
   c. Fred Hampton desired to see lasting social change for African American people
   d. Fred Hampton desiring to see last social change for African American people

2. Which of the following would be the best choice for this sentence (reproduced below)?

   (32) As a result, he became an African American activist during the American Civil Rights Movement and led the Chicago chapter of the Black Panther Party.

   a. NO CHANGE
   b. As a result, he became an African American activist
   c. As a result: he became an African American activist
   d. As a result of, he became an African American activist

3. Which of the following would be the best choice for this sentence (reproduced below)?

(33) Born and raised in Maywood of Chicago, Illinois in 1948 was Hampton.

a. NO CHANGE
b. Hampton was born and raised in Maywood of Chicago, Illinois in 1948.
c. Hampton is born and raised in Maywood of Chicago, Illinois in 1948.
d. Hampton was born and raised in Maywood of Chicago Illinois in 1948.

4. Which of the following would be the best choice for this sentence (reproduced below)?

(34) He was gifted academically and a natural athlete, he became a stellar baseball player in high school.

a. NO CHANGE
b. A natural athlete and gifted though he was,
c. A natural athlete, and gifted,
d. Gifted academically and a natural athlete,

5. Which of the following would be the best choice for this sentence (reproduced below)?

(35) After graduating from Proviso East High School in 1966, he later went on to study law at Triton Junior College.

a. NO CHANGE
b. He later went on to study law at Triton Junior College graduating from Proviso East High School in 1966.
c. Graduating from Proviso East High School and Triton Junior College went to study.
d. Later at Triton Junior College, studying law, from Proviso East High School in 1966.

Answer Explanations

1. C: Choice *C* is the best answer choice here because we have "subject + verb + prepositional phrases" which is usually the most direct sentence combination. Choice *A* is incorrect because the verbs *seeing* and *see* are repetitive. Choice *B* is incorrect because the subject and verb are separated, which is not the best way to divulge information in this particular sentence. Choice *D* is incorrect because the sentence uses a continuous verb form instead of past tense.

2. A: The comma after *result* is necessary for the sentence structure, making it an imperative component. The original sentence is correct, making Choice *A* correct. Choice *B* is incorrect because it lacks the crucial comma that introduces a new idea. Choice *C* is incorrect because a colon is unnecessary, and Choice *D* is incorrect because the addition of the word *of* is unnecessary when applied to the rest of the sentence.

3. B: Hampton was born and raised in Maywood of Chicago, Illinois in 1948. Choice *A* is incorrect because the subject and verb are at the end, and this is not the most straightforward syntax. Choice *C* is incorrect because the rest of the passage is in past tense. Choice *D* is incorrect because there should be a comma between the names of a city and a state.

4. D: Choice *D* is the best answer because it fixes the comma splice in the original sentence by creating a modifying clause at the beginning of the sentence. Choice *A* is incorrect because it contains a comma splice. Choice *B* is too wordy. Choice *C* has an unnecessary comma.

5. A: The sentence is correct as-is. Choice *B* is incorrect because we are missing the chronological signpost *after*. Choice *C* is incorrect because there is no subject represented in this sentence. Choice *D* is incorrect because it also doesn't have a subject.

Social Studies

World History

Political Decentralization in Europe from c. 1200 to c. 1450

From 1200 to 1450, European states were extremely decentralized due to feudalism. In a feudal political system, monarchies granted land rights to nobles (vassals) in exchange for military service and/or tax revenue. As such, the monarchy couldn't enforce laws or otherwise perform most government functions without relying on the local power structures. In contrast, many nobles consolidated power due to their military strength, dominance over the peasant labor force, and control of the food supply. Often, powerful nobles would threaten the monarchy's right to rule, which further incentivized the central government to avoid antagonizing the nobility.

Destabilization in Europe

Europe experienced incredible destabilization during this period. The already dysfunctional English and French monarchies fought the **Hundred Years' War** (1337–1453). Frenchmen originally founded the English monarchy, so the English royal family pressed claims to the French throne based on their substantial landholdings in France. France disputed those land rights, claiming the English royal family were the French monarchy's feudal vassals. The prolonged armed conflict was destabilizing, but it ultimately strengthened the French and English monarchies due to the development of professional armies and rise of nationalism. Consequently, by the end of the Hundred Years' War, both the English and French monarchies were able to consolidate political and economic power at the expense of the nobility, which brought an end to the feudal political system.

The Kingdom of Castile and Kingdom of Aragon spent much of this period fighting back against the Muslim states on the Iberian Peninsula. The two Christian kingdoms united in 1474, establishing the state of Spain. Because the Portuguese Crown consolidated political power in the eleventh century, it had a distinct advantage over their European rivals. During the early fifteenth century, Portugal built a world-renowned armada and launched what quickly would become a vast overseas colonial empire. In Central Europe, the Holy Roman Empire struggled to project authority over the territories it claimed in present-day Germany and Italy. The rest of Italy was divided among relatively small city-states, and the papacy constituted the region's most powerful political force. Eastern Europe was destabilized after the Ottomans sacked Constantinople in 1453, which caused the Byzantine Empire to dissolve into three significantly weaker independent states.

Europe as an Agriculture Society

The European socioeconomic order revolved around agriculture from 1200 to 1450 due to the prevalence of feudalism. The powerful landed class of feudal nobles typically organized their land under the manorial system. The vast majority of European nobles served as the Lord of the Manor, and their extensive landholdings were known as **fiefs**. Peasants who lived on the fief were allowed to farm the land, but they were obligated to pay the Lord of the Manor taxes either in the form of free labor or percentage of crops. Furthermore, the Lord of the Manor held all political power, and even more importantly, created and oversaw the legal system that applied to the fief. Nobles' legal and economic dominance over the peasants is why they held such power under feudalism. Although trade and urban centers generally increased from 1200 to 1450, Europe was still an overwhelmingly agriculture society due to the number and size of fiefs across the continent.

Trading Cities Expand

The **Silk Roads** are a historic system of trade routes that linked the East and West for the first time in world history during the second century BCE. In general, the Silk Roads consisted of overland and maritime trade routes, extending from China to Eastern Europe. Following a relatively dormant period, the Mongol Empire revitalized the Silk Roads during the thirteenth century.

The Mongol Empire increased trade in a number of ways. First, the Mongol Empire was the largest contiguous empire in world history, and its territories incorporated nearly all of the historic Silk Roads in China, Central Asia, the Northern Indian subcontinent, Middle East, and Eastern Europe under a single government and legal system. Second, the Mongol military was a peerless military force, as evidenced by its record-breaking conquests, and its cavalry units were highly effective at preventing and deterring crime. Third, the Mongol Empire's founder, **Genghis Khan**, sponsored merchants as a peaceful means of projecting power and influence over long distances, and this assistance included providing capital, transport, protection, and insurance. Fourth, the Mongol Empire built ties with foreign powers and merchants, such as Mongol elites' relationships with elite Italian merchants and financiers, to gain access to new markets. All of these factors played a role in bringing about the explosion of economic exchanges on the Silk Roads. Historians often refer to this period as **Pax Mongolica**, a play off the term *Pax Romana* used to describe the Roman Empire's stabilizing and stimulating economic influence.

Pax Mongolica resulted in a rapid expansion of cities on the Silk Roads. For some cities, the growth was exponential. Some oasis towns in the Taklamakan Desert (present-day Northwest China), such as Khotan, exploded into bustling multicultural metropolises. Increased trade also revitalized already large cities. Beijing and Shanghai were the most eastern points of the Silk Roads, and as a result, those cities' marketplaces expanded as large merchant caravans arrived and prepared to depart. Similar developments occurred in the Mongol Empire's territory in the northern region of the Indian subcontinent, particularly in Delhi and Patna. One of the Silk Roads' greatest beneficiaries was **Constantinople**, which separated the East and West. Many of the Silk Roads flowed through Constantinople because its ports provided convenient access to European and North African markets. In effect, the Silk Roads functioned as a commercial stimulus program because economic activity naturally increased in cities due to the massive amounts of goods passing through their marketplaces.

A number of cities also expanded due to their proximity to the Maritime Silk Road. Dating back to the second century BCE, the **Maritime Silk Road** linked Chinese, Southeast Asian, Indian, Middle Eastern, and European markets through trade routes on the Pacific and Indian Oceans. The size and frequency of exchanges on the Maritime Silk Road increased during the thirteenth and fourteenth centuries due to the economic growth created from Pax Mongolica. Mongol rulers sought to integrate the terrestrial and maritime Silk Roads, which created tremendous economic growth in cities located on coasts of the Indian subcontinent, Arabian Peninsula, Middle East, and East Africa.

The Mongol city of Hormuz was especially valuable because its territorial waters functioned as a passage from the Persian Gulf to the Gulf of Oman, which carried ships into the Indian Ocean. Similarly, the Mamluk caliphate's city of Acre in present-day Israel served as a major port that connected the European, Asian, and Mediterranean markets. In present-day Malaysia, the city of **Malacca** became a commercial leader on the Maritime Silk Road, serving as a market between India and China. Many North African and East African cities, including the Mamluk Caliphate's capital city, Cairo, and the Hafsid Dynasty's capital city, Tunis, also benefited from the rise of global trade.

Expansion of Empires

Most historians regard the Mongol Empire's expansion and prolonged period of rule as the beginning of the modern world. Through military conquests and diplomacy, Genghis Khan (1162–1227) united the five Mongol confederations in 1206, and he immediately launched a series of aggressive military campaigns. To the east, Genghis Khan's forces fought the Jurchen's Jin Dynasty in modern-day northern China, and to the south, the Mongols engaged the Jin Dynasty's allies and Tibet. However, Genghis Khan was most successful at pushing westward, and his forces devastated states across present Uzbekistan, Tajikistan, Kyrgyzstan, Kazakhstan, Belarus, and Russia. By the time of Genghis Khan's death in 1227, the Mongol Empire extended from the Pacific Ocean to the Caspian Sea, nearly doubling the maximum extent of the Roman Empire in terms of territory.

Ögedei (1186–1241) succeeded Genghis Khan, and he followed a similarly aggressive military strategy. In 1234, Ögedei completed the Mongols' conquest of the Jin Dynasty in northern China. Furthermore, Ögedei's forces defeated several Persian states, invaded Central China, and launched attacks in Central Europe in present-day Poland, Hungary, and Croatia. In the decades following Ögedei's death, the Mongol Empire continued to push deeper into China, and in 1276, the Song Dynasty surrendered to Kublai Khan (1215–1294). Consequently, Kublai Khan claimed the title of emperor of China, declared the start of the Yuan Dynasty, and promptly launched campaigns into Southeast Asia.

Kublai Khan's conquest of China quickened the fragmentation of the Mongol Empire. As Kublai Khan became increasingly more occupied with building the Yuan Dynasty's administrative state to rule China, other Mongol leaders sought to gain more autonomy within the Mongol Empire. Adding fuel to the fire, there were several succession crises over which Mongol leader would serve as the head of the Mongol Empire. Consequently, the Mongol Empire gradually splintered into four smaller and autonomous empires (khanates)—Golden Horde, Chagatai Khanate, Ilkhanate, and Yuan Dynasty—during the latter half of the thirteenth century.

The **Yuan Dynasty** (1271–1368) was the first non-Han dynasty ever established in mainland China. Kublai Khan incorporated a number of traditional practices, such as allowing influential Han leaders to retain their status and maintaining an administrative state consisting of the Three Departments and Six Ministries, to present himself as a legitimate emperor of China. Mongol leaders objected to Kublai Khan's attempts to establish China as the imperial center of the Mongol Empire, and their rebellions resulted in the establishment of three independent khanates. The Yuan Dynasty lasted until the Red Turban Rebellion (1351–1368), an uprising led by a secret society called the **White Lotus**, destabilized the administrative state, triggering the Yuan Dynasty's collapse in 1368. Leaders of the Yuan Dynasty relocated to the Mongolian Plateau and established the Northern Yuan Dynasty, which fought the new Ming Dynasty for centuries. During the seventeenth century, a distant descendant of Genghis Khan united the Manchus and Northern Yuan Dynasty, resulting in the overthrow of the Ming Dynasty.

The **Golden Horde** consolidated control over the northwestern portion of the Mongol Empire, with the territory ranging from Eastern Europe to Siberia. Under the leadership of Khan Uzbeg (1282–1341), the Golden Horde built the largest military in the world and reached its territorial peak after achieving victories over the Ilkhanate, Byzantine Empire, and Bulgarian kingdoms. Although the Golden Horde initially rejected the Yuan Dynasty's claim to the entire Mongol Empire, Uzbeg somewhat relented and agreed to send the Yuan Dynasty tribute in exchange for a share of their common enemies' territories. In addition, Uzbeg converted to Islam and adopted it as the state religion. Following its defeat at the hands of the Timurid Empire in the late fourteenth century, the Golden Horde was besieged with political infighting and civil wars until it officially dissolved in 1502.

The **Chagatai Khanate** controlled the Mongol Empire in Central Asia, including territory located in present-day Mongolia, Xinjiang, Uzbekistan, and Kazakhstan. During the early fourteenth century, the Chagatai Khanate strengthened its ties with the Yuan Dynasty and recognized its economic and military supremacy. In the mid-fourteenth century, the Chagatai Khanate split into two independent empires, and a Turco-Mongol warlord, Timur (1336–1405), conquered the Western Empire in 1370. Historians favorably compare Timur to Genghis Khan, and he is widely recognized as one of the greatest generals in military history, having never been defeated on the battlefield. The Timurid Empire was based in present-day Iran, and it became the undisputed leader of the Muslim world after defeating the Mamluk Caliphate, Ottoman Empire, and Delhi Sultanate. Following Timur's death, political strife destabilized the empire and led to its collapse in the latter half of the fifteenth century.

The **Ilkhanate** governed territory in present-day Iran, Azerbaijan, Armenia, Georgia, Afghanistan, Turkmenistan, Turkey, and Northern India. During the thirteenth century, the Ilkhanate rulers shifted from Tibetan Buddhism to Islam as the state's religion. The Ilkhanate collapsed in the 1330s due to the Black Death, which killed the Ilkahante's last ruler and his line of successors. The Golden Horde initially conquered some of the Ilkhanate's former territory, and Timur later rose to power within its territory in present-day Iran.

Expansion of Empires and Afro-Eurasian Trade

The expansion of the Mongol Empire created a single market across Afro-Eurasia. This meant that a single political entity regulated trade and enforced the same standardized laws under a uniform justice system. Because the Mongol Empire controlled nearly all of the historic Silk Roads, it's not surprising that these trade routes experienced a complete revitalization under their rule, harkening back to their original Golden Age under the Han dynasty. Land-based trade brought an influx of wealth to Afro-Eurasia, and it also encouraged growth in the Maritime Silk Road between China, Southeast Asia, the Indian subcontinent, and East Africa. Overall, economic exchanges, infrastructure projects, and commercial investments increased across Afro-Eurasia due to the Mongol Empire.

Interregional Conflicts Encouraged Technological and Cultural Transfers

Interregional conflicts resulted in technological and cultural transfers across Afro-Eurasia. Numerous military geniuses, including Genghis Khan, Kublai Khan, and Timur, led the Mongol Empire and its successor states. These generals conquered nearly unprecedented amounts of territory by adopting the latest technological innovations, such as combinations of bloomery-based and cast iron-based iron in the production of weapons and light armor. As the Mongols brought most of Afro-Eurasia under their control, these iron techniques became popular across their empire.

Trade also resulted in a variety of technological transfers. For example, the Silk Roads carried gunpowder technologies from China to Afro-Eurasia, altering the course of history. During the late fifteenth and early sixteenth centuries, several Middle Eastern states, such as the Mughal Empire, Ottoman Empire, and Safavid Empire, famously incorporated gunpowder into their militaries. Historians refer to these states as the Islamic Gunpowder Empires because they leveraged gunpowder technologies to rapidly conquer large swathes of territory. Cultural transfers were most evident in the spread of religion. For example, by the early fourteenth century, all of the Mongol successor states except for the Yuan Dynasty had shifted their state religion from Buddhism to Islam.

Causes and Effects of the Columbian Exchange

Connections Between Eastern and Western Hemispheres

Increased contact between Europe and the Americas naturally led to the Columbian Exchange, which involved the transfer of diseases, cash crops, food crops, animals, and agricultural techniques between the Eastern and Western hemispheres. Because Amerindians lacked immunity to Eastern Hemisphere diseases, the spread of these sicknesses decimated their populations. After seizing control over the Amerindians' land, Europeans built massive plantations to grow cash crops, such as tobacco and sugarcane. African slaves, Amerindian captives, and European indentured servants provided the necessary labor for plantations, and these sources of coerced labor further increased cash crops' profitability. The reliance on cash crops incentivized the adoption of unsustainable agricultural practices that led to widespread deforestation and soil depletion. Aside from cash crops, Europeans cultivated Afro-Eurasian fruit trees, grains, and domesticated animals in their American colonies. Some American foods, such as maize, potatoes, sweet potatoes, and tomatoes, were also transported to Afro-Eurasia. These foods supported explosive global population growth due to their nutritional value and high caloric density.

European Colonization of the Americas

European colonization devastated Amerindian populations, primarily due to the spread of new diseases into the Eastern Hemisphere. Diseases such as smallpox, malaria, measles, and influenza were endemic in Eurasia for centuries, but they were relatively unknown in the Americas. These diseases, which were far more common in Eurasia compared to the Americas, were often spread through domesticated animals. Repeated epidemics had ravaged Eurasia due to the prevalence of interregional trade and travel. For example, in the second century, the Antonine Plague traveled along the Eurasian Silk Road, devastating both the Roman Empire and the Han Dynasty. Similarly, in the fourteenth century, the Black Death Plague killed an estimated 100 million people across Eurasia. So, the European colonizers had some immunity to these diseases, while the Amerindians had none. This proved catastrophic. Eastern Hemisphere diseases killed somewhere between 80 and 95 percent of the Amerindians within two centuries of Columbus's first landing in the Caribbean.

The spread of Eastern Hemisphere diseases in the Americas was accelerated by the unintentional transfer of disease vectors. Brown rats traveled to the Americas on ships, and they carried a number of pathogens and parasites. European colonizers also introduced *Aedes aegypti* mosquitoes to the Americas. Unlike native mosquito populations, *Aedes aegypti* mosquitoes are the perfect disease vector because they prefer human blood, and they spread lethal diseases, such as yellow fever and malaria, across the Americas.

Food Crops in the Americas

Europeans exported many traditional American foods to Eurasia, including potatoes, maize, sweet potatoes, and tomatoes. Potatoes were the most successful of these exports. Before the sixteenth century, potatoes were only grown in South America, but within a couple of centuries, they were commonplace in Eurasia and Africa. Many European rulers encouraged local farmers to grow potatoes due to their valuable nutrients and caloric density. By the nineteenth century, potatoes constituted a primary staple crop for farmers throughout Europe, Asia, and Africa. Likewise, maize and sweet potatoes were first exported by the Portuguese during the sixteenth century, and they also became a staple crop across Afro-Eurasia. Tomatoes were also exported from the Americas to Eurasia, though they weren't immediately used in Eurasian cuisine. At first, tomatoes primarily only held aesthetic value to Europeans because they were considered poisonous. European scientists and farmers realized tomatoes were edible several centuries after their initial introduction, and tomato sauce became a staple of Italian cuisine by the nineteenth century.

The food crops described above were generally transferred to Europe as seeds, meaning they were grown locally in Afro-Eurasia. So, the introduction of American food crops had a significant impact on Afro-Eurasians' day-to-day lifestyles; however, they were not a major source of wealth for European governments or merchants. In contrast, **cash crops** were produced in American colonies, and their exportation as raw products drove merchant and government profits. Examples of American cash crops included coffee, cotton, sugarcane, and tobacco. In order to further boost profits, cash crops were grown on huge plantations that relied on large pools of forced or coerced labor.

The overwhelming majority of African slaves and Amerindian captives worked on these plantations. American plantations were able to dominate the Eurasian marketplace due to the mass production of cash crops with coerced labor. For example, it's estimated that Spanish and Portuguese sugarcane plantations produced as much as 90 percent of Europe's sugar. Cash crops also played an important role within colonial societies. Owners of sugarcane and tobacco plantations were among the wealthiest people in the Americas, and they exercised considerable control over local politics. In the Virginia colony, tobacco was so prevalent and valuable, it assumed the status of legal currency, and wealthy plantation owners issued the colony's first paper currencies backed by stores of their cash crops.

Foods Brought to the Americas

Several different types of fruit trees and sugarcane were brought from Afro-Eurasia to the Americas. Citrus fruits and grapes were transported from the Mediterranean to colonies with warmer climates, such as the Caribbean, Florida, the Carolinas, Brazil, Peru, and Mexico. Similarly, the Spanish and Portuguese exported guava and bananas from West Africa to their colonies in Florida, Mexico, and South America. The Portuguese brought sugarcane from the Indian subcontinent to their American colonies. The first sugarcane plantation was established in Brazil during the middle of the sixteenth century, and given its profitability, cultivation spread across South America and the Caribbean.

In order to help feed the European colonizers and their coerced labor force, rice and wheat were brought to the Americas. The most popular strains of rice came from West Africa and Southeast Asia, and the centers of production were located in the Caribbean and the Carolinas. Farther north, European wheat strains were grown in England's **Middle Colonies** (New York, New Jersey, Pennsylvania, and Delaware). The Middle Colonies' colder climate and inhospitable soil made agriculture more difficult, so wheat production helped prevent famine. Barley, oats, and rye were also important grains grown in this area. Spanish and Portuguese colonies were sustained through the production of maize, which predated European colonization of Mexico and South America.

Domesticated animals were also transported overseas, and they similarly altered life in the American colonies. Unlike the Amerindians, Europeans had practiced animal husbandry for centuries, and they transported domesticated animals to be used as sources of meat, dairy, eggs, and labor. Examples of domesticated animals transported from Eurasia to the Americas included cattle, chickens, donkeys, goats, geese, horses, pigs, sheep, and water buffalo. Along with large-scale agriculture, animal husbandry strongly supported European colonization. While Amerindians generally lived closer to a subsistence lifestyle, Europeans were able to feed their populations with less manpower. This allowed for more specialization of professions, facilitating the development of sophisticated militaries, weapons, and navigational techniques.

Although African slaves weren't allowed to transport food or personal belongings to the Americas, slaves' specialized knowledge of farming techniques facilitated the successful cultivation of African crops. The Portuguese initially carried bananas on slave ships, and slaves were forced to help the Portuguese grow the first American crop in Brazil. Slaves were essential to rice production because they were familiar with the required farming techniques, such as irrigation and winnowing. African slaves were also responsible for the success of kola nuts, okra, coffee, and black-eyed peas in the American colonies.

Afro-Eurasia Benefitted Nutritionally from American Food Crops

Afro-Eurasian populations benefited in a variety of ways from the importation of American food crops. First, many of the American food crops had higher caloric density than Afro-Eurasian foods, meaning these foods could support larger populations. Examples included potatoes, sweet potatoes, and maize. Many regions had an absolute dependence on these foods for population growth. This phenomenon is best illustrated by the Great Famine in Ireland during the middle of the nineteenth century. When the potato crop failed, Ireland's population decreased by nearly 25 percent due to mass death and emigration. Second, American food crops constituted a new source of vitamins and added flavor to existing foods. Tomatoes, bananas, and citrus fruits increased the Afro-Eurasian intake of vitamin C, potassium, magnesium, and antioxidants.

The introduction of chili peppers led to the creation of popular dishes, such as Indian curry and Korean kimchee. Third, the American food crops had different soil, weather, and cultivation requirements, which complemented rather than competed with existing food sources. Native Afro-Eurasian food crops could be grown during one season, with American food crops being grown during their off-season; thus, there was an overall increase in the amount of food available to Afro-Eurasians.

More plentiful and nutritional Afro-Eurasian food supplies directly led to explosive global population growth. From 1450 to 1750, the global population nearly doubled, growing from approximately 400 million to 750 million. In addition, American food crops supported widespread industrialization and urbanization during the eighteenth and nineteenth centuries, which marked the most explosive period of population growth in human history.

Deforestation and Soil Depletion from European Colonization

European colonization had a disastrous effect on the physical environment. Prior to colonization, Amerindians had implemented extensive land-use management strategies. Agriculture was more limited, and crops were frequently rotated to prevent soil depletion. Amerindians also used controlled fires to strengthen natural ecosystems. As the fires burned, the resulting ash and biomass enriched the soil. The fires also cleared underbrush and weaker trees, facilitating the development of grasslands and stronger trees, and limited the impact of pests, such as ticks and tree-nut larvae. In general, these land-use management practices were sustainable, ensuring that the environment would support human life for future generations.

In contrast, Europeans exploited the environment for their immediate benefit. Entire forests were cleared to make room for plantations, settlements, and animal husbandry. Centuries-old timberland was chopped down to build ships and infrastructure. Cash crops' profitability incentivized monocropping, which depleted the soil through the continual harvest of the same crop on the same land. For example, tobacco cultivation placed such high demands on the soil that farmland needed to be abandoned after only a few years. Europeans also practiced unsustainable slash-and-burn agriculture. Existing vegetation was burned to enrich the soil, and the resulting field was used for cash crops or animal husbandry until the soil was entirely depleted. Once this happened, the field was abandoned, and agricultural production was moved to the next area. While Aztec, Mayan, and Incan civilizations had used slash-and-burn agriculture for centuries, European practices were far less sustainable due to its larger scale and prioritization of cash crops.

State Building and Expansion from 1450 to 1750

Trading Posts in Africa and Asia

The expansion of the Ottoman Empire reinvigorated interregional trade to an extent not seen since the fall of the Western Roman Empire. The Ottoman Empire began in present-day Turkey, spread into Europe in 1354, and conquered Constantinople in 1453, causing the Byzantine Empire to collapse. By 1683, the Ottomans controlled the Balkans, Greece, Mesopotamia, present-day Syria and Israel, Egypt, and nearly all of the North African coast. As a result, the Ottoman Empire dominated the Mediterranean sea lanes, which were the primary trade routes between Southeastern Europe, North Africa, and West Asia. Like the Roman Empire, the Ottoman administrative state spurred economic growth in the Mediterranean region by connecting ports to overland trade routes, instituting a standardized legal system, establishing an imperial currency, and protecting merchants from piracy.

Similar to its role in the Mediterranean region, the Ottoman Empire's consolidation of economic, military, and political power directly led to increased overland trade across Eurasia, including the Silk Road's reemergence as a major trade route. The **Safavid Empire** in present-day Iran also played a critical role in reviving the Silk Road. When the Safavid Empire fought a prolonged military conflict with the Ottoman Empire, they created a new land route that connected China to Europe through Russia. Consequently, interregional trade between East Asia and Europe boomed throughout this period both on land and through the increasingly prosperous Mediterranean sea lanes.

Interregional maritime trade further increased after Portuguese explorers discovered they could reach the Indian Ocean by sailing around the Cape of Good Hope in 1488. This discovery severely disrupted the Ottoman Empire's monopoly on the Mediterranean sea lanes because it connected Western Europe to East Asia via India. While maritime trade increased throughout the sixteenth and seventeenth centuries, trade was occasionally disrupted by naval wars. Portugal and Spain frequently sparred with the Ottoman Empire in the Mediterranean Sea and Indian Ocean. Similarly, the Safavid Empire fought Portugal to gain control over strategic trade lanes in the Persian Gulf, Gulf of Aden, and Straits of Malacca. With help from the British East India Company, the Safavid Empire was able to fend off the Portuguese. This not only increased English trade with Asia but also opened new economic opportunities for Dutch and French merchants.

India greatly benefited from increased maritime trade based on its geographic position between East Asia and Western Europe. After the **Mughal Empire** consolidated control over the Indian subcontinent, it sought to increase foreign trade. During the late fifteenth century, Emperor Akbar entered into agreements with European trading companies to strengthen economic ties, and as a result, trade steadily increased, particularly in India's textile industry. The Mughal Empire's economy was the largest in the world in the eighteenth century, accounting for an estimated 25 percent of global economic production.

Increased maritime trade greatly disrupted the trans-Saharan caravan networks. As Portugal explored the West African coast, it built military trading posts. Compared to the danger and expense of crossing the Sahara, these trading posts were very attractive. Saharan trading centers further declined after Morocco invaded and conquered the Songhai Empire, which had dominated West African trade. This conquest destroyed much of the infrastructure that had supported the caravan network. In the aftermath of the Songhai Empire's collapse, Portuguese trading posts filled the political and economic vacuum, delivering unprecedented amounts of African slaves and gold to Europe.

New Maritime Empires

European powers raced to establish maritime empires in the Americas to exploit the continents' rich resources. The Portuguese established a new maritime empire in present-day Brazil during the first half of the sixteenth century. Initially, the Portuguese privatized colonization, and the territory was divided into fifteen separate colonies. Portuguese noblemen funded and controlled these colonies, but they failed for a variety of reasons, including agricultural issues, conflicts with Amerindians, and disputes among the noblemen. To save the floundering enterprise, King John III placed Brazil under royal control in 1542. The newly formed central government consolidated its control by conquering the Amerindians and thwarting French attempts at colonizing northern Brazil. In addition, the government converted much of the land into sugarcane plantations, and the intense labor demands led to the importation of African slaves. In the early seventeenth century, the Dutch attempted to seize control over sugarcane production in northwestern Brazil. However, Portuguese forces expelled the Dutch in 1654, leaving Portugal as the uncontested power in Brazil.

The Spanish maritime empire was unprecedented, and it included substantial amounts of territory in the present-day United States, Mexico, Central America, the Caribbean, and South America. In order to construct this empire, Spain had to conquer several advanced Amerindian civilizations. The Spanish conquistador Hernán Cortés launched a campaign against the Aztec Empire in present-day Mexico. Cortés made a strategic alliance with the Tlaxcala city-state, a historic rival of the Aztecs, and the Aztec Empire collapsed in 1521. Another conquistador, Francisco Pizarro, conquered the Inca Empire in 1533. As a result, Spain assumed control over territory in present-day Peru, Bolivia, Argentina, Chile, and Colombia. Other Spanish conquistadors defeated the Muisca Confederation in present-day Colombia as well as numerous Mayan city-states in Central America. Spain divided its massive territories into viceroys, which functioned as provincial administrative states. Along with defending the territories, the viceroys were primarily tasked with mining gold and silver.

The Netherlands relied on the Dutch West India Company to construct its maritime empire. After sponsoring Henry Hudson's voyage to North America, the company established permanent colonies on the Eastern Seaboard of the present-day United States. These colonies were known as **New Netherlands**, and they primarily focused on fur trapping. However, the Dutch had difficulty populating their North American colonies. The most successful colony was New Amsterdam, which was located in present-day New York City. Following their defeat in the Third Anglo-Dutch War, the Dutch ceded New Netherlands to England in 1674. The Dutch also competed with France, England, and Spain in the Caribbean, and they established several successful sugarcane plantations on a series of islands known as the Dutch Antilles.

The French maritime empire stretched from present-day Canada to the Caribbean. France founded numerous settlements in present-day Eastern Canada and Upper Midwestern United States. Those settlements centered on fishing and fur trapping opportunities. French explorers also explored the Mississippi River, and successful settlements were established in St. Louis, Baton Rouge, and New Orleans. However, like the Dutch, the French had difficulty populating their North American territory. France's most successful colony was Saint-Domingue in present-day Haiti. Saint-Domingue's sugarcane plantations outperformed the rest of the Caribbean, but it came at a deadly cost. African slaves' mortality rate on Saint-Domingue was the highest of any European colony in the Americas.

The British maritime empire included thirteen colonies in the present-day United States and numerous settlements in the Caribbean. England's thirteen colonies had incredible economic variation, ranging from shipbuilding in the New England colonies to plantations in the southern colonies. Despite experiencing some initial setbacks, England was markedly successful at attracting European immigrants to the thirteen colonies. For example, in 1750, ten times more Europeans lived in England's thirteen colonies than in all of France's North American colonies. Along with a more favorable climate and geographic position, England was able to attract immigrants by offering extensive rights to self-government. Aside from the thirteen colonies, England established more than a dozen major settlements in the Caribbean between 1623 and 1750. England sent more African slaves to work on Caribbean sugarcane plantations than any other European power.

Growth of States in Africa

Increased maritime trade benefited several African states, such as the Asante Empire and Kingdom of the Kongo. The **Asante Empire** conquered territory in present-day Ghana during the late seventeenth century and established governance by a king who controlled a powerful bureaucracy. One division of the bureaucracy was responsible for negotiating with foreign powers, and the Asante Empire developed a large trade network with European powers, especially the Portuguese. The Asante Empire primarily traded slaves, gold, and ivory in exchange for European manufactured goods and firearms. The **Kingdom of Kongo** held a large swathe of territory on the southwest African coast, and it had extensive trade relationship with European powers. Beginning in the late fifteenth century, Portuguese explorers began arriving in the Kingdom of Kongo, and Portugal established a Catholic Church and slave trading post shortly thereafter. Like the Asante Empire, the Kingdom of Kongo traded slaves and gold to Europeans in exchange for guns, which facilitated the empire's expansion.

Revolutions in the Atlantic World from 1750 to 1900

The Enlightenment laid the foundation for a series of revolutions that altered political systems in the Americas and Europe between 1750 and 1950. Beginning in the early eighteenth century, philosophers began advocating for secular reason, which posed a direct challenge to the Church's authority. In terms of politics, the Enlightenment promoted individualism, freedom, constitutional government, and nationalist ideologies. Most influentially, several Enlightenment philosophers—John Locke, Jean Jacques Rousseau, and Thomas Hobbes—formulated **social contract theories**. Under such theories, the people legitimized the government by sacrificing some freedom to their authority, and in exchange, the government was obligated to protect the people's legal rights. Accordingly, the government's failure to fulfill this obligation would violate the social contract, effectively opening the door to a revolution.

When combined with nationalism, social contract theory functioned as an effective justification for revolution. During the late eighteenth and nineteenth centuries, the Enlightenment influenced revolutions across the Atlantic world, including the American Revolution, French Revolution, and the Latin American wars of independence.

Enlightenment Philosophies

Throughout the 1700s, Enlightenment philosophers created a paradigm shift in thinking, one that upheld reason, nature, happiness, progress, liberty, and inalienable rights as the penultimate objectives of the human experience. Philosophers like Voltaire, Baron de Montesquieu, Jean Jacques Rousseau, and John Locke established a new precedent for human philosophy. Reexamining the role of religion, they encouraged humanist ideals that placed human rights and sentiments at the forefront of existence.

They believed reason should empirically challenge old notions of faith, encouraging scientific discovery as a new lens for understanding the world around them.

New political ideas, particularly those established by John Locke, demanded new social contracts be created between governments and citizens. When these contracts were breached—and individual, natural, or inalienable, rights were compromised—philosophers like Locke and his followers believed citizens had the right to rebel. This new reason-based belief system consequently helped motivate many of the revolutionary efforts of the eighteenth and nineteenth centuries, moving humanity further from monarchy and closer to reimagined classical philosophies such as democracy, liberalism, and republicanism.

Enlightenment Thought Preceded Revolutions

By the 1700s, the New World, disrupted by Enlightenment ideals, was hit with a wave of revolutionary sentiment. American colonists, fueled by new ideological notions of liberty, equality, and human rights, began challenging centralized imperial governments in the region. The colonial powers of the previous century – Britain, France, Spain, and Portugal – were challenged by new colonial-based identities that became culturally distinct from their mother country. **Monarchy**, in particular, was attacked vehemently by Enlightenment thinkers and revolutionary soldiers. Monarchy was seen as a threat because it consolidated political and economic power under the rule of one figurehead. **Republicanism** and **democracy**—classical Roman and Greek political structures—were heralded as the political solutions for the coming age of revolt against monarchy. As colonists gained more independence and power through geographic separation, they sought to decentralize the power of democracy. And, in many cases, they were successful—the American, Haitian, and Latin American independence movements of the era were effective in their efforts to decentralize colonial power.

Nationalism

Nationalism contributed to the development of modern-day states and justified imperial expansion. During the early nineteenth century, German philosopher **G.W. Friedrich Hegel** argued that nationalism bound societies together based on shared language, religion, and culture within a defined territorial area. Hegel's theory influenced **Otto von Bismarck**'s mission to unite Prussian states into a larger German state, which he accomplished in 1871. Similarly, nationalism motivated Italian unification, which gradually occurred between 1815 and 1871. Nationalism also contributed to revolts against nineteenth century imperial powers, which resulted in the creation of newly independent states. For example, Serbian and Greek nationalist movements both succeeded in seceding from the Ottoman Empire in the first half of the nineteenth century. On the other hand, European powers and Japan used nationalism to justify imperial expansion. European powers considered the colonization of Africa as a necessary means of protecting and strengthening their nationalist spirit, and Japan defended the annexation of Korea in the late nineteenth century based on their nationalist claims to superiority.

How the Enlightenment Affected Societies

Reform Movements

In some cases, nationalists challenged boundaries or sought unification of fragmented regions, as in the case of nationalizing Italy in the early nineteenth century. In other cases, nationalists attempted to create homogenous races of people through genocide, as in the Ottoman genocide against the Armenians prior to World War I. In still other cases, some nationalists combined both these trends, as in the British conquest of the Zulus in South Africa. As kingdoms and monarchies were gradually defined as the ruling guards of the past, new nation-states emerged as the primary form of global governance in a world of ever-expanding capitalist endeavors.

Women's Suffrage and an Emergent Feminism

By the end of the nineteenth century, all developed industrial countries had adopted universal male suffrage which gave all men, regardless of race or class, the right to vote. This encouraged women in these societies to fight for their own right to vote. Women in Great Britain and the United States in particular began launching campaigns to earn the right to vote. Women like **Elizabeth Cady Stanton** and **Lucretia Mott** began organizing rallies and conventions in the United States as early as the 1840s. British women also rallied and convened to protest unfair laws and customs that restricted women's rights in the nineteenth century.

One of the biggest breakthroughs for women's suffrage occurred in 1848 when Stanton and Mott led the **Seneca Falls Convention** in Seneca Falls, New York. The convention was the first of many for women's rights in the United States. Borrowing from the Declaration of Independence, the convention eventually released a treatise, known as the **Declaration of Sentiments**, which argued for the equality of women. Resistance shadowed these demands both in the United States and Great Britain. Patriarchal stances argued that women were "unfit" for politics. They saw the Seneca Falls Convention and the women's suffrage movement as too radical a break from traditional values. The Seneca Falls Convention was, therefore, a memorable first attempt at women's suffrage, but it was not an immediate success.

By the 1880s, women's activists across the globe—particularly from Western industrial powers such the United States, Canada, and Europe—began convening at annual events such as the **International Council for Women**. By 1893, large congresses of women were convening at the World's Fair in Chicago, bringing together 27 global territories, including New Zealand, Persia, China, Argentina, and Iceland. By 1893, New Zealand had become the first country to grant universal women's suffrage. Australia (1902), Finland (1906), Norway (1913) soon followed suit as the only pre-World War II universal women's suffrage advocates. As these victories for women's voting rights created a chain reaction across the globe, women's rights activists in the United States and Great Britain became more militant, thanks in large part to the growing militancy of the labor movements in these countries.

After decades of peaceful conventions, women like **Emmeline Pankhurst**, and her daughters Christabel and Sylvia, took a more radical stance against patriarchal government officials in Great Britain, forming the **Women's Social and Political Union (WSPU)**. With the goal of drawing attention to their cause through political stunts, these women heckled government officials in public, committed arson and vandalism, and even threw themselves in front of the king's carriages on parade days. Pankhurst and her daughters were eventually arrested and jailed for their actions. While in jail, they led hunger strikes to garner even more media attention.

Similar radical manifestations of women's rights sprouted in the United States. However, it was not until after World War I that the United States and Great Britain would finally extend universal women's suffrage to their female citizens. These final victories would have not been possible without the work of both the peaceful and radical iterations of the women's suffrage movement of the nineteenth and early twentieth centuries.

Various Revolutions in the Period from 1750 to 1900

The Enlightenment and rise of nationalism heavily contributed to the revolutions that took place between 1750 and 1900, and those revolutions resulted in the dissolution of empires and creation of new nation-states. First, Enlightenment philosophers popularized the concept of a social contract, which created an intellectual justification for self-government. For example, French and American political leaders both claimed violations of the social contract as the basis for their revolutions. Second, nationalism became a popular ideology, especially when communities shared language, religion, social customs, and territory. For example, German and Italian nationalist movements led unification efforts in their respective countries, and both movements succeeded in 1871.

The revolutions had a dramatic effect on several empires as they lost control of their overseas territories. For example, the Spanish Empire and Ottoman Empire were both on the brink of collapse by the beginning of the twentieth century. Another major effect of those revolutions was the creation of dozens of new nation-states governed by constitutional monarchies or representative governments. For example, Romania and Serbia both gained independence from the Ottoman Empire and formed constitutional monarchies during the nineteenth century.

Commonality Based on Language, Religion, Social Customs, and Territory

The development of nation-states was facilitated through people gaining a greater sense of community based on common characteristics, such as language, religion, social customs, and territory. For example, Christianity provided the foundation for nationalist movements in Serbia and Greece that revolted against the Ottoman Empire in 1804 and 1821, respectively. Oftentimes, governments enacted policies intended to emphasize these shared characteristics to create a unifying national identity. For example, after the French Revolution, the new nationalist government formed the Committee for Public Instruction to enlist teachers who instructed communities about the French language and social customs. This initiative was considered an essential part of crafting a national identity that bound the country together. Similarly, governments often sought to establish a national identity by fighting a common enemy, especially when the state's territorial integrity was threatened. For example, in the **Franco-Prussian War** (1870–1871), Prussian forces turned back a French invasion of German territory. This paved the way for the unification of Germany in 1871 under the King of Prussia's leadership.

Revolutions and Rebellions of the 18th Century

During the eighteenth and nineteenth centuries, a series of revolts and revolutions resulted in the creation of new nation-states. In the Americas, the United States revolted and defeated Britain in the **American Revolution** (1776–1783), gaining the right to self-government. Several decades later, the **Haitian Revolution** (1796–1804) against French colonization became the first slave-led rebellion to successfully secure independence. The Haitian Revolution began what's known as the **Latin American wars of independence** that lasted until 1898.

Mexican nationalists defeated Spain in the **Mexican War of Independence** (1810–1821) and established a republican government. Following the loss of its colony in Mexico, Spain issued a proclamation granting independence to Central America in 1821. A revolutionary Venezuelan military leader with European parentage, Simon Bolivar, led nationalist revolts against Spain across South America. Bolivar's forces supported successful revolutions in Bolivia (1825), Colombia (1810), Ecuador (1822), Panama (1821), Peru (1826), and Venezuela (1823). By the end of the twentieth century, nearly all of the Americas had freed themselves from European colonialism and established independent nation-states.

A number of European nation-states formed in the nineteenth century after revolting against the Ottoman Empire. After the **Serbian Revolution** (1804–1817), Serbia received political autonomy from the Ottoman Empire before gaining full independence in 1867. Greek revolutionaries received financial and military support from European powers, including Britain, France, and Russia, and they secured independence from the Ottoman Empire in 1830. Nationalist movements in Bulgaria, Montenegro, and

Romania triggered the **Russo-Turkish War** (1877–1878), and after their victory, all three nation-states gained their full independence from the Ottoman Empire at the Congress of Berlin (1878).

Discontent with Monarchist Imperialism

Monarchical imperialism began to wane as the dominant political ideology in the wake of the Enlightenment and the Industrial Revolution. Throughout the nineteenth century, new political ideologies began to challenge old ways of thinking. These ideologies included democracy, liberalism, socialism, and communism.

Democracy

As a result of both revolutionary and industrial pressures in Europe and the New World, many nations began to "democratize" in the early nineteenth century, which means they quickly began to gravitate toward traditional notions of popular sovereignty, which were first established by Greek philosophers. The democratic fervor of the era helped expand voting rights—first to working-class men, and eventually to all men (black and white) and women. Since **democracy** is based on popular suffrage, or voting, it makes sense that the logical progression globally throughout the nineteenth and early twentieth centuries was a gradual march toward universal suffrage in the Western world. Democracy, often labeled as "chaos" by its challengers, became the dominant political ideology of the United States and Latin America. In the United States, democracy took on a more limited form because of the Electoral College, which placed important voting rights in the hands of appointed officials. In Latin America, many governments adopted popular forms of democracy that elected officials through direct votes.

Liberalism

Classical **liberalism** was not in direct conflict with democracy, but it did not promote popular sovereignty in the same manner. While classical liberalism, which arose from Enlightenment thinking and capitalist enterprises, was largely a political philosophy, much of its philosophical ponderings focused on the role of government in the political economy. Championed by such political thinkers as **John Locke**, author of *Second Treatise of Government* (1689) and **Adam Smith**, author of *The Wealth of Nations* (1776), classical liberalism promoted the free market, noting that freedom was contingent upon the economic self-interests of individuals. Liberalism, therefore, became very critical of extensive government intervention and the **welfare states** (i.e., a government that always watches out for the welfare of its people). Classical liberals championed a "minimal state" that protected the rights of individuals but did not interfere with the free market.

Socialism

Socialism had fundamentally different views from classical liberalism when it came to the political economy. Championed by such political thinkers as **Karl Marx** and **Friedrich Engels**, **socialism** presumes that it is the government's duty to intervene on behalf of the socioeconomic benefit of society. Thanks to the political thought of Marx and Engels, socialism came to be known as a post-capitalist economic system built upon class conflict and the penultimate acquisition of social ownership of the means of production. This brand of socialism was antagonistic to capitalism, paving the way for a more extreme form of social ownership: communism.

Communism

Communism was a radical outgrowth of socialist thought that wanted to eliminate social classes, private property, and the state. It argued that property be distributed equally among people, which would, in turn, eliminate class differences. Thomas More's *Utopia* (1516) is often viewed as a proto-communist text that laid the intellectual foundations for Marx and Engel's radical departure from traditional socialism in *The Communist Manifesto* (1848). *The Communist Manifesto* popularized a new, radical definition of communism that would later stir class conflicts and communist revolutions against capitalist systems, such as the 1917 **October Revolution** in Russia by **Vladimir Lenin** and the **Bolsheviks**. Communism—much like democracy, liberalism, and socialism—was ultimately but one outgrowth of a larger departure away from the autocratic leanings of monarchical political frameworks.

American Revolution, Haitian Revolution, and Latin American Independence Movements

American Revolution

The British Atlantic colonies of North America were not contemplating full-blown independence at the time King George III ascended the throne as the King of England in 1760. In fact, British support was at its highest by the time England was victorious in the **French and Indian War**, or **Seven Years' War** (1754–1763). Many American colonists had fought alongside British troops from the mother country. In spite of the paradigm shifts happening as a result of the spread of Enlightenment ideals, American colonists remained relatively loyal to the British Crown. That all changed in the aftermath of the French and Indian War, which had accrued an excessive amount of debt for Great Britain.

Great Britain, who had fought a significant amount of the war on American soil, believed the Americans had to help the empire emerge from its state of indebtedness. They believed Americans owed the British for their protection. As a result, the British began taxing Americans for certain colonial goods and services. Beginning with the Stamp Act of 1765, Americans, who were economically marginalized in their own right, had to pay taxes for things like wills, deeds, newspapers, and other printed materials. The Americans were outraged; they felt that their rights had been infringed upon. They gravitated toward protest and boycotts, which helped to absolve the Stamp Act by 1766. More taxes followed, however. In 1773, after the British decided to tax tea, the American colonists erupted in protest once again. This time they dumped tea into Boston Harbor, creating anarchical scenes along the Massachusetts coast. The British responded by occupying the American territories with troops.

By 1774, the American colonists had formed the First Continental Congress, which protested British occupation and taxation. By 1775, conflict had erupted in Lexington and Concord. And by 1776, the American colonists had drafted the Declaration of Independence, establishing the thirteen colonies as the United States of America. After years of war against the British, the Americans were finally able to declare victory in the **War for Independence** in 1781. The **American Revolution**, and the consequential **American Republic**, became the first example of a successful revolution in the New World. Other revolutionary nations followed the American example in the years to come.

The Declaration of Independence itself became a source of inspiration for other nations that would eventually declare revolution. The French Revolution (1789–1799), for example, established the **Declaration of the Rights of Man**, a revolutionary document that mixed the values of the Enlightenment and Declaration of Independence. With the success of the American Revolution behind them, and the French Revolution happening across the Atlantic, the early American Republic realized that it had put itself in the precarious position as one of the leaders of the free world. This role became more complicated for the nascent republic as the French were victorious in their efforts, and the Haitians, their colonial neighbors in the Caribbean, took arms against the newly formed French Republic.

Haitian Revolution

Prior to the **Haitian Revolution** of 1791–1804, one-third of the island of Hispaniola was known simply as the French colony of Saint Domingue. The colony, which relied on sugar plantations for economic stability, became a majority slave colony with over 500,000 enslaved Africans. The majority slave population was ruled by a minority of white plantation owners. This engineered inequality, coupled with the successes of the American and French Revolutions, encouraged the slave population of Saint Domingue to rise up against their masters.

Beginning with the rise of a revolutionary leader named **Toussaint L'Ouverture**, the Haitian Revolution focused its energies on creating the first black republic in the Western Hemisphere. Haitian revolutionaries swiftly carried out a successful revolt against their colonial adversaries on the island. French troops swooped in by the tens of thousands to quell the rebellion in 1802. L'Ouverture even temporarily halted his revolutionary efforts in an attempt to compromise with the French government; he tried to negotiate an end to the struggle for national independence in exchange for the abolition of slavery in the region. The negotiations failed, and L'Ouverture was jailed.

He died incarcerated in a cold mountain prison in 1803. The rebellion, however, was continued posthumously by his fellow revolutionaries. One man, **Jean-Jacques Dessalines**, carried the revolution to success in 1804. The success of the Haitian Revolutions sent shockwaves throughout the transatlantic world, including America. Plantation owners in other countries and territories feared that slave populations would revolt on their own soil, and much to their dismay, slave revolts did happen in isolated pockets. But none of these revolts was as successful as the Haitian Revolution, which established the first successful black nation in the New World.

Latin American Revolutions

The first **Latin American revolutions** began in the early 1800s. By that time, Spanish colonial society had become increasingly embroiled in a class struggle between *peninsulares* and *criollos* (sometimes referred to as ***creoles***). ***Peninsulares*** were colonial officials born in Spain. ***Criollos*** were Spaniards born in Latin America. Within the social structure of the Spanish colonies, ***mulattos*** (people of mixed European and African heritage) and ***mestizos*** (people of mixed European and Native American heritage) were believed to be lesser than the *peninsulares* and *criollos*. Native Americans were at the bottom of the socioeconomic ladder because they were considered less economically valuable than African slaves.

This unique social structure gave rise to the Latin American revolutions of the nineteenth century. However, the revolutions would be mostly carried out through the class conflict between *criollos* and *peninsulares*. *Criollos* felt like second-class citizens, though they still maintained a relative amount of power in the colonies. They were allowed to rise to positions as officers in the Spanish armies, but they were not allowed to assume roles as governors. This unique position gave them the military expertise and discontent to spearhead the independence movements. Unlike the relatively isolated American, French, and Haitian revolutions, the Latin American revolutions spread throughout South and Central America.

Following Napoleon's successful destabilization of Spain in 1808, rebellions began to break out in the crumbling empire. Two revolutionaries in particular led the charge: **Simon Bolivar** and **Jose de San Martin**. Bolivar was a wealthy Venezuelan *criollo,* and San Martin was an Argentinian *criollo* who had extensive educational and military experiences in Spain. The revolutions really took hold in 1811 when Bolivar joined with other *criollos* to declare Venezuelan independence from Spain. Both leaders had achieved their own revolutionary victories, and they met in 1822 to discuss combining forces. However, the two could not agree on the best form of government. As a result, San Martin ended up retiring, while Bolivar dealt the Spanish government its final deathblow in South America. In 1824, all Spanish colonies in South America were liberated from Spain.

As this was happening in Spanish South America, revolutionary activity was also spreading in Mexico and Brazil (the lone Portuguese colony in South America). Mexico declared independence after bloody battles waged by **Jose Maria Morelos**, who had carried out **Padre Miguel Hidalgo**'s cry for a mestizo revolution. The Mexico Revolution culminated in 1821. At the same time, a Central American revolution erupted, leading to independence from both Mexico and Spain by 1823. Brazil's revolutionary path was relatively bloodless in comparison. On September 7, 1822, thousands of Brazilian creoles banded together to declare independence from Spain. The Portuguese government relinquished the territory without bloodshed.

Enlightenment Values

Enlightenment values inevitably found their way into some of the most fundamental, paradigm-shifting governance documents of the era. These documents include the American Declaration of Independence, the French "Declaration of the Rights of Man and Citizen," and Bolivar's "Letter from Jamaica." They helped fundamentally alter the governance structures of existing political authorities by disseminating revolutionary and democratic ideals.

American Declaration of Independence

Written by Thomas Jefferson and the Second Continental Congress in 1776, at a key moment in the American Revolution, the **Declaration of Independence** was motivated greatly by the Enlightenment values of John Locke. This document set forth the truths of equality and the inalienable rights of men, including the rights to life, liberty, and the pursuit of happiness. Using Locke's philosophy, the Declaration of Independence asserted the right of the American people to rebel against the king because he had breached his social contract with the American people through tyranny. The document absolves American ties with Great Britain by noting the injustices carried out by King George III of England and his constituents.

French Declaration of the Rights of Man and Citizen

In August 1789, the National Assembly of the French Revolution adopted the **Declaration of the Rights of Man and Citizen**, a document modeled after Enlightenment values and the American Declaration of Independence. Much like the Declaration of Independence, it uses Lockean tenets to argue for the natural rights of men, which included the rights to liberty property, security, and resistance to oppression. Embracing the slogan "Liberty, Equality, and Fraternity," the Declaration of the Rights of Man and Citizen cut ties with the French monarchy, though its rights, like the Declaration of Independence, were extended only to men, not women.

Bolivar's Jamaica Letter

Simon Bolivar's ***Carta de Jamaica***, or **Jamaica Letter**, was written in Kingston in 1815. The letter, like the Declaration of Independence and the Declaration of the Rights of Man and Citizen, employed enlightenment values, justifying the need for Latin American liberation from Spain. The letter called for the union of Latin American states in the battle against colonial aggressors. Like the American and French examples, Bolivar's *Carta de Jamaica* established liberty as an inalienable right of men, which helped further his revolutionary campaigns in the region.

Italian Nationalism

Nationalism destroyed many empires; however, in the case of Italy, it helped build one. **Italian nationalism** emerged from the tumult of the Austrian Empire that had ruled the territory in the early nineteenth century. However, between 1815 and 1848, hundreds of thousands of Italian residents were growing increasingly discontent with Austrian rule; they wanted their own Italian nation, one that would reflect the values of the Italian locals. As a result of this discontent, several Italian nationalist leaders emerged: Giuseppe Mazzini, Camillo di Cavour, and Giuseppe Garibaldi. These men would help lead revolts that would eventually bring about Italian unification.

Efforts for Italian unification really began with **Giuseppe Mazzini** in 1832. Mazzini was a twenty-six-year-old idealist revolutionary who created a group known as **Young Italy**. He wanted to lead a revolution built on youth. In 1848, Mazzini helped lead a brief, violent revolt in the Italian states. As a result, he rose in the ranks to become the head of an Italian republic that was eventually dismantled by foreign influence. Nevertheless, for a brief time, Mazzini was able to unite nine Italian states under his republican rule, which promulgated notions of social justice, peace, and democracy.

Mazzini remained in exile as new revolutionary leaders emerged. One of these leaders was **Camillo di Cavour**, an Italian nationalist from the **Piedmont-Sardinia** region. The largest and most powerful of the Italian states, Piedmont-Sardinia became a hotbed for Italian nationalism even prior to Cavour's revolutionary escapades. During the tumultuous times of 1848, leaders in Piedmont-Sardinia adopted a liberal constitution that championed a more sensible alternative to Italian unification than Mazzini's Young Italy movement. Cavour, a wealthy aristocrat, worked alongside King Victor Emmanuel II to expand the territory and influence of Piedmont Sardinia. Mazzini openly distrusted Cavour, seeing him as a wealthy aristocrat who only had the interests of Piedmont-Sardinia in mind. To some degree, Mazzini was right, but it would be Cavour who would successfully bring about Italian unification.

In an effort to expand, Cavour created strategic alliances so that he could bring territory in Northern Italy within Piedmont-Sardinia's control. With Austria being the greatest adversary of Piedmont-Sardinia in the region, Cavour aligned himself with French troops to successfully annex the territory. An all-out war with Austria ensued as the French-Sardinian alliance took over Lombardi. Only Venetia remained in the possession of Austria following this campaign.

While Cavour campaigned for unification in the north, **Giuseppe Garibaldi** began to unify the southern peninsula. Cavour covertly supported this campaign in the south, hoping that the Italian nationalist rebels there would march north in the name of unification. Garibaldi, leading a nationalist group known as the **Red Shirts**, did just that. After usurping Sicily from foreign control, the Red Shirts marched throughout the southern Italian mainland, conquering territory that bordered the kingdom of Piedmont-Sardinia. Soon thereafter, Cavour facilitated a meeting between King Victor Emmanuel II and Garibaldi. The meeting ended with Garibaldi willingly stepping down as the leader of the south so that Italy could be united as one.

Filipino Nationalism

Filipino nationalism emerged at the end of the nineteenth century due to disruptions caused by the Spanish-American War. As a result of an American victory in the war, the United States took control of a variety of Caribbean and Pacific islands, including Puerto Rico, Guam, and the Philippines. Many Americans were concerned about the prospects of American imperialism, largely due to the nation's own revolutionary efforts against British imperialism. While the Philippines served as a strategic Pacific trade territory, many Americans also feared the diversification of the United States. Staunch imperialists, like President William McKinley, however, saw the acquisition of the Philippines as an opportunity for the United States to further Christianize the world. He hoped to carry out government-supported efforts to assimilate native Filipinos to American culture through education.

Filipino nationalists, however, were not fond of the United States' intrusion on their island. They had previously been fighting against Spain in an effort to decolonize their homeland, and they were hesitant to allow another imperial power on their shores. Filipino nationalists, such as **Emilio Aguinaldo**, promptly criticized the United States, claiming that they promised independence after the war with Spain. The Filipino nationalists declared independence, establishing the Philippine Republic, but the United States rejected these claims. A fierce conflict ensued. The **American War in the Philippines** ended in 1902 with an American victory, but also with a promise of eventual self-rule. The First Philippines Republic was established as an "American Insular Government," which operated much like a protectorate. The United States helped develop railroads and infrastructure in the nation, but also exploited the agricultural capacities of the locals. In 1935, the Philippines became an official American Commonwealth, which ended in 1946 when a new generation of Filipino nationalists declared full sovereignty.

Argentinian Nationalism

The territory of modern-day Argentina had long been under Spanish colonial rule as the Viceroyalty of the Rio de la Plata. The governance of the viceroyalty was controlled by Spanish-born colonialists, known as *peninsulares*. The *peninsulares* did not hold Spanish people born in the Americas, *criollos*, in as high regard as those born in Spain. The resulting tensions between *peninsulares* and *criollos* created the perfect climate for the emergence of an **Argentinian nationalism** that attacked the Spanish Crown for its inability to recognize *criollos* as first-class citizens.

Argentinian nationalism was further fueled by the **Anglo-Spanish War**, a war between Spain and Britain, which gave the *criollos* military experience. Buenos Aires, the largest city of the viceroyalty, became an epicenter of Argentinian nationalism. By 1810, after dueling occupations of the city by the Spanish and British, the *criollos* became increasingly outspoken against colonialism. The result was the **Argentine War for Independence**, which was fought between 1810 and 1818. Under the leadership of soldiers like Manuel Belgrano, Juan Jose Castelli, and Jose de San Martin, the *criollos* were able to finally overthrow Spanish rule in the region, establishing a national constitution in San Miguel de Tucaman in 1816 and winning the entire war for independence by 1818.

Shifting Power by the Century's End

At the outset of the twentieth century, European powers dominated the global political order. Many of these European powers established **maritime empires**, meaning they controlled territory separated by major bodies of water. The **British Empire** is an example of a maritime empire, and it was the largest empire in world history. At its peak in the 1920s, the British Empire controlled nearly twenty-five percent of global territory with extensive territorial holdings in Canada, the Caribbean, North Africa, South Africa, East Africa, the Middle East, the Indian subcontinent, Southeast Asia, and Australia. Prior to World War I, several other European powers sought to compete with the British Empire by seizing territory in Africa, including Belgium, France, Germany, Italy, Portugal, and Spain.

This period is commonly referred to as the **Scramble for Africa**. A similar process occurred in China after the British Empire crushed the **Qing Dynasty** in the Opium Wars (1839–1842; 1856–1860). European powers strong-armed the Qing Dynasty into accepting exploitative trade agreements. As a result, several European powers gained "spheres of influence" in China, meaning they enjoyed political and economic rights over Chinese territory and industries. During the first half of the twentieth century, the British Empire, Germany, the United States, the **Russian Empire**, and Japan all established a sphere of influence in China.

Both world wars dramatically altered the global political order by dissolving empires, which ultimately resulted in the formation of hundreds of new states. World War I triggered the collapse of several land-based empires, such as the Austria-Hungarian Empire and Ottoman Empire. Following its dissolution in 1918, the **Austria-Hungarian Empire** was succeeded by a series of independent countries located in present-day Austria, Hungary, Czech Republic, Poland, Romania, Serbia, Croatia, and Slovenia. The **Ottoman Empire** similarly dissolved due to its defeat in World War I. The present-day country of Turkey succeeded the Ottoman Empire, but all of the empire's former territorial holdings were largely split between the British Empire and France.

European maritime empires mostly collapsed in the aftermath of World War II (1939–1945). World War II essentially bankrupted the British Empire, and it lost control over nearly all of its foreign territory between 1945 and 1965. The **People's Republic of China** was formed in 1949, and it consolidated control over the Chinese mainland. African decolonization occurred during the 1950s and 1960s, resulting in the formation of more than forty newly independent countries. Following the Soviet Union's collapse in 1991, fifteen Eastern European countries gained independence. Overall, the dissolution of European empires resulted in the number of independent countries more than doubling from 1900 to 2000.

Mobilizing Populations for World War I

Combatants in World War I pursued an unprecedented **total war** military strategy, **meaning** that they utilized every possible resource to achieve victory. Many countries instituted **rations** so they could commit the maximum amount of resources to the war effort. Rations placed legal limits on consumption of designated products, which included everything from food to metal. The combatants also mobilized domestic and colonial populations in a variety of ways, including drafts that were instituted to provide more troops for upcoming military actions. As most able-bodied men served in the armed forces, factories hired women in record numbers to sustain the production of weapons and wartime materials.

To increase public support for the rations and draft that were necessary to conduct total war, governments stoked nationalism through political propaganda, art, and media. Nationalist propaganda emphasized that country's superiority and portrayed war as a righteous endeavor. Governments regularly hired artists to glorify their country's historic and wartime achievements, and artists mass-produced posters with heartening slogans and symbols. Political propaganda also appeared in all types of media, including newspapers, radio, and film. Public theaters would show government created films that depicted enemies as hellish monsters, and newspapers published political cartoons with similar themes. Governments heavily censored newspaper and radio reports on the conflict. At times, these reports were often fictional, amounting to what would be called "fake news" in the present day. For example, British media regularly exaggerated or fabricated German war crimes to create the perception that the Central Powers were the aggressor.

New Military Technology in World War I

New military technology had a major impact on World War I. Submarines, naval mines, and modern armored warships transformed naval warfare. In particular, Germany was able to somewhat counter Britain's dominant navy through submarine warfare. However, Britain was able to successfully blockade many German ports, and they adjusted to the threat of submarines by traveling in convoys.

On land, troops were equipped with devastating new weapons, such as grenades, bolt action rifles, light machine guns, fully automatic Maxim guns, and flame throwers. World War I also featured large-scale use of chemical weapons, field artillery, airplanes, and tanks for the first time in history. These new technologies were especially deadly to the powers that continued using outdated military tactics. The most common tactic in World War I was **trench warfare**. Armies dug trenches as natural barriers to protect themselves from the enemies' line of fire, and they defended their own trenches with machine guns. Attacking an entrenched army was incredibly dangerous because the defensive force would be protected by the trench while the offensive force would be exposed. Additionally, the trenches could not protect against artillery fire and chemical attacks. As a result, casualty rates were high in the trenches and in World War I overall. Approximately ten million soldiers died in the conflict with another twenty-three million left wounded. The military conflict also caused widespread famine and disease epidemics, which killed an estimated eight million civilians.

The End of World War I

In June of 1919, the Treaty of Versailles signified an agreement of peace between Germany and the Allied nations, bringing an end to World War I. This treaty identified Germany as the party responsible for the war, and demanded significant reparations from the country, including land and enormous amounts of money that Germany would never be able to pay off. The conditions of the treaty not only put Germany in a bad position economically, but they also instilled fierce discontent and resentment in many Germans. Years later, the fresh leadership within the National Socialist German Workers' Party (more commonly known as the Nazi Party) would appeal to Germans looking for a change. Ultimately, the rise of the Nazis led by Adolf Hitler would contribute to the beginning of World War II.

Mobilizing World War II as a Total War

Similar to World War I, governments deployed the total war strategy in World War II. They formed agencies to disseminate political propaganda, commission artists, censor news outlets, publish media, and intensify nationalism. For example, Britain's Ministry of Information and Germany's Ministry of Public Enlightenment and Propaganda both published informational leaflets, commissioned motivational posters, organized rallies, censored public broadcasting, and glorified their nations while demonizing enemies in film, literature, and artistic works. The propaganda differed from World War I in its use of political ideology to justify the need to wage war. Britain and the United States characterized World War II as a fight to save democracy from totalitarian dictatorships. The Soviet Union championed the need to spread communism across the globe and defend against fascist barbarism. The Axis powers justified their aggression based on their countries' natural superiority and inalienable right to rule over inferior civilizations.

Home countries, colonies, and former colonies also assisted the war effort in a variety of ways, such as enlisting in the armed forces and cooperating with rations. Colonial populations suffered the most from wartime rations, which often caused famine. For example, British wartime policies greatly exacerbated a famine in India, which killed more than two million people. Wartime policies reduced basic freedoms and dominated daily life in every country, but the consequences were the direst for totalitarian societies. Nazi Germany's obsession with racial purity incentivized citizens to report on their fellow citizens, neighbors, and even family members. The Soviet Union sent millions of dissidents and alleged traitors to forced-labor camps (gulags) with minimal due process, and these repressive policies continued long after the Soviet's victory in World War II.

New Military Technology in World War II

Approximately seventy-five million people died in World War II, making it the deadliest armed conflict in human history. Military personnel accounted for an estimated twenty-three million wartime fatalities, while an estimated fifty-two million civilians died during the conflict. The unprecedented death toll can be attributed to new military technologies and tactics. While aircrafts and tanks were first used on a large scale during World War I, updated models deployed during World War II were much more reliable and capable of carrying more lethal weapons. Similarly, combatants were able to mass-produce advanced and deadly small arms, such as the semi-automatic rifle. Naval warfare was transformed through the invention of aircraft carriers, destroyers, cruise missiles, rockets, and anti-submarine radar systems. Nuclear weapons were invented and deployed for the first time in the conflict's final stages. In August 1945, the United States dropped nuclear bombs on Hiroshima and Nagasaki, killing approximately 200,000 Japanese civilians.

Civilians especially suffered in World War II as a result of combatants' total war strategies that disregarded collateral damage, including the new tactic of fire-bombing cities. When bombing urban areas, aircrafts carried both explosive and incendiary bombs, and this deadly combination created uncontrollable fires capable of burning down entire neighborhoods or cities. Between 1940 and 1941, Germany launched the **Blitz** against England. The Blitz targeted industrial centers and large cities, especially London, destroying British property and morale. An estimated forty thousand British civilians died during the Blitz. In the later stages of the war, the United States and Britain regularly fire-bombed Germany, killing approximately 450,000 civilians. The United States also led a fire-bombing campaign against Japan, and the bombing of Tokyo was the most destructive air attack in human history. Overall, American fire-bombing killed approximately 500,000 civilians.

Acts of Genocide and Ethnic Violence

A number of mass atrocities occurred during the twentieth century. During World War I, the Ottoman Empire committed the **Armenian Genocide**. The Ottoman Empire had persecuted Christian Armenians for centuries, but the ethnic violence intensified when Islamic extremists rose to power. Following a series of mass deportations and property confiscations, Armenians were sent to concentration camps where they were denied basic necessities. The mass killing began in 1915 and continued through 1923 with an estimated death toll of 1.5 million Armenians.

The rise of totalitarian governments led by extremist groups increased the frequency and intensity of mass atrocities. When the Soviet Union forcefully took over Ukrainian farmland in an effort to implement communism, the result was a widespread famine called the **Holodomor**. At least 3.9 million Ukrainians died of starvation, and sixteen countries identify this event as a genocide. In addition, the Soviet Union regularly purged ethnic communities, and millions of dissidents died in forced-labor camps (**gulags**).

During the **Great Purge** of 1937 and 1938, Stalin ordered the killing of approximately 750,000 landowners, government officials, and alleged traitors. Fascist Italy committed a variety of war crimes during its imperial conquest of Ethiopia (1935–1937). More than 300,000 Ethiopian civilians died in massacres, poison gas attacks, and concentration camps. Imperial Japan committed extensive war crimes during the Second Sino-Japanese War. Chinese, Korean, and Southeast Asian civilians were killed in massacres, chemical attacks, labor camps, human experiments, and manmade famines. The most infamous mass atrocity is referred to as the **Rape of Nanjing**. After conquering Nanjing, Japanese troops killed an estimated 200,000 Chinese civilians in a month's time.

Nazi Germany committed similarly unthinkable war crimes throughout World War II. Massacres of civilian populations and prisoners of war were commonplace. In the Soviet Union alone, the Nazis killed approximately seven million civilians and three million prisoners of war. Additionally, Adolf Hitler scapegoated Jews for all of Germany's ills, and the persecution of Jews steadily increased throughout his time in power, culminating in the **Holocaust** (1941–1945). Persecution began with boycotting Jewish businesses and segregating Jewish neighborhoods during the 1930s. Following these measures, Jews were rounded up and sent to concentration camps in massive numbers by 1940. Initially, concentration camps were used as a source of coerced labor, but Hitler later adopted what he referred to as the "Final Solution to the Jewish Question." Hitler's "Final Solution" sought to exterminate Jews, and the Nazis began constructing death camps in the early 1940s. Overall, an estimated six million Jews died in the Holocaust.

Numerous mass atrocities occurred after World War II as well. The Indonesian government purged communists from 1965 to 1966, and death squads murdered between 500,000 and three million civilians. The United States and Britain provided the Indonesian government with material support and military training. The **Khmer Rouge**'s communist agrarian policies in Cambodia killed an estimated two million people, which accounted for twenty-five percent of the country's population. The vast majority of victims died from disease and starvation in forced agricultural labor camps, but mass executions in the **Killing Fields** were also common. The **Rwandan Civil War** escalated into a genocide in 1994 when Hutu government officials and militias killed approximately 750,000 Tutsi people.

Emergence of a Cold War

World War II radically transformed the global political order. With the rest of the world still recovering from the conflict's widespread destruction, the United States and Soviet Union emerged as undisputed global superpowers. The World War II alliance between the United States and the Soviet Union would not be permanent. The United States championed democratic values and emphasized individual liberties, such as freedom of expression, freedom of the press, and religious freedom. In contrast, the Soviet Union was authoritarian, and the leader of the Communist Party exercised unilateral control over all political matters.

In addition, the countries had opposing economic systems. The United States was capitalist, while the Soviet Union was communist. **Capitalism** protects private property rights, allows the free market to set prices through supply and demand, and promotes economic competition based on the incentive to maximize profits. **Communism** rejects the concept of private property, empowers a central government to set prices through controlling the supply, and shares the resulting profits among the public. Both the United States and Soviet Union attempted to spread their ideologies across the globe during the post-World War II era, resulting in the **Cold War**.

Almost immediately after World War II, Europe was divided by the metaphorical **Iron Curtain** separating Eastern and Western Europe. Eastern Europe consisted of satellite states under Soviet control, while Western Europe allied itself with the United States. In addition, the Allies had divided Germany into four zones controlled by Britain, France, the United States, and the Soviet Union. The portions of Germany under British, French, and American control were collectively referred to as West Germany, and the Soviets controlled East Germany.

Berlin was divided in the same way, but the city's location in East Germany created logistical problems for the Western Allies. For example, the Soviet Union blockaded Berlin in 1948, and the Western Allies responded by sending supplies into Berlin via airlift. The Soviet Union lifted the blockade less than a year later, but Germany continued to be a source of geopolitical tension throughout the Cold War. Similar tensions simmered around the world as the United States and Soviet Union attempted to disseminate their ideologies across the Middle East, Africa, Asia, and Latin America. Those efforts ranged from instigating regime change to arming insurgent groups. However, the Soviet Union and United States refrained from directly waging war against each other due to the fear of sparking a nuclear war, which is why this conflict is known as the Cold War.

Protection of Sovereignty

As the United States and Soviet Union sought to spread their ideologies around the world, several countries desired to protect their sovereignty. Yugoslavian President **Josip Broz Tito** helped found the **Non-Aligned Movement** in 1961. The Non-Aligned Movement sought to build an international coalition to serve as an alternative to the global political and economic order. Along with promoting economic cooperation, the Non-Aligned Movement vowed to protect its members from foreign interference and aggression. Membership was open to any state that wasn't formally allied to either the United States or Soviet Union. Yugoslavia, India, Egypt, and Cuba played important leadership roles, and nearly one hundred countries joined the Non-Aligned Movement during the Cold War.

More than fifty African countries joined the Non-Aligned Movement, and those countries sought to further strengthen their ties through the **Pan-Africanism movement**. The primary purpose of the Pan-Africanism movement was to collectively oppose European imperialism and recreate the continent's economic, political, and social order by placing control in the hands of Africans. Pan-Africanism led to the formation of the **Organization of African Unity (OAU)** in 1963. The OAU promoted greater economic and political cooperation between member states, and forty-six African nations joined the organization between 1963 and 1975. Egyptian President **Gamal Abdel Nasser** attempted to form a similar Pan-Arabism movement in the 1960s, but the movement failed after Israel defeated an Arab coalition in the **Six-Day War** of 1967.

Military Alliances and Proxy Wars

The United States had developed and deployed nuclear weapons during World War II, and the Soviet Union had successfully developed nuclear weapons in 1949. Beginning in the early 1950s, both superpowers entered into an arms race, and they combined to produce more than 100,000 nuclear weapons during the Cold War. Under the military doctrine of **mutual assured destruction (MAD)**, both countries stockpiled nuclear weapons to deter the other from utilizing theirs. The premise of MAD was that if one country attacked with a nuclear weapon, then the other would retaliate, and both would end up being destroyed.

Although MAD reduced the likelihood of an intentional nuclear exchange, such an exchange nearly occurred during the **Cuban Missile Crisis** of 1962. In response to the United States' nuclear weapons in Turkey, the Soviet Union started constructing nuclear missiles in Cuba. Several high-ranking generals advised **President John F. Kennedy** to launch a nuclear attack against the Soviet Union, but negotiations ultimately prevailed. The United States agreed to remove its missile from Turkey, while the Soviet Union did the same in Cuba.

To amplify their military power, the United States and Soviet Union both entered into new military alliances. The United States, Canada, Iceland, and nine Western European countries formed the **North Atlantic Treaty Organization (NATO)** in 1949. NATO contained a provision that obligated all members to assist any member that came under attack, but it was never triggered during the Cold War. In response to NATO, the Soviet Union and its seven satellite states entered into the **Warsaw Pact** in 1955. While the seven satellite states were already under the Soviet Union's control, the Warsaw Pact provided a greater degree of military integration among the member states. The members of the Warsaw Pact never directly waged war against NATO or the United States, but they did intervene in satellite states to uphold communist rule, most notably in Czechoslovakia in 1968.

Although the United States and Soviet Union never directly entered into a military conflict, the superpowers fought a series of proxy wars as they competed for global influence. **Proxy wars** were conflicts between other opposing countries, each receiving material support from either the United States or the Soviet Union. There were several dozen proxy wars, including the **Korean War**, **Vietnam War**, and the **Soviet-Afghan War**.

After communist North Korea invaded South Korea in 1950, the United States led a UN task force to defend South Korea. The Soviet Union provided North Korea with weapons, but the United States successfully repelled the invasion. Following Chinese intervention on behalf of North Korea, the conflict turned into a stalemate, and the combatants agreed to a ceasefire in 1953.

When North Vietnamese communists challenged American and French imperial control over South Vietnam during the 1950s, the United States began providing South Vietnam with military and financial support. The United States followed a foreign policy of **containment**, which entailed preventing the spread of communism. In 1964, North Vietnamese ships attacked American ones in what is known as the **Gulf of Tonkin** incident. After this, the United States increased its troops from 23,000 to nearly 200,000. The Soviet Union and China functioned as proxies, providing the North Vietnamese with weapons and supplies. Despite achieving significant military victories, the United States was forced to withdraw from Vietnam in 1973 due to lack of domestic support for the war. North Vietnam conquered South Vietnam in 1975, effectively uniting the country under communism.

The Soviet-Afghan War began after the Soviet Union orchestrated a coup and sent troops to consolidate communist control over Afghanistan in 1979. The Soviets faced fierce resistance from guerilla fighters known as the **mujahideen**, and Islamic extremists answered the mujahideen's calls to wage a holy war against the Soviets. The United States provided weapons and training to the mujahideen, and some of those fighters would later launch terrorist attacks against the United States, such as **Osama bin Laden** and **Khalid Sheikh Mohammed**. The Soviet Union was ultimately unable to suppress the mujahideen and officially withdrew its forces from Afghanistan in 1989.

Global Economy from 1900 to Present

The global economy has undergone considerable change between the start of the twentieth century and the present day. During the first half of the twentieth century, the United States, Europe, and Japan dominated the global market. These industrial economies were largely based on manufacturing and steel production. Imperialism continued to play a significant role in the global economy. The United States, Europe, and Japan fueled their industrialization through the extraction of their colonies' raw resources, such as cotton, gold, petroleum, and rubber.

Two major changes occurred in the aftermath of World War II. First, international trade dramatically increased after countries entered into more expansive free trade agreements, including the **General Agreement on Tariffs and Trade** (1947). Greater global integration heavily contributed to the post-war economic boom, which resulted in global economic production increasing by more than five times.

Second, decolonization led to the creation of more independent countries, and major world powers were no longer able to unilaterally dictate economic policies in foreign lands for their own benefit.

As newly independent countries pursued development, they often sought to secure funding from foreign governments, multinational corporations, and supranational organizations. In return for providing the capital, the United States and Europe continued to enjoy access to the developing world's cheap labor and natural resources. Developed countries sought to exploit this cheap source of labor, and they relocated their manufacturing base to Asia and Latin America during the latter half of the twentieth century.

The most recent change to global economies came from the **Information Age of the early twenty-first century**. Computers and automation increased economic productivity, and they enabled the global economy to become even more interconnected through the adoption of communication technologies such as the Internet. Consequently, capital, labor, and technology could be moved anywhere in the world with unprecedented speed and efficiency.

Free-Market Economic Policies and Economic Liberalization in the Twentieth Century

The Soviet Union's collapse was perceived to be a decisive victory for capitalism and free-market economic policies, so many countries entered into free trade agreements during the late twentieth century. The **European Union (EU)** continued to expand in the late 1990s and early 2000s, and it established the **European Economic Area** in 1994, creating a free market for goods, services, labor, and capital for all member states. Similarly, the United States, Canada, and Mexico signed the **North American Free Trade Agreement** to remove trade and investment barriers throughout North America. The **Association of South East Asian Nations (ASEAN)** created the **ASEAN Free Trade Area** in 1992, which removed trade barriers and sought to attract more foreign investments.

The creation of the **World Trade Organization (WTO)** in 1995 further supported international free trade. The WTO is an intergovernmental organization that provides resources for countries seeking to enter free trade agreements or resolve international trade disputes. More than one hundred countries joined the WTO at its founding, and nearly every country in the world has become a member since then.

Knowledge Economies and Industrial Production in the Late Twentieth Century

Groundbreaking information and communication technologies had a profound economic effect during the late twentieth century. Computing power increased during the latter half of the twentieth century, and by the 1990s, most schools and businesses had incorporated computers. Similar advancements were made in automation, which revolutionized manufacturing by further cutting the cost of production. The arrival of the Internet and cell phones in the 1990s allowed people and businesses to instantaneously communicate with anyone in the world. All of these changes led to the Information Age of the twenty-first century. The Information Age spawned a knowledge economy that prioritized skilled workers who provided high-end services, incorporating the latest technological innovations.

As the knowledge economy supplanted the industrial economy in the early twenty-first century, Europe and the United States relocated much of their industrial production to Asia and Latin America. This transition was facilitated through free trade agreements, which removed tariffs, provided access to labor markets, and incentivized greater foreign investment in the developing world. In effect, companies could reduce costs by moving manufacturing production to Asia and Latin America where there would be cheaper labor and less environmental protection. Because free trade agreements remove tariffs, the resulting products can then be shipped back to the home country with minimal economic cost. In return for opening their labor markets, Asian and Latin American countries benefit from foreign companies building commercial infrastructure that boosts employment. However, the relocation of manufacturing has proven controversial. The loss of manufacturing jobs has decimated blue-collar communities in the United States and Europe, and labor conditions in Asian and Latin American countries have been criticized for failing to protect workers from abusive practices and dangerous working conditions.

Free-Market Economic Principles

The free-market economic principles of profit incentive and competition have spread across the global marketplace throughout the late twentieth and early twenty-first centuries. Multinational corporations have prioritized the maximization of profits and shareholders' wealth above all else. In addition, they compete for a global workforce and customer base. Free trade agreements have facilitated economic development by removing barriers to the movement of labor, capital, goods, and services. Governments' economic institutions often explicitly negotiate free trade agreements to benefit specific multinational corporations. Labor unions were one of the most powerful economic institutions in the twentieth century, but their power has markedly declined since the end of the Cold War. Organized labor unions increase business costs, so many governments have enacted pro-business policies to attract more investments.

Social Changes of the Twentieth Century

As the political and social landscape transformed, so did arts and culture. **Modernism** became the dominant art movement in the early twentieth century. Industrialization, urbanization, and the world wars had been wildly disruptive, and artists rejected the realist movement's traditionalism and certainty about the nature of society. Modernism utilized techniques such as cubism and surrealism, incorporated multiple viewpoints, including the subconscious, and often championed anti-war and decolonization movements. The modernist movement also extended into literature and architecture. Modernist architecture innovated with new materials, such as steel and reinforced concrete, and massive high-rise skyscrapers became the definitive feature of cityscapes. Modernist literature similarly experimented with new techniques, such as irony and stream-of-consciousness.

During the latter half of the twentieth century, popular and consumer culture became much more prevalent and global in nature. **Popular culture** refers to the dominating trends within a society at a given time, including anything from entertainment to technology. **Consumer culture** concerns patterns of spending on material goods. As such, popular and consumer culture are intimately related in capitalist economies because popular material goods frequently become ubiquitous throughout society. This relationship deepened in the post-Cold War era due to the rise of global capitalism.

Arts, Entertainment, and Popular Culture

Global integration heavily influenced the development of arts, entertainment, and popular culture in the late twentieth and early twenty-first centuries. Technological innovations, especially television and the Internet, facilitated Increased cultural exchange. **Postmodernism** became the dominant art movement at the end of the twentieth century, and several variants utilized the latest digital technology. As such, postmodern art enjoyed unprecedented reach in terms of how fast and far it could be disseminated. Entertainment and popular culture leveraged technology in similar ways, and the United States particularly benefited from this relationship. Throughout the Cold War, the United States had promoted its art, entertainment, and culture to showcase the virtues of capitalism and liberal society. For example, the CIA secretly financed abstract expressionist artists and organized art shows across Europe to expand the United States' cultural reach.

Following the end of the Cold War, the United States was the lone remaining superpower, and it enjoyed an outsized role in global capitalism. As such, American popular and consumer culture played a dominant role in the newly globalized society. American fast food corporations, movies, television shows, sports, and musicians all achieved a global presence. For example, in 2001, the American film industry accounted for sixty-five percent of the European Union's film consumption. More recently, other countries have broken American hegemony over popular cultural exports. One notable example is Korean pop music (K-pop), which has reached a global audience over the Internet. Similarly, British television programs have captured a sizable global share of television viewers.

Consumer Culture and Global Audience

Multinational corporations have extended consumer culture to a wider audience and have frequently transmitted global advertising campaigns through television and the Internet. For example, McDonald's, an American fast food chain, currently has locations in more than one hundred countries, and its signature Big Mac is sometimes used to evaluate purchasing power parity around the world. Considerable cultural synthesis has also occurred, such as India's Bollywood film industry that produces Americanized movies in the Hindi-language. Other leading capitalist countries have similarly contributed to the newly global consumer culture, such as Britain, France, Germany, Italy, and Japan. As a semi-capitalist rising superpower, China has one of the largest consumer markets in the world, trailing only the United States. However, Chinese consumer culture struggles to transcend national borders due to government censorship.

Advances in Communication, Transportation, Industry, Agriculture, and Medicine

Groundbreaking scientific and technological innovations revolutionized countless fields during the twentieth century. Scientists' discovery of electromagnetic and radio waves paved the way to new forms of communication, including the radio and cell phone. Advances in electromagnetic radiation also allowed doctors to more accurately identify injuries through X-rays. Similarly, the invention of integrated electronic circuits, transistors, and microprocessors led to the development of computers, and the combination of computers and network protocols birthed the Internet. Computing power increased industrial productivity by automating complex processes. The combination of assembly lines and interchangeable parts gave rise to the mass production of cars and industrial products. An improved understanding of air resistance and applied mathematics achieved breakthroughs in aerodynamics.

Scientific discoveries related to plant genetics and breeding resulted in high-yield crops, and advances in chemistry greatly enhanced agrochemicals, such as fertilizers and pesticides. Geological and chemical innovations led to more effective drilling and distillation processes for petroleum. Petroleum and petroleum products, like plastics, amplified industrial output. Greater knowledge of bacteria and viruses helped scientists create antibiotics and vaccines. Furthermore, surgery became safer after doctors discovered techniques for fighting infections, sterilizing conditions, and conducting blood transfusions based on blood group systems.

Canadian History

Precolonial Canada

Aboriginal people lived in Canada for many centuries before the arrival of Europeans. Archaeologists believe that the first Aboriginal peoples traveled from Siberia to present-day Alaska across the Bering land bridge between 20,000 and 12,000 years ago. Small nomadic groups lived in Alaska until the glaciers melted; then, Aboriginal peoples migrated to and settled in present-day Canada. These early migrants relied on hunting big game, such as giant beaver and wooly mammoths, to survive the frigid conditions. After the climate warmed approximately 10,0000 years ago, Aboriginal peoples developed diverse cultural traditions and formed groups that would eventually develop into the First Nations.

Precolonial First Nations shared many similarities. Perhaps most importantly, the First Nations domesticated plants and animals, which paved the way for stable agricultural production. In turn, agriculture facilitated the creation of permanent settlements because Aboriginal societies were no longer completely dependent on tracking big game for survival. Furthermore, agriculture-based settlements led to a more specialized division of labor and supported the development of advanced trading networks. Aboriginal traders regularly exchanged surplus crops, tools, and natural resources over substantial distances. In addition to providing economic benefits, trade networks also enabled the diffusion of technology and culture over vast territory.

As permanent settlements stabilized and trade networks expanded, Aboriginal cultures grew increasingly complex. Aboriginal societies typically emphasized the importance of multi-generational family units and granted elder members much influence. Languages similarly became more nuanced and connected to the local culture. Likewise, Aboriginal societies adopted more elaborate religious traditions and ceremonies, incorporating both anthropomorphic and animist beliefs. Anthropomorphic religious beliefs involve attributing human characteristics to a divine entity, while animistic religious beliefs see divinity in all earthly beings.

Despite economic and cultural similarities, various precolonial First Nations still differed widely based on region. First Nations in the Pacific Northwest, such as the Squamish and Haida, hunted in the temperate rainforest of British Columbia, fished for salmon in the Pacific Ocean, and built relatively large-scale settlements featuring multi-generational homes. The Great Plains of Alberta and Saskatchewan provided First Nations with plentiful bison, so these groups tended to be more nomadic. The Blackfeet, named after the moccasins they wore, dominated the Great Plains for centuries due to a complex alliance system consisting of kinship networks called bands. Similarly, the Iroquois consolidated power in present-day Ontario and Quebec by cultivating the Three Sisters—maize, beans, and squash—to support large-scale settlement and facilitate far-ranging military campaigns.

The Aboriginal peoples living in the Arctic were distinct from the First Nations and are referred to as Inuit. The harsh Arctic climate couldn't support agriculture, so, unlike the First Nations, the Inuit consumed minimal plant-based food and established relatively small-scale settlements. The Inuit relied almost exclusively on hunting, fishing, and whaling. In order to navigate the unforgiving land and sea, the Inuit trained teams of huskies to serve as pack animals and invented highly buoyant vessels called kayaks.

French and British Rule

The colonization of present-day Canada began as an unintended consequence of French interest in finding a northwest passage to Asia. Jacques Cartier did not find the trade route, but he did lay claim to the Gaspé Peninsula on behalf of King Francis I in 1534, and the land mass was christened "Canada." Cartier failed to establish a permanent colony, but in 1604, new French settlers tried again more successfully. A geographer named Samuel de Champlain established Quebec City in 1608, which quickly became France's most stable and successful permanent settlement in North America.

As French settlements, fishing operations, and fur trapping expeditions advanced under Champlain's leadership, the region was renamed New France, with Quebec City as its capital. During the 1630s, the

Roman Catholic Church and Jesuit missionaries assumed control over New France and began converting the Aboriginal peoples. When those efforts failed, the French government intervened to stabilize the settlements, attract immigrants, and generate consistent economic growth. While fishing and fur trapping remained lucrative enterprises, New France struggled to attract immigrants due to the harsh conditions, lack of infrastructure, and hostile relations with the Iroquois. In contrast, Britain enjoyed tremendous success in populating the New England colonies and settlements in Nova Scotia and Newfoundland.

During the late 17th century, the rivalry between France and Britain sparked numerous conflicts in North America as the colonists competed for resources. Between 1688 and 1763, French and British colonists fought six different conflicts with the help of their respective Aboriginal allies. Britain formed a powerful alliance with the Iroquois, while the French allied with a variety of tribes, including the Wabanaki Confederacy, Algonquin, Ottawa, and Shawnee. The Seven Years' War (1754–1763) triggered the final showdown between the imperial rivals, and Britain soundly defeated France. As a result, Britain acquired all of New France, which they renamed Canada.

Britain initially struggled to exercise control over French Canadians, especially in Quebec. To soothe the hostile relations, Britain passed the Quebec Act (1774) to guarantee colonists' rights to speak French, practice Catholicism, and enforce the French Civil Code. This proved critical because the Americans attempted to exploit societal divisions in Quebec during the American Revolutionary War (1776–1783) and the War of 1812. Britain ultimately repelled both American invasions, but the peace was relatively fleeting.

Combined with the rivalry between Francophone and Anglophone colonists, undemocratic governance sparked the Rebellions of 1837–1838. Britain put down the insurrection, and Governor General Lord Durham issued a report that a political union would be necessary to assimilate French Canadians. Soon thereafter, the British Parliament enacted the Act of Union (1840) to divide the colony into Upper Canada and Lower Canada, with equal representation in the Legislative Assembly. This attempt at reconciliation satisfied neither party, and calls for a more cohesive and autonomous union mounted.

Confederation

In 1864, representatives from the British colonies met to come up with a unification agreement. After the Quebec and Charlottetown conferences, the gathered politicians set forth Seventy-Two Resolutions, which outlined the creation of a unified federal system of government. Francophone Canadians hoped this framework would create a more autonomous Quebec, while Anglophone Canadians favored the creation of a united Canada to end the disproportionate representation of Francophones in the Legislative Assembly. After the London Conference of 1866, the British Parliament enacted the British North America Act, otherwise called Constitution Act, 1867. This act united the previously distinct colonies of New Brunswick, Nova Scotia, and Canada, with the latter being split into the provinces of Ontario and Quebec.

Unlike other British colonies, the Dominion of Canada was a self-governing confederation, and it mirrored the American model in its federalized power-sharing arrangement. As such, provinces enjoyed considerable authority and powers protected from federal overreach, such as exclusive control over public lands and the right to establish a legal system. The federal government was modeled after the British government, and it consisted of a bicameral legislature, a justice system, and a taxation system. Despite the unprecedented degree of autonomy granted to the dominion, it was still legally subject to the British Empire's authority.

Canada's reorganization into a united confederation triggered rapid territorial expansion. British Columbia joined the confederation in 1871 after the Canadian Parliament promised to build the Canadian Pacific Railway, a transcontinental railroad. Two years later, Prince Edward Island joined the confederation, largely because it wanted access to Canada's expanding markets. The Canadian Pacific Railway was completed in 1885, and it connected Eastern Canada and British Columbia for the first time. The CPR played a critical role in driving the development of the Canadian frontier because it functioned as the primary means of transporting passengers, goods, and materials into Western Canada. As a result, the Western provinces experienced rapid population growth as Eastern Canadians and immigrants streamed into the region in search of land, gold, and economic opportunities.

Since the confederation consisted of both pre-existing colonies and new provinces, the federal government sought to unify its patchwork legal system. The Parliament of Canada passed the Criminal Code, 1892 to standardize and simplify the criminal justice system. For example, it removed all British criminal laws unless a law was expressly intended to apply to British colonies, dominions, or possessions. The Indian Act of 1876 attempted to clarify the relationship between Aboriginal peoples and the federal government. In general, the Indian Act outlined the laws for reservations and forced some Aboriginal peoples to enfranchise for a variety of reasons, which meant that they lost their Indian Status. First Nations mostly condemned this broad statute for its attempt to coercively assimilate Aboriginal peoples into Canadian society, which led to widespread persecution and cultural suppression.

Canada and the World Wars

The World Wars radically transformed Canada from an imperial outpost to an independent world power. At the outset of World War I (1914–1918), Canada was a British Dominion, and the British Parliament held significant control over Canadian policymaking, especially in terms of foreign policy. Thus, once the United Kingdom declared war on Germany on August 4, 1914, Canada was legally obligated to support the war effort, though the Canadian Parliament enjoyed some autonomy over its wartime commitments and deployments.

Canada successfully created a new army called the Canadian Expeditionary Force that mobilized 630,000 soldiers during World War I. The Expeditionary Force's assault teams gained international recognition for their bravery and ferocity on the Western Front, especially in the crucial Allied victories at the Battles of the Somme and Vimy Ridge. Additionally, the Canadian Royal Navy supported various British blockades and naval campaigns against German U-boats. At the tail end of the war, the Expeditionary Force defeated nearly 25% of Germany's heavy divisions on the Western Front during a series of battles referred to as "Canada's Hundred Days." Canada's commitment came at a heavy price with more than 61,000 deaths.

Unlike in World War I, Canada was not legally obligated to fight on behalf of the United Kingdom in World War II because the Statute of Westminster (1931) had granted Canada near-total autonomy. However, Canada still retained significant links to Britain and followed with its own declaration of war on Germany in 1939 after waiting one week to symbolically emphasize its independence.

Given the nation's staggering sacrifice in World War I, the Canadian Parliament initially restricted its role to supplying the Allied Powers with food and weapons. However, as the conflict escalated, Canada radically increased its commitment of troops. Overall, 1.1 million Canadians served in the military during World War II. Canadian forces fought extensively in both the Atlantic and Pacific theaters, and more than 45,000 Canadians lost their lives with another 55,000 wounded in combat. By the end of the war,

Canada had established itself as one of the largest and most formidable militaries in the world, particularly in terms of its air force and navy.

The Canadian economy also advanced rapidly as a result of wartime production. Prior to World War I, the Canadian economy was primarily agrarian, but the wartime manufacturing boom led to the development of robust automobile and mining industries. Likewise, the wars would forever alter Canadian society. Women capitalized on the newly available economic opportunities and contributed indispensable labor to the wartime mobilization. As a result, white and black women gained full suffrage in 1922 everywhere except for Quebec, which followed suit in 1940. The World Wars also exacerbated divisions between Canada's Anglophone and Francophone citizens. English-speaking Canadians vocally supported both war efforts, largely out of loyalty to Britain. In contrast, French Canadians widely condemned the wars as imperialistic, and several anti-conscription riots occurred in Quebec in 1917, 1918, and 1944.

Postwar Era

Canada emerged from the Second World War as a major world power. Although Canada continued to maintain close ties to the United Kingdom, Canada's relationship to the United States most influenced its trajectory in the postwar era. Along with the Soviet Union, the United States attained global superpower status during this era. Given its geopolitical status as America's northern neighbor, Canada sought to maximize the benefit of having such proximity to the world's most productive economy and advanced military. As a result, Canada adopted capitalist policies domestically, supported the American-led international order, and provided material assistance to American and European interventions against communist threats during the Cold War (1947–1991). Additionally, Canada was a founding member of the North Atlantic Treaty Organization (NATO), an American-led military alliance formed for the explicit purpose of containing the Soviet Union.

Canada enjoyed significant economic growth for much of the postwar era, and the government invested some of its revenue into the creation of a robust, European-style social welfare system. Specifically, the Canadian Parliament passed legislation to guarantee universal healthcare, expand unemployment benefits, and provide pensions to specific groups like retirees and veterans. Canada also expanded its territory during this period. In 1949, Newfoundland held a contentious referendum and ultimately voted to become a Canadian province. Several years later, Canada enforced its territorial claims in the Arctic by relocating Inuit families under false pretenses to desolate areas further north. The Canadian government paid the survivors $10 million CAD in 1993 and officially apologized in 2010.

Two regional movements threatened Canadian national unity during the 1970s and 1980s. First, Francophone separatists advocated for the secession of Quebec, but 59% of Quebecois voters rejected secession in a 1980 referendum. Prime Minister Pierre Trudeau attempted to cool regional tensions by overseeing the passage of constitutional protections for bilingualism and multiculturalism. Second, tensions between the Western provinces and the federal government sparked a phenomenon known as "Western alienation." During the late 1970s, Alberta, British Columbia, Saskatchewan, and Manitoba alleged that they were excluded from federal policymaking, and this caused a strained relationship between the western provinces and the federal government. Alberta has historically been an especially harsh critic of the federal government's taxation policies and regulations of the province's highly lucrative oil and gas industries.

The Constitution Act, 1982 was another major development in the postwar era. Under the Statute of Westminster, 1931, the British Parliament still held the power to amend Canada's constitution. The

Constitution Act, 1982 removed this power and completed the patriation of Canadian constitutional law. The Act also explicitly protected rights of Canadian Aboriginals, established procedures for constitutional amendments, and enacted a bill of rights known as the Canadian Charter of Rights and Freedoms, which is binding on all levels of Canadian government. Quebec never ratified the Constitution Act, 1982, but the Supreme Court ruled that Quebec's consent was unnecessary for enactment.

Recent History

Canadian politics has seen several unprecedented events and shifts in recent years. In 1993, Kim Campbell became the first (and so far only) female prime minister as well as the first to be born in British Columbia. In 2005, Canadian Parliament enacted the Civil Marriage Act to legalize same-sex marriage in every province. In 2003, the rise of a conservative movement challenged historic liberal dominance. Following a prolonged period of division, the Canadian Alliance and Progressive Conservative Party of Canada united to form the Conservative Party of Canada. Under the leadership of Prime Minister Stephen Harper, the Conservative Party elected a minority government in 2006 and 2008 and a majority government in 2011. Pierre Trudeau's son, Justin, led the Liberals back into power in 2015 and has remained in leadership as of 2020.

Recent Canadian history has been marked by increased alignment with the United States. Canada signed the North American Free Trade Agreement (NAFTA) with the United States and Mexico in 1993. In general, NAFTA slashed tariffs and other trade barriers to boost economic productivity and increase investments. Although Canadian consumers have benefited from cheaper consumer goods and more white-collar jobs, NAFTA has been highly controversial due to its negative impact on Canadian manufacturing, labor rights, and the environment. Following the September 11th terrorist attacks, Canada joined the NATO mission to support the American invasion and occupation of Afghanistan. Furthermore, Canada has been a prominent purchaser of American weaponry and has invested heavily in the F-35 fighter jet program.

Canada has continued to struggle with national unity in recent history. Quebecois separatism regained steam in the late 1980s and early 1990s, culminating in a second referendum in 1995. By the extremely narrow margin of 50.6% to 49.4%, Quebecois voters rejected the proposal to secede from Canada and declare independence. Three years later, the Canadian Supreme Court barred provinces from attempting to secede unilaterally, and Canadian Parliament enacted the Clarity Act to establish procedures for provinces seeking independence. The disproportionate attention garnered by Quebec further fueled the anger of Western provinces, especially since Quebec has historically received the greatest share of equalization payments and transfer payments from the federal government. During the late 2010s, secession gained unprecedented momentum in some Western provinces, though it has never reached levels seen in Quebec.

Issues related to Canada's treatment of Aboriginal peoples have also challenged the present-day Canadian identity. Most government actions have centered on apologizing for historic abuses. Most notably, the Truth and Reconciliation Commission of Canada (2008–2015) held extensive hearings to publicize the devastating impact of Canada's Indian residential school system, which was an acculturation program coercively imposed on Aboriginal children for more than a century. High-level Canadian officials have repeatedly discussed the repatriation of Aboriginal land, total autonomy for First Nations, and/or ambitious poverty alleviation programs, though these proposals have been mostly stalled or thwarted.

Geography of Canada

Canada is one of the largest countries in the world, trailing only Russia in terms of total area under territorial control. To the south, Canada borders the contiguous United States, and it shares another land border with the United States (Alaska) in the northwest. Canada borders the Atlantic and Pacific Oceans and shares maritime borders with Greenland to the northeast and Saint Pierre and Miquelon to the southeast.

The physical geography of Canada is immense and incredibly diverse. The Appalachian Mountains and Rocky Mountains extend from the United States northward into Canada, and the Columbia and Coast mountain ranges lie to the west of the Canadian Rockies. In Northern Canada, a series of hills and mountains known as the Canadian Shield has incredibly rich mineral reserves. Other notable geographic features include prairies in central Canada, volcanoes in British Columbia, and Arctic tundra in Nunavut.

Canada also has an abundance of water resources. The Canadian maritime coastline is the longest in the world, and every Canadian province has a maritime border except for Alberta and Saskatchewan. Canada also has more than two million freshwater lakes, which cover approximately 9% of Canadian land. Canada borders four of the five Great Lakes (Erie, Huron, Ontario, and Superior), which form the largest interconnected group of freshwater lakes in the world by total area. In addition, Canada has several enormous drainage basins and numerous rivers, with the two largest being the Mackenzie and St. Lawrence Rivers.

Federal Government of Canada

Modeled after the British system of government, Canada is a Westminster-style parliamentary democracy and constitutional monarchy as established under the Constitution Act, 1867.

As a constitutional monarchy, the British Crown serves as the sovereign and head of state. The Crown's overriding goal is to uphold the principle of responsible government, meaning that government should reflect the will of the people. The monarch is the sole source of the Crown's authority, including the royal prerogative. In general, the royal prerogative involves powers related to the calling of elections, conducting foreign policy, and assenting to legislation. With advice from the Canadian prime minister, the British monarch appoints the Governor General of Canada, who acts as his/her representative.

The Parliament and Supreme Court of Canada make up the rest of the federal government. Canadian Parliament is bicameral, consisting of the Senate (upper house) and House of Commons (lower house). The governor general appoints senators on the advice of the prime minister, and members of the House of Commons are elected by the Canadian people. Following an election, the governor general appoints a prime minister to serve as the head of government. Prime ministers are typically the leader of the winning political party or governing coalition. The prime minister chairs the Cabinet of Canada and selects its ministers. Additionally, the prime minister appoints members to the Supreme Court of Canada, which is the highest level of the Canadian justice system. Supreme Court decisions are binding on all other Canadian courts unless overridden by legislation.

Economy of Canada

The Canadian economy is highly developed, with a free market based on capitalist economic principles, and it regularly ranks as one of the freest marketplaces in the world. Canada's gross domestic product (GDP) was $1.6 trillion in 2020, making it the ninth largest economy in the world. Compared to other highly developed countries, Canada has relatively low income inequality due to its high taxes and generous social welfare system.

The service industry accounts for approximately 70% of Canada's GDP and 75% of its total employment. Some of the largest sectors in Canada's service industry are retail, business services, and high technology. Education and health are also amongst the largest service sectors, but they are publicly owned and operated. In sharp contrast to many other highly developed countries, Canada has maintained robust manufacturing and steel production. Canada is also one of the largest producers and exporters of agricultural products, mineral resources, natural gas, and oil. Lastly, Canada is a world leader in green energy, with 59% of electric generation coming from hydroelectricity.

Canada benefits tremendously from international trade. It is a leading member of the World Trade Organization (WTO), and it has more than a dozen active free trade agreements with its trading partners. One of its most valuable free trade agreements is the Canada–United States–Mexico Agreement, which superseded NAFTA in 2020. While free trade agreements have generally driven economic growth, they have also been heavily criticized for their harmful effects on blue-collar employment, particularly in the manufacturing sector.

Geography

Types of Maps

Geographers utilize a variety of maps in their study of the spatial world. Projections are maps that represent the spherical globe on a flat surface. **Conformal projections** attempt to preserve shape but distort size and area. For example, the most well-known projection, the **Mercator projection**, drastically distorts the size of land areas at the poles. In this particular map, Antarctica, one of the smallest continents, appears massive, while the areas closer to the equator are depicted more accurately. Other projections attempt to lessen the amount of distortion; the **equal-area projection**, for example, attempts to accurately represent the size of landforms. However, equal-area projections alter the shapes and angles of landforms regardless of their positioning on the map. Other projections are hybrids of the two primary models. For example, the **Robinson projection** tries to balance form and area in order to create a more visually accurate representation of the spatial world. Despite the efforts to maintain consistency with shapes, projections cannot provide accurate representations of the Earth's surface due to their flat, two-dimensional nature. In this sense, projections are useful symbols of space, but they do not always provide the most accurate portrayal of reality. The following page displays images of these well-known conformal projections.

Conformal projections

Mercator projection

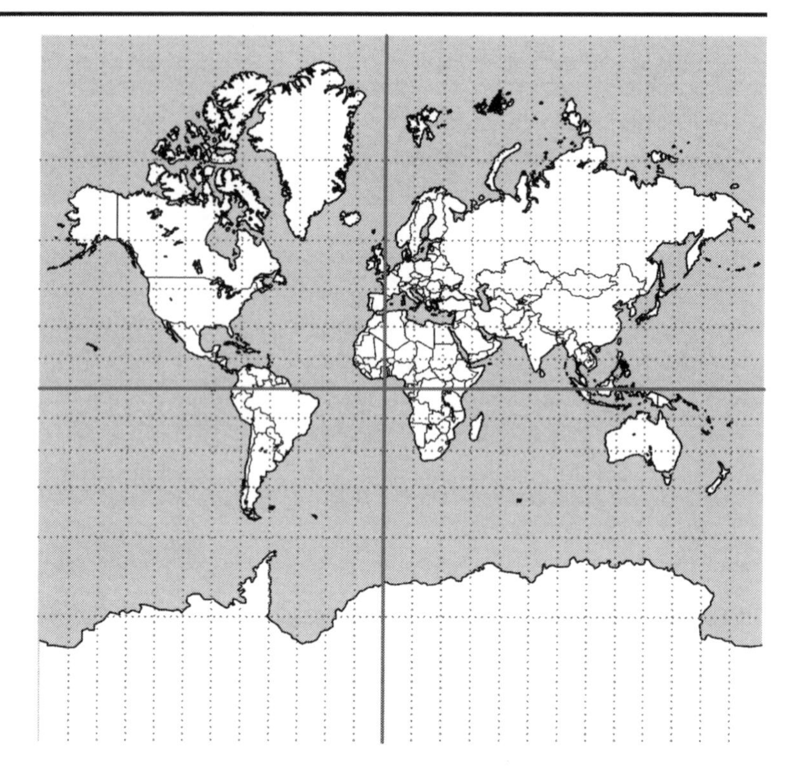

Equal area projection

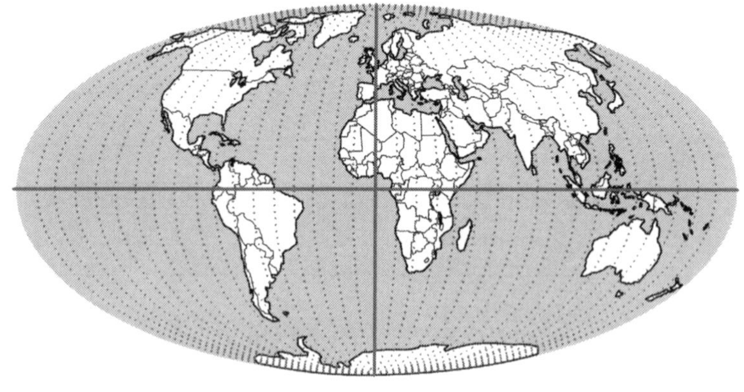

Robinson projection

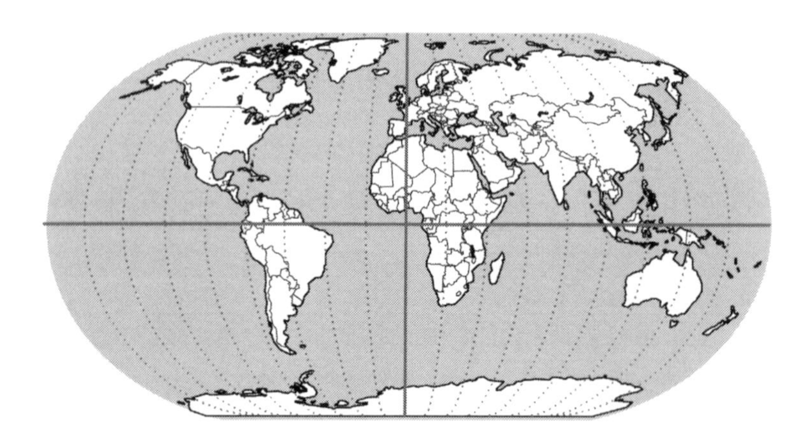

Unlike projections, **topographic maps** display contour lines, which represent the relative elevation of a particular place and are very useful for surveyors, engineers, and/or travelers. For example, hikers may refer to topographic maps to calculate their daily climbs. A section of a topographical map can be viewed below. Where the lines are closer together, the terrain is steeper, and when the lines are more spread out, the terrain is flatter.

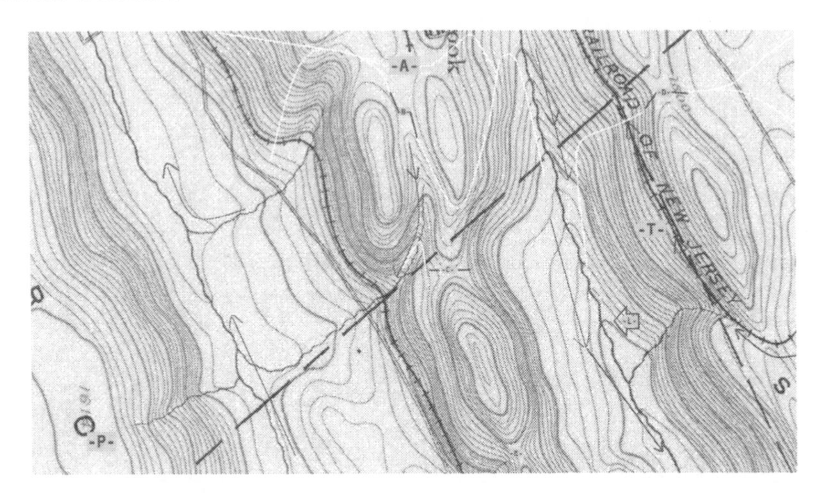

Similar to topographic maps, **isoline maps** are also useful for visualizing and differentiating between data. These maps use symbols to represent values and lines to connect points with the same value. For example, an isoline map could display average temperatures of a given area. The sections which share the same average temperature would be grouped together by lines. Additionally, isoline maps can help geographers study the world by generating questions. For example, is elevation the only reason for differences in temperature? If not, what other factors could cause the disparity between the values?

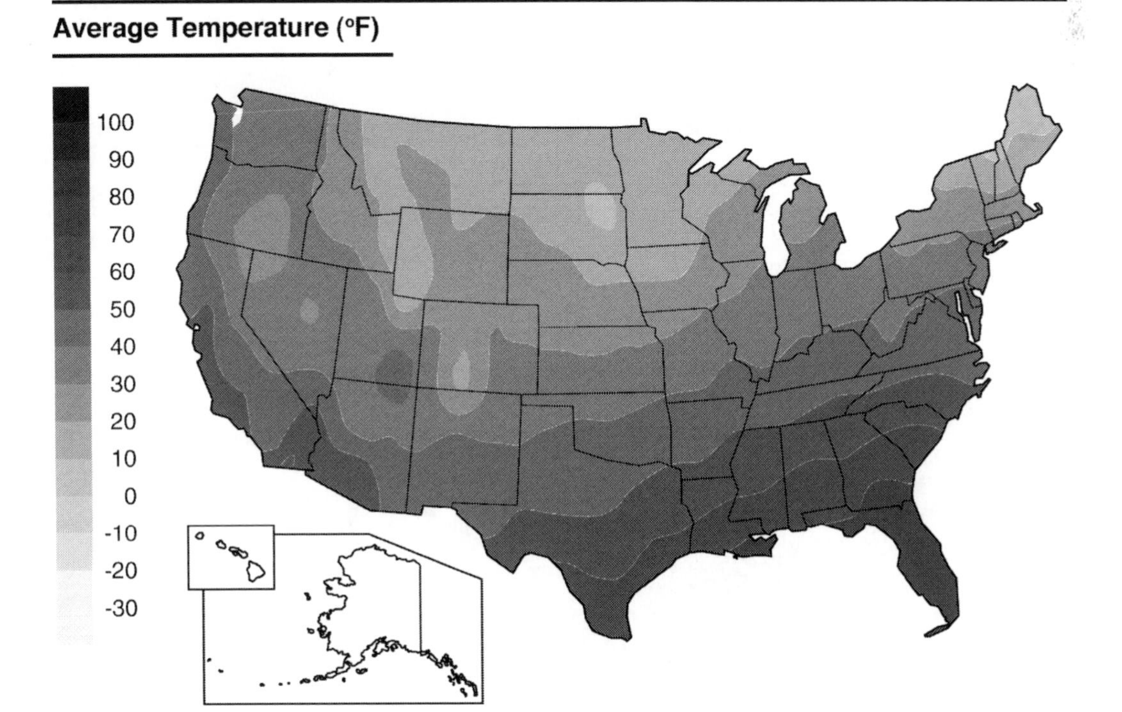

Thematic maps are also quite useful because they display the geographical distribution of complex political, physical, social, cultural, economic, or historical themes. For example, a thematic map could indicate an area's election results using a different color for each candidate. There are several different kinds of thematic maps, including *dot-density maps* and *flow-line maps*. A dot-density map uses dots to illustrate volume and density; these dots could represent a certain population, or the number of specific events that have taken place in an area. Flow-line maps utilize lines of varying thicknesses to illustrate the movement of goods, people, or even animals between two places. Thicker lines represent a greater number of moving elements, and thinner lines represent a smaller number. Below is a dot-density map depicting population density in an urban area.

Interpreting Maps

Geographical concepts are visually conveyed through maps. The map below illustrates some key points about geography.

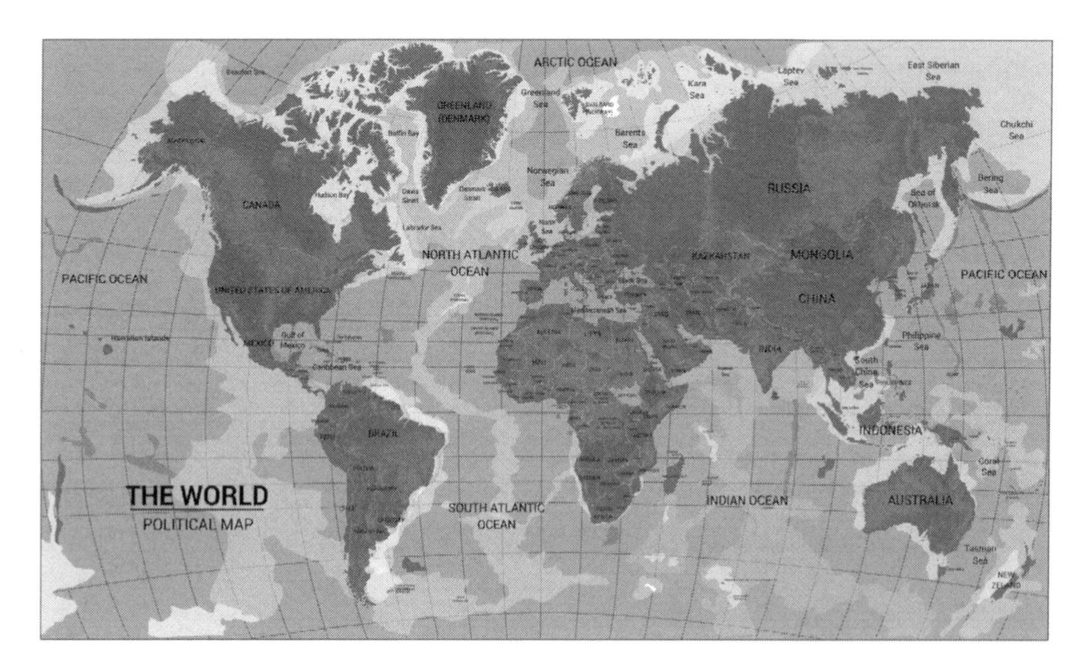

On some colored maps, the oceans, represented in blue between the continents, vary in coloration depending on depth. The differences demonstrate **bathymetry**, which is the study of the ocean floor's depth. Paler areas represent less depth, while darker spots reflect greater depth.

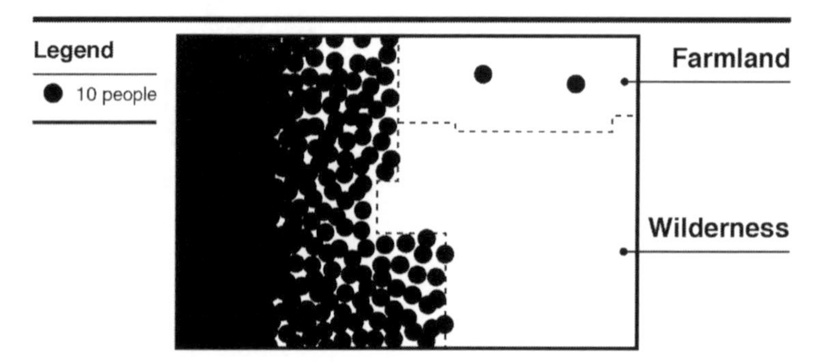

Maps may also display horizonal and vertical lines representing latitude and longitude. The horizontal lines, known as parallels, mark the calculated latitude of those locations and reveal how far north or south these areas are from the equator, which bisects the map horizontally. The vertical lines signify longitude, which determines how far east or west different regions are from each other. The lines of longitude, known as meridians, are also the basis for time zones, which determine the time for different regions. As one travels west between time zones, the given time moves backward accordingly. Conversely, if one travels east, the time moves forward.

There are two particularly significant longitudinal lines. First, the Prime [Greenwich] Meridian marks zero degrees in longitude, and thus determines the other lines. The line circles the globe and divides it into the Eastern and Western hemispheres. Second, the International Date Line represents the change between calendar days. By traveling westward across the International Date Line, a traveler would essentially leap forward a day. For example, a person departing from the United States on Sunday would arrive in Japan on Monday. By traveling eastward across the line, a traveler would go backward a day. For example, a person departing from China on Monday would arrive in Canada on Sunday.

The map below identifies the different time zones.

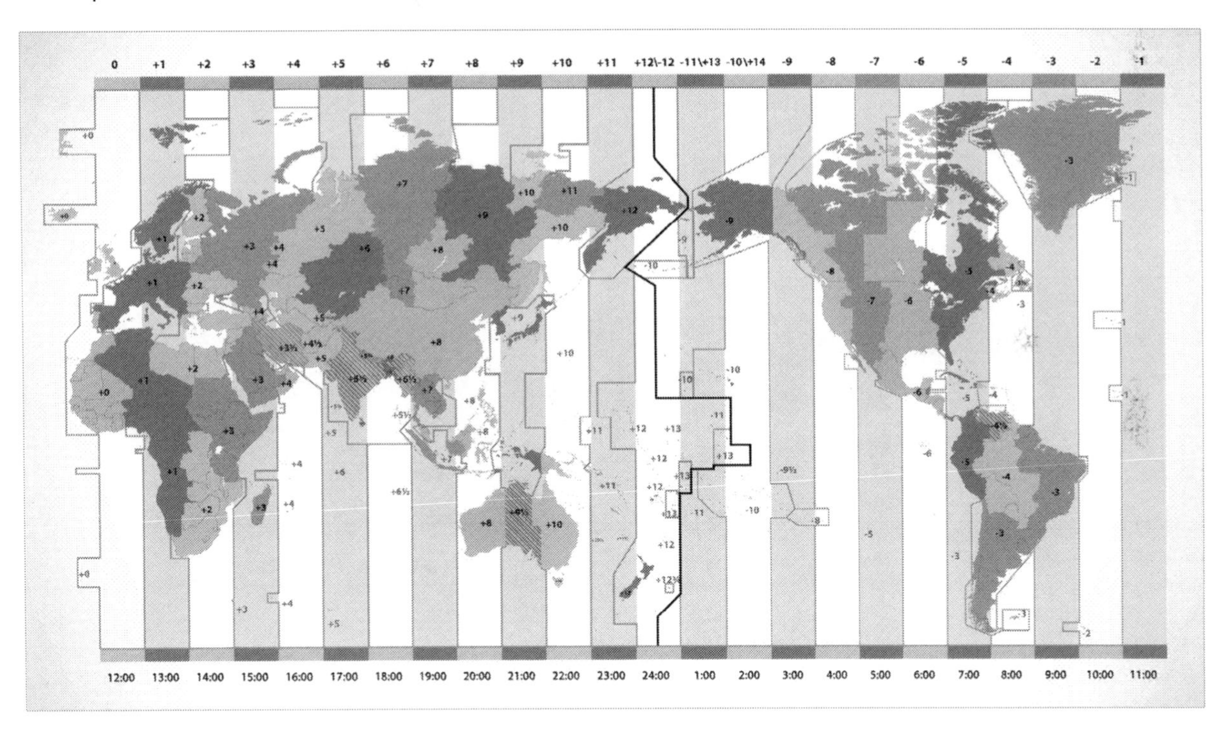

Although world maps are useful in showing the overall arrangement of continents and nations, it is also important at times to look more closely at individual countries because they have unique features that are only visible on more detailed maps.

For example, take the following map of Canada. The country is split into multiple provinces that have their own cultures and localized governments. Other countries are often split into various divisions, such as states, and while these features are ignored for the sake of clarity on larger maps, they are important when studying specific nations. Individual provinces can be further subdivided into counties and municipalities, which may have their own maps that can be examined for closer analysis.

Map of Present-Day Canada

Finally, one of the first steps in examining any map should be to locate its key or legend, which will explain what different symbols represent on the map. As these symbols can be arbitrary depending on the maker, a key will help to clarify the different meanings.

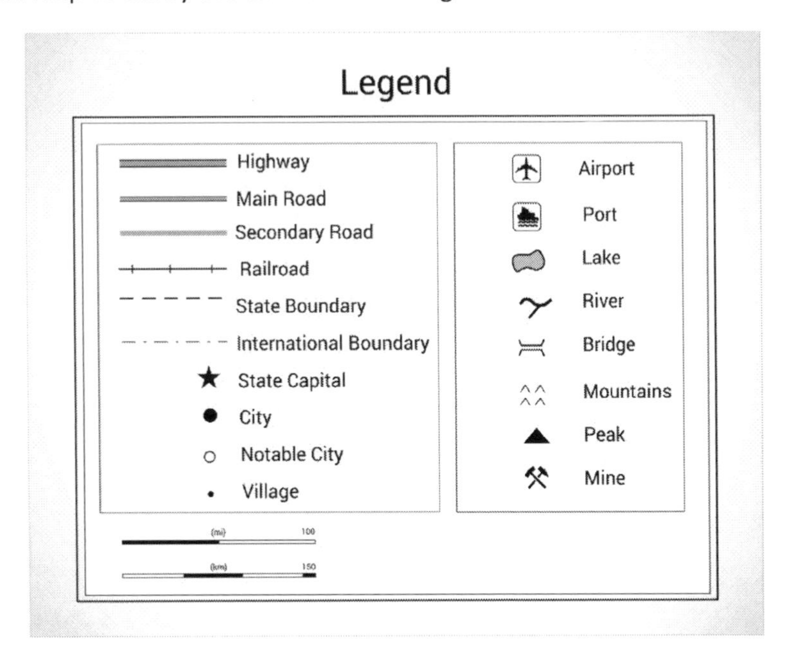

Using Geographic Concepts to Analyze Spatial Phenomena

Using Mental Maps to Organize Spatial Information

Mental maps are exactly what they sound like: maps that exist within someone's mind. The cognitive image of a particular place may differ from person to person, however. Furthermore, mental maps can enable people to travel from point A to point B efficiently. Someone may utilize their mental map to determine the best route on public transit, the least hilly bike path, or the roadways that have the least amount of traffic. Mental maps tend to be more informative when a person has had more experiences in that particular place.

Maps and Scale

Since maps represent a large area on a much smaller two-dimensional space, they must utilize a scale. Scale is simply the ratio of a distance on the ground to the corresponding distance on paper. Geographers and cartographers attempt to make the image on paper representative of the actual place. The United States Geological Survey (USGS) utilizes the mathematical ratio of 1/24,000 in all of its topographical maps. This scale means that one inch on the map is equivalent to 24,000 inches—or nearly two-thirds of a mile—on the ground. Large-scale maps represent a smaller area with greater detail, while small-scale maps are representative of much larger areas with less detail.

Factors Affecting Human Geography

Two primary realms exist within the study of geography. The first, **physical geography**, essentially correlates with the land, water, and foliage of the Earth. The second, **human geography**, is the study of the Earth's people and how they interact with their environment. Several geographical factors impact the human condition, such as access to natural resources. For example, human populations tend to be higher around more reliable sources of fresh water. The metropolitan area of New York City, which has abundant freshwater resources, is home to over 18 million people. Australia, on the other hand, an entire country and continent, has much less accessibility to fresh water and houses only 7 million more people. Although water is not the only factor in this disparity, it certainly plays a role in **population density**—the total number of people in a particular place divided by the total land area, usually in square miles or square kilometers. Australia's population density is about 7 people per square mile, while the most densely populated nation on Earth, Bangladesh, is home to 2,889 people per square mile.

Population density can have a devastating impact on both a physical environment and the humans who live within that environment. For example, Delhi, one of India's most populated cities, is home to nearly five million gasoline-powered vehicles. Each day, those vehicles emit an enormous amount of carbon monoxide into the atmosphere, which directly affects the Delhi citizens' quality of life. In fact, the smog and pollution problems have gotten so severe that many drivers cannot see fifty feet in front of them. Additionally, densely populated areas within third-world nations, or developing nations, struggle significantly in their quest to balance the demands of the modern economy with their nation's lack of infrastructure. For example, nearly as many automobiles operate every day in major American cities like New York and Los Angeles as they do in Delhi, but they create significantly less pollution due to cleaner burning engines, better fuels, and governmental emission regulations.

Although it is a significant factor, population density is not the only source of strain on a place's resources. Historical forces such as civil war, religious conflict, genocide, and government corruption can also profoundly alter the lives of a nation's citizens. For example, the war-torn nation of Somalia has not had a functioning government for nearly three decades. As a result, the nation's citizens have virtually no access to hospital care, vaccinations, or proper facilities for childbirth. Due to these and other factors, the nation's **infant mortality rate**, or the total number of child deaths per 1,000 live births, stands at 74/1000. When compared to Iceland's 2/1000, it's quite evident that Somalia struggles to provide basic services in the realm of childbirth and there is a dire need for humanitarian assistance.

Literacy rates, like infant mortality rates, are also excellent indicators of the relative level of development in a particular place. Many developing nations have both economic and social factors that hinder their ability to educate their own citizens. Due to radical religious factions within some nations like Afghanistan and Pakistan, girls are often denied the ability to attend school, which further reduces the nation's overall literacy rate. For example, girls in Afghanistan have a 24.2 percent literacy rate, one of the lowest rates of any record-keeping nation on Earth.

Although literacy rates are useful in determining a nation's development level, high literacy rates do exist within developing nations. For example, Suriname, which has a significantly lower GDP than Afghanistan, enjoys a 94 percent literacy rate among both sexes. Utilizing this and other data, geographers can form questions and conduct further research about such phenomena. Demographic data, such as population density, the infant mortality rate, and the literacy rate all provide insight into the characteristics of a particular place and help geographers better understand the spatial world.

Locating and Using Sources of Geographic Data

In order to fully understand both geographic realms, geographers must utilize various sources of data. For instance, geographers can use data and comparative analysis to determine the different factors that affect quality of life, such as population density, infant mortality rates, and literacy rates. In addition, organizations such as the *Population Reference Bureau* and the *Central Intelligence Agency* provide incredible amounts of demographic data that are readily accessible for anyone.

The *CIA World Factbook* is an indispensable resource for anyone interested in geography. Providing information about land area, literacy rates, birth rates, and economics, this resource is one of the most comprehensive on the Internet. In addition, the *Population Reference Bureau* (*PRB*) provides students of geography with an abundant supply of information. The *PRB* provides a treasure trove of analyses related to human populations including HIV rates, immigration rates, poverty rates, and more.

Furthermore, the *United States Census Bureau* provides similar information about the dynamics of the American population. Not only does this source focus on the data geographers need to understand the world, but it also provides information about upcoming classes, online workshops, and even includes an online library of resources for both students and teachers.

Websites for each source can be found below:

- Population Reference Bureau: www.prb.org
- United States Census Bureau: www.census.gov
- CIA World Factbook: https://www.cia.gov/library/publications/the-world-factbook/

Spatial Concepts

Location

Location is the central theme in understanding spatial concepts. In geography, there are two primary types of location: relative and absolute. **Relative location** involves locating objects by their proximity to another object. For example, a person giving directions may refer to well-known landmarks, highways, or intersections along the route to provide a better frame of reference. **Absolute location** is the exact latitudinal and longitudinal position on the globe. A common way of identifying absolute location is through the use of digital, satellite-based technologies such as *GPS (Global Positioning System)*, which uses sensors that interact with satellites orbiting the Earth. **Coordinates** correspond with positions on a manmade grid system using imaginary lines known as **latitude** (also known as *parallels*) and **longitude** (also known as *meridians*).

Lines of latitude run parallel to the *Equator* and measure distance from north to south. Lines of longitude run parallel to the *Prime Meridian* and measure distance from east to west.

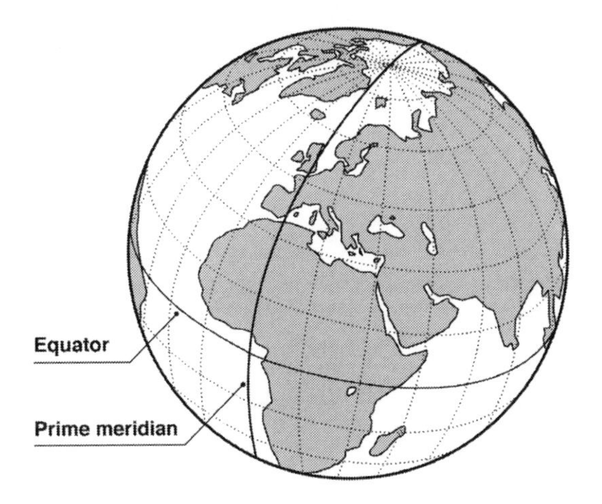

The Equator and the Prime Meridian serve as anchors of the grid system and create the basis for absolute location. They also divide the Earth into **hemispheres.** The Equator divides the Earth into the northern and southern hemispheres, while the Prime Meridian establishes the eastern and western hemispheres.

Lines of latitude are measured by degrees from 0 at the Equator to 90 at the North and South Poles. Lines of longitude are measured by degrees from 0 at the Prime Meridian to 180 at the International Date Line. Coordinates are used to express a specific location using its latitude and longitude and are always expressed in the following format: degree north or south followed by degree east or west (for example, 40°N, 50°E). Since there is great distance between lines of latitude and longitude, absolute locations are often found in between two lines. In those cases, degrees are broken down into *minutes* and *seconds*, which are expressed in this manner: (40° 53′ 44″ N, 50° 22′ 65″ E).

Other major lines of latitude include the *Tropics of Cancer* (23.5 degrees north) and *Capricorn* (23.5 degrees south). These lines correspond with the Earth's *tilt* and mark the positions on the Earth where the sun is directly overhead on the solstices. The tilt and rotation of the earth determine the seasons (or lack thereof) in a given location. For example, the northern hemisphere is tilted directly toward the sun from June 21 to September 21, which creates the summer season in that part of the world. Conversely, the southern hemisphere is tilted away from the direct rays of the sun and experiences winter during those same months. The area between the Tropic of Cancer and the Tropic of Capricorn (called the **tropics**) has more direct exposure to the sun, tends to be warmer year-round, and experiences fewer variations in seasonal temperatures.

Most of the Earth's population lives in the area between the Tropic of Cancer and the *Arctic Circle* (66.5 degrees north), which is one of the **middle latitudes**. In the Southern Hemisphere, the middle latitudes exist between the Tropic of Capricorn and the *Antarctic Circle* (66.5 degrees south). In both of these places, indirect rays of the sun strike the Earth. Therefore, seasons are more pronounced, and milder temperatures generally prevail. The final region, known as the **high latitudes**, is found north of the Arctic Circle and south of the Antarctic Circle. These regions tend to be cold all year, and they experience nearly twenty-four hours of sunlight on their respective *summer solstice* and twenty-four hours of darkness on their *winter solstice*.

Seasons in the Southern Hemispheres are opposite of those in the Northern Hemisphere due to the position of the Earth as it rotates around the sun. An **equinox** occurs when the sun's rays are directly over the Equator, and day and night are of almost equal length throughout the world. Equinoxes occur twice a year; the autumnal equinox occurs around September 22nd, while the spring equinox occurs around March 20th. Since the Northern and Southern hemispheres experience opposite seasons, the season names vary based on location (i.e., when the Northern Hemisphere is experiencing summer, the Southern Hemisphere is in winter).

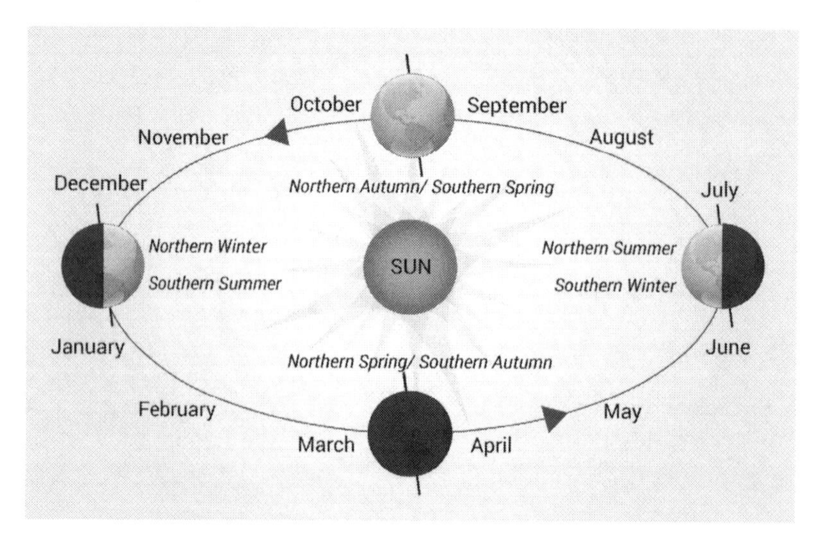

Place

While absolute and relative location identify where something is, the concept of **place** identifies the distinguishing physical and human characteristics of specific locations. People use **toponyms**, names of locations, to define and further orient themselves with their sense of place. Toponyms may be derived from geographical features, important historical figures in the area, or even wildlife commonly found there. For example, many cities in the state of Texas are named in honor of military leaders who fought in the Texas Revolution (such as Houston and Austin), while Mississippi and Alabama got their toponyms from Native American words.

Regions

Geographers divide the world into **regions** to help them understand differences inherent within the world, its people, and its environment. As mentioned previously, lines of latitude and longitude divide the Earth into solar regions relative to the amount of sunlight they receive. Additionally, geographers identify formal and functional regions.

Formal regions are spatially defined areas that have overarching similarities or some level of homogeneity or uniformity. Although not exactly alike, a formal region generally has at least one characteristic that is consistent throughout the entire area. For example, the United States could be classified as one massive formal region because English is the primary language spoken in all fifty states. Even more specifically, the United States is a *linguistic region*—a place where everyone generally speaks the same language.

Functional regions are areas that also have similar characteristics but do not have clear boundaries. Large cities and their metropolitan areas form functional regions, as people from outside the official city limit must travel into the city regularly for work, entertainment, restaurants, etc. Other determining factors of a functional region could be a sports team, a school district, or a shopping center. For example, New York City has two professional baseball, basketball, and football teams. As a result, its citizens may have affinities for different teams even though they live in the same city. Conversely, a citizen in rural Idaho may cheer for the Seattle Seahawks, even though they live over 500 miles from Seattle.

Economic, Political, and Social Factors

Effects of Physical Processes, Climate Patterns, and Natural Hazards on Human Societies

Physical Processes

The Earth's surface, like many other things in the broader universe, does not remain the same for long; in fact, it changes daily. The Earth's surface is subject to a variety of physical processes that continue to shape its appearance. Water, wind, temperature, or sunlight play a role in continually altering the Earth's surface.

Erosion involves the movement of soil from one place to another and can be caused by a variety of stimuli including ice, snow, water, wind, and ocean waves. Wind erosion occurs in generally flat, dry areas with loose topsoil. Over time, the persistent winds can dislodge significant amounts of soil into the air, reshaping the land and wreaking havoc on those who depend on agriculture for their livelihoods. Water can also cause erosion. For example, erosion caused by the Colorado River helped to form the Grand Canyon. Over time, the river moved millions of tons of soil, cutting a huge gorge in the Earth along the way.

In water erosion, material carried by the water is referred to as *sediment*. With time, some sediment can collect at the mouths of rivers, forming *deltas*, which become small islands of fertile soil. This process of detaching loose soils and transporting them to a different location where they remain for an extended period of time is referred to as *deposition*, which is the end result of the erosion process.

In contrast to erosion, **weathering** does not involve the movement of any outside stimuli. Instead, the surface of the Earth is broken down physically or chemically. *Physical weathering* involves the effects of atmospheric conditions such as water, ice, heat, or pressure. For example, when ice forms in the cracks of large rocks or pavement, it can break down or split open the material. *Chemical weathering* generally occurs in warmer climates and involves organic material that breaks down rocks, minerals, or soil. Scientists believe this process led to the creation of fossil fuels such as oil, coal, and natural gas.

Climate Patterns

Weather is the condition of the Earth's atmosphere at a particular time. **Climate** is different; instead of focusing on one particular day, climate is the relative pattern of weather in a place for an extended period of time. Climates are influenced by a variety of factors, including elevation, latitude, proximity to mountains, ocean currents, and wind patterns. For example, the city of Atlanta, Georgia generally has a humid subtropical climate; however, it also occasionally experiences snowstorms in the winter months. Over time, geographers, meteorologists, and other Earth scientists have determined these patterns that are indicative to north Georgia. Almost all parts of the world have predictable climate patterns, which are influenced by the surrounding geography.

The Central Coast of California is an example of a place with a predictable climate pattern. Santa Barbara, California, one of the region's larger cities, has almost the same temperature for most of the year, with only minimal fluctuation during the winter months. The temperatures there, which average between 65 and 75 degrees Fahrenheit regardless of the time of year, are influenced by a variety of different climatological factors including elevation, location relative to the mountains and ocean, and ocean currents. Similarly, Western Europe, which is at the nearly the same latitude as most of Canada, is influenced by the warm waters of the Gulf Stream, an ocean current that acts as a conveyor belt, moving warm tropical waters to the icy north. In fact, the Gulf Stream's influence is so profound that it even keeps Iceland—an island nation in the far North Atlantic—relatively warm.

Natural Hazards

Natural hazards also affect human societies. In tropical and subtropical climates, hurricanes and typhoons that form over warm water can have devastating effects. Additionally, tornadoes, which are powerful cyclonic windstorms, are responsible for widespread destruction in many parts of the world. Earthquakes, caused by shifting plates along faults deep below the Earth's surface, also bring widespread devastation, particularly in nations with poor infrastructure. For example, San Francisco, which experiences earthquakes regularly due to its position near the San Andreas Fault, saw relatively little destruction and death as a result of a major earthquake in 1989. However, in 2010, an earthquake of similar magnitude reportedly killed over 200,000 people in the Western Hemisphere's poorest nation, Haiti. Although a variety of factors may be responsible for the disparity, modern engineering methods and better building materials most likely helped to minimize destruction in San Francisco. Other natural hazards, such as tsunamis, mudslides, avalanches, forest fires, dust storms, flooding, volcanic eruptions, and blizzards, also affect human societies throughout the world.

Characteristics and Spatial Distribution of Earth's Ecosystems

Earth is an incredibly large place filled with a variety of land and water **ecosystems**. *Marine ecosystems* cover over 75 percent of the Earth's surface and contain over 95 percent of the Earth's water. Marine ecosystems can be broken down into two primary subgroups: *freshwater ecosystems*, which only encompass around 2 percent of the earth's surface; and *ocean ecosystems*, which make up over 70 percent. *Terrestrial ecosystems* vary based on latitudinal distance from the equator, elevation, and proximity to mountains or bodies of water. For example, in the high latitudinal regions north of the Arctic Circle and south of the Antarctic Circle, frozen *tundra* dominates. Tundra, which is characterized by low temperatures, short growing seasons, and minimal vegetation, is only found in regions that are far away from the direct rays of the sun.

In contrast, *deserts* can be found throughout the globe and are created by different ecological factors. For example, the world's largest desert, the Sahara, is almost entirely within the tropics; however, other deserts like the Gobi in China, the Mojave in the United States, and the Atacama in Chile, are close to mountain ranges such as the Himalayas, the Sierra Nevada, and the Andes, respectively.

In the United States, *temperate deciduous forests* dominate the southeastern region. The midwestern states such as Nebraska, Kansas, and the Dakotas, are primarily *grasslands*. The states of the Rocky Mountains can have decidedly different climates relative to elevation. Denver, Colorado, will often see snowfalls well into April or May due to colder temperatures, whereas cities in the eastern part of the state, with much lower elevations, may see their last significant snowfall in March.

The tropics generally experience warmer temperatures due to their position on the Earth in relation to the sun. However, like most of the world, the tropics also experience a variety of climatological regions. In Brazil, Southeast Asia, Central America, and even Northern Australia, tropical rainforests are common. These forests, which are known for abundant vegetation, daily rainfall, and a wide variety of animal life, are essential to the health of the world's ecosystems. For example, the Amazon Rain Forest's billions of trees produce substantial amounts of oxygen and absorb an equivalent amount of carbon dioxide—the substance that many climatologists assert is causing climate change or global warming.

Unlike temperate deciduous forests whose trees lose their leaves during the fall and winter months, *tropical rain forests* are always lush, green, and warm. In fact, some rainforests are so dense with vegetation that a few indigenous tribes have managed to exist within them without being influenced by any sort of modern technology, virtually maintaining their ancient way of life in the modern era.

The world's largest land ecosystem, the *taiga*, is found primarily in high latitudinal areas, which receive very little direct sunlight. These forests are generally made up of *coniferous* trees, which do not lose their leaves at any point during the year as *deciduous* trees do. Taigas are cold-climate regions that make up almost 30 percent of the world's land area. These forests dominate the northern regions of Canada, Scandinavia, and Russia, and provide the vast majority of the world's lumber.

Interrelationships Between Humans and Their Environment

Humans both adapt themselves to their environment and adapt their environment to suit their needs. Humans create social systems with the goal of providing people with access to what they need to live more productive, fulfilling, and meaningful lives. Sometimes, humans create destructive systems, but generally speaking, humans tend to leverage their environments to make their lives easier. For example, in warmer climates, people tend to wear lighter clothing such as shorts, linen shirts, and hats. In the excessively sun-drenched nations of the Middle East, both men and women wear flowing white clothing complete with both a head and neck covering in order to prevent the blistering effects of exposure to the sun. Likewise, the native Inuit peoples of northern Canada and Alaska use the thick furs from the animals they kill to insulate their bodies against the bitter cold.

Humans must also manipulate their environments to ensure that they have sufficient access to food and water. In locations where water is not readily available, humans have had to invent ways to redirect water for drinking or agriculture. For example, the city of Los Angeles, America's second most populous city, did not have adequate freshwater resources to sustain its population. However, city and state officials realized that abundant water resources existed approximately three hundred miles to the east. Rather than relocating some of its population to areas with more abundant water resources, the State of California undertook one of the largest construction projects in the history of the world, the Los Angeles Aqueduct, which is a massive water transportation system that connects water-rich areas with the thirsty citizens of Los Angeles.

Farming is another way in which humans use the environment for their advantage. The very first permanent British Colony in North America, Jamestown, VA, was characterized by a hot and humid climate with fertile soil. Consequently, its inhabitants engaged in agriculture for both food and profit. Twelve years after Jamestown's founding in 1607, it was producing millions of dollars of tobacco each year. In order to sustain this booming industry, millions of African slaves and indentured servants from Europe were imported to provide labor.

Conversely, poor soil in the New England colonies did not allow for widespread cash crop production, and the settlers there generally only grew enough food for themselves on small subsistence farms. Due in part to this environmental difference, slavery failed to take a strong foothold in these states, thus creating distinct cultures within the same country.

Renewable and Nonrenewable Resources

Renewable resources are self-replenishing, such as solar, wind, water, and geothermal energy. **Nonrenewable resources**, also known as *fossil fuels,* such as oil, natural gas, and coal, take much longer to replenish but are generally abundant and cheaper to use.

While solar energy is everywhere, the actual means to convert the sun's rays into energy is not. Conversely, coal-fired power plants and gasoline-powered engines, older technologies used during the industrial revolution, remain quite common throughout the world. Reliance on nonrenewable resources continues to grow due to availability and existing infrastructure, but use of renewable energy is also increasing as it becomes more economically competitive with nonrenewable resources.

In addition to sources of energy, nonrenewable resources also include any materials that can be exhausted, such as precious metals, precious stones, and freshwater underground aquifers. Although abundant, most nonrenewable sources of energy are not sustainable because their replenishment takes so long. While renewable resources are sustainable, their use must be properly overseen so that they remain renewable. For example, the beautiful African island of Madagascar is home to some of the most amazing rainforest trees in the world. Logging companies cut, milled, and sold thousands of them in order to make quick profits without planning how to ensure the continued health of the forests. In this way, renewable resources were mismanaged and thus essentially became nonrenewable due to the length of time it takes for replacement trees to grow.

In contrast, many United States paper companies that harvest pine trees must utilize planning techniques to ensure that mature pine trees will always be available. In this manner, these resources remain renewable for human use in a sustainable fashion.

Renewable sources of energy are relatively new in the modern economy. Even though electric cars, wind turbines, and solar panels are becoming more common, they still do not provide enough energy to power the world's economy. As a result, reliance on older forms of energy continues, which can have a devastating effect on the environment. Beijing, China, which has seen a massive boom in industrial jobs, is also one of the most polluted places on Earth. Furthermore, developing nations with very little modern infrastructure also rely heavily on fossil fuels due to the ease in which they are converted into usable energy. Even the United States, which has one of the most developed infrastructures in the world, still relies almost exclusively on fossil fuels, with only ten percent of the required energy coming from renewable sources.

Spatial Patterns of Cultural and Economic Activities

Spatial patterns refer to where things are in the world. Elements of both physical and human geography have spatial patterns regarding where they appear on Earth.

Ethnicity

An ethnic group, or ethnicity, is essentially a group of people with a common language, society, culture, or ancestral heritage. Different ethnicities developed over centuries through historical forces, the impact of religious traditions, and other factors. Thousands of years ago, it was more common for ethnic groups to remain in one area with only occasional interaction with outside groups.

In the modern world, different ethnicities interact on a regular basis due to better transportation resources and the processes of globalization. For example, in countries like the United States and Canada, it is not uncommon for schools, workplaces, or communities to have people of Asian, African, Caucasian, European, Indian, or Native descent.

In less developed parts of the world, travel is limited due to the lack of infrastructure. Consequently, ethnic groups develop in small areas that can differ greatly from other people just a few miles away. For example, on the Balkan Peninsula in southeastern Europe, a variety of different ethnic groups live in close proximity to one another. Croats, Albanians, Serbs, Bosnians, and others all share the same land but have very different worldviews, traditions, and religious influences. Unfortunately, this diversity has not always been a positive characteristic, such as when Bosnia was the scene of a horrible genocide against Albanians in an "ethnic cleansing" effort that continued throughout the late 20th century.

Linguistics

Linguistics, or the study of language, groups certain languages together according to their commonalities. For example, the Romance languages—French, Spanish, Italian, Romanian, and Portuguese—all share language traits from Latin. As people spoke Latin in different regions of Europe, it eventually evolved into different *vernaculars*, or *dialects*, that became their own languages over time. Similarly, the Bantu people of Africa travelled extensively and spread their language, now called Swahili, which became the first Pan-African language.

Since thousands of languages exist, it is important to have a widespread means of communication that can interconnect people from different parts of the world. One way to do this is through a *lingua franca,* or a common language used for business, diplomacy, and other cross-national relationships. English is a primary lingua franca around the world, but there are many others in use as well.

Religion

Religion has played a tremendous role in creating the world's cultures. Devout Christians crossed the Atlantic in hopes of finding religious freedom in New England, Muslim missionaries and traders travelled to the Spice Islands of the East Indies to teach about the Koran, and Buddhist monks traversed the Himalayan Mountains into Tibet to spread their faith.

In some countries, religion helps to shape legal systems. These nations, termed *theocracies*, have no separation of church and state and are more common in Islamic nations such as Saudi Arabia, Iran, and Qatar. In contrast, even though religion has played a tremendous role in the history of the United States, its government remains secular, or nonreligious, due to the influence of European Enlightenment philosophy at the time of its inception.

Like ethnicity and language, religion is a primary way that individuals and people groups self-identify. As a result, religious influences can shape a region's laws, architecture, literature, and music. For example, when the Ottoman Turks, who are Muslim, conquered Constantinople, which was once the home of the Eastern Orthodox Christian Church, they replaced Christian places of worship with mosques. Additionally, they replaced different forms of Roman architecture with those influenced by Arabic traditions.

Economics

Economic activity also has a spatial component. Nations with few natural resources generally tend to import what they need from nations willing to export raw materials to them. Furthermore, areas that are home to certain raw materials generally tend to alter their environment in order to maintain production of those materials. In the San Joaquin Valley of California, an area known for extreme heat and desert-like conditions, local residents have engineered elaborate drip irrigation systems to adequately water lemon, lime, olive, and orange trees, utilizing the warm temperatures to constantly produce citrus fruits. Additionally, other nations with abundant petroleum reserves build elaborate infrastructures in order to pump, house, refine, and transport their materials to nations who require gasoline, diesel, or natural gas.

Essentially, inhabitants of different spatial regions on Earth create jobs, infrastructure, and transportation systems to ensure the continued flow of goods, raw materials, and resources out of their location so long as financial resources keep flowing into the area.

Patterns of Migration and Settlement

Migration is governed by two primary causes: **push factors** that cause someone to leave an area and **pull factors** that lure someone to a particular place. These two factors often work in concert with one another. For example, the United States of America has experienced significant internal migration from the industrial states in the Northeast (such as New York, New Jersey, Connecticut) to the Southern and Western states. Some push factors influencing this migration are high rents in the northeast, dreadfully cold winters, and lack of adequate retirement housing. On the other hand, lower cost of living and warmer climates are pull factors.

International migration also takes place between countries, continents, and other regions. The United States has long been the world's leading nation in regard to **immigration**, the process by which people permanently relocate to a new nation. Conversely, developing nations that suffer from high levels of poverty, pollution, warfare, and other violence all have significant push factors, which cause people to leave and move elsewhere. This process, known as **emigration**, is when people leave a particular area to seek a better life in a different location.

The Development and Changing Nature of Agriculture

Since the genesis of farming as a means of food production, agriculture has been essential to human existence. Humans no longer had to forage and hunt for food, and more consistent food supplies allowed societies to stabilize and grow. In modern times, farming has changed drastically in order to keep up with the increasing world population.

Until the twentieth century, the vast majority of people on Earth engaged in **subsistence farming**, the practice of growing only enough food to feed oneself and one's family. Inventions such as the steel plow, the mechanical reaper, and the seed drill allowed farmers to produce more crops on the same amount of land. As food became cheaper and easier to obtain, populations grew, but fewer people farmed. After the advent of mechanized farming in developed nations, small farms became less common, and many were either abandoned or absorbed by massive commercial farms producing staple crops and cash crops.

In recent years, agricultural practices have undergone further changes in order to keep up with the rapidly growing population. Due in part to the Green Revolution, which introduced the widespread use of fertilizers to produce massive amounts of crops, farming techniques and practices continue to evolve. For example, **genetically modified organisms**, or *GMOs*, are plants or animals whose genetic makeup has been modified using different strands of DNA in hopes of producing more resilient strains of staple crops, livestock, and other foodstuffs. This process, which is a form of biotechnology, attempts to solve the world's food production problems through the use of genetic engineering. Although these crops are abundant and resistant to pests, drought, or frost, they are also the subject of intense scrutiny. For example, the international food company, Monsanto, has faced an incredible amount of criticism regarding its use of GMOs. Many activists assert that such artificial food production processes are inherently problematic and that the resulting food products are dangerous to human health. Despite the controversy, GMOs and biotechnologies continue to change the agricultural landscape and the world's food supply.

Agribusinesses exist throughout the world and produce food for human consumption as well as farming equipment, fertilizers, agrichemicals, and breeding and slaughtering services for livestock. These companies are generally headquartered near the product they produce, like the cereal manufacturer General Mills in the Midwestern United States located near its supply of wheat and corn—the primary ingredients in its cereals.

Contemporary Patterns and Impacts of Development, Industrialization, and Globalization

As mentioned previously, **developing nations** are those that are struggling to modernize their economy, infrastructure, and government systems. Many of these nations may have difficulty providing basic services to their citizens like clean water, adequate roads, or even police protection. Furthermore, government corruption makes life even more difficult for these countries' citizens.

In contrast, **developed nations** are those that have relatively high **Gross Domestic Products** *(GDP)*, or the total value of all goods and services produced in the nation in a given year. The United States, one of the wealthiest nations on Earth, has a GDP of over twenty-one trillion dollars, while Haiti, one of the poorest nations in the Western Hemisphere, has a GDP of over fourteen billion dollars. This comparison is not intended to disparage Haiti or other developing nations, but rather to show that extreme inequities exist in very close proximity to one another, and it may be difficult for developing nations to meet the needs of their citizens and move their economic infrastructure forward toward modernization.

In the modern world, industrialization is the initial key to modernization and development. For developed nations, the process of industrialization took place centuries ago. England, where the Industrial Revolution began, actually began to utilize factories in the early 1700s. Later, the United States and some Western European nations followed suit, using raw materials brought in from their colonies abroad to make finished products. For example, elaborate weaving machines spun cotton into fabric, allowing for the mass production of textiles. As a result, nations that perfected the textile process were able to sell their products around the world, which produced enormous profits. Over time, those nations were able to accumulate wealth, improve their nation's infrastructure, and provide more services for their citizens.

Nations throughout the world are undergoing a similar process in modern times. China exemplifies this concept. While agriculture is still a dominant sector of the Chinese economy, millions of citizens are flocking to major cities like Beijing, Shanghai, and Hangzhou due to the availability of factory jobs that allow workers a certain element of social mobility, or the ability to rise up to a better socioeconomic situation.

Due to improvements in transportation and communication, the world has become figuratively smaller. For example, university students now compete directly with others all over the world to obtain the skills that employers desire. Additionally, many corporations in developed nations have begun to outsource labor to nations with high levels of educational achievement but lower wage expectations. **Globalization**, the process of opening the marketplace to all nations throughout the world, has only just started to take hold in the modern economy. As industrial sites shift to the developing world, more opportunities become available for those nation's citizens as well.

However, due to the massive amounts of pollution produced by factories, the process of globalization also has had significant ecological impacts. The most widely known impact, climate change, which most climatologists assert is caused by an increase of carbon dioxide in the atmosphere, remains a serious problem that has posed challenges for developing nations, who need industries in order to raise their standard of living, and developed nations, whose citizens use a tremendous amount of fossil fuels to run their cars, heat their homes, and maintain their ways of life.

Demographic Patterns and Demographic Change

Demography, the study of human populations, investigates a variety of factors related to the human experience. For instance, several variables impact the geographical movement of people, such as economics, climate, natural disasters, or internal unrest. A recent example of this phenomenon is found in the millions of Syrian immigrants who have moved as far away as possible from the danger in their war-torn homeland.

As previously mentioned, people tend to live near reliable sources of food and water and away from extreme temperatures. Furthermore, the vast majority of people live in the Northern Hemisphere because more land lies in that part of the Earth. In keeping with these factors, human populations tend to be greater where human necessities are easily accessible, or at least more readily available. In other words, such areas have a greater chance of having a higher population density than places without such characteristics.

As push and pull factors fluctuate over time, demographic patterns on Earth will also change. While thousands of Europeans fled their homelands in the 1940s due to the impact of the Second World War, the opposite is true today as thousands of migrants arrive on European shores each month due to conflicts in the Levant and difficult economic conditions in Northern Africa. Furthermore, people tend to migrate to places with a greater economic potential for themselves and their families. As a result, developed nations such as the United States, Germany, Canada, and Australia have a net gain of migrants, while developing nations such as Somalia, Zambia, and Cambodia generally tend to see thousands of their citizens seek better lives elsewhere.

Religion and religious conflict also play a role in determining the composition and location of human populations. For example, the Nation of Israel won its independence in 1948 and has since attracted thousands of Jewish people from all over the world. Additionally, the United States has long been a popular destination due to its promise of religious freedom inherent within its own Constitution. In contrast, nations like Saudi Arabia and Iran do not typically tolerate different religions, resulting in a decidedly uniform religious—and oftentimes ethnic—composition. Other factors such as economic opportunity, social unrest, and cost of living also play a vital role in demographic composition.

Basic Concepts of Political Geography

Nation, state, and **nation-state** are terms with very similar meanings, but knowing the differences aids in a better understanding of geography. A nation is a group of people who share the same cultural, linguistic, and historical heritage. A state is a political unit with sovereignty, or the ability to make its own decisions within defined borders. A nation-state is both a nation and a sovereignly governed state. For example, the province of Quebec is considered its own nation, distinct from the rest of Canada in language and culture, but it is subject to the sovereign state of Canada's governance.

The United Kingdom encompasses four member states: England, Wales, Northern Ireland, and Scotland. Although citizens of those countries may consider themselves to be **sovereign**, or self-governing, the reality is that they cannot make decisions regarding international trade, declarations of war, or other important decisions regarding the rest of the world. Instead, they are **semi-autonomous**, meaning that they can make some decisions regarding how their own state is run but must yield more major powers to a centralized authority. In the United States, this sort of system is called **Federalism**, or the sharing of power among local, state, and federal entities, each of whom is assigned different roles in the overall system of government.

Nation-states and their boundaries are not always permanent. For example, after the fall of the Soviet Union in 1991, new nations emerged that had once been a part of the larger entity called the Union of Soviet Socialists Republics. These formerly sovereign nations were no longer forced to be a part of a unifying communist government, and as a result, they regained their autonomy and became newly independent nations that were no longer satellite nations of the Soviet Union.

In a historical sense, the United States can be seen as a prime example of how national boundaries change. After the conclusion of the American Revolution in 1781, the Treaty of Paris defined the United States' western boundary as the Mississippi River; today, after a series of conflicts with Native American groups, the Mexican government, Hawaiian leadership, the Spanish, and the purchase of Alaska from the Russians, the boundaries of the United States have changed drastically. In a similar fashion, nations in Europe, Africa, and Asia have all shifted their boundaries due to warfare, cultural movements, and language barriers.

In the modern world, boundaries continue to change. For example, the Kurds, an ethnic minority and an excellent example of a nation, are still fighting for control of their right to self-determination, but they have been unsuccessful in establishing a state for themselves. In contrast, the oil-rich region of South Sudan, which has significant cultural, ethnic, and religious differences from Northern Sudan, successfully won its independence in a bloody civil war, which established the nation's newest independent state. In recent years, Russia has made the world nervous by aggressively annexing the Crimean Peninsula, which has been part of Ukraine since the end of the Cold War. Even the United Kingdom and Canada have seen their own people nearly vote for their own rights to self-determination. In 1995, Quebec narrowly voted against becoming a sovereign state through a tightly contested referendum. Similarly, Scotland voted to remain part of the Crown even though many Scots see themselves as inherently different from other regions within the UK. **Decolonization**, or the removal of dependency on colonizers, has altered the political landscape of Africa, allowed more autonomy for the African people, and redefined the boundaries of the entire continent. Essentially, political geography across the globe is constantly changing.

Civics and Government

The Role of the Citizen in a Democratic Society

Citizens express their political beliefs and public opinion through participation in politics. The conventional ways citizens can participate in politics in a democratic state include:

- Obeying laws
- Voting in elections
- Running for public office
- Staying interested in and informed of current events
- Learning Canadian history
- Attending public hearings to be informed and to express opinions on issues, especially on the local level
- Forming interest groups to promote common goals
- Forming political action committees (PACs) that raise money to influence policy decisions
- Petitioning government to create awareness of issues
- Campaigning for a candidate
- Contributing to campaigns
- Using mass media to express political ideas, opinions, and grievances

Obeying Laws

Citizens living in a democracy have several rights and responsibilities to uphold. The first duty is that they uphold the established laws of the government. In a democracy, a system of nationwide laws is necessary to ensure that there is some degree of order. Therefore, citizens must obey the laws and also help enforce them because a law that is inadequately enforced is almost useless. Optimally, a democratic society's laws will be accepted and followed by the community as a whole.

However, conflict can occur when an unjust law is passed. For example, much of the civil rights movement centered around laws that supported segregation between black and whites. Yet these practices were encoded in state laws, which created a dilemma for African Americans who wanted equality but also wanted to respect the law. Fortunately, a democracy offers a system in which government leaders and policies are constantly open to change in accordance with the will of citizens. Citizens can influence the laws that are passed by voting for and electing members of the legislative and executive branches to represent them at the local, provincial, and national levels.

Voting

In a democratic state, the most common way to participate in politics is by voting for candidates in an election. Voting allows the citizens of a state to influence policy by selecting the candidates who share their views and make policy decisions that best suit their interests, or candidates who they believe are most capable of leading the country. In Canada, all citizens over 18—regardless of gender, race, or religion—are allowed to vote.

Citizens can participate by voting in the following types of elections:

- Federal, provincial, territorial, and municipal elections: Citizens elect their representatives in government.
- Referendums: Citizens can vote directly on proposed laws or constitutional amendments.
- Recall elections: Citizens in British Columbia have the unique opportunity to petition the government to remove an official from office before their term ends.
- Voter initiatives: In another process unique to British Columbia, citizens can petition their local government to propose laws that will be approved or rejected by voters.

Running for Public Office

Citizens also have the ability to run for elected office. By becoming leaders in the government, citizens can demonstrate their engagement and help determine government policy. Citizen involvement in the selection of leaders is vital in a democracy because it helps to prevent the formation of an elite group that does not answer to the public. Without the engagement of citizens who run for office, voters are limited in their ability to select candidates that appeal to them. In this case, voting options would become stagnant, inhibiting the nation's ability to grow and change over time. As long as citizens are willing to take a stand for their vision of Canada, the government will remain dynamic and diverse.

Citizen Interest

In order for a democracy to function, it is of the utmost importance that citizens care about the course of politics and be aware of current issues. Apathy among citizens is a constant problem that threatens the endurance of democracies. Citizens should have a desire to take part in the political process, lest they simply accept the status quo and fail to fulfill their civic role. Moreover, they must have acute knowledge of the political processes and the issues that they can address as citizens. Without understanding the world around them, citizens may not fully grasp the significance of political actions and thereby fail to make wise decisions in that regard. Therefore, citizens must stay informed about current affairs, ranging from local to global matters, so that they can properly address them as voters or elected leaders.

Historical Knowledge

Furthermore, knowledge of the nation's history is essential for healthy citizenship. History continues to have an influence on present political decisions. It is especially critical that citizens are aware of the context in which laws were established because it helps clarify the purpose of those laws. In addition, history as a whole shapes the course of societies and the world; therefore, citizens should draw on this knowledge of the past to realize the full consequences of current actions. Issues such as climate change, conflict in the Middle East, and civil rights struggles are rooted in events and cultural developments that reach back centuries and should be addressed.

Therefore, education is a high priority in democracies because it has the potential to instill younger generations of citizens with the right mindset and knowledge required to do their part in shaping the nation. Social studies are especially important because students should understand how democracies function and understand the history of the nation and world. Historical studies should cover national and local events as well because they help provide the basis for the understanding of contemporary politics. Social studies courses should also address the histories of foreign nations because contemporary politics has global consequences. In addition, history lessons should remain open to multiple perspectives, even those that might criticize a nation's past actions, because citizens should be exposed to diverse perspectives that they can apply as voters and leaders.

Purposes and Characteristics of Various Governance Systems

Government is the physical manifestation of the political entity or ruling body of a state. It includes the formal institutions that manage and maintain a society. The form of government does not determine the state's economic system, though these concepts are often closely tied. Many forms of government are based on a society's economic system. However, while the form of government refers to the methods by which a society is managed, the term **economy** refers to the management of resources in a society. Many forms of government exist, often as hybrids of two or more forms of government or economic systems. Forms of government can be distinguished based on protection of civil liberties, protection of rights, distribution of power, power of government, and principles of Federalism.

Regimes

Regime is the term used to describe the ruling body and corresponding political conditions under which citizens live. A regime is defined by the amount of power the government possesses and the number of people who comprise the ruling body. A regime is considered to be ongoing until the culture, priorities, and values of the government are altered, either through a peaceful transition of power or a violent overthrow of the current regime.

Aristocracy

An **aristocracy** is a form of government composed of a small group of wealthy rulers, either holding hereditary titles of nobility or membership in a higher class. Variations of aristocratic governments include:

- Oligarchy: form of government where political power is consolidated in the hands of a small group of people
- Plutocracy: type of oligarchy where a wealthy, elite class dominates the state and society

Though no aristocratic governments exist today, it was the dominant form of government during ancient times, including the vassals and lords during the Middle Ages and the city-state of Sparta in ancient Greece.

Authoritarian

An **authoritarian state** is one in which a single party rules indefinitely. The ruling body operates with unrivaled control and complete power to make policy decisions, including the restriction of denying civil liberties such as freedom of speech, press, religion, and protest. Forms of authoritarian governments include *autocracy*, *dictatorship*, and *totalitarianism*. The Soviet Union, Nazi Germany, and modern-day North Korea are all examples of states with authoritarian governments.

Democracy

Democracy is a form of government in which the people act as the ruling body by electing representatives to voice their views. Forms of democratic governments include:

- Direct democracy: democratic government in which the people make direct decisions on specific policies by majority vote of all eligible voters, like in ancient Athens
- Representative democracy: democratic government in which the people elect representatives to vote in a legislative body. This form of government is also known as a representative republic or indirect democracy. Representative democracy is currently the most popular form of government in the world.

The presidential and parliamentary systems are the most common forms of representative democracy. In the **presidential system**, the executive operates in its own branch distinct from the legislature. In addition, the president is typically both the head of state and head of government. Examples of presidential systems include Brazil, Nigeria, and the United States.

In the **parliamentary system**, the prime minister serves as the head of government. The legislative branch, typically a parliament, elects the prime minister and also has the authority to replace the prime minister with a vote of no confidence. Parliamentary systems often include a president as the head of state, but the office is mostly ceremonial, functioning like a figurehead. Examples of parliamentary systems include Canada, Germany, Australia, and Pakistan.

The presidential system is better designed to distribute power between separate branches of government, which theoretically provides more stability. Presidents serve for a limited number of years, while prime ministers serve until death, resignation, or dismissal.

In the parliamentary system, the interconnectedness between parliament and the prime minister facilitates efficient governance, capable of adjusting to developing situations. In contrast, the presidential system is more prone to political gridlock because there is no direct connection between the legislative and executive branches. The legislature in a presidential system cannot replace the executive, like in the parliamentary system. The separation of powers in a presidential system can lead to disagreement between the executive and legislature, causing gridlock and other delays in governance.

Federalism is a set of principles that divides power between a central government and regional governments. Sovereign states often combine into a federation, and in doing so, they cede some degree of sovereignty to a functional central government that handles broad national policies. The United States and Canada are examples of governments with a Federalist structure.

Monarchy

Monarchy is a form of government in which the state is ruled by a sovereign leader. This leader is called a *monarch* and is typically a hereditary ruler. Monarchs have often justified their power due to some divine right to rule. Types of monarchies include:

- Absolute monarchy: a monarchy in which the monarch has complete power over the people and the state

- Constitutional monarchy: a type of monarchy in which the citizens of the state are protected by a constitution. A separate branch, typically a parliament, makes legislative decisions, and the monarch and legislature share power.

- Crowned republic: a type of monarchy in which the monarch holds only a ceremonial position and the people hold sovereignty over the state. It is defined by the monarch's lack of executive power.

Examples of monarchies:

- Kingdom of Saudi Arabia is an absolute monarchy.
- Canada is a constitutional monarchy (as well as a parliamentary democracy).
- Australia is a crowned republic.

Federal and Unitary Systems

A **unitary government** invests all authority in one central government which oversees subunits or regional bodies for implementation. Conflict between the central and regional authorities is comparatively rare. Most governments today have unitary systems, and they are at least nominally democratic. In unitary governments, there is typically little room for political parties other than the ruling party or for political dissent. Unitary governments that are ethnically homogenous and geographically small in size have the most unity. In a unitary form of government, the central or national government may permit a local or regional government to exercise certain powers but can at any time rescind those powers and dissolve the regional body.

Typically, **federal governments** emerge when a state or a set of states agree to form a union or central government. In this arrangement, states exercise a certain degree of political autonomy, though the central government often has superseding authority on delineated matters. States with federal systems include the United States of America, Brazil, Mexico, and Canada. In federal governments, space is typically made for political parties that participate in elections at the federal, state, and local levels. The division of power in a federal system is constitutionally defined and therefore strictly maintained.

Economics

Fundamental Economic Concepts

Economics is the study of human behavior in response to the production, consumption, and distribution of assets or wealth. Economics can help individuals or societies make decisions for themselves dependent upon their needs, wants, and resources. Economics is divided into two subgroups: microeconomics and macroeconomics.

Microeconomics is the study of individual or small group behaviors related to markets of goods and services. It specifically looks at single factors that could affect these behaviors and decisions. For example, the use of coupons in a grocery store could affect an individual's product choice, quantity purchased, and overall savings that could be directed to a different purchase. Microeconomics encompasses the study of many things, including scarcity, choice, opportunity costs, economics systems, factors of production, supply and demand, market efficiency, the role of government, distribution of income, and product markets.

Macroeconomics examines a much larger scale, analyzing the economy as a whole. It focuses on how aggregate factors such as demand, output, spending habits, unemployment, interest rates, price levels, and national income affect the people in a society or nation. For example, if a national company moves its production overseas to save on costs, how will production, labor, and capital be affected? Governments and corporations use macroeconomic models to help formulate economic policies and strategies.

Microeconomics

Scarcity

When a product is scarce, there is a short supply of it. Limited resources and high demand create **scarcity**. For example, when the newest version of a cellphone is released, people line up to buy the phone or put their name on a wait list if the phone is not immediately available. The new cellphone may become a scarce commodity. In turn, the phone company may raise their prices, knowing that people may be willing to pay more for an item in such high demand.

Factors of Production

There are four factors of production:

- Land: both renewable and nonrenewable resources
- Labor: effort put forth by people to produce goods and services
- Capital: the tools used to create goods and services
- Entrepreneurs: persons who combine land, labor, and capital to create new goods and services

The four factors of production are used to create goods and services to make economic profit. All four factors strongly impact one another.

Supply and Demand

Supply and demand are the most important concepts of economics in a market economy. **Supply** is the amount of a product that a market can offer. **Demand** is the quantity of a product needed or desired by buyers. The price of a product is directly related to supply and demand. The price of a product and the demand for that product go hand in hand in a market economy. For example, when there are a variety of treats at a bakery, certain treats are in higher demand than others. The bakery can raise the cost of the more demanded items as supplies get limited. Conversely, the bakery can sell the less desirable treats by lowering the cost of those items as an incentive for buyers to purchase them.

Product Markets

Product markets are where goods and services are bought and sold. Product markets provide a place for sellers to offer goods and services and for consumers to purchase them. The annual value of goods and services exchanged throughout the year is measured by a nation's Gross Domestic Product (GDP), a monetary measure of goods and services made either quarterly or annually. Department stores, gas stations, grocery stores, and other retail stores are all examples of product markets. However, product markets do not include any raw or unfinished materials.

Theory of the Firm

The behavior of firms is composed of several theories varying between short- and long-term goals. There are four basic firm behaviors: perfect competition, profit maximization, short run, and long run. Each firm follows a pattern, depending on its desired outcome. **Theory of the Firm** posits that firms, after conducting market research, make decisions that will maximize their profits.

- Perfect competition: several businesses are selling the same product simultaneously. There are so many businesses and consumers that none will directly impact the market. Each business and consumer is aware of the competing businesses and markets.
- Profit maximization: Firms decide the quantity of a product that needs to be produced in order to receive maximum profit gains. Profit is the total amount of revenue made after subtracting costs.
- Short run: A short amount of time where fixed prices cannot be adjusted. The quantity of the product depends on the varying amount of labor. Less labor means less product.
- Long run: An amount of time where fixed prices can be adjusted. Firms try to maximize production while minimizing labor costs.

Macroeconomics

Measures of Economic Performance

Measurements of economic performance determine if an economy is growing, stagnant, or deteriorating. To measure the growth and sustainability of an economy, several indicators can be used. Economic indicators provide data that economists can use to determine if there are faulty processes or if some form of intervention is needed.

One of the main indicators of a country's economic performance is the Gross Domestic Product (GDP). GDP growth provides important information that can be used to determine fiscal or financial policies. The GDP does not measure income distribution, quality of life, or losses due to natural disasters. For example, if a community lost everything to a hurricane, it would take a long time to rebuild the community and stabilize its economy. That is why there is a need to take into account more balanced performance measures when factoring overall economic performance.

Other indicators used to measure economic performance are unemployment/employment rates, inflation, savings, investments, surpluses and deficits, debt, labor, trade terms, the HDI (Human Development Index), and the HPI (Human Poverty Index).

Unemployment

Unemployment occurs when an individual does not have a job, is actively trying to find employment, and is not getting paid. *Official* unemployment rates do not factor in the number of people who have stopped looking for work, but *true* unemployment rates do.

There are three types of unemployment: cyclical, frictional, and structural.

Cyclical
Comes as a result of the regular economic cycle and variations in supply and demand. This usually occurs during a recession.
Frictional
When workers voluntarily leave their jobs. An example would be a person changing careers.
Structural
When companies' needs change and a person no longer possesses the skills needed.

Given the nature of a market economy and the fluctuations of the labor market, a 100 percent employment rate is impossible to reach.

Inflation

Inflation is when the value of money decreases and the cost of goods and services increases over time. Supply, demand, and money reserves all affect inflation. Generally, inflation is measured by the Consumer Price Index (CPI), a tool that tracks price changes of goods and services. When the cost of goods and services increase, manufacturers may reduce the quantity they produce due to lower demand. This decreases the purchasing power of the consumer. Basically, as more money is printed, it holds less and less value in purchasing power. When inflation occurs, consumers spend and save less because their currency is worth less. However, if inflation occurs steadily over time, the people can better plan and prepare for future necessities.

Inflation can vary from year to year, usually never fluctuating more than 2 percent. Central banks try to prevent drastic increases or decreases of inflation to prohibit prices from rising or falling too far. Although rare, any country's economy may experience **hyperinflation** (when inflation rates increase to over 50 percent), while other economies may experience **deflation** (when the cost of goods and services decrease over time). Deflation occurs when the inflation rate drops below zero percent.

Business Cycle

A **business cycle** is when the Gross Domestic Product (GDP) moves downward and upward over a long-term growth trend. These cycles help determine where the economy currently stands, as well as where it could be heading. Business cycles usually occur almost every six years and have four phases: expansion, peak, contraction, and trough. Here are some characteristics of each phase:

- Expansion: increased employment rates, production, sales, and economic growth
- Peak: employment rates are at or above full employment and the economy is at maximum productivity
- Contraction: when growth starts slowing and unemployment rises
- Trough: the cycle has hit bottom and is waiting for the next cycle to start again.

When the economy is expanding or "booming," the business cycle is going from a trough to a peak. When the economy is headed down and toward a recession, the business cycle is going from a peak to a trough.

Four phases of a business cycle:

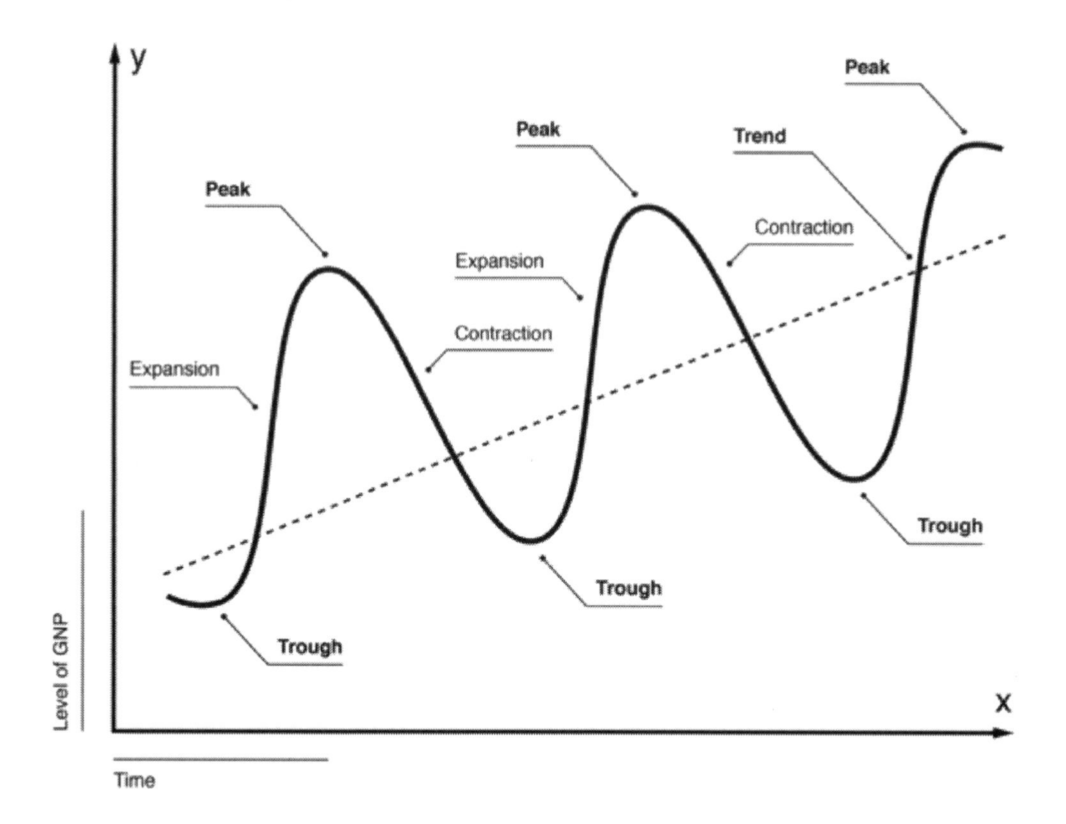

Economic Growth

The most common tool for measuring economic growth is the Gross Domestic Product (GDP). The increase of goods and services over time indicates positive movement in economic growth. The quantity of goods and services produced is not always an indicator of economic growth, however; the value matters more than the quantity.

There are many causes of economic growth, which can be short- or long-term. In the short term, if aggregate demand (the total demand for goods and services produced at a given time) increases, then the overall GDP increases as well. As the GDP increases, interest rates may decrease, which may encourage greater spending and investing. Real estate prices may also rise, and there may be lower income taxes. All of these short-term factors can stimulate economic growth.

In the long term, if aggregate supply (the total supply of goods or services in a given time period) increases, then there is potential for an increase in capital as well. With more working capital, more infrastructure and jobs can be created. New technologies will be developed, and new raw materials may be discovered. All of these long-term factors can also stimulate economic growth.

Other causes of economic growth include low inflation and stability. Lower inflation rates encourage more investing as opposed to higher inflation rates that cause market instability. Stability encourages businesses to continue investing. If the market is unstable, investors may question the volatility of the market.

Potential Costs of Economic Growth:

- Inflation: When economic growth occurs, inflation tends to be high. If supply cannot keep up with demand, then the inflation rate may be unmanageable.
- Economic booms and recessions: The economy goes through cycles of booms and recessions. This causes inflation to fluctuate over time, which puts the economy into a continuous cycle of rising and falling.
- Environmental costs: When the economy is growing, there is an abundance of output, which may result in more pollutants and a reduction in quality of life.
- Inequalities: Growth occurs differently among members of society. While the wealthy may be getting richer, those living in poverty may just be getting on their feet. So, while economic growth is happening, it may happen at very different rates.

While these potential costs could affect economic growth, if the growth is consistent and stable, then it can occur without severe inflation swings. As technology improves, new ways of production can reduce negative environmental factors as well.

Government Involvement in the Economy

Governments have considerable influence over the flow of economies. When a government has full control over the economic decisions of a nation, it is called a command system. This was the case in many absolute monarchies such as eighteenth-century France; King Louis XIV built his economy on the concept of mercantilism, which believed that the state should manage all resources, particularly by accumulating gold and silver. This system of economics discouraged exports and thereby limited trade.

In contrast, the market system is guided by the concept of capitalism, in which individuals and businesses have the freedom to manage their own economic decisions. This allows for private property and increases the opportunities for entrepreneurship and trade. Early proponents of capitalism emphasized *laissez-faire* policies, which means "let it be," and argued that the government should not be involved with the economy at all. They believed that the market is guided by self-interest and that individuals will optimally work for their personal success. However, individuals' interests do not necessarily correlate with the needs of the overall economy. For example, during a financial recession, consumers may decide to save up their money rather than make purchases; doing so helps them in the short run but further reduces demand in a slumping economy. Therefore, most capitalist governments still assert a degree of control over their economies while still allowing for private business.

Likewise, many command system economies have relied heavily on private businesses. Communism has been the primary form of command system economies in the modern era. Communism is a form of socialism that emphasizes communal ownership of property and government control over production. The high degree of government control gives more stability to the economy, but it also creates considerable flaws. The monopolization of the economy by the government limits its ability to respond to local economic conditions because certain regions often have unique resources and needs. With the collapse of the Soviet Union and other communist states, command systems have been largely replaced with market systems.

The Canadian government helps to manage the nation's economy through a market system in several ways. First and foremost, the Bank of Canada, the central bank, is responsible for the production of money for use within the economy; depending on how the government manages the monetary flow, it may lead to a stable economy, deflation, or inflation. Second, provincial and federal governments impose taxes on individuals, corporations, and goods. Third, the government can pass laws that require additional regulation or inspections. In addition, the government has passed competition laws to inhibit the growth of private monopolies, which could limit free growth in the market system. Debates continue over whether the government should take further action to manage private industries or reduce its control over the private sector.

Just as governments can affect the direction of the economy, so can the economy have significant implications on government policies. Financial stability is critical in maintaining a prosperous state. A healthy economy will allow for new developments that contribute to the nation's growth and create jobs. On the other hand, an economic crisis, such as a recession or depression, can gravely damage a government's stability. Without a stable economy, business opportunities plummet, and people begin to lose income and employment. This, in turn, leads to frustration and discontent in the population, which can lead to criticism of the government. This could very well lead to demands for new leadership to resolve the economic crisis.

Economic Systems

Economic systems determine what is being produced, who is producing it, who receives the product, and the money generated by the sale of the product. There are two basic types of economic systems: market economies (including free and competitive markets) and planned or command economies.

Market economies are characterized by:

- Privately owned businesses, groups, or individuals providing goods or services based on demand
- Demand determines the types of goods and services produced (supply)
- Two types: competitive market and free market.

Competitive Market	**Free Market**
Due to the large number of both buyers and sellers, there is no way any one seller or buyer can control the market or price.	Voluntary private trades between buyers and sellers determine markets and prices without government intervention or monopolies.

Planned or command economies are characterized by:

- Government or central authority determines market prices of goods and services
- Government or central authority determines what is being produced and the quantity of production
- Advantage: large number of shared goods such as public services (transportation, schools, or hospitals)
- Disadvantage: wastefulness of resources

Market Efficiency and the Role of Government (Taxes, Subsidies, and Price Controls)

Market efficiency is directly affected by supply and demand. The government can help the market stay efficient by either stepping in when the market is inefficient and/or providing the means necessary for markets to run properly. The government may impose taxes, subsidies, and price controls to increase revenue, lower prices of goods and services, ensure product availability for the government, and maintain fair prices for goods and services.

The Purpose of Taxes, Subsidies, and Price Controls

Taxes	**Subsidies**	**Price Controls**
-Generate government revenue -Discourage purchase or use of "bad" products such as alcohol or cigarettes	-Lower the price of goods and services -Reassure the supply of goods and services -Allow opportunities to compete with overseas vendors	-Act as emergency measures when government intervention is necessary -Set a minimum or maximum price for goods and services

Money and Banking

Money is a means of exchange that provides a convenient way for sellers and consumers to understand the value of their goods and services. As opposed to bartering (when sellers and consumers exchange goods or services as equal trades), money is convenient for both buyers and sellers.

There are three main forms of money: commodity, fiat, and bank. Here are characteristics of each form:

- Commodity money: a valuable good, such as precious metals or tobacco, used as money
- Fiat money: currency that has no intrinsic value but is recognized by the government as valuable for trade, such as paper money
- Bank money: money that is credited by a bank to those who deposit it into bank accounts, such as checking and savings accounts or credit

While price levels within the economy set the demand for money, most countries have central banks that supply the actual money. Essentially, banks buy and sell money. Borrowers can take loans and pay back the bank, with interest, providing the bank with extra capital.

A central bank has control over the printing and distribution of money. Central banks serve three main purposes: manage monetary growth to help steer the direction of the economy, be a backup to commercial banks that are suffering, and provide options and alternatives to government taxation.

The Bank of Canada is the Canada's central bank. The Bank of Canada controls banking systems and determines the value of money in Canada. Basically, it is the bank for banks.

All Western economies have to keep a minimum amount of protected cash called *required reserve*. Once banks meet those minimums, they can then lend or loan the excess to consumers. The required reserves are used within a fractional reserve banking system (fractional because a small portion is kept separate and safe). Not only do banks reserve, manage, and loan money, but they also help form monetary policies.

Monetary Policy

The central bank and other government committees control the amount of money that is made and distributed. The money supply determines monetary policy. Three main features sustain monetary policy.

- Assuring the minimum amount held within banks (bank reserves): when banks are required to hold more money in reserve funds, they are less willing to lend money to help control inflation.
- Adjusting interest rates: raising interest rates makes borrowing more costly, which can slow down unsustainable growth and lower inflation. Lowering interest rates encourages borrowing and can stimulate struggling economies.
- Purchasing and selling bonds (open market operations): Controlling the money supply by buying bonds to increase it and selling bonds to reduce it.

There are two main types of monetary policy: expansionary and contractionary.

Expansionary	Contractionary
• Increases the money supply • Lowers unemployment • Increases consumer spending • Increases private sector borrowing • Possibly decreases interest rates to very low levels, even near zero • Decreases reserve requirements and federal funds	• Decreases the money supply • Helps control inflation • Possibly increases unemployment due to slowdowns in economic growth • Decreases consumer spending • Decreases loans and/or borrowing

The Bank of Canada uses monetary policy to try to achieve maximum employment and secure inflation rates. Because it is the "bank of banks," it truly strives to be the last-resort option for distressed banks. This is because once these kinds of institutions begin to rely on the central bank for help, all parts of the banking industry—including those dealing with loans, bonds, interest rates, and mortgages—are affected.

International Trade and Exchange Rates

International trade is when countries import and export goods and services. Countries often want to deal in terms of their own currency. Therefore, when importing or exporting goods or services, consumers and businesses need to enter the market using the same form of currency. For example, if Canada would like to trade with China, then Canada may have to trade in China's form of currency, the *Yuan*, versus the dollar, depending on the business.

The exchange rate is what one country's currency will exchange for another. There are two forms of exchange rates: fixed and floating. Fixed exchange rates involve government interventions (like central banks) to help keep the exchange rates stable. Floating, or "flexible," exchange rates constantly change because they rely on supply and demand needs. While each type of exchange rate has advantages and disadvantages, the rate truly depends on the current state of each country's economy. Therefore, each exchange rate may differ from country to country.

Fixed Versus Floating Exchange Rates			
Fixed Exchange Rate: government intervenes to keep exchange rates stable		Floating or "Flexible" Exchange Rate: Supply and demand determines the exchange rate	
Advantages -Stable prices -Exports are more competitive and in turn more profitable -Helps keep inflation low	*Disadvantages* -Requires a large amount of reserve funds -Possible mispricing of currency values	*Advantages* -Central bank involvement is unnecessary -Facilitates free trade	*Disadvantages* -Currency speculation -Exchange rate risks -Inflation increases

Countries may have differing economic statuses and exchange rates, but they rely on one another for goods and services. Prices of imports and exports are affected by the strength of another country's currency. For example, if the Canadian dollar is at a higher value than another country's currency, imports will be less expensive because the dollar will have more value than that of the country selling its good or service. On the other hand, if the dollar is at a low value compared to the currency of another country, Canadian importers will tend to avoid buying international items from that country. However, exporters to that country could benefit from the low value of the dollar.

Fiscal Policy

Fiscal policy refers to how the government adjusts spending and tax rates to influence the functions of the economy. Fiscal policies can either increase or decrease tax rates and spending. These policies represent a tricky balancing act, because if the government increases taxes too much, consumer spending and monetary value will decrease. Conversely, if the government lowers taxes, consumers will have more money in their pockets to buy more goods and services, which increases demand and the need for companies to supply those goods and services. Due to the higher demand, suppliers can add jobs to fulfill that demand. While increases in supply, demand, and jobs are positive for the overall economy, they may result in a devaluation of the dollar and less purchasing power.

Consumer Economics

Economics is closely linked with the flow of resources, technology, and population in societies. The use of natural resources, such as water and fossil fuels, has always depended in part on the pressures of the economy. A supply of a specific good may be limited in the market, but with sufficient demand, sellers are incentivized to increase the available quantity. Unfortunately, the demand for certain objects can often be unlimited, and a high price or limited supply may prevent consumers from obtaining the product or service. If the sellers succumb to the consumers' demand and continue to exploit a scarce resource, supply could potentially be exhausted.

The resources for most products, both renewable and nonrenewable, are finite. This is a particularly difficult issue with nonrenewable resources, but even renewable resources often have limits: organic products such as trees and animals require stable populations and sufficient habitats to support those populations. Furthermore, the costs of certain decisions can have detrimental effects on other resources. For example, industrialization provides economic benefits in many countries but also has had the negative effect of polluting surrounding environments; the pollution, in turn, often eliminates or harms fish, plants, and other potential resources.

The control of resources within an economy is particularly important in determining how resources are used. While demand may change with the consumers' choices and preferences, supply depends on the objectives of the producers. They determine how much of their supply they allot for sale, and in the case of monopolies, they might have sole access to the resource. They might limit their use of resources or gather more to meet the demand. Consumers can choose which sellers they rely on for their supply, except in the case of a monopoly because there is no alternative supplier. Therefore, the function of supply within an economy can drastically influence how resources are exploited.

The availability of resources, in turn, affects the human population. Humans require basic resources such as food and water for survival, as well as additional resources for healthy lifestyles. Therefore, access to these resources helps determine the survival rate of humans. For much of human existence, economies have had limited ability to extract resources from the natural world, which restricted the growth rate of populations. However, the development of new technologies, combined with increasing demand for certain products, has pushed resource use to a new level. On one hand, this led to higher living standards and lower death rates. On the other hand, the increasing exploitation of resources has increased the world's population as a whole to unsustainable levels. The rising population leads to higher demand for resources that cannot be met. This creates poverty, reduced living conditions, and higher death rates. As a result, economics can significantly influence local and world population levels.

Technology is also intricately related to population, resources, and economics. The role of demand within economies has incentivized people to innovate new technologies that enable societies to have a higher quality of life and greater access to resources. Entrepreneurs expand technologies by finding ways to create new products for the market. The Industrial Revolution, in particular, illustrates the relationship between economics and technology because the ambitions of businessmen led to new infrastructure that enabled more efficient and sophisticated use of resources. Many of these inventions reduced the amount of work necessary for individuals and allowed the development of leisure activities, which in turn created new economic markets. However, economic systems can also limit the growth of technology. In the case of monopolies, the lack of alternative suppliers reduces the incentive to meet and exceed consumer expectations. Moreover, as demonstrated by the effects of economics on resources, technology's increasing ability to extract resources can lead to their depletion and create significant issues that need to be addressed.

Distribution of Income

Distribution of income refers to how wages are spread across a society or segments of a society. If everyone made the same amount of money, the distribution of income would be equal. That is not the case in most societies. Wealth varies among people and companies. Income inequality gaps are present in Canada and many other nations. Taxes provide an option to redistribute income or wealth because they provide revenue to build new infrastructure and provide cash benefits to some of the poorest members in society.

Choice and Opportunity Costs

When an individual decides between possibilities, that individual is making a choice. Choices allow people to compare opportunity costs. Opportunity costs are benefits that a person could have received, but gave up, in choosing another course of action. What is an individual willing to trade or give up for a different choice? For example, if an individual pays someone to mow the lawn because he or she would rather spend that time doing something else, then the opportunity cost of paying someone to mow the lawn is worth the time gained from not doing the job himself or herself.

On a larger scale, governments have to assess different opportunity costs when it comes to using taxpayer money. Should the government build a new school, repair roads, or allocate funds to local hospitals? Each choice has a tradeoff, and decision makers must choose which option they think is best.

Practice Questions

1. Which of the following best summarizes the Black Death's impact on European society?
   a. The Black Death increased the labor supply, resulting in wage increases for urban workers.
   b. The Black Death allowed governments to consolidate economic and political power.
   c. The Black Death reduced agricultural production and resulted in the Great Famine.
   d. The Black Death undermined governments and led to the scapegoating of vulnerable populations.

Questions 2–5 refer to the passage below:

> The trade which was thus opened between the dominions of the sultan and those of Genghis Khan was not, however, wholly in the hands of merchants coming from the former country. Soon after the coming of the caravan last mentioned, Genghis Khan fitted out a company of merchants from his own country, who were to go into the country of the sultan, taking with them such articles, the products of the country of the Monguls, as they might hope to find a market for there. There were four principal merchants, but they were attended by a great number of assistants, servants, camel-drivers, etc., so that the whole company formed quite a large caravan. Genghis Khan sent with them three embassadors, who were to present to the sultan renewed assurances of the friendly feelings which he entertained for him, and of his desire to encourage and promote as much as possible the commercial intercourse between the two countries which had been so happily begun.

Excerpt from *Genghis Khan* by Jacob Abbott, 1901

2. Genghis Khan founded which of the following empires?
   a. Golden Horde
   b. Mongol Empire
   c. Timurid Empire
   d. Yuan dynasty

3. The Mongol Empire was one of the largest empires in human history. Which of the following is the only empire that was larger?
   a. The Roman Empire
   b. The Persian Empire
   c. The Ottoman Empire
   d. The British Empire

4. According to the passage, why did Genghis Khan send a company of merchants to the sultan?
   a. Genghis Khan hoped to strengthen his empire's commercial ties with the sultan's country.
   b. Genghis Khan believed a commercial relationship would pave the way for a military conquest.
   c. Genghis Khan always favored commercial relationships over military conquests.
   d. Genghis Khan viewed the financing of a merchant caravan to be merely a diplomatic obligation.

5. Assuming the passage is referring to a sultan in the Middle East, which trade route did the caravan likely travel along?
   a. Indian Ocean trade routes
   b. Maritime Silk Road
   c. Silk Roads
   d. Trans-Saharan trade routes

Answer Explanations

1. D: The Black Death had severe and long-lasting consequences due to the chaos it wreaked on European society. In the aftermath of this human tragedy, Europeans sought someone to blame. Given their poor performance in preventing the outbreak and spread of the disease, European governments suffered a fierce backlash from the incensed citizenry. For many European governments, the struggle to regain their legitimacy took nearly a century. Many vulnerable communities were also scapegoated and accused of maliciously spreading the Black Death. For example, vigilantes regularly expelled and/or massacred Jewish and Romani minority communities. Thus, Choice *D* is the correct answer. The Black Death caused a steep decline in the labor supply, which led to wage increases for urban workers. So, Choice *A* is incorrect. Likewise, Choice *B* is incorrect because the Black Death led to an even greater decentralization of economic and political power as angry communities agitated against the government. The Great Famine (1315–1317) occurred several decades before the Black Death (1347–1351), so Choice *C* is incorrect.

2. B: Genghis Khan was one of the most accomplished military generals in history. During the early thirteenth century, he united the five Mongol confederations and successfully invaded territory in present-day China, Central Asia, Eastern Europe, and the Middle East. These conquests marked the beginning of the Mongol Empire. Thus, Choice *B* is the correct answer. Following the Mongol Empire's fragmentation at the end of the thirteenth century, a Mongol faction known as the Golden Horde assumed control of the Mongol Empire's former territory in Eastern Europe and Siberia. So, Choice *A* is incorrect. Kublai Khan completed the Mongol Empire's conquest of China. After declaring himself the emperor of China, he founded the Yuan Dynasty in 1271. So, Choice *C* is incorrect. Timur conquered the western half of the Chagatai Khanate and established the Timurid Empire in 1370. Therefore, Choice *D* is incorrect.

3. A: In the last sentence of the passage, the author explains how Genghis Khan sent the company of merchants in order to encourage and promote a commercial relationship with the sultan's country. In addition to his military conquests, Genghis Khan was famous for using diplomacy and state-funded economic policies to expand his territories. Prior to an invasion, Genghis Khan would usually offer his target the option of paying tribute and/or opening their markets to the Mongols in exchange for peace. If refused, the Mongols would conduct terror campaigns and launch aggressive military actions to force the state into submission. Thus, Choice *A* is the correct answer. The passage doesn't reference Genghis Khan considering or planning an invasion of the sultan's country, so Choice *B* is incorrect. Genghis Khan seemingly favored a commercial relationship to a military conquest in regard to the sultan's country (however, this wasn't always the case, as evidenced by Genghis Khan's prolific military career). So, Choice *C* is incorrect. Genghis Khan considered the financing of merchants and investments in commercial infrastructure to be an invaluable piece of his foreign policy, so Choice *D* is incorrect.

4. D: The Mongol Empire is recognized as the largest contiguous empire in world history. However, the British Empire surpassed the Mongol Empire in terms of total land area, though its land was more spread out. Therefore, Choice *D* is the correct answer. The Mongol Empire nearly doubled the Roman Empire's territory, making Choice *A* incorrect. Likewise, the Mongol Empire was significantly larger than both the Persian and Ottoman Empires, so Choices *B* and *C* are also incorrect.

5. C: The caravan would have been traveling from Central Asia to the Middle East. As such, the caravan most likely traveled along the Silk Roads, a historic series of land-based travel routes that ran from East Asia to Eastern Europe. Middle Eastern and Central Asian cities were some of the most important destinations on the Silk Roads. Thus, Choice *C* is the correct answer. The Indian Ocean trade routes were a maritime trading network that primarily connected East Africa, the Indian subcontinent, and Southeast Asian states. The caravan almost certainly wouldn't have traveled from Central Asia to the Middle East by ship, so Choice *A* is incorrect. Choice *B* is incorrect for the same reason. The Maritime Silk Road was a maritime trade route that combined the Indian Ocean trade network with an additional trade route to Chinese markets. Choice *D* is incorrect because the trans-Saharan trade routes were in Africa, which wouldn't have been on the caravan's most direct path to the Middle East.

Science

Physical Science (Physics and Chemistry)

Structure of Matter

Elements, Compounds, and Mixtures

Everything that takes up space and has mass is composed of **matter**. Understanding the basic characteristics and properties of matter helps with classification and identification.

An **element** is a substance that cannot be chemically decomposed to a simpler substance, while still retaining the properties of the element.

Compounds are composed of two or more elements that are chemically combined. The constituent elements in the compound are in constant proportions by mass.

When a material can be separated by physicals means (such as sifting it through a colander), it is called a **mixture**. Mixtures are categorized into two types: *heterogeneous* and *homogeneous*. Heterogeneous mixtures have physically distinct parts, which retain their different properties. A mix of salt and sugar is an example of a heterogeneous mixture. With heterogenous mixtures, it is possible that different samples from the same parent mixture may have different proportions of each component in the mixture. For example, in the sugar and salt mixture, there may be uneven mixing of the two, causing one random tablespoon sample to be mostly salt, while a different tablespoon sample may be mostly sugar.

A homogeneous mixture, also called a **solution,** has uniform properties throughout a given sample. An example of a homogeneous solution is salt fully dissolved in warm water. In this case, any number of samples taken from the parent solution would be identical.

Atoms, Molecules, and Ions

The basic building blocks of matter are **atoms,** which are extremely small particles that retain their identity during chemical reactions. Atoms can be singular or grouped to form elements. Elements are composed of one type of atom with the same properties.

Molecules are a group of atoms—either the same or different types—that are chemically bonded together by attractive forces. For example, hydrogen and oxygen are both atoms but, when bonded together, form water.

Ions are electrically charged particles that are formed from an atom or a group of atoms via the loss or gain of electrons.

Basic Properties of Solids, Liquids, and Gases

Matter exists in certain **states**, or physical forms, under different conditions. These states are called *solid*, *liquid*, or *gas*.

A solid has a rigid, or set, form and occupies a fixed shape and volume. Solids generally maintain their shape when exposed to outside forces.

Liquids and gases are considered fluids, which have no set shape. Liquids are fluid yet are distinguished from gases by their incompressibility (incapable of being compressed) and set volume. Liquids can be transferred from one container to another but cannot be forced to fill containers of different volumes via compression without causing damage to the container. For example, if one attempts to force a given volume or number of particles of a liquid, such as water, into a fixed container, such as a small water bottle, the container would likely explode from the extra water.

A gas can easily be compressed into a confined space, such as a tire or an air mattress. Gases have no fixed shape or volume. They can also be subjected to outside forces, and the number of gas molecules that can fill a certain volume vary with changes in temperature and pressure.

Basic Structure of an Atom

Atomic Models

Theories of the atomic model have developed over the centuries. The most commonly referenced model of an atom was proposed by Niels Bohr. Bohr studied the models of J.J. Thomson and Ernest Rutherford and adapted his own theories from these existing models. Bohr compared the structure of the atom to that of the Solar System, where there is a center, or nucleus, with various sized orbitals circulating around this nucleus. This is a simplified version of what scientists have discovered about atoms, including the structures and placements of any orbitals. Modern science has made further adaptations to the model, including the fact that orbitals are actually made of electron "clouds."

Atomic Structure: Nucleus, Electrons, Protons, and Neutrons

Following the Bohr model of the atom, the **nucleus**, or *core*, is made up of positively charged **protons** and neutrally charged **neutrons**. The neutrons are theorized to be in the nucleus with the protons to provide greater "balance" at the center of the atom. The nucleus of the atom makes up the majority (more than 99%) of the mass of an atom, while the orbitals surrounding the nucleus contain negatively charged **electrons**. The entire structure of an atom is incredibly small.

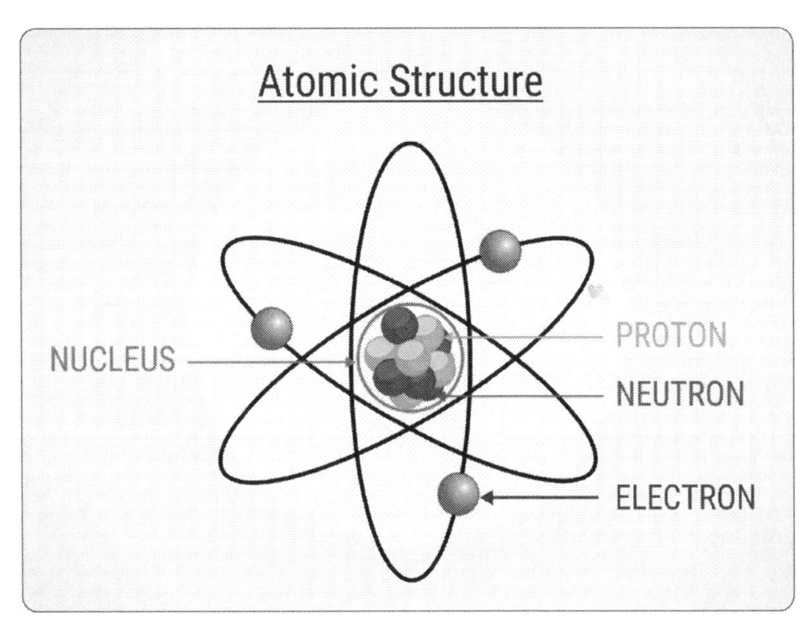

Atomic Number, Atomic Mass, and Isotopes

The **atomic number** of an atom is determined by the number of protons within the nucleus. When a substance is composed of atoms that all have the same atomic number, it is called an **element**. Elements are arranged by atomic number and grouped by properties in the **periodic table**.

An atom's **mass number** is determined by the sum of the total number of protons and neutrons in the atom. Most nuclei have a net neutral charge, and all atoms of one type have the same atomic number. However, there are some atoms of the same type that have a different mass number, due to an imbalance of neutrons. These are called **isotopes**. In isotopes, the atomic number, which is determined by the number of protons, is the same, but the mass number, which is determined by adding the protons and neutrons, is different due to the irregular number of neutrons.

Electron Arrangements

Electrons are most easily organized into distributions of subshells called **electron configurations**. Subshells fill from the inside (closest to the nucleus) to the outside. Therefore, once a subshell is filled, the next shell farther from the nucleus begins to fill, and so on. Atoms with electrons on the outside of a noble gas core (an atom with an electron inner shell that corresponds to the configuration of one of the noble gases, such as Neon) and pseudo-noble gas core (an atom with an electron inner shell that is similar to that of a noble gas core along with (n -1) d^{10} electrons), are called **valence** electrons. Valence electrons are primarily the electrons involved in chemical reactions. The similarities in their configurations account for similarities in properties of groups of elements. Essentially, the groups (vertical columns) on the periodic table all have similar characteristics, such as solubility and reactivity, due to their similar electron configurations.

Basic Characteristics of Radioactive Materials

Radioisotopes

As mentioned, an isotope is a variation of an element with a different number of neutrons in the nucleus, causing the nucleus to be unstable. When an element is unstable, it will go through decay or disintegration. All manmade elements are unstable and will break down. The length of time for an unstable element to break down is called the **half-life**. As an element breaks down, it forms other elements, known as daughters. Once a stable daughter is formed, the radioactive decay stops.

Characteristics of Alpha Particles, Beta Particles, and Gamma Radiation

As radioactive decay is occurring, the unstable element emits *alpha*, *beta*, and *gamma* radiation. Alpha and beta radiation are not as far-reaching or as powerful as gamma radiation. Alpha radiation is caused by the emission of two protons and two neutrons, while beta radiation is caused by the emission of either an electron or a positron. In contrast, gamma radiation is the release of photons of energy, not particles. This makes it the farthest-reaching and the most dangerous of these emissions.

Fission and Fusion

The splitting of an atom is referred to as fission, whereas the combination of two atoms into one is called fusion. To achieve fission and break apart an isotope, the unstable isotope is bombarded with high-speed particles. This process releases a large amount of energy and is what provides the energy in a nuclear power plant. Fusion occurs when two nuclei are merged to form a larger nucleus. The action of fusion also creates a tremendous amount of energy. To put the difference in the levels of energy between fission and fusion into perspective, the level of energy from fusion is what provides energy to the Earth's sun.

Basic Concepts and Relationships Involving Energy and Matter

The study of energy and matter, including heat and temperature, is called **thermodynamics**. There are four fundamental laws of thermodynamics, but the first two are the most commonly discussed.

First Law of Thermodynamics

The first law of thermodynamics is also known as the **conservation of energy**. This law states that energy cannot be created or destroyed, but is just transferred or converted into another form through a thermodynamic process. For example, if a liquid is boiled and then removed from the heat source, the liquid will eventually cool. This change in temperature is not because of a loss of energy or heat, but from a transfer of energy or heat to the surroundings. This can include the heating of nearby air molecules, or the transfer of heat from the liquid to the container or to the surface where the container is resting.

This law also applies to the idea of perpetual motion. A self-powered perpetual motion machine cannot exist. This is because the motion of the machine would inevitably lose some heat or energy to friction, whether from materials or from the air.

Second Law of Thermodynamics

The second law of thermodynamics is also known as the **law of entropy**. Entropy means chaos or disorder. In simple terms, this law means that all systems tend toward chaos. When one or more systems interacts with another, the total entropy is the sum of the interacting systems, and this overall sum also tends toward entropy.

Conservation of Matter in Chemical Systems

The conservation of energy is seen in the conservation of matter in chemical systems. This is helpful when attempting to understand chemical processes, since these processes must balance out. This means that extra matter cannot be created or destroyed, it must all be accounted for through a chemical process.

Kinetic and Potential Energy

The conservation of energy also applies to the study of energy in physics. This is clearly demonstrated through the kinetic and potential energy involved in a system.

The energy of motion is called **kinetic energy**. If an object has height, or is raised above the ground, it has **potential energy**. The total energy of any given system is the sum of the potential energy and the kinetic energy of the subject (object) in the system.

Potential energy is expressed by the equation:

$$PE = mgh$$

Where *m* equals the object's mass, *g* equals acceleration caused by the gravitational force acting on the object, and *h* equals the height of the object above the ground.

Kinetic energy is expressed by the following equation:

$$KE = \frac{1}{2} mv^2$$

Where *m* is the mass of the object and *v* is the velocity of the object.

Conservation of energy allows the total energy for any situation to be calculated by the following equation:

$$KE + PE$$

For example, a roller coaster poised at the top of a hill has all potential energy, and when it reaches the bottom of that hill, as it is speeding through its lowest point, it has all kinetic energy. Halfway down the hill, the total energy of the roller coaster is about half potential energy and half kinetic energy. Therefore, the total energy is found by calculating both the potential energy and the kinetic energy and then adding them together.

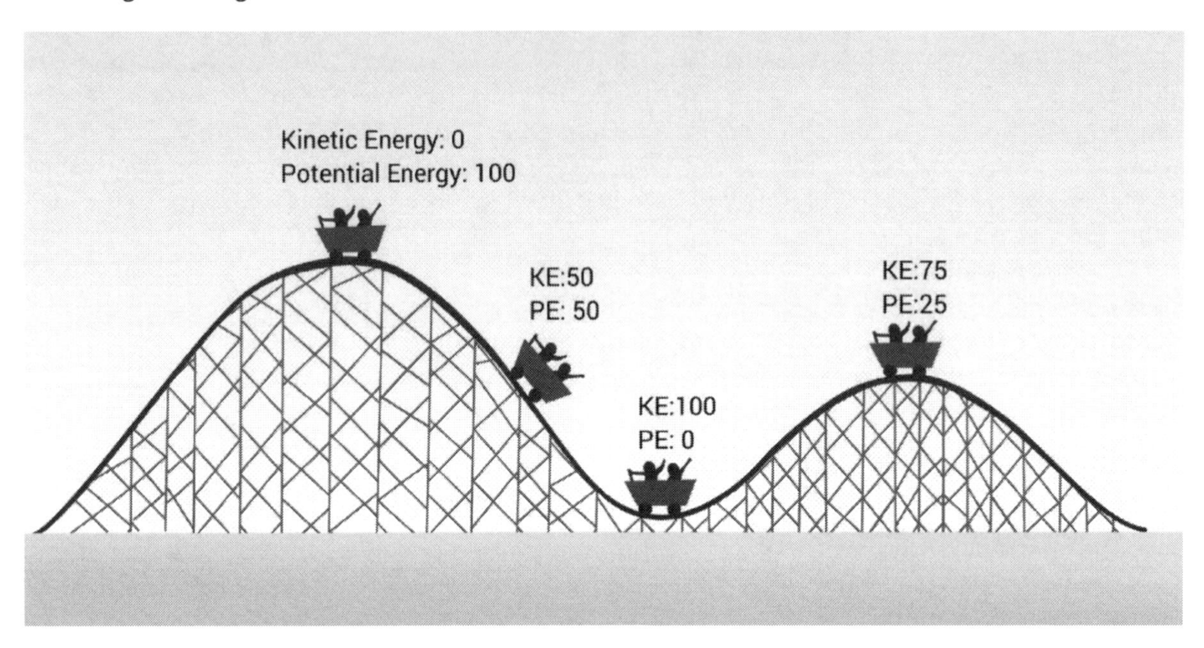

Transformations Between Different Forms of Energy

As stated by the conservation of energy, energy cannot be created or destroyed. If a system gains or loses energy, it is transformed within a single system from one type of energy to another or transferred from one system to another. For example, if the roller coaster system has potential energy that transfers to kinetic energy, the kinetic energy can then be transferred into thermal energy or heat released through braking as the coaster descends the hill. Energy can also transform from the chemical energy inside of a battery into the electrical energy that lights a train set. The energy released through nuclear fusion (when atoms are joined together, they release heat) is what supplies power plants with the energy for electricity. All energy is transferred from one form to another through different reactions. It can also be transferred through the simple action of atoms bumping into each other, causing a transfer of heat.

Differences Between Chemical and Physical Properties/Changes

A change in the physical form of matter, but not in its chemical identity, is known as a **physical change**. An example of a physical change is tearing a piece of paper in half. This changes the shape of the matter, but it is still paper.

Conversely, a **chemical change** alters the chemical composition or identity of matter. An example of a chemical change is burning a piece of paper. The heat necessary to burn the paper alters the chemical composition of the paper. This chemical change cannot be easily undone, since it has created at least one form of matter different than the original matter.

Temperature Scales

There are three main temperature scales used in science. The **Fahrenheit** scale is based on the measurement of water freezing at 32^0 F and water boiling at 212^0 F. The **Celsius** scale uses 0^0 C as the temperature for water freezing and 100^0 C for water boiling. The Celsius scale is the most widely used in the scientific community. The accepted measurement by the International System of Units (from the French Système international d'unités), or SI, for temperature is the **Kelvin** scale. This is the scale employed in thermodynamics, since its zero is the basis for absolute zero, or the unattainable temperature, when matter no longer exhibits degradation.

The conversions between the temperature scales are as follows:

^{0}Fahrenheit to ^{0}Celsius: $^0C = \frac{5}{9}(^0F - 32)$

^{0}Celsius to ^{0}Fahrenheit: $^0F = \frac{9}{5}(^0C) + 32$

^{0}Celsius to Kelvin: $K = {}^0C + 273.15$

Transfer of Thermal Energy and Its Basic Measurement

There are three basic ways in which energy is transferred. The first is through **radiation**. Radiation is transmitted through electromagnetic waves and it does not need a medium to travel (it can travel in a vacuum). This is how the sun warms the Earth, and typically applies to large objects with great amounts of heat or objects with a large difference in their heat measurements.

The second form of heat transfer is **convection**. Convection involves the movement of "fluids" from one place to another. (The term *fluid* does not necessarily apply to a liquid, but any substance in which the molecules can slide past each other, such as gases.) It is this movement that transfers the heat to or from an area. Generally, convective heat transfer occurs through diffusion, which is when heat moves from areas of higher concentrations of particles to those of lower concentrations of particles and less heat. This process of flowing heat can be assisted or amplified through the use of fans and other methods of forcing the molecules to move.

The final process is called **conduction**. Conduction involves transferring heat through the touching of molecules. Molecules can either bump into each other to transfer heat, or they may already be touching each other and transfer the heat through this connection. For example, imagine a circular burner on an electric stove top. The coil begins to glow orange near the base of the burner that is connected to the stove because it heats up first. Since the burner is one continuous piece of metal, the molecules are touching each other. As they pass heat along the coil, it begins to glow all the way to the end.

To determine the amount of heat required to warm the coil in the above example, the type of material from which the coil is made must be known. The quantity of heat required to raise one gram of a substance one degree Celsius (or Kelvin) at a constant pressure is called *specific heat*. This measurement can be calculated for masses of varying substances by using the following equation:

$$q = s \times m \times \Delta t$$

Where *q* is the specific heat, *s* is the specific heat of the material being used, *m* is the mass of the substance being used, and Δt is the change in temperature.

A calorimeter is used to measure the heat of a reaction (either expelled or absorbed) and the temperature changes in a controlled system. A simple calorimeter can be made by using an insulated coffee cup with a thermometer inside. For this example, a lid of some sort would be preferred to prevent any escaping heat that could be lost by evaporation or convection.

Applications of Energy and Matter Relationships

When considering the cycling of matter in ecosystems, the flow of energy and atoms is from one organism to another. The **trophic level** of an organism refers to its position in a food chain. The level shows the relationship between it and other organisms on the same level and how they use and transfer energy to other levels in the food chain. This includes consumption and decomposition for the transfer of energy among organisms and matter. The sun provides energy through radiation to the Earth, and plants convert this light energy into chemical energy, which is then released to fuel the organism's activities.

Naturally occurring elements deep within the Earth's mantle release heat during their radioactive decay. This release of heat drives convection currents in the Earth's magma, which then drives plate tectonics. The transfer of heat from these actions causes the plates to move and create convection currents in the oceans. This type of cycling can also be seen in transformations of rocks. Sedimentary rocks can undergo significant amounts of heat and pressure to become metamorphic rocks. These rocks can melt back into magma, which then becomes igneous rock or, with extensive weathering and erosion, can revert to sediment and form sedimentary rocks over time. Under the right conditions (weathering and erosion), igneous rocks can also become sediment, which eventually compresses into sedimentary rock. Erosion helps the process by redepositing rocks into sediment on the sea floor.

All of these cycles are examples of the transfer of energy from one type into another, along with the conservation of mass from one level to the next.

Periodicity and States of Matter

Periodic Table of the Elements

Using the periodic table, elements are arranged by atomic number, similar characteristics, and electron configurations in a tabular format. The columns, called *groups*, are sorted by similar chemical properties and characteristics such as appearance and reactivity. This can be seen in the shiny texture of metals, the high melting points of alkali Earth metals, and the softness of post-transition metals. The rows are arranged by electron valance configurations and are called *periods*.

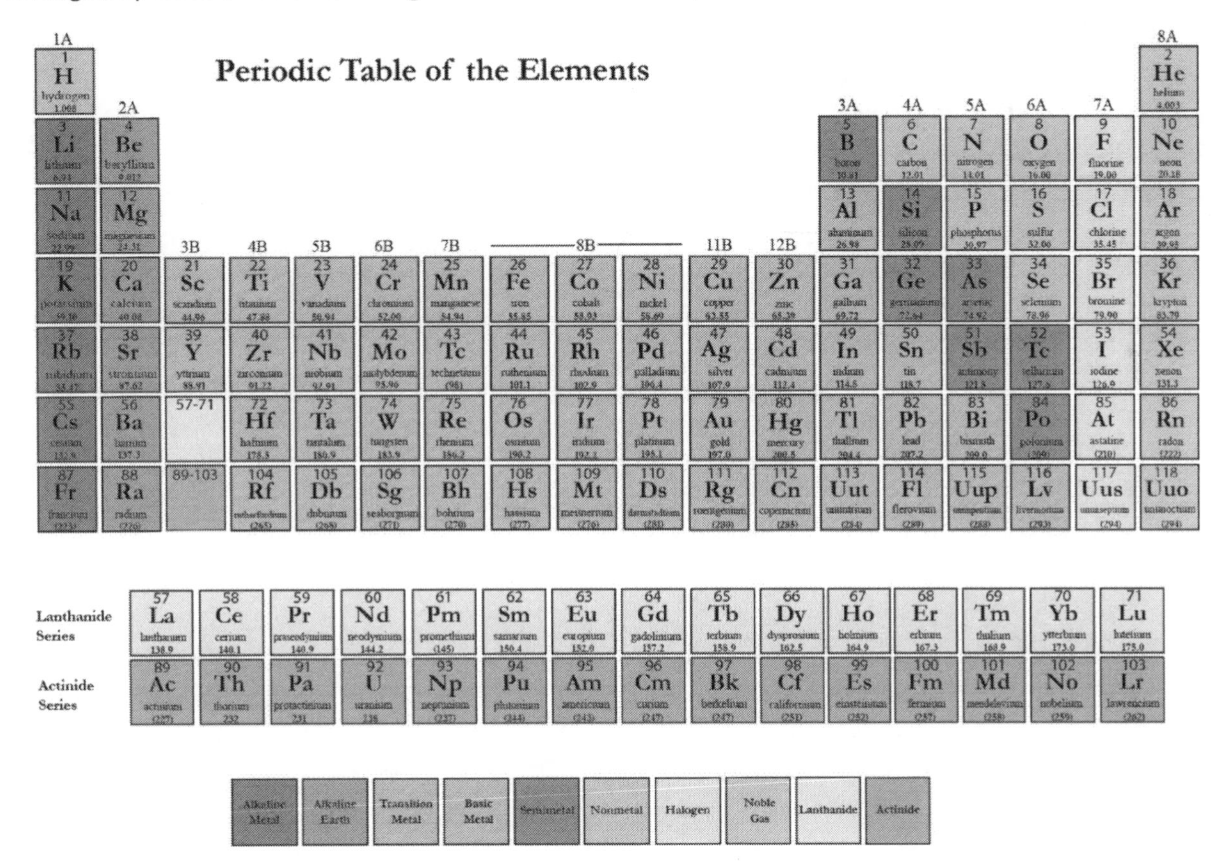

The elements are set in ascending order from left to right by atomic number. As mentioned, the atomic number is the number of protons contained within the nucleus of the atom. For example, the element helium has an atomic number of 2 because it has two protons in its nucleus.

An element's mass number is calculated by adding the number of protons and neutrons of an atom together, while the atomic mass of an element is the weighted average of the naturally occurring atoms of a given element, or the relative abundance of isotopes that might be used in chemistry. For example, the atomic (mass) number of chlorine is 35; however, the atomic mass of chlorine is 35.5 amu (atomic mass unit). This discrepancy exists because there are many isotopes (meaning the nucleus could have 36 instead of 35 protons) occurring in nature. Given the prevalence of the various isotopes, the average of all of the atomic masses turns out to be 35.5 amu, which is slightly higher than chlorine's number on the periodic table. As another example, carbon has an atomic number of 12, but its atomic mass is 12.01 amu because, unlike chlorine, there are few naturally occurring isotopes to raise the average number.

Elements are arranged according to their valance electron configurations, which also contribute to trends in chemical properties. These properties help to further categorize the elements into blocks, including metals, non-metals, transition metals, alkali metals, alkali earth metals, metalloids, lanthanides, actinides, diatomics, post-transition metals, polyatomic non-metals, and noble gases. Noble gases (the far-right column) have a full outer electron valence shell. The elements in this block possess similar characteristics such as being colorless, odorless, and having low chemical reactivity. Another block, the metals, tend to be shiny, highly conductive, and easily form alloys with each other, non-metals, and noble gases.

The symbols of the elements on the periodic table are a single letter or a two-letter combination that is usually derived from the element's name. Many of the elements have Latin origins for their names, and their atomic symbols do not match their modern names. For example, iron is derived from the word *ferrum*, so its symbol is Fe, even though it is now called iron. The naming of the elements began with those of natural origin and their ancient names, which included the use of the ending "ium." This naming practice has been continued for all elements that have been named since the 1940s. Now, the names of new elements must be approved by the International Union of Pure and Applied Chemistry.

The elements on the periodic table are arranged by number and grouped by trends in their physical properties and electron configurations. Certain trends are easily described by the arrangement of the periodic table, which includes the increase of the atomic radius as elements go from right to left and from top to bottom on the periodic table. Another trend on the periodic table is the increase in ionization energy (or the tendency of an atom to attract and form bonds with electrons). This tendency increases from left to right and from bottom to top of the periodic table—the opposite directions of the trend for the atomic radius. The elements on the right side and near the bottom of the periodic table tend to attract electrons with the intent to gain, while the elements on the left and near the top usually lose, or give up, one or more electrons in order to bond. The only exceptions to this rule are the noble gases. Since the noble gases have full valence shells, they do not have a tendency to lose or gain electrons.

Chemical reactivity is another trend identifiable by the groupings of the elements on the periodic table. The chemical reactivity of metals decreases from left to right and while going higher on the table. Conversely, non-metals increase in chemical reactivity from left to right and while going lower on the table. Again, the noble gases present an exception to these trends because they have very low chemical reactivity.

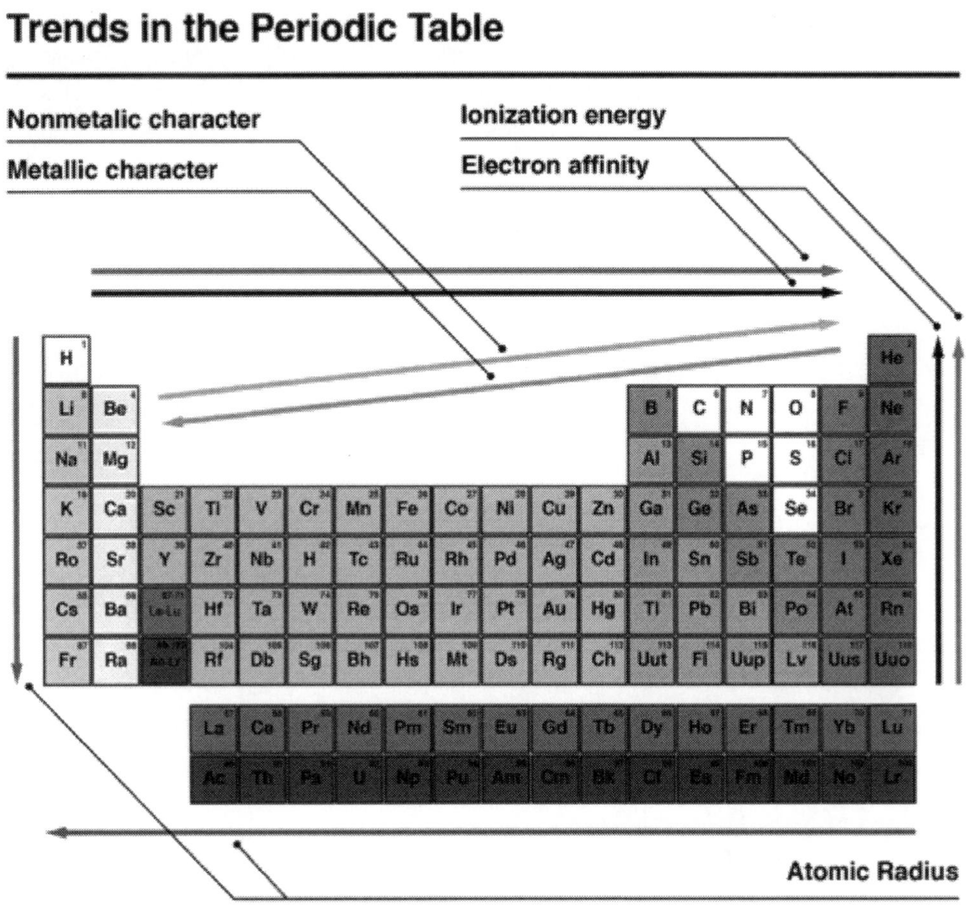

Trends in the Periodic Table

Nonmetalic character

Metallic character

Ionization energy

Electron affinity

Atomic Radius

States of Matter and Factors that Affect Phase Changes

Matter is most commonly found in three distinct states: solid, liquid, and gas. A solid has a distinct shape and a defined volume. A liquid has a more loosely defined shape and a definite volume, while a gas has no definite shape or volume. The **Kinetic Theory of Matter** states that matter is composed of a large number of small particles (specifically, atoms and molecules) that are in constant motion. The distance between the separations in these particles determines the state of the matter: solid, liquid, or gas. In gases, the particles have a large separation and no attractive forces. In liquids, there is moderate separation between particles and some attractive forces to form a loose shape. Solids have almost no separation between their particles, causing a defined and set shape. The constant movement of particles causes them to bump into each other, thus allowing the particles to transfer energy between each other. This bumping and transferring of energy helps explain the transfer of heat and the relationship between pressure, volume, and temperature.

The **Ideal Gas Law** states that pressure, volume, and temperature are all related through the equation: $PV = nRT$, where P is pressure, V is volume, n is the amount of the substance in moles, R is the gas constant, and T is temperature.

Through this relationship, volume and pressure are both proportional to temperature, but pressure is inversely proportional to volume. Therefore, if the equation is balanced, and the volume decreases in the system, pressure needs to proportionately increase to keep both sides of the equation balanced. In contrast, if the equation is unbalanced and the pressure increases, then the temperature would also increase, since pressure and temperature are directly proportional.

When pressure, temperature, or volume change in matter, a change in state can occur. Changes in state include solid to liquid (melting), liquid to gas (evaporation), solid to gas (sublimation), gas to solid (deposition), gas to liquid (condensation), and liquid to solid (freezing). There is one other state of matter called *plasma*, which is seen in lightning, television screens, and neon lights. Plasma is most commonly converted from the gas state at extremely high temperatures.

The amount of energy needed to change matter from one state to another is labeled by the terms for phase changes. For example, the temperature needed to supply enough energy for matter to change from a liquid to a gas is called the *heat of vaporization*. When heat is added to matter in order to cause a change in state, there will be an increase in temperature until the matter is about to change its state. During its transition, all of the added heat is used by the matter to change its state, so there is no increase in temperature. Once the transition is complete, then the added heat will again yield an increase in temperature.

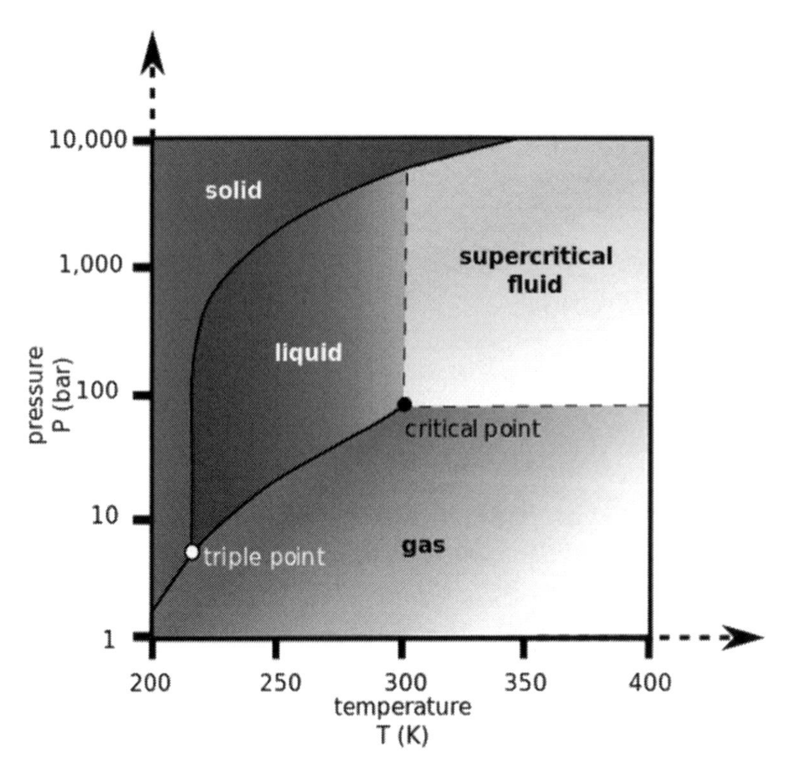

Each state of matter is considered to be a phase, and changes between phases are represented by phase diagrams. These diagrams show the effects of changes in pressure and temperature on matter. The states of matter fall into areas on these charts called *heating curves*.

Chemical Nomenclature, Composition, and Bonding

Simple Compounds and Their Chemical Formulas

Chemical formulas represent the proportion of the number of atoms in a chemical compound. Chemical symbols are used for the elements present and numerical values. Parentheses are also sometimes used to show the number of combinations of the elements in relation to their ionic charges. An element's ionic charge can be determined by its location on the periodic table. This information is then used to correctly combine its atoms in a compound.

For example, the chemical formula for sodium chloride (table salt) is the combination of sodium (Na, ionic charge of +1) and chlorine (Cl, ionic charge of -1). From its placement on the periodic table, the electron valence of an outer shell can be determined: sodium has an ionic charge of +1, while chlorine has an ionic charge of -1. Since these two elements have an equal and opposite amount of charge, they combine in a neutral one-to-one ratio: NaCl. The naming of compounds depends mainly on the second element in a chemical compound. If it is a non-metal (such as chlorine), it is written with an "ide" at the end. The compound NaCl is called "sodium chloride."

If the elements forming a compound do not have equal and opposite ionic charges, there will be an unequal number of each element in the compound to balance the ionic charge. This situation happens with many elements, for example, in the combination of nickel and oxygen into nickel oxide (Ni_2O_3). Nickel has a +3 ionic charge and oxygen has a -2 ionic charge, so when forming a compound, there must be two nickel atoms for every three oxygen atoms (a common factor of 6) to balance the charge of the compound. This compound is called "nickel oxide."

A chemical formula can also be written from a compound's name. For instance, the compound carbon dioxide is formed by the combination of carbon and oxygen. The word "dioxide" means there are two oxygen atoms for every carbon atom, so it is written as CO_2.

To better represent the composition of compounds, structural formulas are used. The combination of atoms is more precisely depicted by lining up the electron configuration of the outer electron shell through a Lewis dot diagram.

The Lewis dot diagram, named for Gilbert N. Lewis, shows the arrangement of the electrons in the outer shell and how these electrons can pair/bond with the outer shell electrons of other atoms when forming compounds. The diagram is created by writing the symbol of an element and then drawing dots to represent the outer shell of valence electrons around what would be an invisible square surrounding the symbol. The placement of the first two dots can vary based on the school of teaching. For the given example, the first dot is placed on the top and then the next dot is placed beside it, since it represents the pair of electrons in the 1s valence shell. The next dots (electrons) are placed one at a time on each side—right, bottom, left, right bottom left, etc.—of the element symbol until all of the valence shell electrons are represented, or the structure has eight dots (electrons), which means it is full. This method gives a more specific picture of compounds, how they are structured, and what electrons are available for bonding, sharing, and forming new compounds. For example, the compound sodium chloride is written separately with sodium having one valence electron and chlorine having seven valence electrons. Then, combined with a total of eight electrons, it is written with two dots being shared between the two elements.

$$Na + \cdot\ddot{\underset{\cdot\cdot}{Cl}}: \rightarrow Na^{+} + :\ddot{\underset{\cdot\cdot}{Cl}}:^{-}$$

Types of Chemical Bonding

A chemical bond is a strong attractive force that can exist between atoms. The bonding of atoms is separated into two main categories. The first category, **ionic bonding**, primarily describes the bonding that occurs between oppositely charged ions in a regular crystal arrangement. It primarily exists between salts, which are known to be ionic. Ionic bonds are held together by the electrostatic attraction between oppositely charged ions. This type of bonding involves the transfer of electrons from the valence shell of one atom to the valence shell of another atom. If an atom loses an electron from its valence shell, it becomes a positive ion, or *cation*. If an atom gains an electron, it becomes a negative ion, or an *anion*. The Lewis electron-dot symbol is used to more simply express the electron configuration of atoms, especially when forming bonds.

The second type of bonding is **covalent bonding**. This bonding involves the sharing of a pair of electrons between atoms. There are no ions involved in covalent bonding, but the force holding the atoms together comes from the balance between the attractive and repulsive forces involving the shared electron and the nuclei. Atoms frequently engage is this type of bonding when it enables them to fill their outer valence shell.

Mole Concept and Its Applications

The calculation of mole ratios of reactants and products involved in a chemical reaction is called *stoichiometry*. To find these ratios, one must first find the proportion of the number of molecules in one mole of a substance. This relates the molar mass of a compound to its mass and this relationship is a constant known as *Avogadro's number* (6.23×10^{23}). Since it is a ratio, there are no dimensions (or units) for Avogadro's number.

Molar Mass and Percent Composition

The molar mass of a substance is the measure of the mass of one mole of the substance. For pure elements, the molar mass is also known as the atomic mass unit (amu) of the substance. For compounds, it can be calculated by adding the molar masses of each substance in the compound. For example, the molar mass of carbon is 12.01 g/mol, while the molar mass of water (H_2O) requires finding the sum of the molar masses of the constituents ((1.01 x 2 = 2.02 g/mol for hydrogen) + (16.0 g/mol for oxygen) = 18.02 g/mol).

The percentage of a compound in a composition can be determined by taking the individual molar masses of each component divided by the total molar mass of the compound, multiplied by 100.

Determining the percent composition of carbon dioxide (CO_2) first requires the calculation of the molar mass of CO_2.

molar mass of carbon = 12.01 x 1 atom = 12.01 g/mol

molar mass of oxygen = 16.0 × 2 atoms = 32.0 g/mol

molar mass of CO_2 = 12.01 g/mol + 32.0 g/mol = 44.01 g/mol

Next, each individual mass is divided by the total mass and multiplied by 100 to get the percent composition of each component.

12.01/44.01 = (0.2729 × 100) = 27.29% carbon

32.0/44.01 = (0.7271 × 100) = 72.71% oxygen

(A quick check in the addition of the percentages should always yield 100%.)

Chemical Reactions

Basic Concepts of Chemical Reactions

Chemical reactions rearrange the initial atoms of the reactants into different substances. These types of reactions can be expressed through the use of balanced chemical equations. A **chemical equation** is the symbolic representation of a chemical reaction through the use of chemical terms. The reactants at the beginning (or on the left side) of the equation must equal the products at the end (or on the right side) of the equation.

For example, table salt (NaCl) forms through the chemical reaction between sodium (Na) and chlorine (Cl) and is written as: $Na + Cl_2 \rightarrow NaCl$.

However, this equation is not balanced because there are two sodium atoms for every pair of chlorine atoms involved in this reaction. So, the left side is written as: $2Na + Cl_2 \rightarrow NaCl$.

Next, the right side needs to balance the same number of sodium and chlorine atoms. So, the right side is written as: $2Na + Cl_2 \rightarrow 2NaCl$. Now, this is a balanced chemical equation.

Chemical reactions typically fall into two types of categories: endothermic and exothermic.

An **endothermic reaction** absorbs heat, whereas an **exothermic reaction** releases heat. For example, in an endothermic reaction, heat is drawn from the container holding the chemicals, which cools the container. Conversely, an exothermic reaction emits heat from the reaction and warms the container holding the chemicals.

Factors that can affect the rate of a reaction include temperature, pressure, the physical state of the reactants (e.g., surface area), concentration, and catalysts/enzymes.

The formula $PV = nRT$ shows that an increase in any of the variables (pressure, volume, or temperature) affects the overall reaction. The physical state of two reactants can also determine how much interaction they have with each other. If two reactants are both in a fluid state, they may have the capability of interacting more than if solid. The addition of a catalyst or an enzyme can increase the rate of a chemical reaction, without the catalyst or enzyme undergoing a change itself.

Le Chatelier's principle describes factors that affect a reaction's equilibrium. Essentially, when introducing a "shock" to a system (or chemical reaction), a positive feedback/shift in equilibrium is often the response. In accordance with the second law of thermodynamics, this imbalance will eventually even itself out, but not without counteracting the effects of the reaction.

There are many different types of chemical reactions. A **synthesis reaction** is the combination of two or more elements into a compound. For example, the synthesis reaction of hydrogen and oxygen forms water.

$$2\ H_2(g) + O_2(g) \rightarrow 2\ H_2O(g)$$

A **decomposition reaction** is the breaking down of a compound into its more basic components. For example, the decomposition, or electrolysis, of water results in it breaking down into oxygen and hydrogen gas.

$$2\ H_2O \rightarrow 2\ H_2 + O_2$$

A **combustion reaction** is similar to a decomposition reaction, but it requires oxygen and heat for the reaction to occur. For example, the burning of a candle requires oxygen to ignite and the reaction forms carbon dioxide during the process.

$$CH_4(g) + 2O_2(g) \rightarrow CO_2(g) + 2H_2O(g)$$

There are also single and double replacement reactions where compounds swap components with each other to form new compounds. In the **single replacement reaction**, a single element will swap into a compound, thus releasing one of the compound's elements to become the new single element. For example, the reaction between iron and copper sulfate will create copper and iron sulfate.

$$1Fe(s) + 1CuSO_4(aq) \rightarrow 1FeSO_4(aq) + 1Cu(s)$$

In a **double replacement reaction**, two compounds swap components to form two new compounds. For example, the reaction between sodium sulfide and hydrochloric acid forms sodium chloride and hydrogen sulfide.

$$Na_2S + HCl \rightarrow NaCl + H_2S$$

After balancing the reaction, we get:

$$Na2S + 2HCl \rightarrow 2NaCl + H2S$$

An organic reaction is a chemical reaction involving the components of carbon and hydrogen.

Finally, there are oxidation/reduction (redox or half) reactions. These reactions involve the loss of electrons from one species (oxidation), and the gain of electrons to the other species (reduction). For example, the oxidation of magnesium is as follows:

$$2\ Mg(\boldsymbol{s}) + O_2(\boldsymbol{g}) \rightarrow 2\ MgO(\boldsymbol{s})$$

Acid-Base Chemistry

Simple Acid-Base Chemistry

If something has a sour taste, it is acidic, and if something has a bitter taste, it is basic. Unfortunately, it can be extremely dangerous to ingest chemicals in an attempt to classify them as an acid or a base. Therefore, acids and bases are generally identified by the reactions they have when combined with water. An acid will increase the concentration of the hydrogen ion (H^+), while a base will increase the concentration of the hydroxide ion (OH^-).

To better categorize the varying strengths of acids and bases, the pH scale is used. The pH scale provides a logarithmic (base 10) grading to acids and bases based on their strength. The pH scale contains values from 0 through 14, with 7 being neutral. If a solution registers below 7 on the pH scale, it is considered an acid. If it registers higher than 7, it is considered a base. To perform a quick test on a solution, litmus paper can be used. A base will turn red litmus paper blue, whereas an acid will turn blue litmus paper red. To gauge the strength of an acid or base, a test of phenolphthalein can be used. An acid will turn red phenolphthalein colorless, and a base will turn colorless phenolphthalein pink. As demonstrated with these types of tests, acids and bases neutralize each other. When acids and bases react with one another, they produce salts (also called ionic substances).

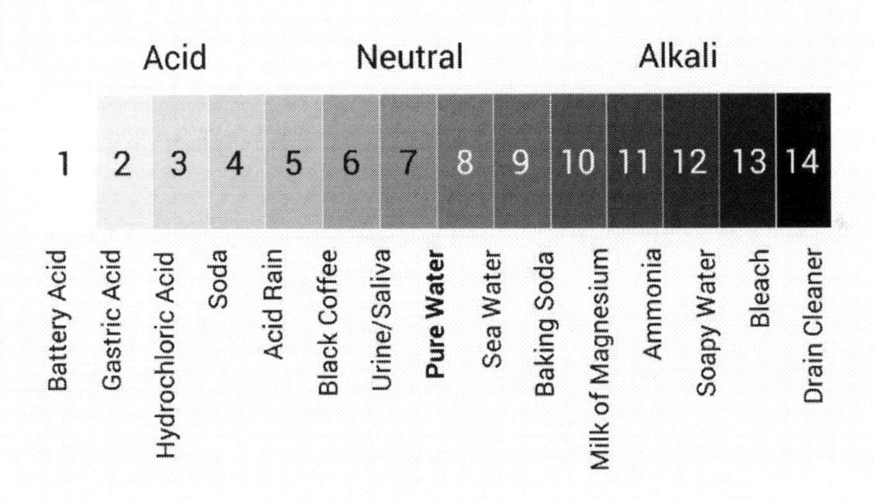

Solutions and Solubility

Different Types of Solutions

A **solution** is a homogenous mixture of more than one substance. A **solute** is another substance that can be dissolved into a substance called a **solvent**. If only a small amount of solute is dissolved in a solvent, the solution formed is said to be *diluted*. If a large amount of solute is dissolved into the solvent, then the solution is said to be *concentrated*. For example, water from a typical, unfiltered household tap is diluted because it contains other minerals in very small amounts.

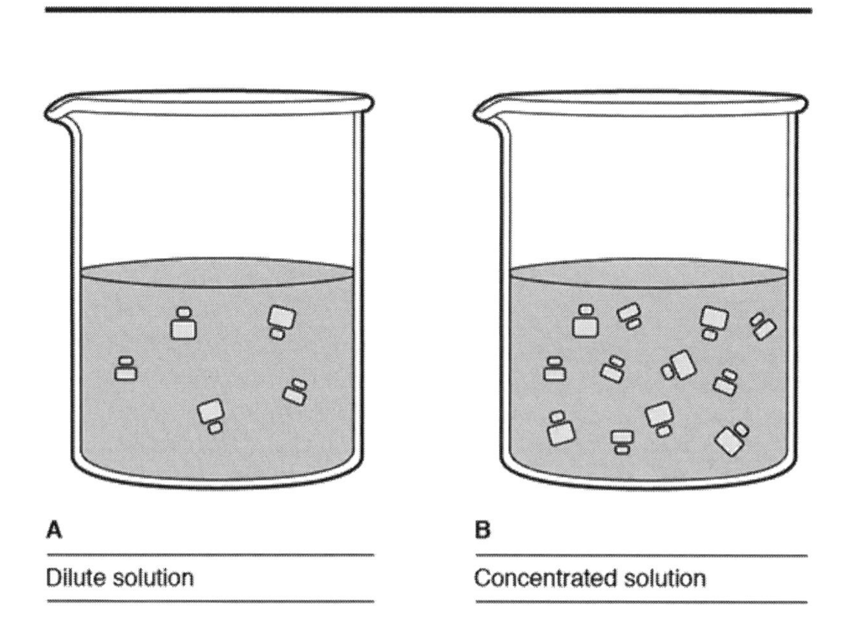

If more solute is being added to a solvent, but not dissolving, the solution is called *saturated*. For example, when hummingbirds eat sugar-water from feeders, they prefer it as sweet as possible. When trying to dissolve enough sugar (solute) into the water (solvent), there will be a point where the sugar crystals will no longer dissolve into the solution and will remain as whole pieces floating in the water. At this point, the solution is considered saturated and cannot accept more sugar. This level, at which a solvent cannot accept and dissolve any more solute, is called its *saturation point*. In some cases, it is possible to force more solute to be dissolved into a solvent, but this will result in crystallization. The state of a solution on the verge of crystallization, or in the process of crystallization, is called a *supersaturated* solution. This can also occur in a solution that seems stable, but if it is disturbed, the change can begin the crystallization process.

Although the terms *dilute*, *concentrated*, *saturated*, and *supersaturated* give qualitative descriptions of solutions, a more precise quantitative description needs to be established for the use of chemicals. This holds true especially for mixing strong acids or bases. The method for calculating the concentration of a solution is done through finding its molarity. In some instances, such as environmental reporting, molarity is measured in parts per million (ppm). Parts per million, is the number of milligrams of a substance dissolved in one liter of water. To find the *molarity*, or the amount of solute per unit volume of solution, for a solution, the following formula is used:

$$c = \frac{n}{V}$$

In this formula, c is the molarity (or unit moles of solute per volume of solution), n is the amount of solute measured in moles, and V is the volume of the solution, measured in liters.

Example:

What is the molarity of a solution made by dissolving 2.0 grams of NaCl into enough water to make 100 mL of solution?

To solve this, the number of moles of NaCl needs to be calculated:

First, to find the mass of NaCl, the mass of each of the molecule's atoms is added together as follows:

23.0g (Na) + 35.5g (Cl) = 58.5g NaCl

Next, the given mass of the substance is multiplied by one mole per total mass of the substance:

2.0g NaCl × (1 mol NaCl/58.5g NaCl) = 0.034 mol NaCl

Finally, the moles are divided by the number of liters of the solution to find the molarity:

(0.034 mol NaCl)/(0.100L) = 0.34 M NaCl

To prepare a solution of a different concentration, the *mass solute* must be calculated from the molarity of the solution. This is done via the following process:

Example:

How would you prepare 600.0 mL of 1.20 M solution of sodium chloride?

To solve this, the given information needs to be set up:

1.20 M NaCl =1.20 mol NaCl/1.00 L of solution

0.600 L solution × (1.20 mol NaCl/1.00 L of solution) = 0.72 moles NaCl

0.72 moles NaCl × (58.5g NaCl/1 mol NaCl) = 42.12 g NaCl

This means that one must dissolve 42.12 g NaCl in enough water to make 600.0 L of solution.

Factors Affecting the Solubility of Substances and the Dissolving Process

Certain factors can affect the rate in dissolving processes. These include temperature, pressure, particle size, and agitation (stirring). As mentioned, the *ideal gas law* states that $PV = nRT$, where P equals pressure, V equals volume, and T equals temperature. If the pressure, volume, or temperature are affected in a system, it will affect the entire system. Specifically, if there is an increase in temperature, there will be an increase in the dissolving rate. An increase in the pressure can also increase the dissolving rate. Particle size and agitation can also influence the dissolving rate, since all of these factors contribute to the breaking of intermolecular forces that hold solute particles together. Once these forces are broken, the solute particles can link to particles in the solvent, thus dissolving the solute.

A *solubility curve* shows the relationship between the mass of solute that a solvent holds at a given temperature. If a reading is on the solubility curve, the solvent is *full* (*saturated*) and cannot hold anymore solute. If a reading is above the curve, the solvent is *unstable* (*supersaturated*) from holding more solute than it should. If a reading is below the curve, the solvent is *unsaturated* and could hold more solute.

If a solvent has different electronegativities, or partial charges, it is considered to be *polar*. Water is an example of a polar solvent. If a solvent has similar electronegativities, or lacking partial charges, it is considered to be *non-polar*. Benzene is an example of a non-polar solvent. Polarity status is important when attempting to dissolve solutes. The phrase "like dissolves like" is the key to remembering what will happen when attempting to dissolve a solute in a solvent. A polar solute will dissolve in a like, or polar solvent. Similarly, a non-polar solute will dissolve in a non-polar solvent. When a reaction produces a solid, the solid is called a *precipitate.* A precipitation reaction can be used for removing a salt (an ionic compound that results from a neutralization reaction) from a solvent, such as water. For water, this process is called ionization. Therefore, the products of a neutralization reaction (when an acid and base react) are a salt and water. Therefore, the products of a neutralization reaction (when an acid and base react) are a salt and water.

When a solute is added to a solvent to lower the freezing point of the solvent, it is called *freezing point depression*. This is a useful process, especially when applied in colder temperatures. For example, the addition of salt to ice in winter allows the ice to melt at a much lower temperature, thus creating safer road conditions for driving. Unfortunately, the freezing point depression from salt can only lower the melting point of ice so far and is ineffectual when temperatures are too low. This same process, with a mix of ethylene glycol and water, is also used to keep the radiator fluid (antifreeze) in an automobile from freezing during the winter.

Mechanics

Description of Motion in One and Two Dimensions

The description of motion is known as **kinetics**, and the causes of motion are known as **dynamics**. Motion in one dimension is known as a *scalar* quantity. It consists of one measurement such as length (length or distance is also known as displacement), speed, or time. Motion in two dimensions is known as a *vector* quantity. This would be a speed with a direction, or velocity.

Velocity is the measure of the change in distance over the change in time. All vector quantities have a direction that can be relayed through the sign of an answer, such as -5.0 m/s or +5.0 m/s. The objects registering these velocities would be in opposite directions, where the change in distance is denoted by Δx and the change in time is denoted by Δt:

$$v = \frac{\Delta x}{\Delta t}$$

Acceleration is the measure of the change in an object's velocity over a change in time, where the change in velocity, $v_2 - v_1$, is denoted by Δv and the change in time, $t_1 - t_2$, is denoted by Δt:

$$a = \frac{\Delta v}{\Delta t}$$

The linear momentum, p, of an object is the result of the object's mass, m, multiplied by its velocity, v, and is described by the equation:

$$p = mv$$

This aspect becomes important when one object hits another object. For example, the linear momentum of a small sports car will be much smaller than the linear momentum of a large semi-truck. Thus, the semi-truck will cause more damage to the car than the car to the truck.

Newton's Three Laws of Motion

Sir Isaac Newton summarized his observations and calculations relating to motion into three concise laws.

First Law of Motion: Inertia

This law states that an object in motion tends to stay in motion or an object at rest tends to stay at rest, unless the object is acted upon by an outside force.

For example, a rock sitting on the ground will remain in the same place, unless it is pushed or lifted from its place.

The First Law also includes the relation of weight to gravity and force between objects relative to the distance separating them.

$$Weight = G\frac{Mm}{r^2}$$

In this equation, *G* is the gravitational constant, *M* and *m* are the masses of the two objects, and *r* is the distance separating the two objects.

Second Law of Motion: F = ma

This law states that the force on a given body is the result of the object's mass multiplied by any acceleration acting upon the object. For objects falling on Earth, an acceleration is caused by gravitational force (9.8 m/s^2).

Third Law of Motion: Action-Reaction

This law states that for every action there is an equal and opposite reaction. For example, if a person punches a wall, the wall exerts a force back on the person's hand equal and opposite to his or her punching force. Since the wall has more mass, it absorbs the impact of the punch better than the person's hand.

Mass, Weight, and Gravity

Mass is a measure of how much of a substance exists, or how much inertia an object has. The mass of an object does not change based on the object's location, but the weight of an object does vary with its location.

For example, a 15-kg mass has a weight that is determined by acceleration from the force of gravity here on Earth. However, if that same 15-kg mass were to be weighed on the moon, it would weigh much less, since the acceleration force from the moon's gravity is approximately one-sixth of that on Earth.

Weight = mass × acceleration

$$W_{Earth} = 15\text{ kg} \times 9.8\text{ m/s}^2 \quad > \quad W_{Moon} = 15\text{ kg} \times 1.62\text{ m/s}^2$$

$$W_{Earth} = 147\text{N} \quad > \quad 24.3\text{N}$$

Analysis of Motion and Forces

Projectile Motion describes the path of an object in the air. Generally, it is described by two-dimensional movement, such as a stone thrown through the air. This activity maps to a parabolic curve. However, the definition of projectile motion also applies to free fall, or the non-arced motion of an object in a path straight up and/or straight down. When an object is thrown horizontally, it is subject to the same influence of gravity as an object that is dropped straight down. The farther the projectile motion, the farther the distance of the object's flight.

Friction is a force that opposes motion. It can be caused by a number of materials; there is even friction caused by air. Whenever two differing materials touch, rub, or pass by each other, it will create friction, or an oppositional force, unless the interaction occurs in a true vacuum. To move an object across a floor, the force exerted on the object must overcome the frictional force keeping the object in place. Friction is also why people can walk on surfaces. Without the oppositional force of friction to a shoe pressing on the floor, a person would not be able to grip the floor to walk—similar to the challenge of walking on ice. Without friction, shoes slip and are unable to help people propel forward and walk.

When calculating the effects of objects hitting (or colliding with) each other, several things are important to remember. One of these is the definition of momentum: the mass of an object multiplied by the object's velocity. As mentioned, it is expressed by the following equation:

$$p = mv$$

Here, p is equal to an object's momentum, m is equal to the object's mass, and v is equal to the object's velocity.

Another important thing to remember is the principle of the conservation of linear momentum. The total momentum for objects in a situation will be the same before and after a collision. There are two primary types of collisions: elastic and inelastic. In an elastic collision, the objects collide and then travel in different directions. During an inelastic collision, the objects collide and then stick together in their final direction of travel. The total momentum in an elastic collision is calculated by using the following formula:

$$m_1v_1 + m_2v_2 = m_1v_1 + m_2v_2$$

Here, m_1 and m_2 are the masses of two separate objects, and v_1 and v_2 are the velocities, respectively, of the two separate objects.

The total momentum in an inelastic collision is calculated by using the following formula:

$$m_1v_1 + m_2v_2 = (m_1 + m_2)v_f$$

Here, v_f is the final velocity of the two masses after they stick together post-collision.

Example:

If two bumper cars are speeding toward each other, head-on, and collide, they are designed to bounce off of each other and head in different directions. This would be an elastic collision.

If real cars are speeding toward each other, head-on, and collide, there is a good chance their bumpers might get caught together and their direction of travel would be together in the same direction.

An **axis** is an invisible line on which an object can rotate. This is most easily observed with a toy top. There is actually a point (or rod) through the center of the top on which the top can be observed to be spinning. This is called the axis.

When objects move in a circle by spinning on their own axis, or because they are tethered around a central point (also an axis), they exhibit circular motion. Circular motion is similar in many ways to linear (straight line) motion; however, there are a few additional points to note. A spinning object is always accelerating because it is always changing direction. The force causing this constant acceleration on or around an axis is called *centripetal force* and is often associated with centripetal acceleration. Centripetal force always pulls toward the axis of rotation. An imaginary reactionary force, called *centrifugal force*, is the outward force felt when an object is undergoing circular motion. This reactionary force is not the real force; it just feels like it is there. For this reason, it has also been referred to as a "fictional force." The true force is the one pulling inward, or the centripetal force.

The terms *centripetal* and *centrifugal* are often mistakenly interchanged. If the centripetal force acting on an object moving with circular motion is removed, the object will continue moving in a straight line tangent to the point on the circle where the object last experienced the centripetal force. For example, when a traditional style washing machine spins a load of clothes to expunge the water from the load, it rapidly spins the machine barrel. A force is pulling in toward the center of the circle (centripetal force). At the same time, the wet clothes, which are attempting to move in a straight line, are colliding with the outer wall of the barrel that is moving in a circle. The interaction between the wet clothes and barrel wall causes a reactionary force to the centripetal force and this expels the water out of the small holes that line the outer wall of the barrel.

Conservation of Angular Momentum

An object moving in a circular motion also has momentum; for circular motion, it is called **angular momentum**. This is determined by rotational inertia, rotational velocity, and the distance of the mass from the axis or center of rotation. When objects exhibit circular motion, they also demonstrate the **conservation of angular momentum**, meaning that the angular momentum of a system is always constant, regardless of the placement of the mass. Rotational inertia can be affected by how far the mass of the object is placed with respect to the axis of rotation. The greater the distance between the mass and the axis of rotation, the slower the rotational velocity. Conversely, if the mass is closer to the axis of rotation, the rotational velocity is faster. A change in one affects the other, thus conserving the angular momentum. This holds true as long as no external forces act upon the system.

For example, ice skaters spinning in on one ice skate extends their arms out for a slower rotational velocity. When skaters bring their arms in close to their bodies (which lessens the distance between the mass and the axis of rotation), their rotational velocity increases and they spin much faster. Some skaters extend their arms straight up above their head, which causes an extension of the axis of rotation, thus removing any distance between the mass and the center of rotation, which maximizes their rotational velocity.

Another example is when a person selects a horse on a merry-go-round: the placement of their horse can affect their ride experience. All of the horses are traveling with the same rotational speed, but in order to travel along the same plane as the merry-go-round turns, a horse on the outside will have a greater linear speed because it is further away from the axis of rotation. Essentially, an outer horse has to cover a lot more ground than a horse on the inside in order to keep up with the rotational speed of the merry-go-round platform. Thrill seekers should always select an outer horse.

The center of mass is the point that provides the average location for the total mass of a system. The word "system" can apply to just one object/particle or to many. The center of mass for a system can be calculated by finding the average of the mass of each object and multiplying by its distance from an origin point using the following formula:

$$x_{center of mass} = \frac{m_1 x_1 + m_2 x_2}{m_1 + m_2}$$

In this case, *x* is the distance from the point of origin for the center of mass and each respective object, and *m* is the mass of each object.

To calculate for more than one object, the pattern can be continued by adding additional masses and their respective distances from the origin point.

Simple Machines

A simple machine is a mechanical device that changes the direction or magnitude of a force. There are six basic types of simple machines: lever, wedge, screw, inclined plane, wheel and axle, and pulley.

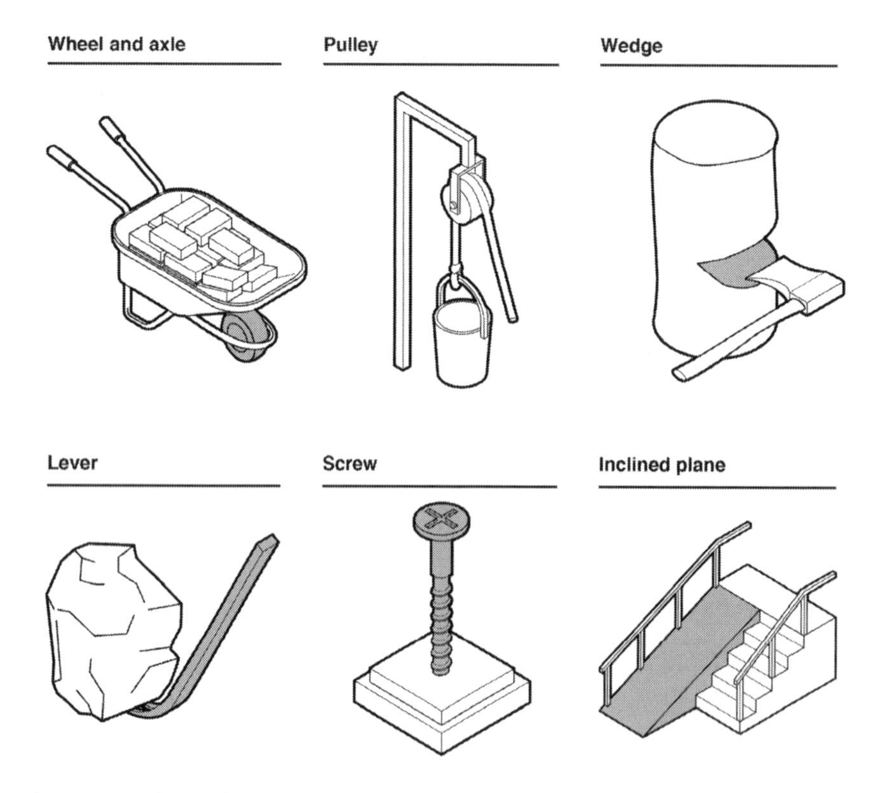

Here is how each type works and an example:

- A lever helps lift heavy items higher with less force, such as a crowbar lifting a large cast iron lid.
- A wedge helps apply force to a specific area by focusing the pressure, such as an axe splitting a tree.
- An inclined plane, such as a loading dock ramp, helps move heavy items up vertical distances with less force.

- A screw is an inclined plane wrapped around an axis and allows more force to be applied by extending the distance of the plane. For example, a screw being turned into a piece of wood provides greater securing strength than hitting a nail into the wood.

- A wheel and axle allows the use of rotational force around an axis to assist with applying force. For example, a wheelbarrow makes it easier to haul large loads by employing a wheel and axle at the front.

- A pulley is an application of a wheel and axle with the addition of cords or ropes and it helps move objects vertically. For example, pulling a bucket out of a well is easier with a pulley and ropes.

Using a simple machine employs an advantage to the user. This is referred to as the mechanical advantage. It can be calculated by comparing the force input by the user to the simple machine with the force output from the use of the machine (also displayed as a ratio).

$$MechanicalAdvantage = \frac{outputforce}{inputforce}$$

$$MA = \frac{F_{out}}{F_{in}}$$

In the following instance of using a lever, it can be helpful to calculate the torque, or circular force, necessary to move something. This is also employed when using a wrench to loosen a bolt.

$$Torque = F \times distanceofleverarmfromtheaxisofrotation \text{ (called the moment arm)}$$

$$T = F \times d$$

Electricity and Magnetism

Electrical Nature of Common Materials

Generally, an atom carries no net charge because the positive charges of the protons in the nucleus balance the negative charges of the electrons in the outer shells of the atom. This is considered to be electrically neutral. However, since electrons are the only portion of the atom known to have the freedom to "move," this can cause an object to become electrically charged. This happens either through a gain or a loss of electrons. Electrons have a negative charge, so a gain creates a net negative charge for the object. On the contrary, a loss of electrons creates a positive charge for the object. This charge can also be focused on specific areas of an object, causing a notable interaction between charged objects. For example, if a person rubs a balloon on a carpet, the balloon transfers some of is electrons to the carpet. So, if that person were to hold a balloon near his or her hair, the electrons in the "neutral" hair would make the hair stand on end. This is due to the electrons wanting to fill the deficit of electrons on the balloon. Unless electrically forced into a charged state, most natural objects in nature tend toward reestablishing and maintaining a neutral charge.

When dealing with charges, it is easiest to remember that *like charges repel* each other and *opposite charges attract* each other. Therefore, negatives and positives attract, while two positives or two negatives will repel each other. Similarly, when two charges come near each other, they exert a force on one another.

This is described through **Coulomb's Law**:

$$F = k\frac{q_1 q_2}{r^2}$$

In this equation, *F* is equal to the force exerted by the interaction, *k* is a constant ($k = 8.99 \times 10^9$ N m^2/C^2), q_1 and q_2 are the measure of the two charges, and *r* is the distance between the two charges.

When materials readily transfer electricity or electrons, or can easily accept or lose electrons, they are considered to be good conductors. The transferring of electricity is called **conductivity**. If a material does not readily accept the transfer of electrons or readily loses electrons, it is considered to be an **insulator**. For example, copper wire easily transfers electricity because copper is a good conductor. However, plastic does not transfer electricity because it is not a good conductor. In fact, plastic is an insulator.

Basic Electrical Concepts

In an electrical circuit, the flow from a power source, or the voltage, is "drawn" across the components in the circuit from the positive end to the negative end. This flow of charge creates an electric current (*I*), which is the time (*t*) rate of flow of net charge (*q*). It is measured with the formula:

$$I = \frac{q}{t}$$

Current is measured in amperes (amps). There are two main types of currents:

1. *Direct current* (DC): a unidirectional flow of charges through a circuit

2. *Alternating current* (AC): a circuit with a changing directional flow of charges or magnitude

Every circuit will show a loss in voltage across its conducting material. This loss of voltage is from resistance within the circuit and can be caused by multiple factors, including resistance from wiring and components such as light bulbs and switches. To measure the resistance in a given circuit, Ohm's law is used:

$$Resistance = \frac{Voltage}{current} = R = \frac{V}{I}$$

Resistance (*R*) is measured in Ohms (Ω).

Components in a circuit can be wired *in series* or *in parallel*. If the components are wired in series, a single wire connects each component to the next in line. If the components are wired in parallel, two wires connect each component to the next. The main difference is that the voltage across those in series is directly related from one component to the next. Therefore, if the first component in the series becomes inoperable, no voltage can get to the other components. Conversely, the components in parallel share the voltage across each other and are not dependent on the prior component wired to allow the voltage across the wire.

To calculate the resistance of circuit components wired in series or parallel, the following equations are used:

Resistance in series:

$$R_{total} = R_1 + R_2 + R_3 + \cdots$$

Resistance in parallel:

$$R_{total} = \frac{1}{R_1} + \frac{1}{R_2} + \frac{1}{R_3} + \cdots$$

To make electrons move so that they can carry their charge, a change in voltage must be present. On a small scale, this is demonstrated through the electrons traveling from the light switch to a person's finger. This might happen in a situation where a person runs his or her socks on a carpet, touches a light switch, and receives a small jolt from the electrons that run from the switch to the finger. This minor jolt is due to the deficit of electrons created by rubbing the socks on the carpet, and then the electrons going into the ground. The difference in charge between the switch and the finger caused the electrons to move.

If this situation were to be created on a larger and more sustained scale, the factors would need to be more systematic, predictable, and harnessed. This could be achieved through batteries/cells and generators. Batteries or cells have a chemical reaction that occurs inside, causing energy to be released and charges to be able to move freely. Batteries generally have nodes (one positive and one negative), where items can be hooked up to complete a circuit and allow the charge to travel freely through the item. Generators convert mechanical energy into electric energy using power and movement.

Basic Properties of Magnetic Fields and Forces

Consider two straight rods that are made from magnetic material. They will naturally have a negative end (pole) and a positive end (pole). These charged poles react just like any charged item: opposite charges attract and like charges repel. They will attract each other when arranged positive pole to negative pole. However, if one rod is turned around, the two rods will now repel each other due to the alignment of negative to negative and positive to positive. These types of forces can also be created and amplified by using an electric current. For example, sending an electric current through a stretch of wire creates an electromagnetic force around the wire from the charge of the current. This force exists as long as the flow of electricity is sustained. This magnetic force can also attract and repel other items with magnetic properties. Depending on the strength of the current in the wire, a greater or smaller magnetic force can be generated around the wire. As soon as the current is stopped, the magnetic force also stops.

Optics and Waves

Electromagnetic Spectrum

The movement of light is described like the movement of waves. Light travels with a wave front, has an amplitude (height from the neutral), a cycle or wavelength, a period, and energy. Light travels at approximately 3.00×10^8 m/s and is faster than anything created by humans thus far.

Light is commonly referred to by its measured wavelengths, or the distance between two successive crests or troughs in a wave. Types of light with the longest wavelengths include radio, TV, and micro, and infrared waves. The next set of wavelengths are detectable by the human eye and create the *visible spectrum*. The visible spectrum has wavelengths of 10^{-7} m, and the colors seen are red, orange, yellow, green, blue, indigo, and violet. Beyond the visible spectrum are shorter wavelengths (also called the *electromagnetic spectrum*) containing ultraviolet light, X-rays, and gamma rays. The wavelengths outside of the visible light range can be harmful to humans if they are directly exposed or are exposed for long periods of time.

Basic Characteristics and Types of Waves

A **mechanical wave** is a type of wave that passes through a medium (solid, liquid, or gas). There are two basic types of mechanical waves: longitudinal and transverse.

A **longitudinal wave** has motion that is parallel to the direction of the wave's travel. This can best be visualized by compressing one side of a tethered spring and then releasing that end. The movement travels in a bunching/un-bunching motion across the length of the spring and back.

A **transverse wave** has motion that is perpendicular to the direction of the wave's travel. The particles on a transverse wave do not move across the length of the wave; instead, they oscillate up and down, creating peaks and troughs.

A wave with a combination of both longitudinal and transverse motion can be seen through the motion of a wave on the ocean—with peaks and troughs, and particles oscillating up and down.

Mechanical waves can carry energy, sound, and light, but they need a medium through which transport can occur. An electromagnetic wave can transmit energy without a medium, or in a vacuum.

A more recent addition in the study of waves is the **gravitational wave**. Its existence has been proven and verified, yet the details surrounding its capabilities are still somewhat under inquiry. Gravitational waves are purported to be ripples that propagate as waves outward from their source and travel in the curvature of space/time. They are thought to carry energy in a form of radiant energy called *gravitational radiation*.

Basic Wave Phenomena

When a wave crosses a boundary or travels from one medium to another, certain things occur. If the wave can travel through one medium into another medium, it experiences **refraction**. This is the bending of the wave from one medium to another due to a change in density of the mediums, and thus, the speed of the wave changes. For example, when a pencil is sitting in half of a glass of water, a side view of the glass makes the pencil appear to be bent at the water level. What the viewer is seeing is the refraction of light waves traveling from the air into the water. Since the wave speed is slowed in water, the change makes the pencil appear bent.

When a wave hits a medium that it cannot penetrate, it is bounced back in an action called **reflection**. For example, when light waves hit a mirror, they are reflected, or bounced, off the mirror. This can cause it to seem like there is more light in the room, since there is a "doubling back" of the initial wave. This same phenomenon also causes people to be able to see their reflection in a mirror.

When a wave travels through a slit or around an obstacle, it is known as **diffraction**. A light wave will bend around an obstacle or through a slit and cause what is called a *diffraction pattern*. When the waves bend around an obstacle, it causes the addition of waves and the spreading of light on the other side of the opening.

Dispersion is used to describe the splitting of a single wave by refracting its components into separate parts. For example, if a wave of white light is sent through a dispersion prism, the light appears as its separate rainbow-colored components, due to each colored wavelength being refracted in the prism.

When wavelengths hit boundaries, different things occur. Objects will absorb certain wavelengths of light and reflect others, depending on the boundaries. This becomes important when an object appears to be a certain color. The color of an object is not actually within that object, but rather, in the wavelengths being transmitted by that object. For example, if a table appears to be red, that means the table is absorbing all other wavelengths of visible light except those of the red wavelength. The table is reflecting, or transmitting, the wavelengths associated with red back to the human eye, and so it appears red.

Interference describes when an object affects the path of a wave, or another wave interacts with a wave. Waves interacting with each other can result in either *constructive interference* or *destructive interference*, based on their positions. With constructive interference, the waves are in sync with each other and combine to reinforce each other. In the case of deconstructive interference, the waves are out of sync and reduce the effect of each other to some degree. In *scattering*, the boundary can change the direction or energy of a wave, thus altering the entire wave. *Polarization* changes the oscillations of a wave and can alter its appearance in light waves. For example, polarized sunglasses remove the "glare" from sunlight by altering the oscillation pattern observed by the wearer.

When a wave hits a boundary and is completely reflected, or if it cannot escape from one medium to another, it is called **total internal reflection**. This effect can be seen in the diamonds with a brilliant cut.

The angle cut on the sides of the diamond causes the light hitting the diamond to be completely reflected back inside the gem, making it appear brighter and more colorful than a diamond with different angles cut into its surface.

The **Doppler effect** applies to situations with both light and sound waves. The premise of the Doppler effect is that, based upon the relative position or movement of a source and an observer, waves can seem shorter or longer than they actually are. When the Doppler effect is noted with sound, it warps the noise being heard by the observer. This makes the pitch or frequency seem shorter or higher as the source is approaching, and then longer or lower as the source is getting farther away. The frequency/pitch of the source never actually changes, but the sound in respect to the observer makes it seem like the sound has changed. This can be observed when a siren passes by an observer on the road. The siren sounds much higher in pitch as it approaches the observer and then lower after it passes and is getting farther away.

The Doppler effect also applies to situations involving light waves. An observer in space would see light approaching as being shorter wavelengths than the light actually is, causing it to look blue. When the light wave gets farther away, the light would appear red because of the apparent elongation of the wavelength. This is called the *red-blue shift*.

Basic Optics

When reflecting light, a mirror can be used to observe a virtual (not real) image. A *plane mirror* is a piece of glass with a coating in the background to create a reflective surface. An image is what the human eye sees when light is reflected off the mirror in an unmagnified manner. If a *curved mirror* is used for reflection, the image seen will not be a true reflection. Instead, the image will either be enlarged or miniaturized compared to its actual size. Curved mirrors can also make the object appear closer or farther away than the actual distance the object is from the mirror.

Lenses can be used to refract or bend light to form images. Examples of lenses are the human eye, microscopes, and telescopes. The human eye interprets the refraction of light into images that humans understand to be actual size. *Microscopes* allow objects that are too small for the unaided human eye to be enlarged enough to be seen. *Telescopes* allow objects to be viewed that are too far away to be seen with the unaided eye. *Prisms* are pieces of glass that can have a wavelength of light enter one side and appear to be divided into its component wavelengths on the other side. This is due to the ability of the prism to slow certain wavelengths more than others.

Sound

Sound travels in waves and is the movement of vibrations through a medium. It can travel through air (gas), land, water, etc. For example, the noise a human hears in the air is the vibration of the waves as they reach the ear. The human brain translates the different frequencies (pitches) and intensities of the vibrations to determine what created the noise.

A tuning fork has a predetermined frequency because of the length and thickness of its tines. When struck, it allows vibrations between the two tines to move the air at a specific rate. This creates a specific tone, or note, for that size of tuning fork. The number of vibrations over time is also steady for that tuning fork and can be matched with a frequency. All pitches heard by the human ear are categorized by using frequency and are measured in Hertz (cycles per second).

The level of sound in the air is measured with sound level meters on a decibel (dB) scale. These meters respond to changes in air pressure caused by sound waves and measure sound intensity. One decibel is

1/10th of a *bel*, named after Alexander Graham Bell, the inventor of the telephone. The decibel scale is logarithmic, so it is measured in factors of 10. This means, for example, that a 10 dB increase on a sound meter equates to a 10-fold increase in sound intensity.

Life Science

Structure and Function of Animal and Plant Cell Organelles

Animal and plant cells contain many of the same or similar **organelles**, which are membrane enclosed structures that each have a specific function; however, there are a few organelles that are unique to either one or the other general cell type. The following cell organelles are found in both animal and plant cells, unless otherwise noted in their description:

- *Nucleus*: The nucleus consists of three parts: the nuclear envelope, the nucleolus, and chromatin. The *nuclear envelope* is the double membrane that surrounds the nucleus and separates its contents from the rest of the cell. The *nucleolus* produces ribosomes. *Chromatin* consists of DNA and protein, which form chromosomes that contain genetic information. Most cells have only one nucleus; however, some cells, such as skeletal muscle cells, have multiple nuclei.

- *Endoplasmic reticulum (ER)*: The ER is a network of membranous sacs and tubes that is responsible for membrane synthesis. It is also responsible for packaging and transporting proteins into vesicles that can move out of the cell. It folds and transports other proteins to the Golgi apparatus. It contains both smooth and rough regions; the rough regions have ribosomes attached, which are the sites of protein synthesis.

- *Flagellum*: Flagellum are found only in animal cells. They are made up of a cluster of microtubules projected out of the plasma membrane, and they aid in cell mobility.

- *Centrosome*: The centrosome is the area of the cell where *microtubules*, which are filaments that are responsible for movement in the cell, begin to be formed. Each centrosome contains two centrioles. Each cell contains one centrosome.

- *Cytoskeleton*: The cytoskeleton in animal cells is made up of microfilaments, intermediate filaments, and microtubules. In plant cells, the cytoskeleton is made up of only microfilaments and microtubules. These structures reinforce the cell's shape and aid in cell movement.

- *Microvilli*: Microvilli are found only in animal cells. They are protrusions in the cell membrane that increase the cell's surface area. They have a variety of functions, including absorption, secretion, and cellular adhesion. They are found on the apical surface of epithelial cells, such as in the small intestine. They are also located on the plasma surface of a female's eggs to help anchor sperm that are attempting fertilization.

- *Peroxisome*: A peroxisome contains enzymes that are involved in many of the cell's metabolic functions, one of the most important being the breakdown of very long chain fatty acids. Peroxisomes produces hydrogen peroxide as a byproduct of these processes and then converts the hydrogen peroxide to water. There are many peroxisomes in each cell.

- *Mitochondrion*: The mitochondrion is often called the powerhouse of the cell and is one of the most important structures for maintaining regular cell function. It is where aerobic cellular respiration occurs and where most of the cell's adenosine triphosphate (ATP) is generated. The number of mitochondria in a cell varies greatly from organism to organism, and from cell to cell. In human cells, the number of mitochondria can vary from zero in a red blood cell, to 2000 in a liver cell.
- *Lysosome*: Lysosomes are responsible for digestion and can hydrolyze macromolecules. There are many lysosomes in each cell.
- *Golgi apparatus*: The Golgi apparatus is responsible for the composition, modification, organization, and secretion of cell products. Because of its large size, it was actually one of the first organelles to be studied in detail. There are many Golgi apparatuses in each cell.
- *Ribosomes*: Ribosomes are found either free in the cytosol, bound to the rough ER, or bound to the nuclear envelope. They manufacture proteins within the cell.
- *Plasmodesmata*: The plasmodesmata are found only in plant cells. They are cytoplasmic channels, or tunnels, that go through the cell wall and connect the cytoplasm of adjacent cells.
- *Chloroplast*: Chloroplasts are found only in plant cells. They are responsible for *photosynthesis*, which is the process of converting sunlight to chemical energy that can be stored and used later to drive cellular activities.
- *Central vacuole*: A central vacuole is found only in plant cells. It is responsible for storing material and waste. This is the only vacuole found in a plant cell.
- *Plasma membrane*: The plasma membrane is a phospholipid bilayer that encloses the cell.

- *Cell wall*: Cell walls are only present in plant cells. The cell wall is made up of strong fibrous substances including cellulose and other polysaccharides, and protein. It is a layer outside of the plasma membrane, which protects the cell from mechanical damage and helps maintain the cell's shape.

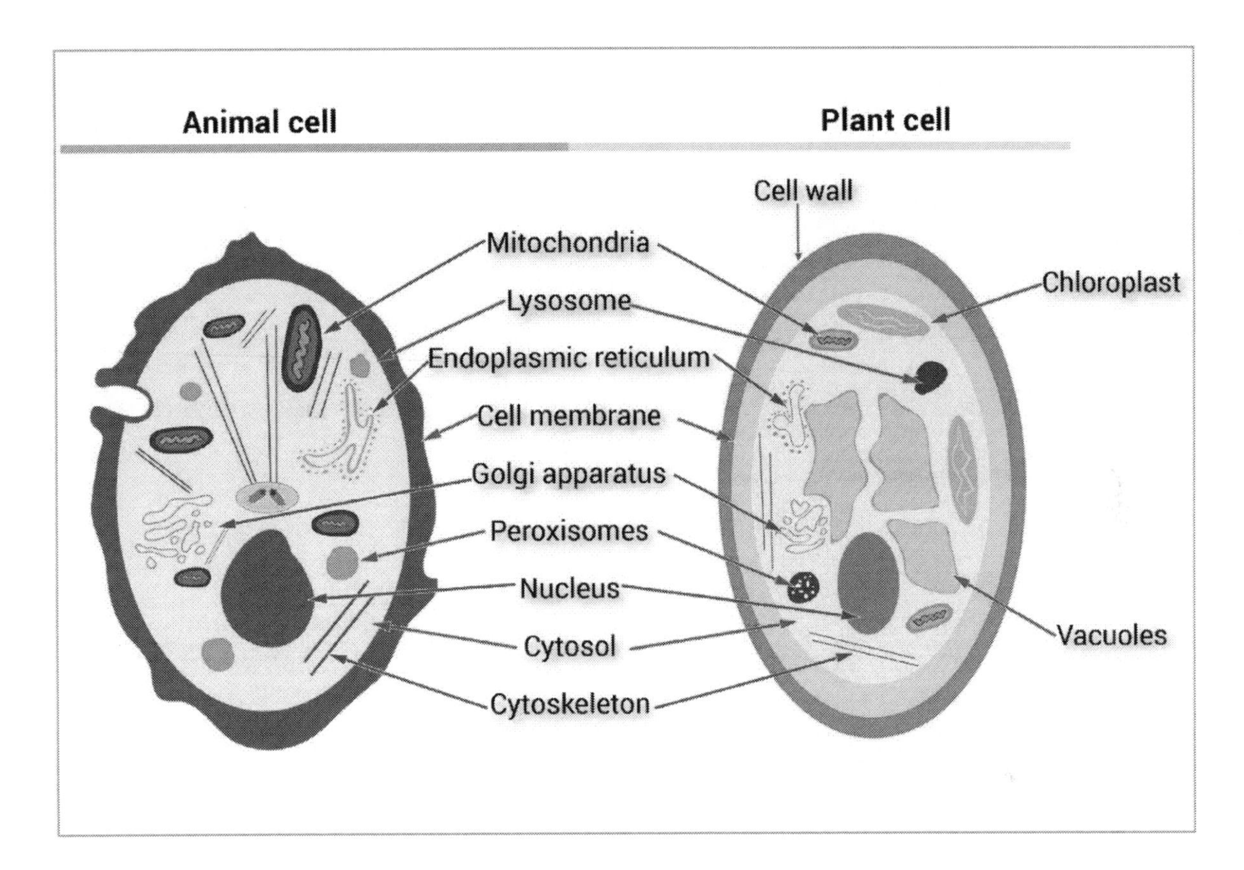

Levels of Organization

There are about two hundred different types of cells in the human body. Cells group together to form *biological* tissues, and tissues combine to form organs, such as the heart and kidneys. Organs that work together to perform vital functions of the human body form organ systems. There are eleven organ systems in the human body: skeletal, muscular, urinary, nervous, digestive, endocrine, reproductive, respiratory, cardiovascular, integumentary, and lymphatic. Although each system has its own unique function, they all rely on each other, either directly or indirectly, to operate properly.

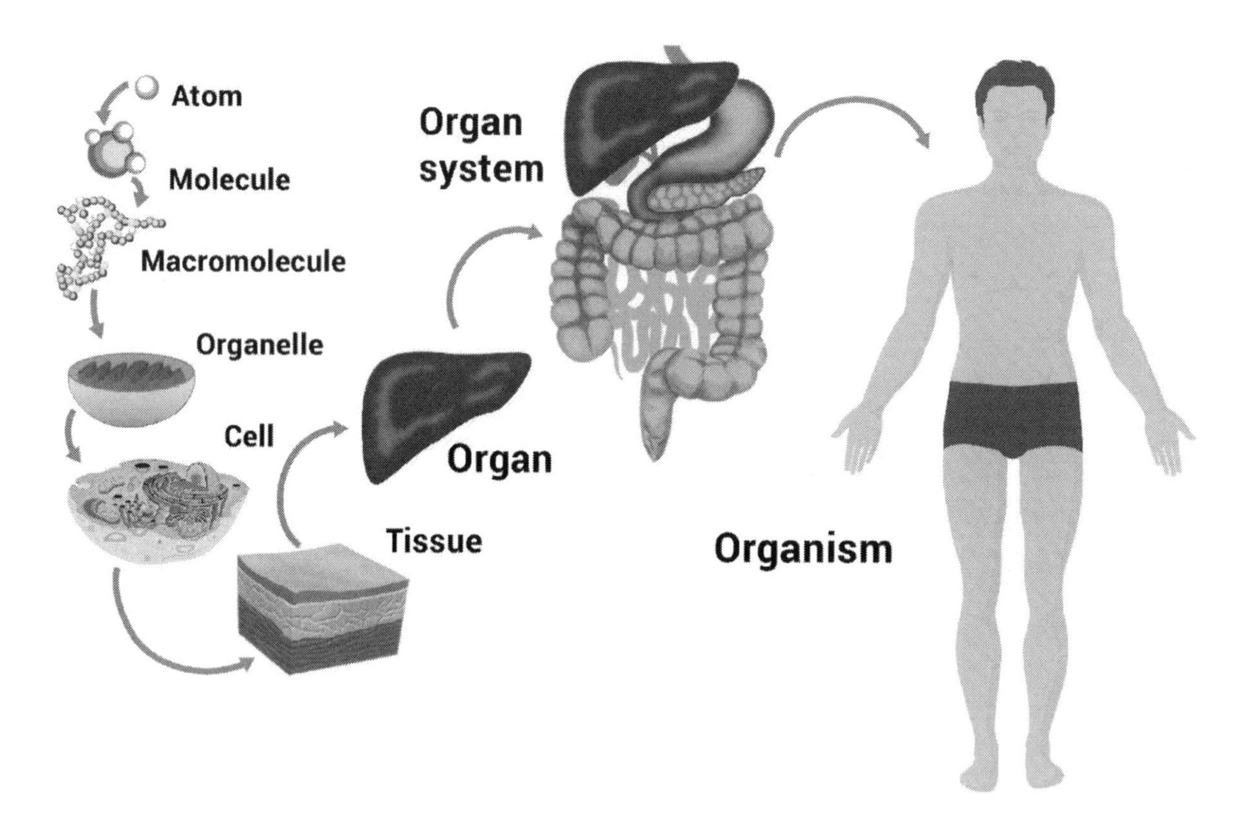

Major Features of Common Animal Cell Types

The most common animal cell types are blood, muscle, nerve, epithelial, and gamete cells. The three main blood cells are *red blood cells (RBCs), white blood cells (WBCs),* and *platelets*. RBCs transport oxygen and carbon dioxide through the body. They do not have a nucleus and they live for about 120 days in the blood. WBCs defend the body against diseases. They do have a nucleus and live for only three to four days in the human body. Platelets help with the formation of blood clots following an injury. They do not have a nucleus and live for about eight days after formation. *Muscle cells* are long, tubular cells that form muscles, which are responsible for movement in the body. On average, they live for about fifteen years, but this number is highly dependent on the individual body. There are three main types of muscle tissue: skeletal, cardiac, and smooth. *Skeletal muscle cells* have multiple nuclei and are the only voluntary muscle cell, which means that the brain consciously controls the movement of skeletal muscle. *Cardiac muscle cells* are only found in the heart; they have a single nucleus and are involuntary. *Smooth muscle cells* make up the walls of the blood vessels and organs. They have a single nucleus and are involuntary. *Nerve cells* conduct electrical impulses that help send information and instructions from the brain to the rest of the body. They contain a single nucleus and have a specialized membrane that allows for this electrical signaling between cells. *Epithelial* cells cover exposed surfaces, and line internal cavities and passageways. *Gametes* are specialized cells that are responsible for reproduction. In the human body, the gametes are the egg and the sperm.

Prokaryotes and Eukaryotes

There are two distinct types of cells that make up most living organisms: *prokaryotic* and *eukaryotic*. Both types of cells are enclosed by a cell membrane, which is selectively permeable. Selective permeability means essentially that it is a gatekeeper, allowing certain molecules and ions in and out, and keeping unwanted ones at bay, at least until they are ready for use. They both contain *ribosomes* and DNA. One major difference between these types of cells is that in eukaryotic cells, the cell's DNA is enclosed in a membrane-bound nucleus, whereas in prokaryotic cells, the cell's DNA is in a region—called the *nucleoid*—that is not enclosed by a membrane. Another major difference is that eukaryotic cells contain organelles, while prokaryotic cells do not have organelles.

Prokaryotic cells include *bacteria* and archaea. They do not have a nucleus or any membrane-bound organelles, are unicellular organisms, and are generally very small in size. Eukaryotic cells include animal, plant, fungus, and protist cells. *Fungi* are unicellular microorganisms such as yeasts, molds, and mushrooms. Their distinguishing characteristic is the chitin that is in their cell walls. *Protists* are organisms that are not classified as animals, plants, or fungi; they are unicellular; and they do not form tissues.

Classification Schemes

Taxonomy is the science behind the biological names of organisms. Biologists often refer to organisms by their Latin scientific names to avoid confusion with common names, such as with fish. Jellyfish, crayfish, and silverfish all have the word "fish" in their name, but they belong to three different species. In the eighteenth century, Carl Linnaeus invented a naming system for species that included using the Latin scientific name of a species, called the *binomial*, which has two parts: the *genus*, which comes first, and the *specific epithet*, which comes second. Similar species are grouped into the same genus. The Linnaean system is the commonly used taxonomic system today and, moving from comprehensive similarities to more general similarities, classifies organisms into their species, genus, family, order, class, phylum, and kingdom. *Homo sapiens* is the Latin scientific name for humans.

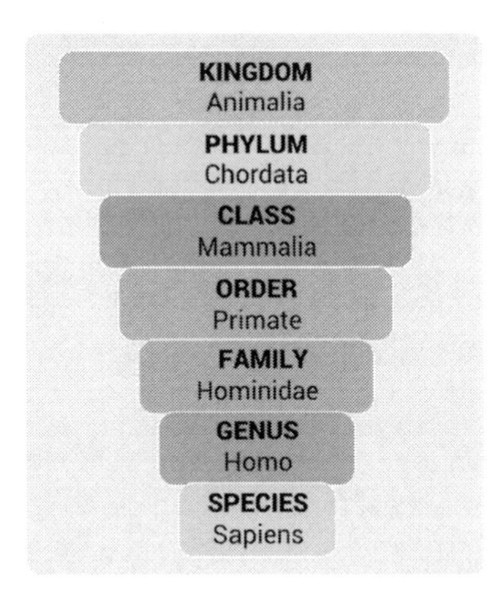

Phylogenetic trees are branching diagrams that represent the proposed evolutionary history of a species. The branch points most often match the classification groups set forth by the Linnaean system.

Using this system helps elucidate the relationship between different groups of organisms. The diagram below is that of an empty phylogenetic tree:

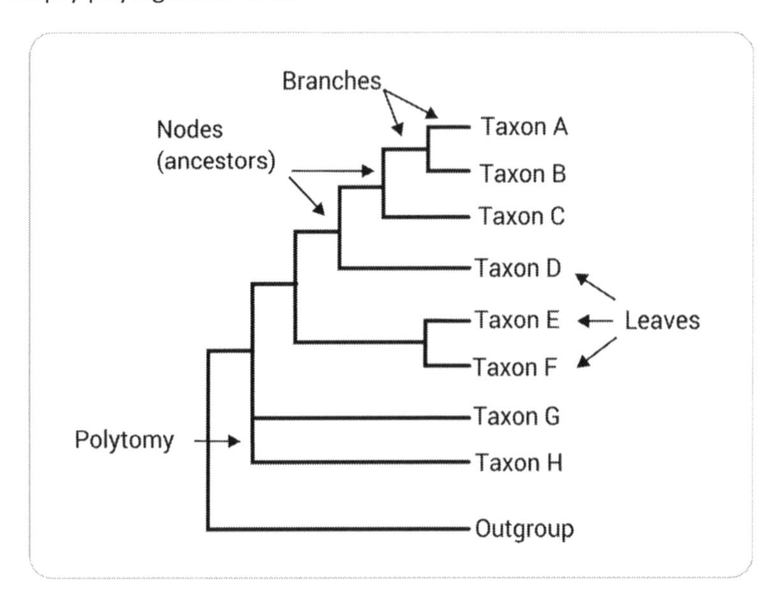

Each branch of the tree represents the divergence of two species from a common ancestor. For example, the coyote is known as Canis latrans and the gray wolf is known as Canis lupus. Their common ancestor, the Canis lepophagus, which is now extinct, is where their shared genus derived.

Characteristics of Bacteria, Animals, Plants, Fungi, and Protists

As discussed earlier, there are two distinct types of cells that make up most living organisms: prokaryotic and eukaryotic. Bacteria (and archaea) are classified as prokaryotic cells, whereas animal, plant, fungi, and protist cells are classified as eukaryotic cells.

Although animal cells and plant cells are both eukaryotic, they each have several distinguishing characteristics. *Animal cells* are surrounded by a plasma membrane, while *plant cells* have a cell wall made up of cellulose that provides more structure and an extra layer of protection for the cell. Animals use oxygen to breathe and give off carbon dioxide, while plants do the opposite—they take in carbon dioxide and give off oxygen. Plants also use light as a source of energy. Animals have highly developed sensory and nervous systems and the ability to move freely, while plants lack both abilities. Animals, however, cannot make their own food and must rely on their environment to provide sufficient nutrition, whereas plants do make their own food.

Fungal cells are typical eukaryotes, containing both a nucleus and membrane-bound organelles. They have a cell wall, similar to plant cells; however, they use oxygen as a source of energy and cannot perform photosynthesis. They also depend on outside sources for nutrition and cannot produce their own food. Of note, their cell walls contain *chitin.*

Protists are a group of diverse eukaryotic cells that are often grouped together because they do not fit into the categories of animal, plant, or fungal cells. They can be categorized into three broad categories: protozoa, protophyta, and molds. These three broad categories are essentially "animal -like," "plant-like," and "fungus-like," respectively. All of them are unicellular and do not form tissues. Besides this simple similarity, protists are a diverse group of organisms with different characteristics, life cycles, and cellular structures.

Types of Anatomy and Physiology

Anatomy is the study of external and internal body parts and structures, and their physical relationships to each other. Physiology is the study of the function of living organisms and their body parts. Understanding these areas of study is easier after learning the specific terms relating to the field.

Gross Anatomy: The study of large structures and features visible to the naked eye. It is also known as macroscopic anatomy.

Microscopic: The study of structures and their functions that can only be seen with magnification, such as through a magnifying glass or microscope.

Systemic: The study of a major organ system.

Developmental: The study of the physical changes that occur between conception and physical maturity. It includes gross anatomy, microscopic anatomy and physiology, because of the broad range of anatomical structures that undergo changes during this time.

Comparative: The study of structural organization and their functions among different types of animals. Animals that look different now but have the same internal structures with similar functions are proposed to have an evolutionary relationship.

Pathophysiology: The study of the effects of diseases on how an organ or organ system functions.

Integumentary System (Skin)

Skin consists of three layers: epidermis, dermis, and the hypodermis. There are four types of cells that make up the keratinized stratified squamous epithelium in the epidermis. They are keratinocytes, melanocytes, Merkel cells, and Langerhans cells. Skin is composed of many layers, starting with a basement membrane. On top of that sits the stratum germinativum, the stratum spinosum, the stratum granulosum, the stratum lucidum, and then the stratum corneum at the outer surface. Skin can be classified as thick or thin. These descriptions refer to the epidermis layer. Most of the body is covered with thin skin, but areas such as the palm of the hands are covered with thick skin. The dermis consists of a superficial papillary layer and a deeper reticular layer. The papillary layer is made of loose connective tissue, containing capillaries and the axons of sensory neurons. The reticular layer is a meshwork of tightly packed irregular connective tissue, containing blood vessels, hair follicles, nerves, sweat glands, and sebaceous glands. The hypodermis is a loose layer of fat and connective tissue. Since it is the third layer, if a burn reaches this third degree, it has caused serious damage.

Sweat glands and sebaceous glands are important exocrine glands found in the skin. Sweat glands regulate temperature, and remove bodily waste by secreting water, nitrogenous waste, and sodium salts to the surface of the body. Some sweat glands are classified as apocrine glands. Sebaceous glands are holocrine glands that secrete sebum, which is an oily mixture of lipids and proteins. Sebum protects the skin from water loss, as well as bacterial and fungal infections.

The three major functions of skin are protection, regulation, and sensation. Skin acts as a barrier and protects the body from mechanical impacts, variations in temperature, microorganisms, and chemicals. It regulates body temperature, peripheral circulation, and fluid balance by secreting sweat. It also contains a large network of nerve cells that relay changes in the external environment to the body.

Skeletal System

The skeletal system consists of the 206 bones that make up the skeleton, as well as the cartilage, ligaments, and other connective tissues that stabilize them. Bone is made of collagen fibers and calcium inorganic minerals, mostly in the form of hydroxyapatite, calcium carbonate, and phosphate salts. The inorganic minerals are strong but brittle, and the collagen fibers are weak but flexible, so the combination makes bone resistant to shattering. There are two types of bone: compact and spongy. Compact bone has a basic functional unit, called the Haversian system. Osteocytes, or bone cells, are arranged in concentric circles around a central canal, called the Haversian canal, which contains blood vessels. While Haversian canals run parallel to the surface of the bone, perforating canals, also known as the canals of Volkmann, run perpendicularly between the central canal and the surface of the bone. The concentric circles of bone tissue that surround the central canal within the Haversian system are called lamellae. The spaces that are found between the lamellae are called lacunae. The Haversian system is a reservoir for calcium and phosphorus for blood. Spongy bone, in contrast to compact bone, is lightweight and porous. It covers the outside of the bone and it gives it a shiny, white appearance. It has a branching network of parallel lamellae, called trabeculae.

Although spongy bone forms an open framework around the compact bone, it is still quite strong. Different bones have different ratios of compact-to-spongy bone, depending on their functions. The outside of the bone is covered by a periosteum, which has four major functions. It isolates and protects bones from the surrounding tissue; provides a place for attachment of the circulatory and nervous system structures; participates in growth and repair of the bone; and attaches the bone to the deep fascia. An endosteum is found inside the bone, covers the trabeculae of the spongy bone and lines the inner surfaces of the central canals.

One major function of the skeletal system is to provide structural support for the entire body. It provides a framework for the soft tissues and organs to attach to. The skeletal system also provides a reserve of important nutrients, such as calcium and lipids. Normal concentrations of calcium and phosphate in body fluids are partly maintained by the calcium salts stored in bone. Lipids that are stored in yellow bone marrow can be used as a source of energy. Red bone marrow produces red blood cells, white blood cells, and platelets that circulate in the blood. Certain groups of bones form protective barriers around delicate organs. The ribs, for example, protect the heart and lungs, the skull encloses the brain, and the vertebrae cover the spinal cord.

Muscular System

The muscular system of the human body is responsible for all movement that occurs. There are approximately 700 muscles in the body that are attached to the bones of the skeletal system and that make up half of the body's weight. Muscles are attached to the bones through tendons. Tendons are made up of dense bands of connective tissue and have collagen fibers that firmly attach to the bone on one side and the muscle on the other. Their fibers are actually woven into the coverings of the bone and muscle so they can withstand the large forces that are put on them when muscles are moving. There are three types of muscle tissue in the body: Skeletal muscle tissue pulls on the bones of the skeleton and causes body movement; cardiac muscle tissue helps pump blood through veins and arteries; and smooth muscle tissue helps move fluids and solids along the digestive tract and contributes to movement in other body systems. All of these muscle tissues have four important properties in common: They are excitable, meaning they respond to stimuli; contractile, meaning they can shorten and pull on connective tissue; extensible, meaning they can be stretched repeatedly, but maintain the ability to contract; and elastic, meaning they rebound to their original length after a contraction.

Muscles begin at an origin and end at an insertion. Generally, the origin is proximal to the insertion and the origin remains stationary while the insertion moves. For example, when bending the elbow and moving the hand up toward the head, the part of the forearm that is closest to the wrist moves and the part closer to the elbow is stationary. Therefore, the muscle in the forearm has an origin at the elbow and an insertion at the wrist.

Body movements occur by muscle contraction. Each contraction causes a specific action. Muscles can be classified into one of three muscle groups based on the action they perform. Primary movers, or agonists, produce a specific movement, such as flexion of the elbow. Synergists are in charge of helping the primary movers complete their specific movements. They can help stabilize the point of origin or provide extra pull near the insertion. Some synergists can aid an agonist in preventing movement at a joint. Antagonists are muscles whose actions are opposite of the agonist's. If an agonist is contracting during a specific movement, the antagonist is stretched. During flexion of the elbow, the biceps' brachii muscle contracts and acts as an agonist, while the triceps' brachii muscle on the opposite side of the upper arm acts as an antagonist and stretches.

Skeletal muscle tissue has several important functions. It causes movement of the skeleton by pulling on tendons and moving the bones. It maintains body posture through the contraction of specific muscles responsible for the stability of the skeleton. Skeletal muscles help support the weight of internal organs and protect these organs from external injury. They also help to regulate body temperature within a normal range. Muscle contractions require energy and produce heat, which heats the body when cold.

Nervous System

Although the nervous system is one of the smallest organ systems in the human body, it is the most complex. It consists of all of the neural tissue and is in charge of controlling and adjusting the activities of all of the other systems of the body. Neural responses to stimuli are often fast but disappear quickly once the neural activity stops. Neural tissue contains two types of cells: neurons and neuroglia. Neurons, or nerve cells, are the main cells responsible for transferring and processing information in the nervous system. Neuroglia support the neurons by providing a framework around them and isolating them from the surrounding environment. They also act as phagocytes and protect neurons from harmful substances.

The nervous system is made of the central nervous system (CNS) and the peripheral nervous system (PNS). The CNS includes the brain and the spinal cord, while the PNS includes the rest of the neural tissue not included in the CNS. The CNS is where intelligence, memory, learning, and emotions are processed. It is responsible for processing and coordinating sensory data and motor commands. The PNS is responsible for relaying sensory information and motor commands between the CNS and peripheral tissues and systems. The PNS has two subdivisions, known as the afferent and efferent divisions. While the afferent division relays sensory information to the CNS, the efferent division transmits motor commands to muscles and glands. The efferent division consists of the somatic nervous system (SNS), which controls skeletal muscle contractions, and the autonomic nervous system (ANS), which regulates activity of smooth muscle, cardiac muscle, and glands.

Two types of pathways are used to communicate information between the brain and the peripheral tissues. Sensory pathways start in a peripheral system and end in the brain. Motor pathways carry information from the brain to peripheral systems. Motor commands often occur in response to the information transmitted through a sensory pathway. Processing in both pathways happens at several points along the way, where neurons pass the information to each other.

The nervous system is responsible for processing both general senses and specialized senses. General senses include temperature, pain, touch, pressure, vibration and proprioception. Specialized senses include olfaction (smell), gustation (taste), equilibrium, hearing, and vision. The information from each sense is processed through a specific receptor for that sense. A receptor that is sensitive to touch may not be responsive to chemical stimuli, for example. The specificity of the receptor is developed either from its individual structure or from accessory cells or structures creating a shield against other senses.

Endocrine System

The endocrine system is made of the ductless tissues and glands that secrete hormones into the interstitial fluids of the body. Interstitial fluid is the solution that surrounds tissue cells within the body. This system works closely with the nervous system to regulate the physiological activities of the other systems of the body to maintain homeostasis. While the nervous system provides quick, short-term responses to stimuli, the endocrine system acts by releasing hormones into the bloodstream that get distributed to the whole body. The response is slow but long-lasting, ranging from a few hours to a few weeks.

Hormones are chemical substances that change the metabolic activity of tissues and organs. While regular metabolic reactions are controlled by enzymes, hormones can change the type, activity, or quantity of the enzymes involved in the reaction. They bind to specific cells and start a biochemical chain of events that changes the enzymatic activity. Hormones can regulate development and growth, digestive metabolism, mood, and body temperature, among other things. Often small amounts of hormone will lead to large changes in the body.

Major Endocrine Glands

Hypothalamus: A part of the brain, the hypothalamus connects the nervous system to the endocrine system via the pituitary gland. Although it is considered part of the nervous system, it plays a dual role in regulating endocrine organs.

Pituitary Gland: A pea-sized gland found at the bottom of the hypothalamus. It has two lobes, called the anterior and posterior lobes. It plays an important role in regulating the function of other endocrine glands. The hormones released control growth, blood pressure, certain functions of the sex organs, salt concentration of the kidneys, internal temperature regulation, and pain relief.

Thyroid Gland: This gland releases hormones, such as thyroxine, that are important for metabolism, growth and development, temperature regulation, and brain development during infancy and childhood. Thyroid hormones also monitor the amount of circulating calcium in the body.

Parathyroid Glands: These are four pea-sized glands located on the posterior surface of the thyroid. The main hormone secreted is called parathyroid hormone (PTH) and helps with the thyroid's regulation of calcium in the body.

Thymus Gland: The thymus is located in the chest cavity, embedded in connective tissue. It produces several hormones important for development and maintenance of normal immunological defenses. One hormone promotes the development and maturation of lymphocytes, which strengthens the immune system.

Adrenal Gland: One adrenal gland is attached to the top of each kidney. It produces adrenaline and is responsible for the "fight or flight" reactions in the face of danger or stress. The hormones epinephrine and norepinephrine cooperate to regulate states of arousal.

Pancreas: The pancreas is an organ that has both endocrine and exocrine functions. The endocrine functions are controlled by the pancreatic islets of Langerhans, which are groups of beta cells scattered throughout the gland that secrete insulin to lower blood sugar levels in the body. Neighboring alpha cells secrete glucagon to raise blood sugar.

Pineal Gland: The pineal gland secretes melatonin, a hormone derived from the neurotransmitter serotonin. Melatonin can slow the maturation of sperm, oocytes, and reproductive organs. It also regulates the body's circadian rhythm, which is the natural awake/asleep cycle. It also serves an important role in protecting the CNS tissues from neural toxins.

Testes and Ovaries: These glands secrete testosterone and estrogen, respectively, and are responsible for secondary sex characteristics, as well as reproduction.

Circulatory System

The circulatory system is composed of the heart and blood vessels of the body. The heart is the main organ of the circulatory system. It acts as a pump and works to circulate blood throughout the body. Gases, nutrients, and waste are constantly exchanged between the circulating blood and interstitial fluid, keeping tissues and organs alive and healthy. The circulatory system is divided into the pulmonary and systemic circuits. The pulmonary circuit is responsible for carrying carbon dioxide-rich blood to the lungs and returning oxygen-rich blood to the heart. The systemic circuit transports the oxygen-rich blood to the rest of the body and returns carbon dioxide-rich blood to the heart.

Heart

The heart is located posterior to the sternum, on the left side, in the front of the chest. The heart wall is made of three distinct layers. The outer layer, the epicardium, is a *serous* membrane that is also known as the visceral pericardium. The middle layer is called the myocardium, and contains connective tissue, blood vessels, and nerves within its layers of cardiac muscle tissue. The inner layer is the endocardium and is made of a simple squamous epithelium. This layer includes the heart valves and is continuous with the endothelium of the attached blood vessels.

The heart has four chambers: the right atrium, the right ventricle, the left atrium, and the left ventricle. An interatrial septum, or wall, separates the right and left atria, and the right and left ventricles are separated by an interventricular septum. The atrium and ventricle on the same side of the heart have an opening between them that is regulated by a valve. The valve maintains blood flow in only one direction, moving from the atrium to the ventricle, and prevents backflow. The systemic circuit pumps oxygen-poor blood into the right atrium, then pumps it into the right ventricle. From there, the blood enters the pulmonary trunk and then flows into the pulmonary arteries, where it can become re-oxygenated. Oxygen-rich blood from the lungs flows into the left atrium and then passes into the left ventricle. From there, blood enters the aorta and is pumped to the entire systemic circuit.

Blood

Blood circulates throughout the body in a system of vessels that includes arteries, veins, and capillaries. It distributes oxygen, nutrients, and hormones to all the cells in the body. The vessels are muscular tubes that allow gas exchange to occur. Arteries carry oxygen-rich blood from the heart to the other tissues of the body. The largest artery is the aorta. Veins collect oxygen-depleted blood from tissues and organs and return it to the heart. The walls of veins are thinner and less elastic than arteries, because the blood pressure in veins is lower than in arteries. Capillaries are the smallest of the blood vessels and do not function individually; instead, they work together in a unit, called a capillary bed. This network of capillaries provides oxygen-rich blood from arterioles to tissues and feeds oxygen-poor blood from tissues back to venules.

Blood comprises plasma and formed elements, which include red blood cells (RBCs), white blood cells (WBCs), and platelets. Plasma is the liquid matrix of the blood and contains dissolved proteins. RBCs transport oxygen and carbon dioxide. WBCs are part of the immune system and help fight diseases. Platelets contain enzymes and other factors that help with blood clotting.

Respiratory System

The respiratory system mediates the exchange of gas between the air and the blood, mainly by the act of breathing. This system is divided into the upper respiratory system and the lower respiratory system. The upper system comprises the nose, the nasal cavity and sinuses, and the pharynx. The lower respiratory system comprises the larynx (voice box), the trachea (windpipe), the small passageways leading to the lungs, and the lungs. The upper respiratory system is responsible for filtering, warming, and humidifying the air that gets passed to the lower respiratory system, protecting the lower respiratory system's more delicate tissue surfaces.

The Lungs

The right lung is divided into three lobes: superior, middle, and inferior. The left lung is divided into two lobes: superior and inferior. The left lung is smaller than the right, likely because it shares space in the chest cavity with the heart. Together, the lungs contain approximately 1500 miles of airway passages. The bronchi, which carry air into the lungs, branch into bronchioles and continue to divide into smaller and smaller passageways, until they become alveoli, which are the smallest passages. Most of the gas exchange in the lungs occurs between the blood-filled pulmonary capillaries and the air-filled alveoli.

Functions of the Respiratory System

The respiratory system has many functions. Most importantly, it provides a large area for gas exchange between the air and the circulating blood. It protects the delicate respiratory surfaces from environmental variations and defends them against pathogens. It is responsible for producing the sounds that the body makes for speaking and singing, as well as for non-verbal communication. It also helps regulate blood volume, blood pressure, and body fluid pH.

Breathing

When a breath of air is inhaled, oxygen enters the nose or mouth, and passes into the sinuses, where the temperature and humidity of the air get regulated. The air then passes into the trachea and is filtered. From there, the air travels into the bronchi and reaches the lungs. Bronchi are tubes that lead from the trachea to each lung and are lined with cilia and mucus that collect dust and germs along the way. Within the lungs, oxygen and carbon dioxide are exchanged between the air in the alveoli and the blood in the pulmonary capillaries. Oxygen-rich blood returns to the heart and is pumped through the systemic circuit. Carbon dioxide-rich air is exhaled from the body.

Breathing is possible due to the muscular diaphragm pulling on the lungs, increasing their volume and decreasing their pressure. Air flows from the external high-pressure system to the low-pressure system inside the lungs. When breathing out, the diaphragm releases its pressure difference, decreases the lung volume, and forces the stale air back out.

Digestive System

The digestive system is a group of organs that work together to transform food and liquids into energy, which can then be used by the body as fuel. Food is ingested and then passes through the alimentary canal, or GI tract, which comprises the mouth, pharynx, esophagus, stomach, small intestine, and large intestine. The digestive system has accessory organs, including the liver, gallbladder, and pancreas, that help with the processing of food and liquids, but do not have food pass directly through them. These accessory organs and the digestive system organs work together in the following functions:

Ingestion: Food and liquids enter the alimentary canal through the mouth.

Introductory Mechanical and Chemical Processing: Teeth grind the food, and the tongue swirls it to facilitate swallowing. Enzymes in saliva begin chemical digestion.

Advanced Mechanical and Chemical Digestion: The muscular stomach uses physical force and enzymes, which function at low pH levels, to break down the food and liquid's complex molecules, such as sugars, lipids, and proteins, into smaller molecules that can be absorbed by the small intestine.

Secretion: Most of the acids, buffers, and enzymes that aid in digestion are secreted by the accessory organs, but some are provided by the digestive tract. Bile from the liver facilitates fat digestion.

Absorption: Vitamins, electrolytes, organic molecules, and water are absorbed by the villi and microvilli lining in the small intestine and are moved to the interstitial fluid of the digestive tract.

Compaction: Indigestible materials and organic wastes are dehydrated in the large intestine and compacted before elimination from the body.

Excretion: Waste products are excreted from the digestive tract.

Major Organs of the Alimentary Canal

Stomach: This organ stores food so the body has time to digest large meals. Its highly acidic environment and enzyme secretions, such as pepsin and trypsin, aid in digestion. It also aids in mechanical processing through muscular contractions.

Small Intestine: This organ is a thin tube that is approximately ten feet long. It secretes enzymes to aid in digestion and has many folds that increase its surface area and allows for maximum absorption of nutrients from the digested food.

Large Intestine: This organ is a long thick tube that is about five feet long. It absorbs water from the digested food and transports waste to be excreted from the body. It also contains symbiotic bacteria that further breaks down the waste products, allowing for any extra nutrients to be absorbed.

Major Accessory Organs

Liver: The liver produces and secretes bile, which is important for the digestion of lipids. It also plays a large role in the regulation of circulating levels of carbohydrates, amino acids, and lipids in the body. Excess nutrients are removed by the liver and deficiencies are corrected with its stored nutrients.

Gallbladder: This organ is responsible for storing and concentrating bile before it gets secreted into the small intestine. While the gallbladder is storing bile, it can regulate the bile's composition by absorbing water, thereby increasing the concentration of bile salts and other components.

Pancreas: The Pancreas has exocrine cells that secrete buffers and digestive enzymes. It contains specific enzymes for each type of food molecule, such as carbohydrases for carbohydrates, lipases for lipids, and proteinases for proteins.

Urinary System

The urinary system is made up of the kidneys, ureters, urinary bladder, and the urethra. It is the main system responsible for getting rid of the organic waste products, excess water and electrolytes are generated by the body's other systems. The kidneys are responsible for producing urine, which is a fluid waste product containing water, ions, and small soluble compounds. The urinary system has many important functions related to waste excretion. It regulates the concentrations of sodium, potassium, chloride, calcium, and other ions in the plasma by controlling the amount of each that is excreted in urine. This also contributes to the maintenance of blood pH. It regulates blood volume and pressure by controlling the amount of water lost in the urine and releasing erythropoietin and renin. It eliminates toxic substances, drugs, and organic waste products, such as urea and uric acid. It also synthesizes calcitriol, which is a hormone derivative of vitamin D3 that aids in calcium ion absorption by the intestinal epithelium.

The Kidneys

Under normal circumstances, humans have two functioning kidneys. They are the main organs are responsible for filtering waste products out of the blood and transferring them to urine. Every day, the kidneys filter approximately 120 to 150 quarts of blood and produce one to two quarts of urine. Kidneys are made of millions of tiny filtering units, called nephrons. Nephrons have two parts: a glomerulus, which is the filter, and a tubule. As blood enters the kidneys, the glomerulus allows fluid and waste products to pass through it and enter the tubule. Blood cells and large molecules, such as proteins, do not pass through and remain in the blood. The filtered fluid and waste then pass through the tubule, where any final essential minerals are sent back to the bloodstream. The final product at the end of the tubule is called urine.

Waste Excretion

Once urine accumulates, it leaves the kidneys. The urine travels through the ureters into the urinary bladder, a muscular organ that is hollow and elastic. As more urine enters the urinary bladder, its walls stretch and become thinner so there is no significant difference in internal pressure. The urinary bladder stores the urine until the body is ready for urination, at which time the muscles contract and force the urine through the urethra and out of the body.

Reproductive System

The reproductive system is responsible for producing, storing, nourishing, and transporting functional reproductive cells, or gametes, in the human body. It includes the reproductive organs, also known as gonads, the reproductive tract, the accessory glands and organs that secrete fluids into the reproductive tract, and the perineal structures, which are the external genitalia. The human male and female reproductive systems are very different from each other.

The Male System

The male gonads are called testes. The testes secrete androgens, mainly testosterone, and produce and store 500 million sperms cells, which are the male gametes, each day. An androgen is a steroid hormone that controls the development and maintenance of male characteristics. Once the sperm are mature, they move through a duct system, where they mix with additional fluids secreted by accessory glands, forming a mixture called semen. The sperm cells in semen are responsible for fertilization of the female gametes to produce offspring.

The Female System

The female gonads are the ovaries. Ovaries generally produce one immature gamete, or oocyte, per month. They are also responsible for secreting the hormones estrogen and progesterone. When the oocyte is released from the ovary, it travels along the uterine tubes, or Fallopian tubes, and then into the uterus. The uterus opens into the vagina. When sperm cells enter the vagina, they swim through the uterus and may fertilize the oocyte in the Fallopian tubes. The resulting zygote travels down the tube and implants into the uterine wall. The uterus protects and nourishes the developing embryo for nine months until it is ready for the outside environment. If the oocyte is not fertilized, it is released in the uterine, or menstrual, cycle. The menstrual cycle occurs monthly and involves the shedding of the functional part of the uterine lining.

Mammary glands are a specialized accessory organ of the female reproductive system. The mammary glands are located in the breast tissue, and during pregnancy begin to grow, and the cells proliferate in preparation for lactation. After pregnancy, the cells begin to secrete nutrient-filled milk, which is transferred into a duct system and out through the nipple for nourishment of the baby.

Response to Stimuli and Homeostasis

A *stimulus* is a change in the environment, either internal or external, around an organism that is received by a sensory receptor and causes the organism to react. *Homeostasis* is the stable state of an organism. When an organism reacts to stimuli, it works to counteract the change in order to reach homeostasis again.

Exchange with the Environment

Animals exchange gases and nutrients with the environment through several different organ systems. The *respiratory system* mediates the exchange of gas between the air and the circulating blood, mainly through the act of breathing. It filters, warms, and humidifies the air that gets inhaled and then passes it into the blood stream. The main function of the *excretory system* is to eliminate excess material and fluids in the body. The kidneys and bladder work together to filter organic waste products, excess water, and electrolytes from the blood that are generated by the other physiologic systems and excrete them from the body. The *digestive system* is a group of organs that work together to transform ingested food and liquid into energy, which can then be used by the body as fuel. Once all of the nutrients are absorbed, the waste products are excreted from the body.

Characteristics of Vascular and Nonvascular Plants

Plants that have an extensive vascular transport system are called *vascular plants*. Those plants without a transport system are called *nonvascular plants*. Approximately ninety-three percent of plants that are currently living and reproducing are vascular plants. The cells that comprise the vascular tissue in vascular plants form tubes that transport water and nutrients through the entire plant. Nonvascular plants include mosses, liverworts, and hornworts. They do not retain any water; instead, they transport water using other specialized tissue. They have structures that look like leaves but are actually just single sheets of cells without a cuticle or stomata.

Structure and Function of Roots, Leaves, and Stems

Roots are responsible for anchoring plants in the ground. They absorb water and nutrients and transport them up through the plant. *Leaves* are the main location of photosynthesis. They contain *stomata*, which are pores used for gas exchange, on their underside to take in carbon dioxide and release oxygen. *Stems* transport materials through the plant and support the plant's body. They contain *xylem*, which conducts water and dissolved nutrients upward through the plant, and *phloem*, which conducts sugars and metabolic products downward through the leaves.

Earth and Space Science

Major Features of the Solar System

Structure of the Solar System

The **solar system** is an elliptical planetary system with a large sun in the center that provides gravitational pull on the planets.

Laws of Motion

Planetary motion is governed by three scientific laws called **Kepler's laws**:

1. The orbit of a planet is elliptical in shape, with the Sun as one focus.

2. An imaginary line joining the center of a planet and the center of the Sun sweeps out equal areas during equal intervals of time.

3. For all planets, the ratio of the square of the orbital period is the same as the cube of the average distance from the Sun.

The most relevant of these laws is the first. Planets move in elliptical paths because of gravity; when a planet is closer to the Sun, it moves faster because it has built up gravitational speed. As illustrated in the diagram below, the second law states that it takes planet 1 the same time to travel along the A1 segment as the A2 segment, even though the A2 segment is shorter.

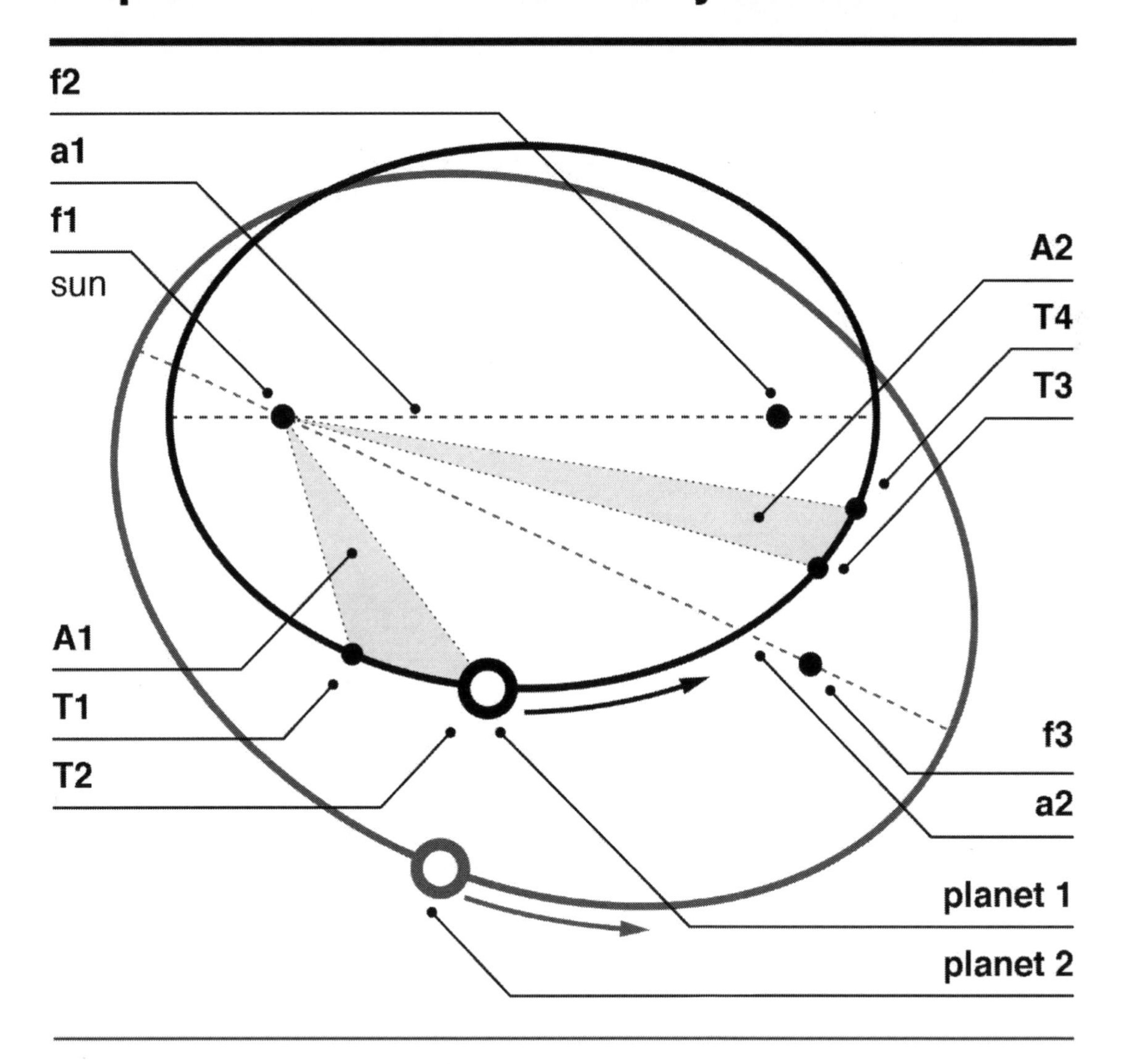

Characteristics of the Sun, Moon, and Planets

The Sun is comprised mainly of hydrogen and helium. Metals make up only about 2% of its total mass. The Sun is 1.3 million kilometers wide, weighs 1.989 x 10^{30} kilograms, and has temperatures of 5,800 Kelvin (9980 °F) on the surface and 15,600,000 Kelvin (28 million °F) at the core. The Sun's enormous size and gravity give it the ability to provide sunlight. The gravity of the Sun compresses hydrogen and helium atoms together through nuclear fusion and releases energy and light.

The Moon has a distinct core, mantle, and crust. It has elevations and craters created by impacts with large objects in the solar system. The Moon makes a complete orbit around the Earth every 27.3 days. It's relatively large compared to other moons in the Solar System, with a diameter one-quarter of the Earth and a mass 1/81 of the Earth.

The eight planets of the Solar System are divided into four inner (or terrestrial) planets and four outer (or Jovian) planets. In general, terrestrial planets are small, and Jovian planets are large and gaseous. The planets in the Solar System are listed below from nearest to farthest from the Sun:

- Mercury: the smallest planet in the Solar System; it only takes about 88 days to completely orbit the Sun
- Venus: around the same size, composition, and gravity as Earth and orbits the Sun every 225 days
- Earth: the only known planet with life
- Mars: called the Red Planet due to iron oxide on the surface; takes around 687 days to complete its orbit
- Jupiter: the largest planet in the system; made up of mainly hydrogen and helium
- Saturn: mainly composed of hydrogen and helium along with other trace elements; has 61 moons; has beautiful rings, which may be remnants of destroyed moons
- Uranus: the coldest planet in the system, with temperatures as low as -224.2 °Celsius (-371.56 °F)
- Neptune: the last and third-largest planet; also, the second-coldest planet

Asteroids, Meteoroids, Comets, and Dwarf/Minor Planets

Several other bodies travel through the universe. *Asteroids* are orbiting bodies composed of minerals and rock. They're also known as *minor planets*—a term given to any astronomical object in orbit around the Sun that doesn't resemble a planet or a comet. *Meteoroids* are mini asteroids with no specific orbiting pattern. *Meteors* are meteoroids that have entered the Earth's atmosphere and started melting from contact with greenhouse gases. *Meteorites* are meteors that have landed on Earth. *Comets* are composed of dust and ice and look like a comma with a tail from the melting ice as they streak across the sky.

Theories of Origin of the Solar System

One theory of the origins of the Solar System is the *nebular hypothesis*, which posits that the Solar System was formed by clouds of extremely hot gas called a *nebula*. As the nebula gases cooled, they became smaller and started rotating. Rings of the nebula left behind during rotation eventually condensed into planets and their satellites. The remaining nebula formed the Sun.

Another theory of the Solar System's development is the *planetesimal hypothesis*. This theory proposes that planets formed from cosmic dust grains that collided and stuck together to form larger and larger bodies. The larger bodies attracted each other, growing into moon-sized protoplanets and eventually planets.

Interactions of the Earth-Moon-Sun System

The Earth's Rotation and Orbital Revolution Around the Sun

Besides revolving around the Sun, the Earth also spins like a top. It takes one day for the Earth to complete a full spin, or rotation. The same is true for other planets, except that their "days" may be shorter or longer. One Earth day is about 24 hours, while one Jupiter day is only about nine Earth hours, and a Venus day is about 241 Earth days. Night occurs in areas that face away from the Sun, so one side of the planet experiences daylight and the other experiences night. This phenomenon is the reason that the Earth is divided into time zones. The concept of time zones was created to provide people around the world with a uniform standard time, so the Sun would rise around 7:00 AM, regardless of location.

Effect on Seasons

The Earth's tilted axis creates the seasons. When Earth is tilted toward the Sun, the Northern Hemisphere experiences summer while the Southern Hemisphere has winter—and vice versa. As the Earth rotates, the distribution of direct sunlight slowly changes, explaining how the seasons gradually change.

Phases of the Moon

The Moon goes through two phases as it revolves around Earth: waxing and waning. Each phase lasts about two weeks:

- Waxing—the right side of the Moon is illuminated
- New moon (dark): the Moon rises and sets with the Sun
- Crescent: a tiny sliver of illumination on the right
- First quarter: the right half of the Moon is illuminated
- Gibbous: more than half of the Moon is illuminated
- Full moon: the Moon rises at sunset and sets at sunrise
- Waning—the left side of the Moon is illuminated
- Gibbous: more than half is illuminated, only here it is the left side that is illuminated
- Last quarter: the left half of the Moon is illuminated
- Crescent: a tiny sliver of illumination on the left
- New moon (dark)—the Moon rises and sets with the Sun

Effect on Tides

Although the Earth is much larger, the Moon still has a significant gravitational force that pulls on Earth's oceans. At its closest to Earth, the Moon's gravitation pull is greatest and creates high tide. The opposite is true when the Moon is farthest from the Earth: less pull creates low tide.

Solar and Lunar Eclipses

Eclipses occur when the Earth, the Sun, and the Moon are all in line. If the three bodies are perfectly aligned, a total eclipse occurs; otherwise, it's only a partial eclipse. A *solar eclipse* occurs when the Moon is between the Earth and the Sun, blocking sunlight from reaching the Earth. A *lunar eclipse* occurs when the Earth interferes with the Sun's light reflecting off the full moon. The Earth casts a shadow on the Moon, but the particles of the Earth's atmosphere refract the light, so some light reaches the Moon, causing it to look yellow, brown, or red.

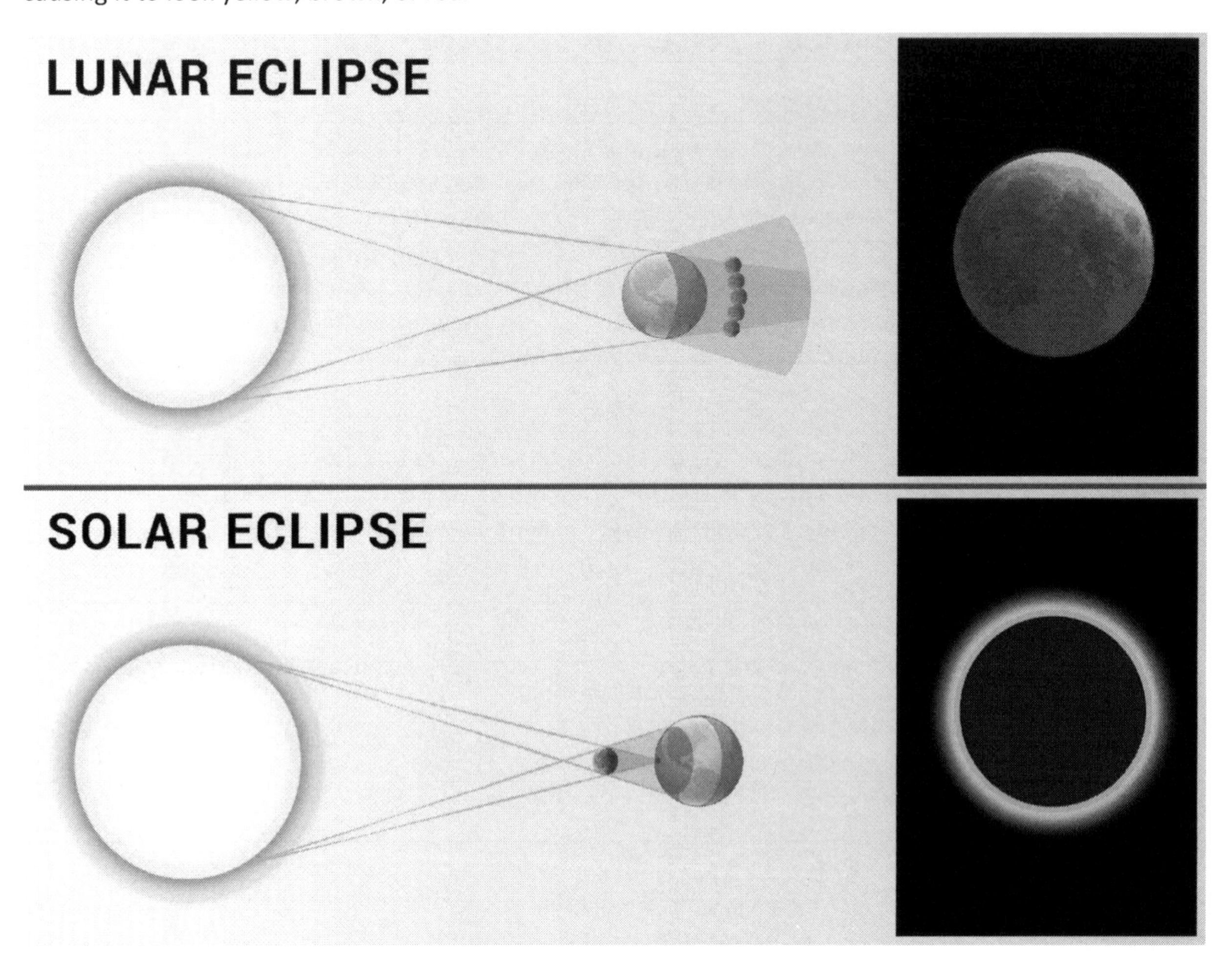

Time Zones

Longitudinal, or vertical, lines determine how far east or west different regions are from each other. These lines, also known as *meridians,* are the basis for time zones, which allocate different times to regions depending on their position eastward and westward of the prime meridian.

Effect of Solar Wind on the Earth

Solar winds are streams of charged particles emitted by the Sun, consisting of mostly electrons, protons, and alpha particles. The Earth is largely protected from solar winds by its magnetic field. However, the winds can still be observed, as they create phenomena like the beautiful Northern Lights (or Aurora Borealis).

Major Features of the Universe

Galaxies

Galaxies are clusters of stars, rocks, ice, and space dust. Like everything else in space, the exact number of galaxies is unknown, but there could be as many as a hundred billion. There are three types of galaxies: spiral, elliptical, and irregular. Most galaxies are *spiral galaxies*; they have a large, central galactic bulge made up of a cluster of older stars. They look like a disk with spinning arms. *Elliptical galaxies* are groups of stars with no pattern of rotation. They can be spherical or extremely elongated, and they don't have arms. *Irregular galaxies* vary significantly in size and shape.

To say that galaxies are large is an understatement. Most galaxies are 1,000 to 100,000 parsecs in diameter, with one *parsec* equal to about 19 trillion miles. The Milky Way is the galaxy that contains Earth's Solar System. It's one of the smaller galaxies that has been studied. The diameter of the Milky Way is estimated to be between 31,000 to 55,000 parsecs.

Characteristics of Stars and Their Life Cycles

Life Cycle of Stars

All stars are formed from nebulae. Depending on their mass, stars take different pathways during their life. Low- and medium-mass stars start as nebulae and then become red giants and white dwarfs. High-mass stars become red supergiants, supernovas, and then either neutron stars or black holes. Official stars are born as red dwarves because they have plentiful amounts of gas—mainly hydrogen—to undergo nuclear fusion. Red dwarves mature into white dwarves before expending their hydrogen fuel source. When the fuel is spent, it creates a burst of energy that expands the star into a red giant. Red giants eventually condense to form white dwarves, which is the final stage of a star's life.

Stars that undergo nuclear fusion and energy expenditure extremely quickly can burst in violent explosions called *supernovas*. These bursts can release as much energy in a few seconds as the Sun can release in its entire lifetime. The particles from the explosion then condense into the smallest type of star—a neutron star—and eventually form a *blackhole*, which has such a high amount of gravity that not even light energy can escape. The Sun is currently a red dwarf, early in its life cycle.

Color, Temperature, Apparent Brightness, Absolute Brightness, and Luminosity

The color of a star depends on its surface temperature. Stars with cooler surfaces emit red light, while the hottest stars give off blue light. Stars with temperatures between these extremes, such as the Sun, emit white light. The *apparent brightness* of a star is a measure of how bright a star appears to an observer on the Earth. The *absolute brightness* is a measure of the intrinsic brightness of a star and is measured at a distance of exactly 10 parsecs away. The *luminosity* of a star is the amount of light emitted from its surface.

Hertzsprung-Russell Diagrams

Hertzsprung-Russell diagrams are scatterplots that show the relationship of a star's brightness and temperature, or color. The general layout shows stars of greater luminosity toward the top of the diagram. Stars with higher surface temperatures appear toward the left side of the diagram. The diagonal area from the top-left of the diagram to the bottom-right is called the *main sequence*.

Dark Matter

Dark matter is an unidentified type of matter that comprises approximately 27% of the mass and energy in the observable universe. As the name suggests, dark matter is so dense and small that it doesn't emit or interact with electromagnetic radiation, such as light, making it electromagnetically invisible. Although dark matter has never been directly observed, its existence and properties can be inferred from its gravitational effects on visible objects as well as the cosmic microwave background. Patterns of movement have been observed in visible objects that would only be possible if dark matter exerted a gravitational pull.

Theory About the Origin of the Universe

The *Big Bang theory* is a proposed cosmological model for the origin of the universe. It theorizes that the universe expanded from a high-density and high-temperature state. The theory offers comprehensive explanations for a wide range of astronomical phenomena, such as the cosmic microwave background and Hubble's Law. From detailed measurements of the expansion rate of the universe, Big Bang theorists estimate that the Big Bang occurred approximately 13.8 billion years ago, which is considered the age of the universe. The theory states that after the initial expansion, the universe cooled enough for subatomic particles and atoms to form and aggregate into giant clouds. These clouds coalesced through gravity and formed the stars and galaxies. If this theory holds true, it's predicted that the universe will reach a point where it will stop expanding and start to pull back toward the center due to gravity.

Types and Basic Characteristics of Rocks and Minerals and Their Formation Processes

The Rock Cycle

Although it may not always be apparent, rocks are constantly being destroyed while new ones are created in a process called the *rock cycle*. This cycle is driven by plate tectonics and the water cycle, which are discussed in detail later. The rock cycle starts with *magma*, the molten rock found deep within the Earth. As magma moves toward the Earth's surface, it hardens and transforms into igneous rock. Then, over time, igneous rock is broken down into tiny pieces called *sediment* that are eventually deposited all over the surface. As more and more sediment accumulates, the weight of the newer sediment compresses the older sediment underneath and creates sedimentary rock. As sedimentary rock is pushed deeper below the surface, the high pressure and temperature transform it into metamorphic rock. This metamorphic rock can either rise to the surface again or sink even deeper and melt back into magma, thus starting the cycle again.

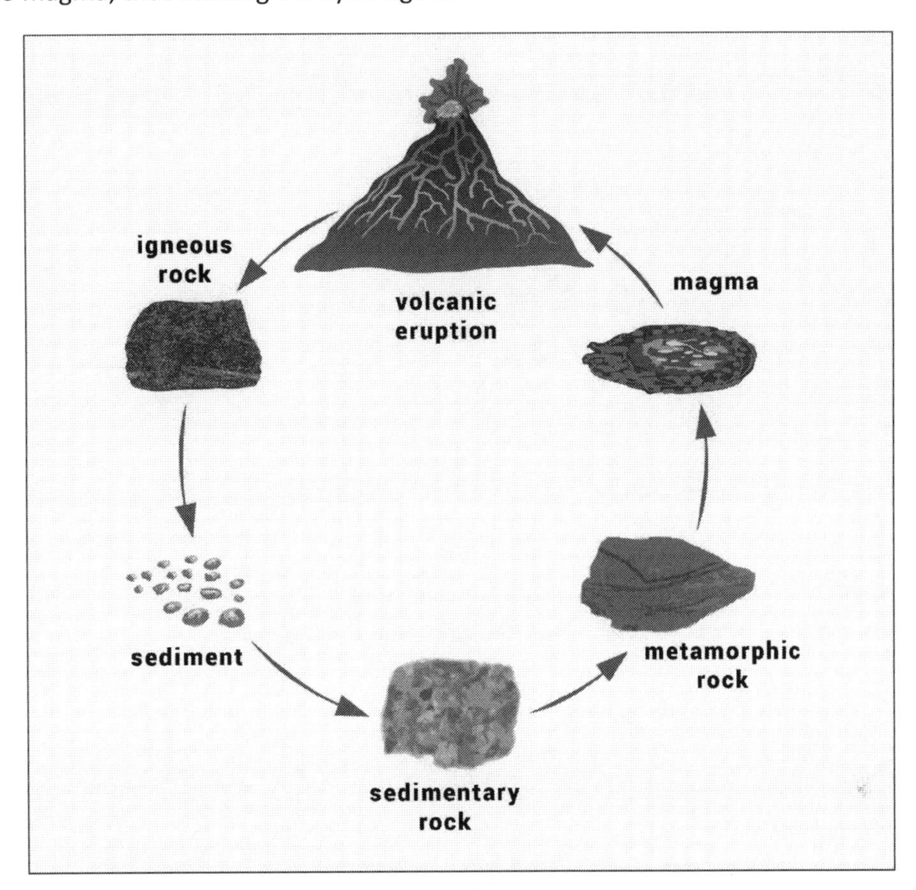

Characteristics of Rocks and Their Formation Processes

There are three main types of rocks: sedimentary, igneous, and metamorphic. Aside from physical characteristics, one of their main differences is how they are created. *Sedimentary rocks* are formed at the surface, on land and in bodies of water, through processes called deposition and cementation. They can be classified as clastic, biochemical, and chemical. *Clastic rocks*, such as sandstone, are composed of other pieces of inorganic rocks and sediment. *Biochemical rocks* are created from an organic material, such as coal, forming from dead plant life. *Chemical rocks* are created from the deposition of dissolved minerals, such as calcium salts that form stalagmites and stalactites in caves.

Igneous rocks are created when magma solidifies at or near the Earth's surface. When they're formed at the surface, (i.e., from volcanic eruption), they are *extrusive*. When they form below the surface, they're called *intrusive*. Examples of extrusive rocks are obsidian and tuff, while rocks like granite are intrusive.

Metamorphic rocks are the result of a transformation from other rocks. Based on appearance, these rocks are classified as foliated or non-foliated. *Foliated rocks* are created from compression in one direction, making them appear layered or folded like slate. *Non-foliated rocks* are compressed from all directions, giving them a more homogenous appearance, such as marble.

Characteristics of Minerals and Their Formation Processes

A *mineral*, such as gold, is a naturally occurring inorganic solid composed of one type of molecule or element that's organized into a crystalline structure. Rocks are aggregates of different types of minerals. Depending on their composition, minerals can be mainly classified into one of the following eight groups:

- *Carbonates*: formed from molecules that have either a carbon, nitrogen, or boron atom at the center.
- *Elements*: formed from single elements that occur naturally; includes metals such as gold and nickel, as well as metallic alloys like brass.
- *Halides:* formed from molecules that have halogens; halite, which is table salt, is a classic example.
- *Oxides*: formed from molecules that contain oxygen or hydroxide and are held together with ionic bonds; encompasses the phosphates, silicates, and sulfates.
- *Phosphates*: formed from molecules that contain phosphates; the apatite group minerals are in this class.
- *Silicates*: formed from molecules that contain silicon, silicates are the largest class and usually the most complex minerals; topaz is an example of a silicate.
- *Sulfates*: formed from molecules that contain either sulfur, chromium, tungsten, selenium, tellurium, and/or molybdenum.
- *Sulfides*: formed from molecules that contain sulfide (S^{2-}); includes many of the important metal ores, such as lead and silver.

One important physical characteristic of a mineral is its *hardness*, which is defined as its resistance to scratching. When two crystals are struck together, the harder crystal will scratch the softer crystal. The most common measure of hardness is the Mohs Hardness Scale, which ranges from 1 to 10, with 10 being the hardest. Diamonds are rated 10 on the Mohs Hardness Scale, and talc, which was once used to make baby powder, is rated 1. Other important characteristics of minerals include *luster* or shine, *color,* and *cleavage*, which is the natural plane of weakness at which a specific crystal breaks.

Erosion, Weathering, and Deposition of Earth's Surface Materials and Soil Formation

Erosion and Deposition

Erosion is the process of moving rock and occurs when rock and sediment are picked up and transported. Wind, water, and ice are the primary factors for erosion. *Deposition* occurs when the particles stop moving and settle onto a surface, which can happen through gravity or involve processes such as precipitation or flocculation. *Precipitation* is the solidification or crystallization of dissolved ions that occurs when a solution is oversaturated. *Flocculation* is similar to coagulation and occurs when colloid materials (materials that aren't dissolved but are suspended in the medium) aggregate or clump until they are too heavy to remain suspended.

Chemical and Physical (Mechanical) Weathering

Weathering is the process of breaking down rocks through mechanical or chemical changes. Mechanical forces include animal contact, wind, extreme weather, and the water cycle. These physical forces don't alter the composition of rocks. In contrast, chemical weathering transforms rock composition. When water and minerals interact, they can start chemical reactions and form new or secondary minerals from the original rock. In chemical weathering, the processes of oxidation and hydrolysis are important. When rain falls, it dissolves atmospheric carbon dioxide and becomes acidic. With sulfur dioxide and nitrogen oxide in the atmosphere from volcanic eruptions or burning fossil fuels, the rainfall becomes even more acidic and creates acid rain. Acidic rain can dissolve the rock that it falls upon.

Characteristics of Soil

Soil is a combination of minerals, organic materials, liquids, and gases. There are three main types of soil, as defined by their compositions, going from coarse to fine: sand, silt, and clay. Large particles, such as those found in sand, affect how water moves through the soil, while tiny clay particles can be chemically active and bind with water and nutrients. An important characteristic of soil is its ability to form a crust when dehydrated. In general, the finer the soil, the harder the crust, which is why clay (and not sand) is used to make pottery.

There are many different classes of soil, but the components are always sand, silt, or clay. Below is a chart used by the United States Department of Agriculture (USDA) to define soil types:

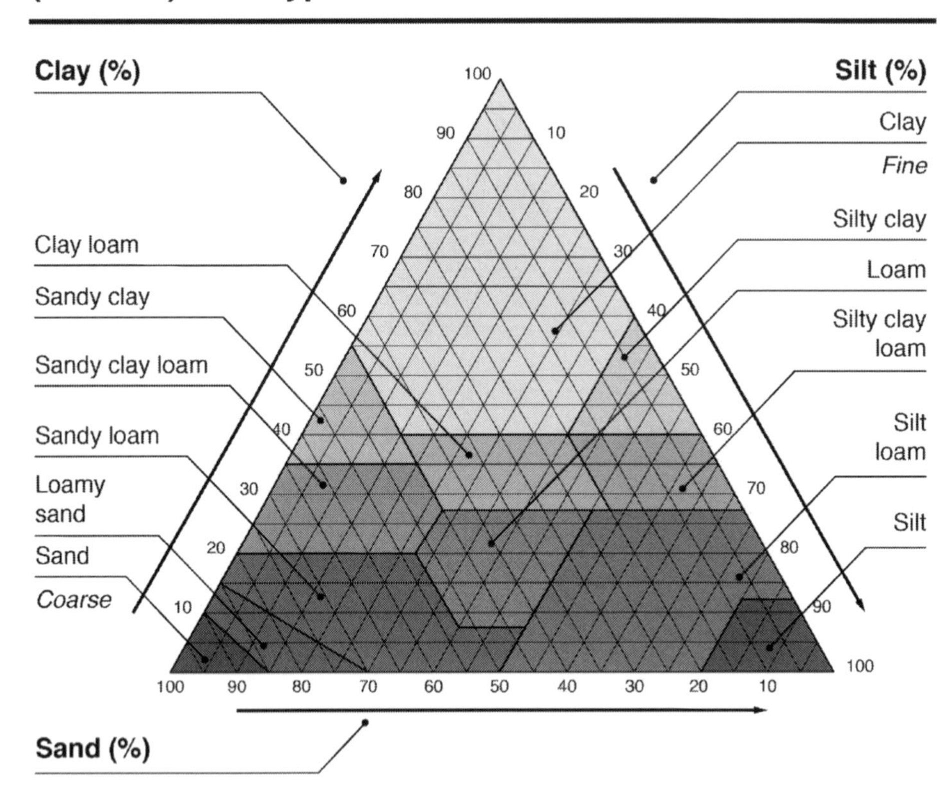

Loam is a term for soil that is a mixture of sand, silt, and clay. It's also the soil most commonly used for agriculture and gardening.

Porosity and Permeability

Porosity and permeability refer to how water moves through rock and soil underground. *Porosity* is a measure of the open space in a rock. This space can be between grains or within cracks and cavities in the rock. *Permeability* is a measure of the ease with which water can move through a porous rock. Therefore, rock that is more porous is also more permeable. When a rock is more permeable, it's less effective as a water purifier because dirty particles in the water can pass through porous rock.

Runoff and Infiltration

An important function of soil is to absorb water to be used by plants or released into groundwater. *Infiltration capacity* is the maximum amount of water that can enter soil at any given time and is regulated by the soil's porosity and composition. For example, sandy soils have larger pores than clays, allowing water to infiltrate them easier and faster. *Runoff* is water that moves across land's surface and may end up in a stream or a rut in the soil. Runoff generally occurs after the soil's infiltration capacity is reached. However, during heavy rainfalls, water may reach the soil's surface at a faster rate than infiltration can occur, causing runoff without soil saturation. In addition, if the ground is frozen and the soil's pores are blocked by ice, runoff may occur without water infiltrating the soil.

Earth's Basic Structure and Internal Processes

Earth's Layers

Earth has three major layers: a thin solid outer surface or *crust*, a dense *core,* and a *mantle* between them that contains most of the Earth's matter. This layout resembles an egg, where the eggshell is the crust, the mantle is the egg white, and the core is the yolk. The outer crust of the Earth consists of igneous or sedimentary rocks over metamorphic rocks. Together with the upper portion of the mantle, it forms the *lithosphere*, which is broken into tectonic plates.

Major plates of the lithosphere

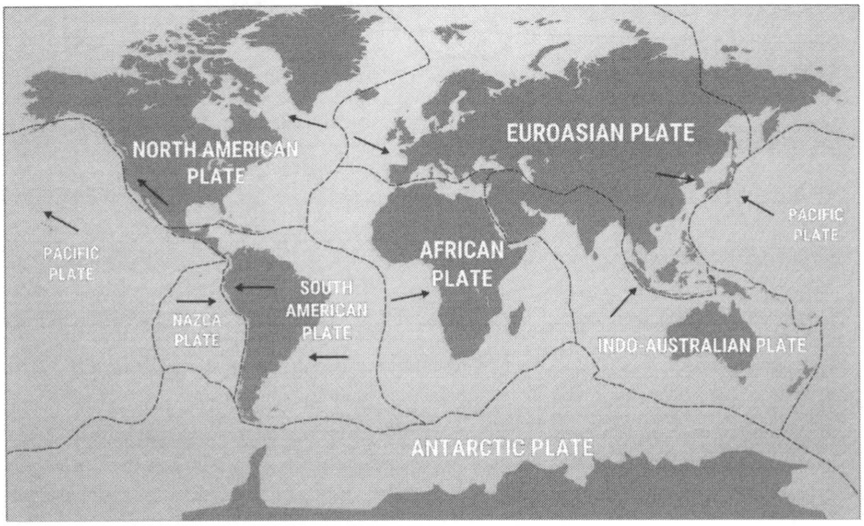

The mantle can be divided into three zones. The *upper mantle* is adjacent to the crust and composed of solid rock. Below the upper mantle is the *transition zone*. The *lower mantle* below the transition zone is a layer of completely solid rock. Underneath the mantle is the molten *outer core* followed by the compact, solid *inner core*. The inner and outer cores contain the densest elements, consisting of mostly iron and nickel.

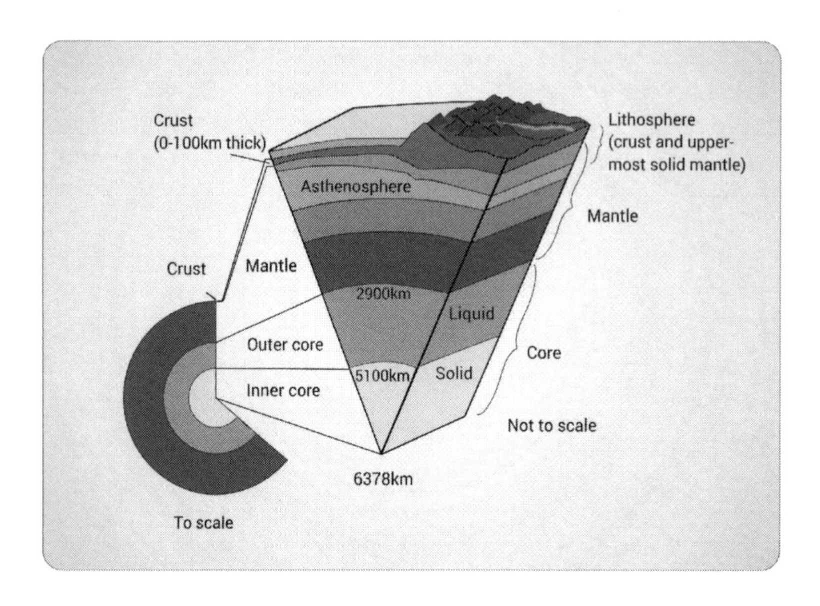

Shape and Size of the Earth

The Earth isn't a perfect sphere; it's slightly elliptical. From center to surface, its radius is almost 4,000 miles, and its circumference around the equator is about 24,902 miles. In comparison, the Sun's radius is 432,288 miles—over 1,000 times larger than the Earth's—and the Moon's radius is about 1,000 miles.

Geographical Features

The Earth's surface is dynamic and consists of various landforms. As tectonic plates are pushed together, *mountains* are formed. *Canyons* are deep trenches that are usually created by plates moving apart but can also be created by constant weathering and erosion from rivers and runoff. *Deltas* are flat, triangular stretches of land formed by rivers that deposit sediment and water into the ocean. *Sand dunes* are mountains of sand located in desert areas or the bottom of the ocean. They are formed by wind and water movement when there's an absence of plants or other features that would otherwise hold the sand in place.

The Earth's Magnetic Field

The Earth's magnetic field is created by the magnetic forces that extend from the Earth's interior to outer space. It can be modeled as a magnetic dipole tilted about 10 degrees from the Earth's rotational axis, as if a bar magnet was placed at an angle inside the Earth's core. The geomagnetic pole located near Greenland in the northern hemisphere is actually the south pole of the Earth's magnetic field, and vice versa for the southern geomagnetic pole. The *magnetosphere* is the Earth's magnetic field, which extends tens of thousands of kilometers into space and protects the Earth and the atmosphere from damaging solar wind and cosmic rays.

Plate Tectonics Theory and Evidence

The theory of *plate tectonics* hypothesizes that the continents weren't always separated like they are today but were once joined and slowly drifted apart. Evidence for this theory is based upon the fossil record. Fossils of one species were found in regions of the world now separated by an ocean. It's unlikely that a single species could have travelled across the ocean.

Folding and Faulting

The exact number of tectonic plates is debatable, but scientists estimate there are around nine to fifteen major plates and almost 40 minor plates. The line where two plates meet is called a *fault*. The San Andreas Fault is where the Pacific and North American plates meet. Faults or boundaries are classified depending on the interaction between plates. Two plates collide at *convergent boundaries. Divergent boundaries* occur when two plates move away from each other. Tectonic plates can move vertically and horizontally.

Continental Drift, Seafloor Spreading, Magnetic Reversals

The movement of tectonic plates is similar to pieces of wood floating in a pool of water. They can bob up and down as well as bump, slide, and move away from each other. These different interactions create the Earth's landscape. The collision of plates can create mountain ranges, while their separation can create canyons or underwater chasms. One plate can also slide atop another and push it down into the Earth's hot mantle, creating magma and volcanoes, in a process called *subduction*.

Subduction

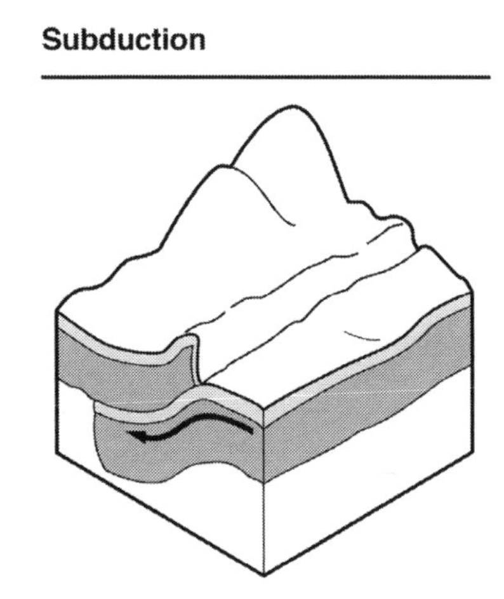

Unlike a regular magnet, the Earth's magnetic field changes over time because it's generated by the motion of molten iron alloys in the outer core. Although the magnetic poles can wander geographically, they do so at such a slow rate that they don't affect the use of compasses in navigation. However, at irregular intervals that are several hundred thousand years long, the fields can reverse, with the north and south magnetic poles switching places.

Characteristics of Volcanoes

Volcanoes are mountainous structures that act as vents to release pressure and magma from the Earth's crust. During an *eruption*, the pressure and magma are released, and volcanoes smoke, rumble, and throw ash and *lava*, or molten rock, into the air. *Hot spots* are volcanic regions of the mantle that are hotter than surrounding regions.

Characteristics of Earthquakes

Earthquakes occur when tectonic plates slide or collide as a result of the crust suddenly releasing energy. Stress in the Earth's outer layer pushes together two faults. The motion of the planes of the fault continues until something makes them stop. The *epicenter* of an earthquake is the point on the surface directly above where the fault is slipping. If the epicenter is located under a body of water, the earthquake may cause a *tsunami*, a series of large, forceful waves.

Seismic Waves and Triangulation

Earthquakes cause *seismic waves*, which travel through the Earth's layers and give out low-frequency acoustic energy. Triangulation of seismic waves helps scientists determine the origin of an earthquake.

Historical Geology

Principle of Uniformitarianism

Uniformitarianism is the assumption that natural laws and processes haven't changed and apply everywhere in the universe. In geology, uniformitarianism includes the *gradualist model*, which states that "the present is the key to the past" and claims that natural laws functioned at the same rates as observed today.

Basic Principles of Relative Age Dating

Relative age dating is the determination of the relative order of past events without determining absolute age. The Law of Superposition states that older geological layers are deeper than more recent layers. Rocks and fossils can be used to compare one stratigraphic column with another. A *stratigraphic column* is a drawing that describes the vertical location of rocks in a cliff wall or underground. Correlating these columns from different geographic areas allows scientists to understand the relationships between different areas and strata. Before the discovery of radiometric dating, geologists used this technique to determine the ages of different materials. Relative dating can only determine the sequential order of events, not the exact time they occurred. The Law of Fossil Succession states that when the same kinds of fossils are found in rocks from different locations, the rocks are likely the same age.

Absolute (Radiometric) Dating

Absolute or *radiometric dating* is the process of determining age on a specified chronology in archaeology and geology. It provides a numerical age by measuring the radioactive decay of elements (such as carbon-14) trapped in rocks or minerals and using the known rate of decay to determine how much time has passed. *Uranium-lead dating* can be used to date some of the oldest rocks on Earth.

Fossil Record as Evidence of the Origin and Development of Life

The *fossil record* is the location of fossils throughout the Earth's surface layers. Deeper fossils are older than the fossils above. Scientists use the fossil record to determine when certain organisms existed and how they evolved. There are several ways a fossil can form:

- *Permineralization:* when an organism is buried and its empty spaces fill with mineral-rich groundwater
- *Casts:* when the original remains are completely destroyed and an organism-shaped hole is left in the existing rock
- *Replacement or recrystallization:* when shell or bone is replaced with another mineral

Basic Structure and Composition of the Earth's Atmosphere

Layers

The Earth's atmospheric layers are determined by their temperatures but are reported by their distance above sea level. Listed from closest to sea level on upward, the levels are:

- Troposphere: sea level to 11 miles above sea level
- Stratosphere: 11 miles to 31 miles above sea level
- Mesosphere: 31 miles to 50 miles above sea level
- Ionosphere: 50 miles to 400 miles above sea level
- Exosphere: 400 miles to 800 miles above sea level

The ionosphere and exosphere are together considered the thermosphere. The ozone layer is in the stratosphere and weather experienced on Earth's surface is a product of factors in the troposphere.

Composition of the Atmosphere

The Earth's atmosphere is composed of gas particles: 78% nitrogen, 21% oxygen, 1% other gases such as argon, and 0.039% carbon dioxide. The atmospheric layers are created by the number of particles in the air and gravity's pull upon them.

Atmospheric Pressure and Temperature

The lower atmospheric levels have higher atmospheric pressures due to the mass of the gas particles located above. The air is less dense (it contains fewer particles per given volume) at higher altitudes. The temperature changes from the bottom to top of each atmospheric layer. The tops of the troposphere and mesosphere are colder than their bottoms, but the reverse is true for the stratosphere and thermosphere. Some of the warmest temperatures are actually found in the thermosphere because of a type of radiation that enters that layer.

Basic Concepts of Meteorology

Relative Humidity

Relative humidity is the ratio of the partial pressure of water vapor to water's equilibrium vapor pressure at a given temperature. At low temperatures, less water vapor is required to reach a high relative humidity. More water vapor is needed to reach a high relative humidity in warm air, which has a greater capacity for water vapor. At ground level or other areas of higher pressure, relative humidity increases as temperatures decrease because water vapor condenses as the temperature falls below the dew point. As relative humidity cannot be greater than 100%, the dew point temperature cannot be greater than the air temperature.

Dew Point

The *dew point* is the temperature at which water vapor in the air condenses into liquid water due to saturation. At temperatures below the dew point, the rate of condensation will be greater than the rate of evaporation, forming more liquid water. When condensed water forms on a surface, it's called *dew*; when it forms in the air, it's called *fog* or *clouds*, depending on the altitude.

Wind

Wind is the movement of gas particles across the Earth's surface. Winds are generated by differences in atmospheric pressure. Air inherently moves from areas of higher pressure to lower pressure, which is what causes wind to occur. Surface friction from geological features, such as mountains or man-made features can decrease wind speed. In meteorology, winds are classified based on their strength, duration, and direction. *Gusts* are short bursts of high-speed wind, *squalls* are strong winds of intermediate duration (around one minute), and winds with a long duration are given names based on their average strength. *Breezes* are the weakest, followed by *gales*, *storms*, and *hurricanes*.

Types of Precipitation

There are three distinct processes by which precipitation occurs. *Convection precipitation* occurs when air rises vertically in a forceful manner, quickly overturning the atmosphere and resulting in heavy precipitation. It's generally more intense and shorter in duration than *stratiform precipitation*, which occurs when large masses of air move over each other. *Orographic precipitation* occurs when moist air is forced upwards over rising terrain, such as a mountain. Most storms are a result of convection precipitation.

Precipitation can fall in liquid or solid phases, as well as any form in between. Liquid precipitation includes rain and drizzle. Frozen precipitation includes snow, sleet, and hail. Intensity is classified by rate of fall or visibility restriction. The forms of precipitation are:

- *Rain*: water vapor that condenses on dust particles in the troposphere until it becomes heavy enough to fall to Earth
- *Sleet*: rain that freezes on its way down; it starts as ice that melts and then freezes again before hitting the ground
- *Hail*: balls of ice thrown up and down several times by turbulent winds, so that more and more water vapor can condense and freeze on the original ice; hail can be as large as golf balls or even baseballs
- *Snow*: loosely packed ice crystals that fall to Earth

Cloud Types and Formation

Water in the atmosphere can exist as visible masses called *clouds* composed of water droplets, tiny crystals of ice, and various chemicals. Clouds exist primarily in the troposphere and can be classified based on altitude.

High Clouds (5,000 - 13,000 meters above sea level)	**Middle Clouds** (2,000 - 7,000 meters above sea level)	**Low Clouds** (below 2,000 meters above sea level)
• Cirrus: thin and wispy • Cirrocumulus: rows of small, puffy "pillows" • Cirrostratus: thin sheets that cover the sky	• Altocumulus: gray and white, made up of water droplets • Altostratus: grayish/bluish gray clouds	• Stratus: gray clouds that can cover the sky • Stratocumulus: gray, lumpy, low-lying clouds • Nimbostratus: thick and dark gray, typical of rain or snow clouds

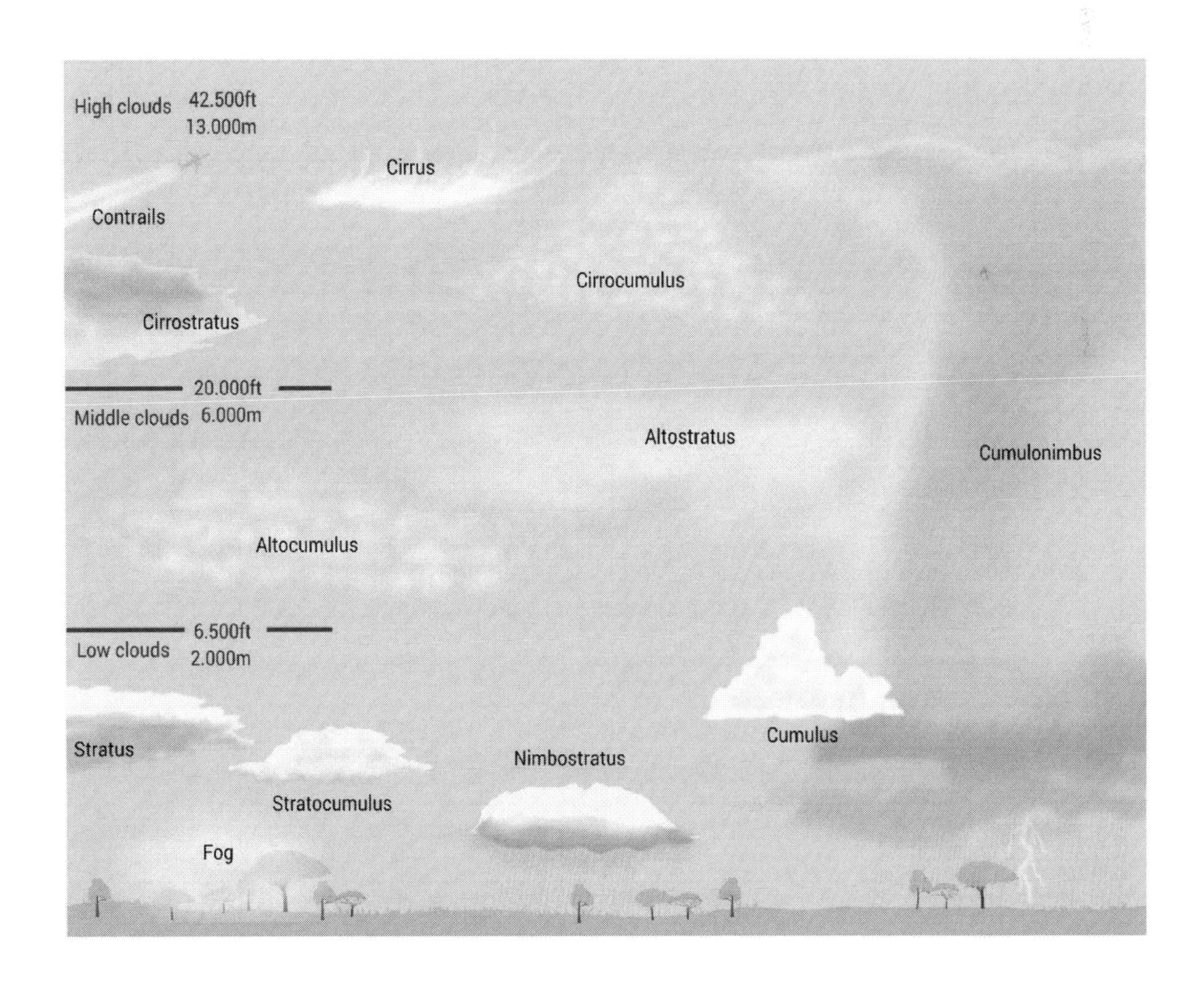

Air Masses, Fronts, Storms, and Severe Weather

Air masses are volumes of air defined by their temperature and the amount of water vapor they contain. A *front* is where two air masses of different temperatures and water vapor content meet. Fronts can be the site of extreme weather, such as thunderstorms, which are caused by water particles rubbing against each other. When they do so, electrons are transferred, and energy and electrical currents accumulate. When enough energy accumulates, thunder and lightning occur. *Lightning* is a massive electric spark created by a cloud, and *thunder* is the sound created by an expansion of air caused by the sudden increase in pressure and temperature around lightning.

Extreme weather includes tornadoes and hurricanes. *Tornadoes* are created by changing air pressure and winds that can exceed 300 miles per hour. *Hurricanes* occur when warm ocean water quickly evaporates and rises to a colder, low-pressure portion of the atmosphere. Hurricanes, typhoons, and tropical cyclones are all created by the same phenomena, but they occur in different regions. *Blizzards* are similar to hurricanes in that they're created by the clash of warm and cold air, but they only occur when cold Arctic air moves toward warmer air. They usually involve large amounts of snow.

Development and Movement of Weather Patterns

A *weather pattern* is weather that's consistent for a period of time. Weather patterns are created by fronts. A *cold front* is created when two air masses collide in a high-pressure system. A *warm front* is created when a low-pressure system results from the collision of two air masses; they are usually warmer and less dense than high-pressure systems. When a cold front enters an area, the air from the warm front is forced upwards. The temperature of the warm front's air decreases, condenses, and often creates clouds and precipitation. When a warm front moves into an area, the warm air moves slowly upwards at an angle. Clouds and precipitation form, but the precipitation generally lasts longer because of how slowly the air moves.

The Water Cycle

Evaporation and Condensation

The *water cycle* is the cycling of water between its three physical states: solid, liquid, and gas. The Sun's thermal energy heats surface water so it evaporates. As water vapor collects in the atmosphere from evaporation, it eventually reaches a saturation level where it condenses and forms clouds heavy with water droplets.

Precipitation

When the droplets condense as clouds get heavy, they fall as different forms of precipitation, such as rain, snow, hail, fog, and sleet. *Advection* is the process of evaporated water moving from the ocean and falling over land as precipitation.

Runoff and Infiltration

Runoff and *infiltration* are important parts of the water cycle because they provide water on the surface available for evaporation. Runoff can add water to oceans and aid in the advection process. Infiltration provides water to plants and aids in the transpiration process.

Transpiration

Transpiration is an evaporation-like process that occurs in plants and soil. Water from the stomata of plants and from pores in soil evaporates into water vapor and enters the atmosphere.

Major Factors that Affect Climate and Seasons

Effects of Latitude, Geographical Location, and Elevation

The climate and seasons of different geographical areas are primarily dictated by their sunlight exposure. Because the Earth rotates on a tilted axis while travelling around the Sun, different latitudes get different amounts of direct sunlight throughout the year, creating different climates. Polar regions experience the greatest variation, with long periods of limited or no sunlight in the winter and up to 24 hours of daylight in the summer. Equatorial regions experience the least variance in direct sunlight exposure. Coastal areas experience breezes in the summer as cooler ocean air moves ashore, while areas southeast of the Great Lakes can get "lake effect" snow in the winter, as cold air travels over the warmer water and creates snow on land. Mountains are often seen with snow in the spring and fall. Their high elevation causes mountaintops to stay cold. The air around the mountaintop is also cold and holds less water vapor than air at sea level. As the water vapor condenses, it creates snow.

Effects of Atmospheric Circulation

Global winds are patterns of wind circulation and they have a major influence on global weather and climate. They help influence temperature and precipitation by carrying heat and water vapor around the Earth. These winds are driven by the uneven heating between the polar and equatorial regions created by the Sun. Cold air from the polar regions sinks and moves toward the equator, while the warm air from the equator rises and moves toward the poles. The other factor driving global winds is the *Coriolis Effect*. As air moves from the North Pole to the equator, the Earth's rotation makes it seem as if the wind is also moving to the right, or westbound, and eastbound from South Pole to equator.

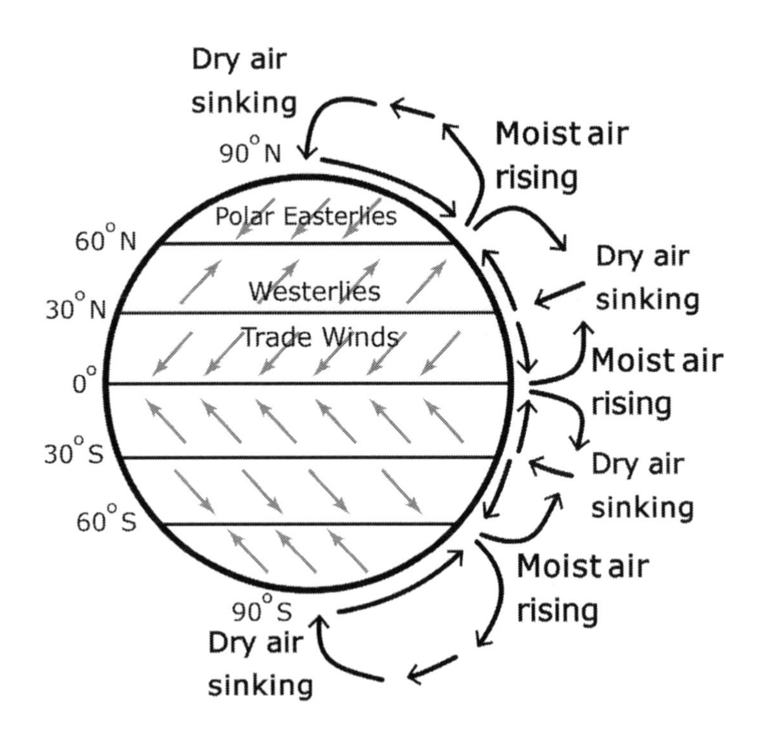

Global wind patterns are given names based on which direction they blow. There are three major wind patterns in each hemisphere. Notice the image above diagramming the movement of warm (dry) air and moist (cold) air.

Tradewinds—easterly surface winds found in the troposphere near the equator—blow predominantly from the northeast in the Northern Hemisphere and from the southeast in the Southern Hemisphere. These winds direct the tropical storms that develop over the Atlantic, Pacific, and Indian Oceans and land in North America, Southeast Asia, and eastern Africa, respectively. *Jet streams* are westerly winds that follow a narrow, meandering path. The two strongest jet streams are the polar jets and the subtropical jets. In the Northern Hemisphere, the polar jet flows over the middle of North America, Europe, and Asia, while in the Southern Hemisphere, it circles Antarctica.

Effects of Ocean Circulation

Ocean currents are similar to global winds because winds influence how the oceans move. Ocean currents are created by warm water moving from the equator towards the poles while cold water travels from the poles to the equator. The warm water can increase precipitation in an area because it evaporates faster than the colder water.

Characteristics and Locations of Climate Zones

Climate zones are created by the Earth's tilt as it travels around the Sun. These zones are delineated by the equator and four other special latitudinal lines: the Tropic of Cancer or Northern Tropic at 23.5° North; the Tropic of Capricorn or Southern Tropic at 23.5° South; the Arctic Circle at 66.5° North; and the Antarctic Circle at 66.5° South. The areas between these lines of latitude represent different climate zones. Tropical climates are hot and wet, like rainforests, and tend to have abundant plant and animal life, while polar climates are cold and usually have little plant and animal life. Temperate zones can vary and experience the four seasons.

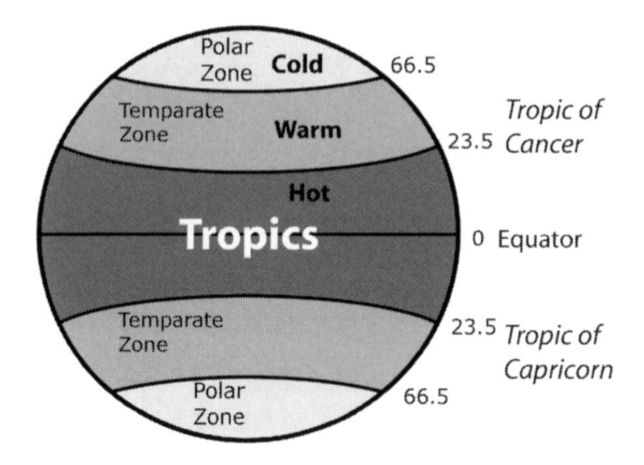

Effects of Natural Phenomena

Natural phenomena can have a sizeable impact on climate and weather. Chemicals released from volcanic eruptions can fall back to Earth in acid rain. In addition, large amounts of carbon dioxide released into the atmosphere can warm the climate. Carbon dioxide creates the *greenhouse effect* by trapping solar energy from sunlight reflected off the Earth's surface within the atmosphere. The amount of solar radiation emitted from the Sun varies and has recently been discovered to be cyclical.

El Niño and La Niña

El Niño and *La Niña* are terms for severe weather anomalies associated with torrential rainfall in the Pacific coastal regions, mainly in North and South America. These events occur irregularly every few years, usually around December, and are caused by a band of warm ocean water that accumulates in the central Pacific Ocean around the equator. The warm water changes the wind patterns over the

Pacific and stops cold water from rising toward the American coastlines. The rise in ocean temperature also leads to increased evaporation and rain. These events are split into two phases—a warm, beginning phase called El Niño and a cool end phase called La Niña.

Characteristics and Processes of the Earth's Oceans and Other Bodies of Water

Distribution and Location of the Earth's Water

A *body of water* is any accumulation of water on the Earth's surface. It usually refers to oceans, seas, and lakes, but also includes ponds, wetlands, and puddles. Rivers, streams, and canals are bodies of water that involve the movement of water.

Most bodies of water are naturally occurring geographical features, but they can also be artificially created like lakes created by dams. Saltwater oceans make up 96% of the water on the Earth's surface. Freshwater makes up 2.5% of the remaining water.

Seawater Composition

Seawater is water from a sea or ocean. On average, seawater has a salinity of about 3.5%, meaning every kilogram of seawater has approximately 35 grams of dissolved sodium chloride salt. The average density of saltwater at the surface is 1.025 kg/L, making it denser than pure or freshwater, which has a density of 1.00 kg/L. Because of the dissolved salts, the freezing point of saltwater is also lower than that of pure water; saltwater freezes at −2 °C (28 °F). As the concentration of salt increases, the freezing point decreases. Thus, it's more difficult to freeze water from the Dead Sea—a saltwater lake known to have water with such high salinity that swimmers cannot sink.

Coastline Topography and the Topography of Ocean Floor

Topography is the study of natural and artificial features comprising the surface of an area. *Coastlines* are an intermediate area between dry land and the ocean floor. The ground progressively slopes from the dry coastal area to the deepest depth of the ocean floor. At the continental shelf, there's a steep descent of the ocean floor. Although it's often believed that the ocean floor is flat and sandy like a beach, its topography includes mountains, plateaus, and valleys.

Tides, Waves, and Currents

Tides are caused by the pull of the Moon and the Sun. When the Moon is closer in its orbit to the Earth, its gravity pulls the oceans away from the shore. When the distance between the Moon and the Earth is greater, the pull is weaker, and the water on Earth can spread across more land. This relationship creates low and high tides. Waves are influenced by changes in tides as well as the wind. The energy transferred from wind to the top of large bodies of water creates *crests* on the water's surface and *waves* below. Circular movements in the ocean are called *currents*. They result from the Coriolis Effect, which is caused by the Earth's rotation. Currents spin in a clockwise direction above the equator and counterclockwise below the equator.

Estuaries and Barrier Islands

An *estuary* is an area of water located on a coast where a river or stream meets the sea. It's a transitional area that's partially enclosed, has a mix of salty and fresh water, and has calmer water than the open sea. *Barrier islands* are coastal landforms created by waves and tidal action parallel to the mainland coast. They usually occur in chains, and they protect the coastlines and create areas of protected waters where wetlands may flourish.

Islands, Reefs, and Atolls

Islands are land that is completely surrounded by water. *Reefs* are bars of rocky, sandy, or coral material that sit below the surface of water. They may form from sand deposits or erosion of underwater rocks. An *atoll* is a coral reef in the shape of a ring (but not necessarily circular) that encircles a lagoon. In order for an atoll to exist, the rate of its erosion must be slower than the regrowth of the coral that composes the atoll.

Polar Ice, Icebergs, Glaciers

Polar ice is the term for the sheets of ice that cover the poles of a planet. *Icebergs* are large pieces of freshwater ice that break off from glaciers and float in the water. A *glacier* is a persistent body of dense ice that constantly moves because of its own weight. Glaciers form when snow accumulates at a faster rate than it melts over centuries. They form only on land, in contrast to *icecaps*, which can form from sheets of ice in the ocean. When glaciers deform and move due to stresses created by their own weight, they can create crevasses and other large distinguishing land features.

Lakes, Ponds, and Wetlands

Lakes and *ponds* are bodies of water that are surrounded by land. They aren't part of the ocean and don't contain flowing water. Lakes are larger than ponds, but otherwise the two bodies don't have a scientific distinction. *Wetlands* are areas of land saturated by water. They have a unique soil composition and provide a nutrient-dense area for vegetation and aquatic plant growth. They also play a role in water purification and flood control.

Streams, Rivers, and River Deltas

A *river* is a natural flowing waterway usually consisting of freshwater that flows toward an ocean, sea, lake, or another river. Some rivers flow into the ground and become dry instead of reaching another body of water. Small rivers are usually called *streams* or *creeks*. River *deltas* are areas of land formed from the sediment carried by a river and deposited before it enters another body of water. As the river reaches its end, the flow of water slows, and the river loses the power to transport the sediment, so it falls out of suspension.

Geysers and Springs

A *spring* is a natural occurrence where water flows from an aquifer to the Earth's surface. A *geyser* is a spring that intermittently and turbulently discharges water. Geysers form only in certain hydrogeological conditions. They require proximity to a volcanic area or magma to provide enough heat to boil or vaporize the water. As hot water and steam accumulate, pressure grows and creates the spraying geyser effect.

Properties of Water that Affect Earth Systems

Water is a chemical compound composed of two hydrogen atoms and one oxygen atom (H_2O) and has many unique properties. In its solid state, water is less dense than its liquid form; therefore, ice floats in water. Water also has a very high heat capacity, allowing it to absorb a high amount of the Sun's energy without getting too hot or evaporating. Its chemical structure makes it a polar compound, meaning one side has a negative charge while the other is positive. This characteristic—along with its ability to form strong intermolecular hydrogen bonds with itself and other molecules—make water an effective solvent for other chemicals.

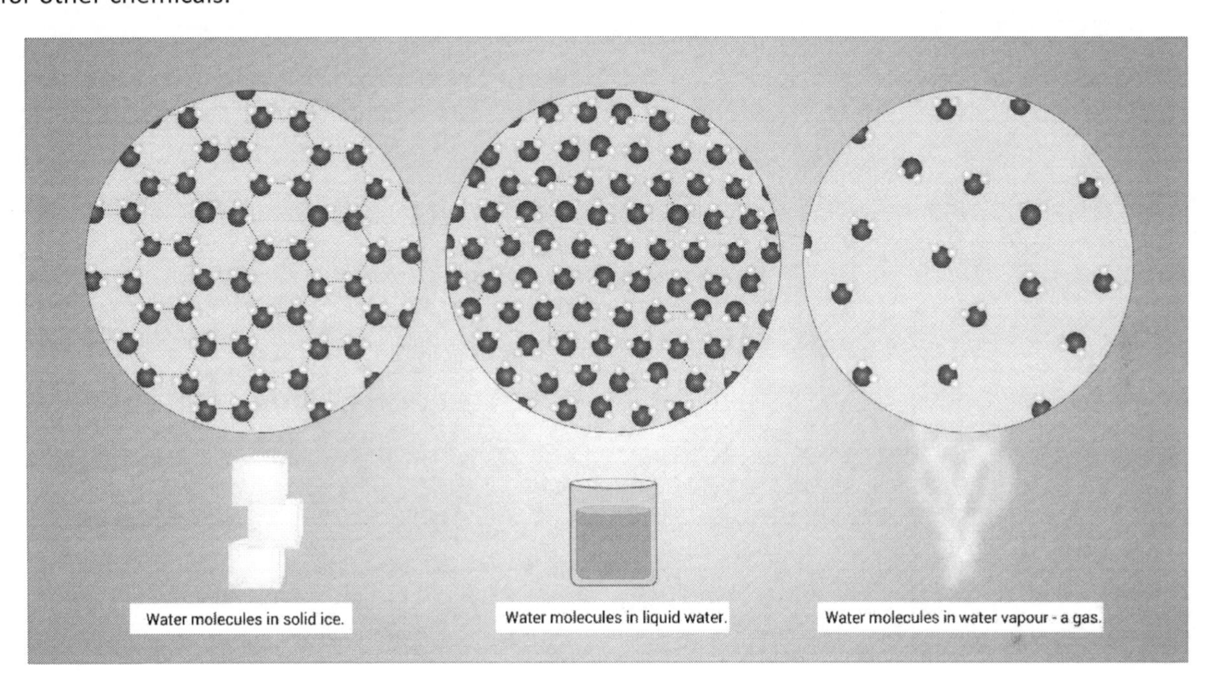

Practice Questions

Questions 1–5 pertain to Passage 1 and the figures that follow:

Predators are animals that eat other animals. Prey are animals that are eaten by a predator. Predators and prey have a distinct relationship. Predators rely on the prey population for food and nutrition. They evolve physically to catch their prey. For example, they develop a keen sense of sight, smell, or hearing. They may also be able to run very fast or camouflage to their environment in order to sneak up on their prey. Likewise, the prey population may develop these features to escape and hide from their predators. As predators catch more prey, the prey population dwindles. With fewer prey to catch, the predator population also dwindles. This happens in a cyclical manner over time.

Figure 1 below shows the cyclical population growth in a predator-prey relationship.

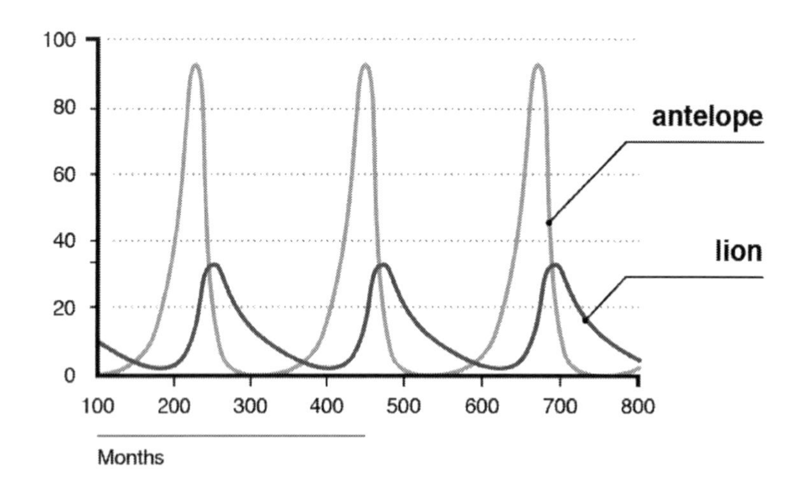

Figure 2 below shows a predator-prey cycle in a circular picture diagram.

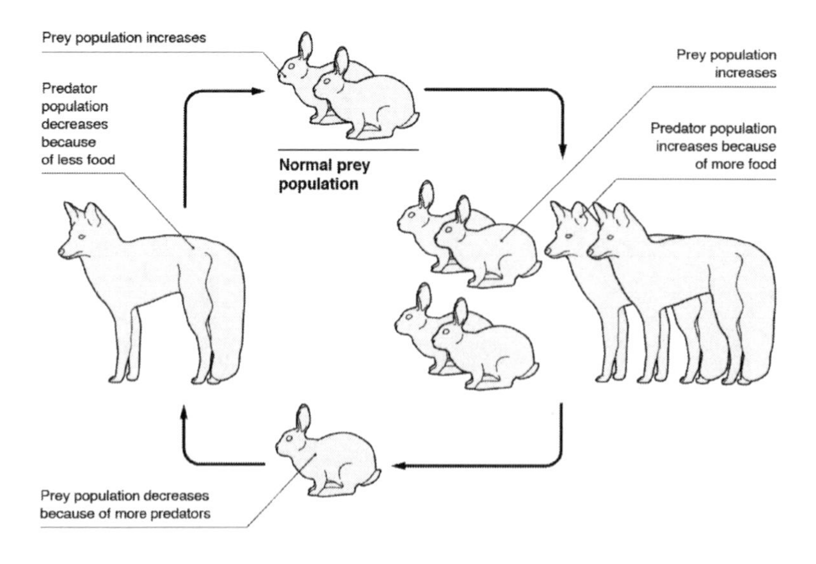

1. Looking at Figure 1, approximately how long is one cycle of the prey population, which includes the population being low, reaching a peak, and then becoming low again?
   a. 300 months
   b. 200 months
   c. 800 months
   d. 400 months

2. In Figure 2, which animal is the predator?
   a. Both the fox and rabbit
   b. Fox only
   c. Rabbit only
   d. Neither the fox nor the rabbit

3. What causes the predator population to decrease?
   a. When there's an increase in the prey population
   b. When winter arrives
   c. When the prey start attacking the predators
   d. When there are fewer prey to find

4. What causes the prey population to increase?
   a. When the predator population decreases, so more prey survive and reproduce.
   b. When there's an increase in the predator population
   c. When there's more sunlight
   d. The prey population always remains the same size.

5. Which is NOT a feature that a prey population can develop to hide from their predator?
   a. Keen sense of smell
   b. Camouflage ability
   c. A loud voice
   d. Keen sense of hearing

Answer Explanations

1. A: One cycle takes 300 months. It starts with the population being low, rising and reaching a peak, and then falling again. It takes 200 months to reach the peak, Choice *B*. Three cycles could be completed in 800 months, Choice *C*, and the second cycle would have started by 400 months, Choice *D*.

2. B: Looking at Figure 2, the fox is the predator. When the diagram notes that the predator population decreases on the left side, there is only one fox left. As it increases, as noted on the right side, there are two foxes drawn. Foxes are also much larger than rabbits and would be able to catch them much easier than the other way around.

3. D: When the prey population decreases, the predators have less food, i.e., prey, to feed on. This causes the predator population to dwindle. An increase in the prey population, Choice *A*, would actually increase the predator population because they would have more food, which would lengthen survival and increase reproduction. Seasons do not affect the predator population in this situation, Choice *B*. Generally, prey do not have the ability to attack their predators, Choice *C*, due to physical constraints, such as differences in size.

4. A: When the predator population decreases, the rate of survival of the prey population increases, and they can then also reproduce more. An increase in the predator population, Choice *B*, would cause the prey population to decrease. Weather and amount of sunlight, Choice *C*, does not affect the growth of the prey population. The prey population is cyclical and does not remain the same size, Choice *D*.

5. C: Prey populations can develop different features to try and hide from and escape the predator population. The features help them blend into their environment, such as Choice *B*, or help them identify predators early and quickly, Choices *A* and *D*. Choice *C* would just allow the predators to hear the prey easily.

Practice Test

Mathematics

Calculator Questions

1. At the beginning of the day, Xavier has 20 apples. At lunch, he meets his sister Emma and gives her half of his apples. After lunch, he stops by his neighbor Jim's house and gives him 6 of his apples. He then uses $\frac{3}{4}$ of his remaining apples to make an apple pie for dessert at dinner. At the end of the day, how many apples does Xavier have left?

   a. 4
   b. 6
   c. 2
   d. 1
   e. 3

2. What is the product of two irrational numbers?

   a. Irrational
   b. Rational
   c. Irrational or rational
   d. Complex and imaginary
   e. Imaginary

3. The graph shows the position of a car over a 10-second time interval. Which of the following is the correct interpretation of the graph for the interval 1 to 3 seconds?

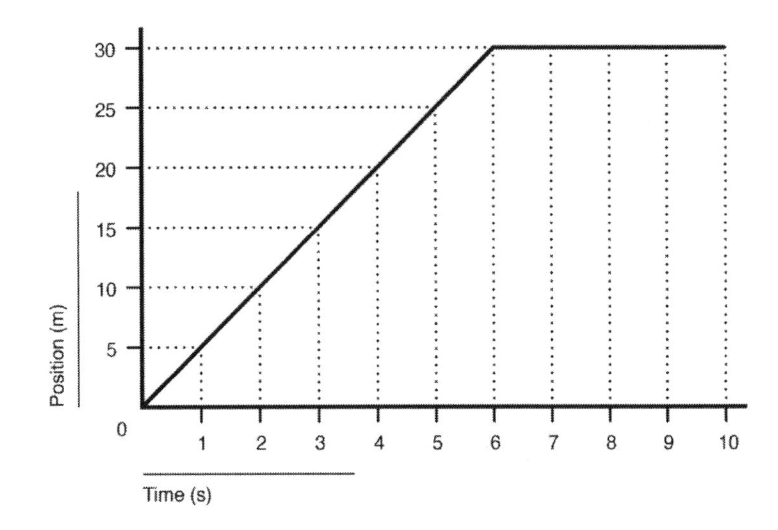

   a. The car remains in the same position.
   b. The car is traveling at a speed of 5 m/s.
   c. The car is traveling up a hill.
   d. The car is traveling at 5 mph.
   e. The car accelerates at a rate of 5 m/s.

4. Being as specific as possible, how is the number -4 classified?
   a. Real, rational, integer, whole, natural
   b. Real, rational, integer, natural
   c. Real, rational, integer
   d. Real, irrational, complex
   e. Real, irrational, whole

5. After a 20% discount, Frank purchased a new refrigerator for $850. How much did he save from the original price?
   a. $170
   b. $212.50
   c. $105.75
   d. $200
   e. $187.50

6. Store brand coffee beans cost $1.23 per pound. A local coffee bean roaster charges $1.98 per 1 ½ pounds. How much more would 5 pounds from the local roaster cost than 5 pounds of the store brand?
   a. $0.55
   b. $1.55
   c. $1.45
   d. $0.45
   e. $0.35

7. What is the solution to the following problem in decimal form?

$$\frac{3}{5} \times \frac{7}{10} \div \frac{1}{2}$$

   a. 0.042
   b. 84%
   c. 0.84
   d. 0.42
   e. 42%

8. Dwayne has received the following scores on his math tests: 78, 92, 83, and 97. What score must Dwayne get on his next math test to have an overall average of 90?
   a. 89
   b. 98
   c. 95
   d. 94
   e. 100

9. What are all the factors of 12?
   a. 12, 24, 36
   b. 1, 2, 4, 6, 12
   c. 12, 24, 36, 48
   d. 1, 2, 3, 4, 6, 12
   e. 0, 1, 12

10. Which of the following augmented matrices represents the system of equations below?

$$2x - 3y + z = -5$$
$$4x - y - 2z = -7$$
$$-x + 2z = -1$$

a. $\begin{bmatrix} 2 & -3 & 1 & 5 \\ 4 & -1 & 0 & -7 \\ -1 & 0 & 2 & 1 \end{bmatrix}$

b. $\begin{bmatrix} 2 & 4 & -1 \\ -3 & -1 & 0 \\ 1 & -2 & 2 \\ -5 & -7 & -1 \end{bmatrix}$

c. $\begin{bmatrix} 2 & 4 & -1 & -5 \\ -3 & -1 & 0 & -7 \\ 2 & -2 & 2 & -1 \end{bmatrix}$

d. $\begin{bmatrix} 2 & -3 & 1 \\ 4 & -1 & -2 \\ -1 & 0 & 2 \end{bmatrix}$

e. $\begin{bmatrix} 2 & -3 & 1 & -5 \\ 4 & -1 & -2 & -7 \\ -1 & 0 & 2 & -1 \end{bmatrix}$

11. What are the zeros of the function: $f(x) = x^3 + 4x^2 + 4x$?
a. -2
b. 0, -2
c. 2
d. 0, 2
e. 0

12. If $g(x) = x^3 - 3x^2 - 2x + 6$ and $f(x) = 2$, then what is $g(f(x))$?
a. -26
b. 6
c. $2x^3 - 6x^2 - 4x + 12$
d. -2
e. $2x^2 - 6$

13. What is the solution to the following system of equations?

$$x^2 - 2x + y = 8$$
$$x - y = -2$$

a. $(-2, 3)$
b. There is no solution.
c. $(-2, 0)\ (1, 3)$
d. $(-2, 0)\ (3, 5)$
e. $(2, 0)\ (-1, 3)$

14. Which of the following is the result after simplifying the expression:

$$(7n + 3n^3 + 3) + (8n + 5n^3 + 2n^4)?$$

a. $9n^4 + 15n - 2$
b. $2n^4 + 5n^3 + 15n - 2$
c. $9n^4 + 8n^3 + 15n$
d. $2n^4 + 8n^3 + 15n + 3$
e. $2n^4 + 5n^3 + 15n - 3$

15. What is the product of the following expression?

$$(4x - 8)(5x^2 + x + 6)$$

a. $20x^3 - 36x^2 + 16x - 48$
b. $6x^3 - 41x^2 + 12x + 15$
c. $20x^4 + 11x^2 - 37x - 12$
d. $2x^3 - 11x^2 - 32x + 20$
e. $20x^3 - 40x^2 + 24x - 48$

16. How could the following equation be factored to find the zeros?

$$y = x^3 - 3x^2 - 4x$$

a. $0 = x^2(x - 4), x = 0, 4$
b. $0 = 3x(x + 1)(x + 4), x = 0, -1, -4$
c. $0 = x(x + 1)(x + 6), x = 0, -1, -6$
d. $0 = 3x(x + 1)(x - 4), x = 0, 1, -4$
e. $0 = x(x + 1)(x - 4), x = 0, -1, 4$

17. The hospital has a nurse-to-patient ratio of 1:25. If there is a maximum of 325 patients admitted at a time, how many nurses are there?

a. 13 nurses
b. 25 nurses
c. 325 nurses
d. 12 nurses
e. 5 nurses

18. Which of the following is the solution for the given equation?

$$\frac{x^2 + x - 30}{x - 5} = 11$$

a. $x = -6$
b. All real numbers.
c. $x = 16$
d. $x = 5$
e. There is no solution.

19. Mom's car drove 72 miles in 90 minutes. How fast did she drive in feet per second?
    a. 0.8 feet per second
    b. 48.9 feet per second
    c. 0.009 feet per second
    d. 70.4 feet per second
    e. 21.3 feet per second

20. Solve $V = lwh$ for h.
    a. $lwV = h$
    b. $h = \frac{V}{lw}$
    c. $h = \frac{Vl}{w}$
    d. $h = \frac{Vw}{l}$
    e. $h = \frac{Vl}{w}$

21. What is the domain for the function $y = \sqrt{x}$?
    a. All real numbers
    b. $x \geq 0$
    c. $x > 0$
    d. $y \geq 0$
    e. $x < 0$

22. If Sarah reads at an average rate of 21 pages in four nights, how long will it take her to read 140 pages?
    a. 6 nights
    b. 26 nights
    c. 8 nights
    d. 27 nights
    e. 21 nights

23. The phone bill is calculated each month using the equation $c = 50g + 75$. The cost of the phone bill per month is represented by c, and g represents the gigabytes of data used that month. Identify and interpret the slope of this equation.
    a. 75 dollars per day
    b. 75 gigabytes per day
    c. 50 dollars per day
    d. 50 dollars per gigabyte
    e. The slope cannot be determined

24. What is the function that forms an equivalent graph to $y = \cos(x)$?
    a. $y = \tan(x)$
    b. $y = \csc(x)$
    c. $y = \sin\left(x + \frac{\pi}{2}\right)$
    d. $y = \sin\left(x - \frac{\pi}{2}\right)$
    e. $y = \tan\left(x + \frac{\pi}{2}\right)$

25. If $\sqrt{1+x} = 4$, what is x?
   a. 10
   b. 15
   c. 20
   d. 25
   e. 36

No Calculator Questions

26. What is the inverse of the function $f(x) = 3x - 5$?
   a. $f^{-1}(x) = \frac{x}{3} + 5$
   b. $f^{-1}(x) = \frac{5x}{3}$
   c. $f^{-1}(x) = 3x + 5$
   d. $f^{-1}(x) = \frac{x+5}{3}$
   e. $f^{-1}(x) = \frac{x}{3} - 5$

27. What are the zeros of $f(x) = x^2 + 4$?
   a. $x = -4$
   b. $x = \pm 2i$
   c. $x = \pm 2$
   d. $x = \pm 4i$
   e. $x = 2, 4$

28. Twenty is 40% of what number?
   a. 60
   b. 8
   c. 200
   d. 70
   e. 50

29. What is the simplified form of the expression $1.2 * 10^{12} \div 3.0 * 10^8$?
   a. $0.4 * 10^4$
   b. $4.0 * 10^4$
   c. $4.0 * 10^3$
   d. $3.6 * 10^{20}$
   e. $4.0 * 10^2$

30. You measure the width of your door to be 36 inches. The true width of the door is 35.75 inches. What is the relative error in your measurement?
   a. 0.7%
   b. 0.007%
   c. 0.99%
   d. 0.1%
   e. 7.0%

31. What is the y-intercept for $y = x^2 + 3x - 4$?
   a. $y = 1$
   b. $y = -4$
   c. $y = 3$
   d. $y = 4$
   e. $y = -3$

32. Is the following function even, odd, neither, or both?

$$y = \frac{1}{2}x^4 + 2x^2 - 6$$

   a. Even
   b. Odd
   c. Neither
   d. Both
   e. Even for all negative x-values and odd for all positive x-values

33. Which equation is not a function?
   a. $y = |x|$
   b. $y = \sqrt{x}$
   c. $x = 3$
   d. $y = 4$
   e. $y = 3x$

34. How could the following function be rewritten to identify the zeros?

$$y = 3x^3 + 3x^2 - 18x$$

   a. $y = 3x(x + 3)(x - 2)$
   b. $y = x(x - 2)(x + 3)$
   c. $y = 3x(x - 3)(x + 2)$
   d. $y = (x + 3)(x - 2)$
   e. $y = 3x(x + 3)(x + 2)$

35. A six-sided die is rolled. What is the probability that the roll is 1 or 2?
   a. $\frac{1}{6}$
   b. $\frac{1}{4}$
   c. $\frac{1}{3}$
   d. $\frac{1}{2}$
   e. $\frac{1}{36}$

36. A line passes through the origin and through the point (-3, 4). What is the slope of the line?

a. $-\frac{4}{3}$

b. $-\frac{3}{4}$

c. $\frac{4}{3}$

d. $\frac{3}{4}$

e. $\frac{1}{3}$

37. What type of function is modeled by the values in the following table?

x	f(x)
1	2
2	4
3	8
4	16
5	32

a. Linear
b. Exponential
c. Quadratic
d. Cubic
e. Logarithmic

38. An investment of $2,000 is made into an account with an annual interest rate of 5%, compounded continuously. What is the total value of the investment after eight years?

a. $4,707
b. $3,000
c. $2,983.65
d. $10,919.63
e. $1,977.61

39. A ball is drawn at random from a ball pit containing 8 red balls, 7 yellow balls, 6 green balls, and 5 purple balls. What's the probability that the ball drawn is yellow?

a. $^1/_{26}$

b. $^{19}/_{26}$

c. $^{14}/_{26}$

d. 1

e. $^7/_{26}$

40. Two cards are drawn from a shuffled deck of 52 cards. What's the probability that both cards are Kings if the first card isn't replaced after it's drawn?

a. $^1/_{169}$
b. $^1/_{221}$
c. $^1/_{13}$
d. $^4/_{13}$
e. $^1/_{104}$

41. Gary is driving home to see his parents for Christmas. He travels at a constant speed of 60 miles per hour for a total of 350 miles. How many minutes will it take him to travel home if he takes a break for 10 minutes every 100 miles?

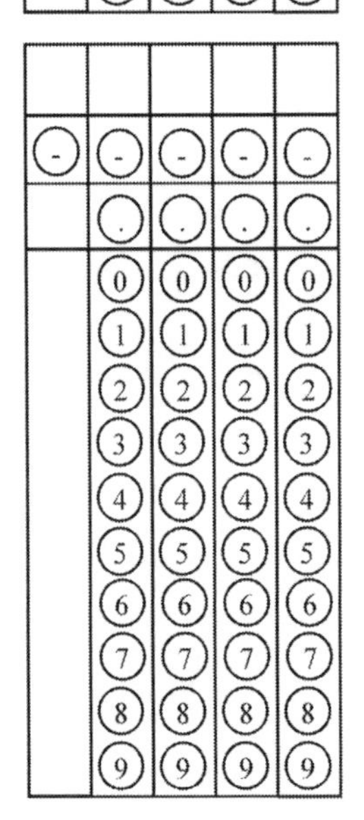

42. Kelly is selling cookies to raise money for the chorus. She has 500 cookies to sell. She sells 45% of the cookies to the sixth graders. At the second lunch, she sells 40% of what's left to the seventh graders. If she sells 60% of the remaining cookies to the eighth graders, how many cookies does Kelly have left at the end of all lunches?

43. Find the value of x.

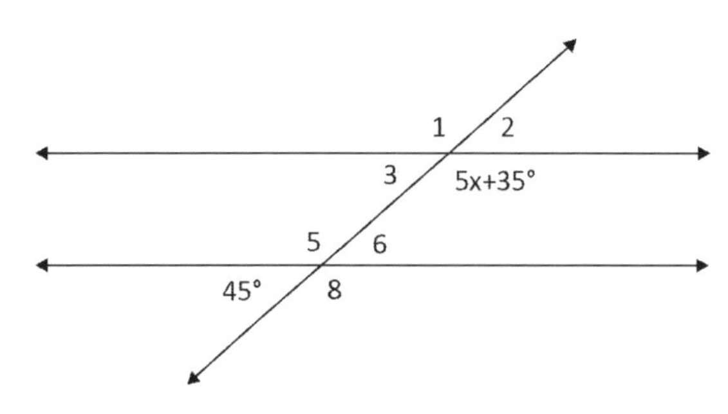

44. What is the value of the following expression?

$$\sqrt{8^2 + 6^2}$$

45. $864 \div 36 =$

46. Sam is twice as old as his sister, Lisa. Their oldest brother, Ray, will be 25 in three years. If Lisa is 13 years younger than Ray, how old is Sam?

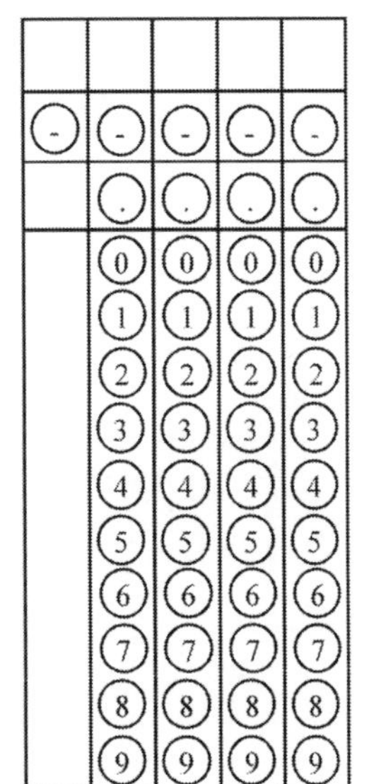

47. What is the perimeter of the following figure rounded to the nearest tenth?

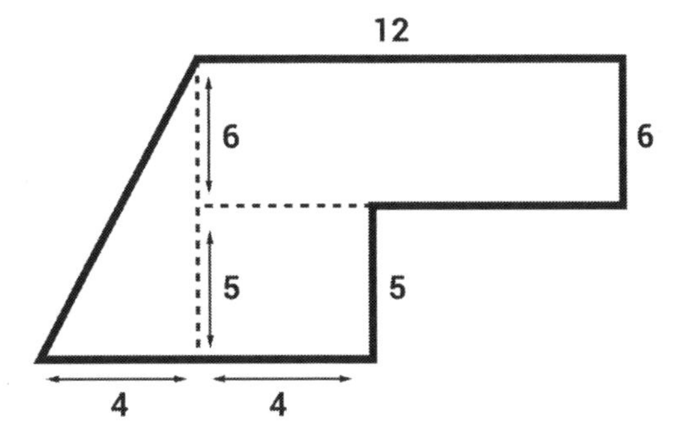

48. Solve the following:

$$(\sqrt{36} \times \sqrt{16}) - 3^2$$

49. What is the overall median of Dwayne's current scores: 78, 92, 83, 97?

50. The total perimeter of a rectangle is 36 cm. If the length is 12 cm, what is the width?

Reading

Questions 1–5 are based on the following passage:

As long ago as 1860 it was the proper thing to be born at home. At present, so I am told, the high gods of medicine have decreed that the first cries of the young shall be uttered upon the anesthetic air of a hospital, preferably a fashionable one. So young Mr. and Mrs. Roger Button were fifty years ahead of style when they decided, one day in the summer of 1860, that their first baby should be born in a hospital. Whether this anachronism had any bearing upon the astonishing history I am about to set down will never be known.

I shall tell you what occurred, and let you judge for yourself.

The Roger Buttons held an enviable position, both social and financial, in ante-bellum Baltimore. They were related to the This Family and the That Family, which, as every Southerner knew, entitled them to membership in that enormous peerage which largely populated the Confederacy. This was their first experience with the charming old custom of having babies—Mr. Button was naturally nervous. He hoped it would be a boy so that he could be sent to Yale College in Connecticut, at which institution Mr. Button himself had been known for four years by the somewhat obvious nickname of "Cuff."

On the September morning consecrated to the enormous event he arose nervously at six o'clock, dressed himself, adjusted an impeccable stock, and hurried forth through the streets of Baltimore to the hospital, to determine whether the darkness of the night had borne in new life upon its bosom.

When he was approximately a hundred yards from the Maryland Private Hospital for Ladies and Gentlemen, he saw Doctor Keene, the family physician, descending the front steps, rubbing his hands together with a washing movement—as all doctors are required to do by the unwritten ethics of their profession.

Mr. Roger Button, the president of Roger Button & Co., Wholesale Hardware, began to run toward Doctor Keene with much less dignity than was expected from a Southern gentleman of that picturesque period. "Doctor Keene!" he called. "Oh, Doctor Keene!"

The doctor heard him, faced around, and stood waiting, a curious expression settling on his harsh, medicinal face as Mr. Button drew near.

"What happened?" demanded Mr. Button, as he came up in a gasping rush. "What was it? How is she? A boy? Who is it? What—"

"Talk sense!" said Doctor Keene sharply. He appeared somewhat irritated.

"Is the child born?" begged Mr. Button.

Doctor Keene frowned. "Why, yes, I suppose so—after a fashion." Again he threw a curious glance at Mr. Button.

From The Curious Case of Benjamin Button by F.S. Fitzgerald, 1922.

1. According to the passage, what major event is about to happen in this story?
   a. Mr. Button is about to go to a funeral.
   b. Mr. Button's wife is about to have a baby.
   c. Mr. Button is getting ready to go to the doctor's office.
   d. Mr. Button is about to go shopping for new clothes.
   e. Mr. Button is about to be promoted to the president of his company.

2. What kind of tone does the above passage have?
   a. Irate and Belligerent
   b. Sad and Angry
   c. Shameful and Confused
   d. Grateful and Joyous
   e. Nervous and Excited

3. As it is used in the fourth paragraph, the word *consecrated* most nearly means:
   a. Numbed
   b. Chained
   c. Dedicated
   d. Moved
   e. Exonerated

4. What does the author mean to do by adding the following statement?

   "rubbing his hands together with a washing movement—as all doctors are required to do by the unwritten ethics of their profession."

   a. Suggesting that Mr. Button is tired of the doctor.
   b. Trying to explain the detail of the doctor's profession.
   c. Hinting to readers that the doctor is an unethical man.
   d. Giving readers a visual picture of what the doctor is doing.
   e. Crediting the doctor with authorship of medical ethics guides.

5. Which of the following best describes the development of this passage?
   a. It starts in the middle of a narrative in order to transition smoothly to a conclusion.
   b. It is a chronological narrative from beginning to end.
   c. The sequence of events is backwards—we go from future events to past events.
   d. To introduce the setting of the story and its characters.
   e. To provide a testimonial that leads into a persuasive argument.

Questions 6–10 are based upon the following passage:

The taxi's radio was tuned to a classical FM broadcast. Janáček's *Sinfonietta*—probably not the ideal music to hear in a taxi caught in traffic. The middle-aged driver didn't seem to be listening very closely, either. With his mouth clamped shut, he stared straight ahead at the endless line of cars stretching out on the elevated expressway, like a veteran fisherman standing in the bow of his boat, reading the ominous confluence of two currents. Aomame settled into the broad back seat, closed her eyes, and listened to the music.

How many people could recognize Janáček's *Sinfonietta* after hearing just the first few bars? Probably somewhere between "very few" and "almost none." But for some reason, Aomame was one of the few who could.

Janáček composed his little symphony in 1926. He originally wrote the opening as a fanfare for a gymnastics festival. Aomame imagined 1926 Czechoslovakia: The First World War had ended, and the country was freed from the long rule of the Hapsburg Dynasty. As they enjoyed the peaceful respite visiting central Europe, people drank Pilsner beer in cafés and manufactured handsome light machine guns. Two years earlier, in utter obscurity, Franz Kafka had left the world behind. Soon Hitler would come out of nowhere and gobble up this beautiful little country in the blink of an eye, but at the time no one knew what hardships lay in store for them. This may be the most important proposition revealed by history: "At the time, no one knew what was coming." Listening to Janáček's music, Aomame imagined the carefree winds sweeping across the plains of Bohemia and thought about the vicissitudes of history.

In 1926 Japan's Taisho Emperor died, and the era name was changed to Showa. It was the beginning of a terrible, dark time in this country, too. The short interlude of modernism and democracy was ending, giving way to fascism.

Aomame loved history as much as she loved sports. She rarely read fiction, but history books could keep her occupied for hours. What she liked about history was the way all its facts were linked with particular dates and places. She did not find it especially difficult to remember historical dates. Even if she did not learn them by rote memorization, once she grasped the relationship of an event to its time and to the events preceding and following it, the date would come to her automatically. In both middle school and high school, she had always gotten the top grade on history exams. It puzzled her to hear someone say he had trouble learning dates. How could something so simple be a problem for anyone?

"Aomame" was her real name. Her grandfather on her father's side came from some little mountain town or village in Fukushima Prefecture, where there were supposedly a number of people who bore the name, written with exactly the same characters as the word for "green peas" and pronounced with the same four syllables, "Ah-oh-mah-meh." She had never been to the place, however. Her father had cut his ties with his family before her birth, just as her mother had done with her own family, so she had never met any of her grandparents. She didn't travel much, but on those rare occasions when she stayed in an unfamiliar city or town, she would always open the hotel's phone book to see if there were any Aomames in the area. She had never found a single one, and whenever she tried and failed, she felt like a lonely castaway on the open sea.

Excerpt from *1Q84* by Haruki Murakami, translated by Jay Rubin, 2011

6. In context, "reading the ominous confluence of two currents" is a metaphor that most nearly represents:
   a. Seeking to catch a fish for dinner
   b. Watching the merging of traffic
   c. Eyeing a wreck on the highway
   d. Enjoying an afternoon drive
   e. Managing a line of cars through a tunnel

7. In relation to the last two paragraphs, the rest of the passage serves primarily to:
   a. Explain why the First World War occurred
   b. Depict a taxi driver at a certain point in history
   c. Engage the reader in an introduction to classical music
   d. Show two characters' inability to socially interact.
   e. Highlight the accuracy and depth of a character's knowledge.

8. The narrator's perspective throughout the passage might be described as that of:
   a. A neutral observer
   b. An intrigued storyteller
   c. An excited historian
   d. A disillusioned journalist
   e. A pedantic professor

9. The statement "Aomame imagined the carefree winds sweeping across the plains of Bohemia" serves as:
   a. Imagery to symbolize the blissful ignorance of the upcoming World War II in Eastern Europe
   b. Simile to denote that our perspective of history is mostly taken for granted
   c. Synecdoche to indicate that the plains and winds stand for bad weather in general
   d. Rhetorical question to call attention to the issue of the upcoming war and its prevention
   e. Personification to indicate that the wind and plains assume human characteristics

10. From the information in the passage, the setting of the story is located in:
   a. Germany
   b. Czech Republic
   c. Japan
   d. Austria
   e. Switzerland

Questions 11–15 are based on the following passage:

> *2. Chap.*
>
> *Of their departure into Holland and their troubls ther aboute, with some of the many difficulties they found and mete withall.*
>
> Being thus constrained to leave their native country, their lands and livings, and all their friends and familiar acquaintance, it was much, and thought marvellous by many. But to go into a country they knew not, but by hearsay, where they must learn a new language, and get their livings they knew not how, it being a dear place, and subject to the miseries of war, it was by many thought an adventure almost desperate, a case intolerable, and a misery worse than death; especially seeing they were not acquainted with trades nor traffic, (by which the country

doth subsist) but had only been used to a plain country life and the innocent trade of husbandry. But these things did not dismay them, (although they did sometimes trouble them,) for their desires were set on the ways of God, and to enjoy his ordinances. But they rested on his providence, and knew whom they had believed. Yet this was not all. For although they could not stay, yet were they not suffered to go; but the ports and havens were shut against them, so as they were fain to seek secret means of conveyance, and to fee the mariners, and give extraordinary rates for their passages. And yet were they oftentimes betrayed, many of them, and both they and their goods intercepted and surprised, and thereby put to great trouble and charge; of which I will give an instance or two, and omit the rest."

"There was a great company of them purposed to get passage at Boston, in Lincolnshire; and for that end had hired a ship wholly to themselves, and made agreement with the master to be ready at a certain day, and take them and their goods in, at a convenient place, where they accordingly would all attend in readiness. So after long waiting and large expenses, though he kept not the day with them, yet he came at length, and took them in, in the night. And when he had them and their goods aboard, he betrayed them, having beforehand complotted with the searchers and other officers so to do; who took them and put them into open boats, and there rifled and ransacked them, searching them to their shirts for money, yae, even the women, further than became modesty; and then carried them back into the town, and made them a spectacle and wonderment to the multitude, which came flocking on all sides to behold them. Being thus by the catchpole officers riffled and stripped of their money, books and much other goods, they were presented to the magistrates, and messengers sent to inform the Lords of the Council of them; and so they were committed to ward. Indeed the magistrates used them courteously, and showed them what favor they could; but could not deliver them until order came from the Council table. But the issue was, that after a month's imprisonment the greatest part were dismissed, and sent to the places from whence they came; but seven of the principal men were still kept in prison and bound over to the assizes."

In the spring of 1608 another attempt was made to embark and another Dutch shipmaster engaged. This second party assembled at a point between Grimsby and Hull not far from the mouth of the Humber. The women and children arrived in a small bark which became grounded at low water and while some of the men on shore were taken off in the ship's boat they were again apprehended. And to quote again:

"But after the first boat-full was got aboard, and she was ready to go for more, the master espied a great company, both horse and foot, with bills and guns and other weapons: for the country was raised to take them."

"But the poor men which were got on board were in great distress for their wives and children, which they saw thus to be taken, and were left distitute of their helps, and themselves also not having a cloth to shift them with, more than they had on their backs, and some scarce a penny about them, all they had being on the bark. It drew tears from their eyes, and anything they had they would have given to have been on shore again. But all in vain; there was no remedy; they must thus sadly part.

Excerpt from History of Plymouth Plantation by William Bradford, written between 1630 and 1651

11. Which of the following was NOT a difficulty faced by the group of travelers?
    a. They didn't know the new country's language.
    b. They weren't allowed to bring their wives or children to the new country.
    c. They didn't know how to make a living in the new country.
    d. Ships charged exorbitant rates for the voyage to the new country.
    e. They were met with hostility and prejudice by others in their travels.

12. What type of passage is this?
    a. Historical memoir or journal
    b. Passage from a textbook
    c. Poem or Epic Poem
    d. An instructional/technical document
    e. A religious text

13. What went wrong with the ship hired at Boston in Lincoln-shire?
    a. The women got sick due to the rough seas.
    b. The men were separated from their wives and children.
    c. The group couldn't get out of jail in time to board the ship.
    d. The shipmaster betrayed them.
    e. The ship got caught in a storm which caused a shipwreck.

14. Based on the passage, why did the men leave their wives and children?
    a. Armed government officials arrived on the shore.
    b. The government refused to let them out of prison.
    c. The sea was too rough to risk going back for them.
    d. The women were too sick to make the voyage.
    e. The women and children had contracted a disease.

15. Ships charged the group of travelers extra fees because:
    a. they were a vulnerable minority group.
    b. the voyage was particularly long and dangerous.
    c. the government outlawed their departure.
    d. they were wealthy enough to afford it.
    e. the group was so large and rowdy.

Questions 16–20 are based on the following poem. Read it carefully then answer the questions.

I sit and sew—a useless task it seems,
My hands grown tired, my head weighed down with dreams—
The panoply of war, the martial tred of men,
Grim-faced, stern-eyed, gazing beyond the ken
Of lesser souls, whose eyes have not seen Death, *5*
Nor learned to hold their lives but as a breath—
But—I must sit and sew.

I sit and sew—my heart aches with desire—
That pageant terrible, that fiercely pouring fire
On wasted fields, and writhing grotesque things *10*
Once men. My soul in pity flings

Appealing cries, yearning only to go
There in that holocaust of hell, those fields of woe—
But—I must sit and sew.

The little useless seam, the idle patch; *15*
Why dream I here beneath my homely thatch,
When there they lie in sodden mud and rain,
Pitifully calling me, the quick ones and the slain?
You need me, Christ! It is no roseate dream
That beckons me—this pretty futile seam, *20*
It stifles me—God, must I sit and sew?

Poem "I Sit and Sew" by Alice Moore Dunbar-Nelson, 1918

16. In line 9, the speaker mentions a "pageant." What is she referring to?
    a. A beauty pageant
    b. The current war
    c. A popular play
    d. A wedding celebration
    e. A sewing jubilee

17. In the first stanza, who are the "martial tred of men, / Grim-faced, stern-eyed" that the speaker mentions in lines 3 and 4?
    a. Children
    b. Enemies
    c. Soldiers
    d. Neighbors
    e. Family

18. What idea does the speaker effectively contrast in this poem?
    a. The idea between the usefulness of sewing and the uselessness of war, namely that men's bodies are literally being wasted on the battlefield while sewing gives the opportunity of creating clothes for those same bodies.
    b. The idea between right and wrong, specifically that the war and everything relating to it is immoral, and the domestic side of life, including sewing, can be seen as doing good.
    c. The idea between sacrifice and selfishness; the speaker is admitting that she is being selfish for wanting to pursue her passion of sewing rather than helping out with the war.
    d. The idea between cold and warmth; the speaker contrasts cold and warm imagery and sees the war as something cold and distant and the home as something warm and intimate.
    e. The idea between activeness and passiveness in the sense that the speaker views sewing as passiveness and longs to do something active in order to help out in the war.

19. What type of poetic lines are included in this poem?
    a. English sonnets
    b. Alternating tercets
    c. Rhyming couplets
    d. Syllabic haikus
    e. Iambic trimeter

20. How many stanzas does this poem have?
    a. 1
    b. 2
    c. 3
    d. 4
    e. 5

Questions 21–25 are based on the following passage:

> Learning how to write a ten-minute play may seem like a monumental task at first; but, if you follow a simple creative writing strategy, similar to writing a narrative story, you will be able to write a successful drama. The first step is to open your story as if it is a puzzle to be solved. This will allow the reader a moment to engage with the story and to mentally solve the story with you, the author. Immediately provide descriptive details that steer the main idea, the tone, and the mood according to the overarching theme you have in mind. For example, if the play is about something ominous, you may open Scene One with a thunderclap. Next, use dialogue to reveal the attitudes and personalities of each of the characters who have a key part in the unfolding story. Keep the characters off balance in some way to create interest and dramatic effect. Maybe what the characters say does not match what they do. Show images on stage to speed up the narrative; remember, one picture speaks a thousand words. As the play progresses, the protagonist must cross the point of no return in some way; this is the climax of the story. Then, as in a written story, you create a resolution to the life-changing event of the protagonist. Let the characters experience some kind of self-discovery that can be understood and appreciated by the patient audience. Finally, make sure all things come together in the end so that every detail in the play makes sense right before the curtain falls.

21. Based on the passage above, which of the following statements is false?
    a. Writing a ten-minute play may seem like an insurmountable task.
    b. Providing descriptive details is not necessary until after the climax of the story line.
    c. Engaging the audience by jumping into the story line immediately helps them solve the story's developing ideas with you, the writer.
    d. Descriptive details give clues to the play's intended mood and tone.
    e. The introduction of a ten-minute play does not need to open with a lot of coffee pouring or cigarette smoking to introduce the scenes. The action can get started right away.

22. Based on the passage above, which of the following is true?
    a. The class of eighth graders quickly learned that it is not that difficult to write a ten-minute play.
    b. The playwrights of the twenty-first century all use the narrative writing basic feature guide to outline their initial scripts.
    c. In order to follow a simple structure, a person can write a ten-minute play based on some narrative writing features.
    d. Women find playwriting easier than men because they are used to communicating in writing.
    e. The structure of writing a poem is similar to that of play writing and of narrative writing.

23. Based on your understanding of the passage, it can be assumed that which of the following statements are true?
    a. One way to reveal the identities and nuances of the characters in a play is to use dialogue.
    b. Characters should follow predictable routes in the challenge presented in the unfolding narrative, so the audience may easily follow the sequence of events.
    c. Using images in the stage design is a detrimental element of creating atmosphere and meaning for the drama.
    d. There is no need for the protagonist to come to terms with a self-discovery; he or she simply needs to follow the prescription for life lived as usual.
    e. It is perfectly fine to avoid serious consequences for the actors of a ten-minute play because there is not enough time to unravel perils.

24. In the passage, why does the writer suggest that writing a ten-minute play is accessible for a novice playwright?
    a. It took the author of the passage only one week to write his first play.
    b. The format follows similar strategies of writing a narrative story.
    c. There are no particular themes or points to unravel; a playwright can use a stream of consciousness style to write a play.
    d. Dialogue that reveals the characters' particularities is uncommonly simple to write.
    e. The characters of a ten-minute play wrap up the action simply by revealing their ideas in a monologue.

25. Based on the passage, which basic feature of narrative writing is NOT mentioned with respect to writing a ten-minute play?
    a. Character development
    b. Descriptive details
    c. Dialogue
    d. Mood and tone
    e. Style

Questions 26–30 are based on the following passage.

> Dana Gioia argues in his article that poetry is dying, now little more than a limited art form confined to academic and college settings. Of course poetry remains healthy in the academic setting, but the idea of poetry being limited to this academic subculture is a stretch. New technology and social networking alone have contributed to poets and other writers' work being shared across the world. YouTube has emerged to be a major asset to poets, allowing live performances to be streamed to billions of users. Even now, poetry continues to grow and voice topics that are relevant to the culture of our time. Poetry is not in the spotlight as it may have been in earlier times, but it's still a relevant art form that continues to expand in scope and appeal.
>
> Furthermore, Gioia's argument does not account for live performances of poetry. Not everyone has taken a poetry class or enrolled in university—but most everyone is online. The Internet is a perfect launching point to get all creative work out there. An example of this was the performance of Buddy Wakefield's *Hurling Crowbirds at Mockingbars*. Wakefield is a well-known poet who has published several collections of contemporary poetry. One of my favorite works by Wakefield is *Crowbirds*, specifically his performance at New York University in 2009. Although

his reading was a campus event, views of his performance online number in the thousands. His poetry attracted people outside of the university setting.

Naturally, the poem's popularity can be attributed both to Wakefield's performance and the quality of his writing. *Crowbirds* touches on themes of core human concepts such as faith, personal loss, and growth. These are not ideas that only poets or students of literature understand, but all human beings: "You acted like I was hurling crowbirds at mockingbars / and abandoned me for not making sense. / Evidently, I don't experience things as rationally as you do" (Wakefield 15-17). Wakefield weaves together a complex description of the perplexed and hurt emotions of the speaker undergoing a separation from a romantic interest. The line "You acted like I was hurling crowbirds at mockingbars" conjures up an image of someone confused, seemingly out of their mind . . . or in the case of the speaker, passionately trying to grasp at a relationship that is fading. The speaker is looking back and finding the words that described how he wasn't making sense. This poem is particularly human and gripping in its message, but the entire effect of the poem is enhanced through the physical performance.

At its core, poetry is about addressing issues/ideas in the world. Part of this is also addressing the perspectives that are exiguously considered. Although the platform may look different, poetry continues to have a steady audience due to the emotional connection the poet shares with the audience.

26. Which one of the following best explains how the passage is organized?
   a. The author begins with a long definition of the main topic, and then proceeds to prove how that definition has changed over the course of modernity.
   b. The author presents a puzzling phenomenon and uses the rest of the passage to showcase personal experiences in order to explain it.
   c. The author contrasts two different viewpoints, then builds a case showing preference for one over the other.
   d. The passage is an analysis of another theory that the author has no stake in.
   e. The passage is a summary of a main topic from its historical beginnings to its contemplated end.

27. The author of the passage would likely agree most with which of the following?
   a. Buddy Wakefield is a genius and is considered at the forefront of modern poetry.
   b. Poetry is not irrelevant; it is an art form that adapts to the changing time while containing its core elements.
   c. Spoken word is the zenith of poetic forms and the premier style of poetry in this decade.
   d. Poetry is on the verge of vanishing from our cultural consciousness.
   e. Poetry is a writing art. While poetry performances are useful for introducing poems, the act of reading a poem does not contribute to the piece overall.

28. Which one of the following words, if substituted for the word *exiguously* in the last paragraph, would LEAST change the meaning of the sentence?
   a. Indolently
   b. Inaudibly
   c. Interminably
   d. Infrequently
   e. Impecunious

29. Which of the following is most closely analogous to the author's opinion of Buddy Wakefield's performance in relation to modern poetry?
   a. Someone's refusal to accept that the Higgs Boson will validate the Standard Model.
   b. An individual's belief that soccer will lose popularity within the next fifty years.
   c. A professor's opinion that poetry contains the language of the heart, while fiction contains the language of the mind.
   d. An individual's assertion that video game violence was the cause of the Columbine shootings.
   e. A student's insistence that psychoanalysis is a subset of modern psychology.

30. What is the primary purpose of the passage?
   a. To educate readers on the development of poetry and describe the historical implications of poetry in media.
   b. To disprove Dana Gioia's stance that poetry is becoming irrelevant and is only appreciated in academia.
   c. To inform readers of the brilliance of Buddy Wakefield and to introduce them to other poets that have influenced contemporary poetry.
   d. To prove that Gioia's article does have some truth to it and to shed light on its relevance to modern poetry.
   e. To recount the experience of watching a live poetry performance and to look forward to future performances.

Questions 31–35 are based on the following passage:

> Becoming a successful leader in today's industry, government, and nonprofit sectors requires more than a high intelligence quotient (IQ). Emotional Intelligence (EI) includes developing the ability to know one's own emotions, to regulate impulses and emotions, and to use interpersonal communication skills with ease while dealing with other people. A combination of knowledge, skills, abilities, and mature emotional intelligence (EI) reflects the most effective leadership recipe. Successful leaders sharpen more than their talents and IQ levels; they practice the basic features of emotional intelligence. Some of the hallmark traits of a competent, emotionally intelligent leader include self-efficacy, drive, determination, collaboration, vision, humility, and openness to change. An unsuccessful leader exhibits opposite leadership traits: unclear directives, inconsistent vision and planning strategies, disrespect for followers, incompetence, and an uncompromising transactional leadership style. There are ways to develop emotional intelligence for the person who wants to improve his or her leadership style. For example, an emotionally intelligent leader creates an affirmative environment by incorporating collaborative activities, using professional development training for employee self-awareness, communicating clearly about the organization's vision, and developing a variety of resources for working with emotions. Building relationships outside the institution with leadership coaches and with professional development trainers can also help leaders who want to grow their leadership success. Leaders in today's work environment need to strive for a combination of skill, knowledge, and mature emotional intelligence to lead followers to success and to promote the vision and mission of their respective institutions.

31. The passage suggests that the term *emotional intelligence (EI)* can be defined as which of the following?

a. A combination of knowledge, skills, abilities, and mature emotional intelligence reflects the most effective EI leadership recipe.
b. An emotionally intelligent leader creates an affirmative environment by incorporating collaborative activities, using professional development training for employee self-awareness, communicating clearly about the organization's vision, and developing a variety of resources for working with emotions.
c. EI includes developing the ability to know one's own emotions, to regulate impulses and emotions, and to use interpersonal communication skills with ease while dealing with other people.
d. Becoming a successful leader in today's industry, government, and nonprofit sectors requires more than a high IQ.
e. An EI leader exhibits the following leadership traits: unclear directives, inconsistent vision and planning strategies, disrespect for followers, incompetence, and uncompromising transactional leadership style.

32. Based on the information in the passage, a successful leader must have a high EI quotient.

a. The above statement can be supported by the fact that Daniel Goldman conducted a scientific study.
b. The above statement can be supported by the example that emotionally intelligent people are highly successful leaders.
c. The above statement is not supported by the passage.
d. The above statement is supported by the illustration that claims, "Leaders in today's work environment need to strive for a combination of skill, knowledge, and mature emotional intelligence to lead followers to success and to promote the vision and mission of their respective institutions."
e. The above statement can be inferred because emotionally intelligent people obviously make successful leaders.

33. According to the passage, some of the characteristics of an unsuccessful leader include which of the following?

a. Talent, IQ level, and abilities
b. Humility, knowledge, and skills
c. Loud, demeaning actions toward female employees
d. Outdated technological resources and strategies
e. Transactional leadership style

34. According to the passage, which of the following must be true?

a. The leader exhibits a healthy work/life balance lifestyle.
b. The leader is uncompromising in transactional directives for all employees, regardless of status.
c. The leader learns to strategize using future trends analysis to create a five-year plan.
d. The leader uses a combination of skill, knowledge, and mature reasoning to make decisions.
e. The leader continually tries to improve their EI test quotient by studying the intelligence quotient of other successful leaders.

35. According to the passage, which one of the following choices are true?
    a. To be successful, leaders in the nonprofit sector need to develop academic intelligence.
    b. It is not necessary for military leaders to develop emotional intelligence because they prefer a transactional leadership style.
    c. Leadership coaches cannot add value to someone who is developing his or her emotional intelligence.
    d. Humility is a valued character value; however, it is not necessarily a trademark of an emotionally intelligent leader.
    e. If a leader does not have the level of emotional intelligence required for a certain job, they are capable of increasing emotional intelligence.

Questions 36 through 40 refer to the following passage:

ACT II SCENE II

Capulet's orchard.

[Enter ROMEO]

ROMEO He jests at scars that never felt a wound.

[JULIET appears above at a window]

But, soft! what light through yonder window breaks?

It is the east, and Juliet is the sun.

Arise, fair sun, and kill the envious moon,

Who is already sick and pale with grief,

That thou her maid art far more fair than she:

Be not her maid, since she is envious;

Her vestal livery is but sick and green

And none but fools do wear it; cast it off.

It is my lady, O, it is my love! 10

O, that she knew she were!

She speaks yet she says nothing: what of that?

Her eye discourses; I will answer it.

I am too bold, 'tis not to me she speaks:

Two of the fairest stars in all the heaven,

Having some business, do entreat her eyes

To twinkle in their spheres till they return.

What if her eyes were there, they in her head?

The brightness of her cheek would shame those stars,

As daylight doth a lamp; her eyes in heaven 20

Would through the airy region stream so bright

That birds would sing and think it were not night.

See, how she leans her cheek upon her hand!

O, that I were a glove upon that hand,

That I might touch that cheek!

JULIET Ay me!

ROMEO She speaks:

O, speak again, bright angel! for thou art

As glorious to this night, being o'er my head

As is a winged messenger of heaven

Unto the white-upturned wondering eyes

Of mortals that fall back to gaze on him 30

When he bestrides the lazy-pacing clouds

And sails upon the bosom of the air.

JULIET O Romeo, Romeo! wherefore art thou Romeo?

Deny thy father and refuse thy name;

Or, if thou wilt not, be but sworn my love,

And I'll no longer be a Capulet.

ROMEO *[Aside]* Shall I hear more, or shall I speak at this?

JULIET 'Tis but thy name that is my enemy;

Thou art thyself, though not a Montague.

What's Montague? it is nor hand, nor foot, 40

Nor arm, nor face, nor any other part

Belonging to a man. O, be some other name!

What's in a name? that which we call a rose

By any other name would smell as sweet;

So Romeo would, were he not Romeo call'd,

Retain that dear perfection which he owes

Without that title. Romeo, doff thy name,

And for that name which is no part of thee

Take all myself.

Excerpt from Shakespeare's *Romeo and Juliet*

36. In the passage that begins, "What's Montague? It is nor hand, nor foot, / Nor arm, nor face, nor any other part…," what is Juliet essentially saying?
    a. That she isn't special
    b. That Romeo shouldn't care about her
    c. That his name means they cannot be together
    d. That she would love Romeo no matter what his name is
    e. That her name is more important than his

37. Which statement best describes Romeo's intent when he says the moon is envious?
    a. The sun is rising.
    b. The sun and stars are shining.
    c. Juliet is more beautiful.
    d. The moon is sick.
    e. The moon is paler than the sun.

38. Which of the following describes the tone of lines 10 through 25?
    a. Desirous
    b. Remorseful
    c. Respectful
    d. Tentative
    e. Hesitant

39. Which of the following lines indicates the famous balcony scene from the play?
    a. "And sails upon the bosom of the air."
    b. "Two of the fairest stars in all the heaven,"
    c. "As glorious to this night, being o'er my head"
    d. "He jests at scars that never felt a wound."
    e. "When he bestrides the lazy-pacing clouds"

40. The line, "As glorious to this night, being o'er my head / As is a winged messenger of heaven," is an example of which of the following?

a. Metaphor
b. Simile
c. Personification
d. Imagery
e. Theme

Writing

Part I

Questions 1–9 are based on the following passage:

While all dogs (1) descend through gray wolves, it's easy to notice that dog breeds come in a variety of shapes and sizes. With such a (2) drastic range of traits, appearances and body types dogs are one of the most variable and adaptable species on the planet. (3) But why so many differences. The answer is that humans have actually played a major role in altering the biology of dogs. (4) This was done through a process called selective breeding.

(5) Selective breeding which is also called artificial selection is the processes in which animals with desired traits are bred in order to produce offspring that share the same traits. In natural selection, (6) animals must adapt to their environments increase their chance of survival. Over time, certain traits develop in animals that enable them to thrive in these environments. Those animals with more of these traits, or better versions of these traits, gain an (7) advantage over others of their species. Therefore, the animal's chances to mate are increased and these useful (8) genes are passed into their offspring. With dog breeding, humans select traits that are desired and encourage more of these desired traits in other dogs by breeding dogs that already have them.

The reason for different breeds of dogs is that there were specific needs that humans wanted to fill with their animals. For example, sent hounds are known for their extraordinary ability to track game through scent. These breeds are also known for their endurance in seeking deer and other prey. Therefore, early hunters took dogs that displayed these abilities and bred them to encourage these traits. Later, these generations took on characteristics that aided these desired traits. (9) For example, Bloodhounds have broad snouts and droopy ears that fall to the ground when they smell. These physical qualities not only define the look of the Bloodhound, but also contribute to their amazing tracking ability. The broad snout is able to define and hold onto scents longer than many other breeds. The long floppy hears serve to collect and hold the scents the earth holds so that the smells are clearer and able to be distinguished.

1. Which of the following would be the best choice for this sentence (reproduced below)?

While all dogs (1) descend through gray wolves, it's easy to notice that dog breeds come in a variety of shapes and sizes.

a. NO CHANGE
b. descend by gray wolves
c. descend from gray wolves
d. descended through gray wolves

2. Which of the following would be the best choice for this sentence (reproduced below)?

With such a (2) drastic range of traits, appearances and body types, dogs are one of the most variable and adaptable species on the planet.

a. NO CHANGE
b. drastic range of traits, appearances, and body types,
c. drastic range of traits and appearances and body types,
d. drastic range of traits, appearances, as well as body types,

3. Which of the following would be the best choice for this sentence (reproduced below)?

(3) But why so many differences.

a. NO CHANGE
b. But are there so many differences?
c. But why so many differences are there.
d. But why so many differences?

4. Which of the following would be the best choice for this sentence (reproduced below)?

(4) This was done through a process called selective breeding.

a. NO CHANGE
b. This was done, through a process called selective breeding.
c. This was done, through a process, called selective breeding.
d. This was done through selective breeding, a process.

5. Which of the following would be the best choice for this sentence (reproduced below)?

(5) Selective breeding which is also called artificial selection is the processes in which animals with desired traits are bred in order to produce offspring that share the same traits.

a. NO CHANGE
b. Selective breeding, which is also called artificial selection is the processes
c. Selective breeding which is also called, artificial selection, is the processes
d. Selective breeding, which is also called artificial selection, is the processes

6. Which of the following would be the best choice for this sentence (reproduced below)?

In natural selection, (6) animals must adapt to their environments increase their chance of survival.

a. NO CHANGE
b. animals must adapt to their environments to increase their chance of survival.
c. animals must adapt to their environments, increase their chance of survival.
d. animals must adapt to their environments, increasing their chance of survival.

7. Which of the following would be the best choice for this sentence (reproduced below)?

Those animals with more of these traits, or better versions of these traits, gain an (7) advantage over others of their species.

a. NO CHANGE
b. advantage over others, of their species.
c. advantages over others of their species.
d. advantage over others.

8. Which of the following would be the best choice for this sentence (reproduced below)?

Therefore, the animal's chances to mate are increased and these useful (8) genes are passed into their offspring.

a. NO CHANGE
b. genes are passed onto their offspring.
c. genes are passed on to their offspring.
d. genes are passed within their offspring.

9. Which of the following would be the best choice for this sentence (reproduced below)?

(9) For example, Bloodhounds have broad snouts and droopy ears that fall to the ground when they smell.

a. NO CHANGE
b. For example, Bloodhounds,
c. For example Bloodhounds
d. For example, bloodhounds

Questions 10–18 are based on the following passage:

I'm not alone when I say that it's hard to pay attention sometimes. I can't count how many times I've sat in a classroom, lecture, speech, or workshop and (10) been bored to tears or rather sleep. (11) Usually I turn to doodling in order to keep awake. This never really helps; I'm not much of an artist. Therefore, after giving up on drawing a masterpiece, I would just concentrate on keeping my eyes open and trying to be attentive. This didn't always work because I wasn't engaged in what was going on.

(12) Sometimes in particularly dull seminars, I'd imagine comical things going on in the room or with the people trapped in the room with me. Why? (13) Because I wasn't invested in what was

going on I wasn't motivated to listen. I'm not going to write about how I conquered the difficult task of actually paying attention in a difficult or unappealing class—it can be done, sure. I have sat through the very epitome of boredom (in my view at least) several times and come away learning something. (14) Everyone probably has had to at one time do this. What I want to talk about is that profound moment when curiosity is sparked (15) in another person drawing them to pay attention to what is before them and expand their knowledge.

What really makes people pay attention? (16) Easy it's interest. This doesn't necessarily mean (17) embellishing subject matter drawing people's attention. This won't always work. However, an individual can present material in a way that is clear to understand and actually engages the audience. Asking questions to the audience or class will make them a part of the topic at hand. Discussions that make people think about the content and (18) how it applies to there lives world and future is key. If math is being discussed, an instructor can explain the purpose behind the equations or perhaps use real-world applications to show how relevant the topic is. When discussing history, a lecturer can prompt students to imagine themselves in the place of key figures and ask how they might respond. The bottom line is to explore the ideas rather than just lecture. Give people the chance to explore material from multiple angles, and they'll be hungry to keep paying attention for more information.

10. Which of the following would be the best choice for this sentence (reproduced below)?

I can't count how many times I've sat in a classroom, lecture, speech, or workshop and (10) been bored to tears or rather sleep.

a. NO CHANGE
b. been bored to, tears, or rather sleep.
c. been bored, to tears or rather sleep.
d. been bored to tears or, rather, sleep.

11. Which of the following would be the best choice for this sentence (reproduced below)?

(11) Usually I turn to doodling in order to keep awake.

a. NO CHANGE
b. Usually, I turn to doodling in order to keep awake.
c. Usually I turn to doodling, in order, to keep awake.
d. Usually I turned to doodling in order to keep awake.

12. Which of the following would be the best choice for this sentence (reproduced below)?

(12) Sometimes in particularly dull seminars, I'd imagine comical things going on in the room or with the people trapped in the room with me.

a. NO CHANGE
b. Sometimes, in particularly, dull seminars,
c. Sometimes in particularly dull seminars
d. Sometimes in particularly, dull seminars,

13. Which of the following would be the best choice for this sentence (reproduced below)?

(13) Because I wasn't invested in what was going on I wasn't motivated to listen.

a. NO CHANGE
b. Because I wasn't invested, in what was going on, I wasn't motivated to listen.
c. Because I wasn't invested in what was going on. I wasn't motivated to listen.
d. I wasn't motivated to listen because I wasn't invested in what was going on.

14. Which of the following would be the best choice for this sentence (reproduced below)?

(14) Everyone probably has had to at one time do this.

a. NO CHANGE
b. Everyone probably has had to, at one time. Do this.
c. Everyone's probably had to do this at some time.
d. At one time everyone probably has had to do this.

15. Which of the following would be the best choice for this sentence (reproduced below)?

What I want to talk about is that profound moment when curiosity is sparked (15) in another person drawing them to pay attention to what is before them and expand their knowledge.

a. NO CHANGE
b. in another person, drawing them to pay attention
c. in another person; drawing them to pay attention to what is before them.
d. in another person, drawing them to pay attention to what is before them.

16. Which of the following would be the best choice for this sentence (reproduced below)?

(16) Easy it's interest.

a. NO CHANGE
b. Easy it is interest.
c. Easy. It's interest.
d. Easy—it's interest.

17. Which of the following would be the best choice for this sentence (reproduced below)?

This doesn't necessarily mean (17) embellishing subject matter drawing people's attention.

a. NO CHANGE
b. embellishing subject matter which draws people's attention.
c. embellishing subject matter to draw people's attention.
d. embellishing subject matter for the purpose of drawing people's attention.

18. Which of the following would be the best choice for this sentence (reproduced below)?

Discussions that make people think about the content and (18) how it applies to there lives world and future is key.

a. NO CHANGE
b. how it applies to their lives, world, and future is key.
c. how it applied to there lives world and future is key.
d. how it applies to their lives, world and future is key.

Questions 19–27 are based on the following passage:

Since the first discovery of dinosaur bones, (19) scientists has made strides in technological development and methodologies used to investigate these extinct animals. We know more about dinosaurs than ever before and are still learning fascinating new things about how they looked and lived. However, one has to ask, (20) how if earlier perceptions of dinosaurs continue to influence people's understanding of these creatures? Can these perceptions inhibit progress towards further understanding of dinosaurs?

(21) The biggest problem with studying dinosaurs is simply that there are no living dinosaurs to observe. All discoveries associated with these animals are based on physical remains. To gauge behavioral characteristics, scientists cross-examine these (22) finds with living animals that seem similar in order to gain understanding. While this method is effective, these are still deductions. Some ideas about dinosaurs can't be tested and confirmed simply because humans can't replicate a living dinosaur. For example, a Spinosaurus has a large sail, or a finlike structure that grows from its back. Paleontologists know this sail exists and have ideas for the function of (23) the sail however they are uncertain of which idea is the true function. Some scientists believe (24) the sail serves to regulate the Spinosaurus' body temperature and yet others believe its used to attract mates. Still, other scientists think the sail is used to intimidate other predatory dinosaurs for self-defense. These are all viable explanations, but they are also influenced by what scientists know about modern animals. (25) Yet, it's quite possible that the sail could hold a completely unique function.

While it's (26) plausible, even likely that dinosaurs share many traits with modern animals, there is the danger of overattributing these qualities to a unique, extinct species. For much of the early nineteenth century, when people first started studying dinosaur bones, the assumption was that they were simply giant lizards. (27) For the longest time this image was the prevailing view on dinosaurs, until evidence indicated that they were more likely warm blooded. Scientists have also discovered that many dinosaurs had feathers and actually share many traits with modern birds.

19. Which of the following would be the best choice for this sentence (reproduced below)?

Since the first discovery of dinosaur bones, (19) scientists has made strides in technological development and methodologies used to investigate these extinct animals.

a. NO CHANGE
b. scientists has made strides in technological development, and methodologies, used to investigate
c. scientists have made strides in technological development and methodologies used to investigate
d. scientists, have made strides in technological development and methodologies used, to investigate

20. Which of the following would be the best choice for this sentence (reproduced below)?

However, one has to ask, (20) how if earlier perceptions of dinosaurs continue to influence people's understanding of these creatures?

a. NO CHANGE
b. how perceptions of dinosaurs
c. how, if, earlier perceptions of dinosaurs
d. whether earlier perceptions of dinosaurs

21. Which of the following would be the best choice for this sentence (reproduced below)?

(21) The biggest problem with studying dinosaurs is simply that there are no living dinosaurs to observe.

a. NO CHANGE
b. The biggest problem with studying dinosaurs is simple, that there are no living dinosaurs to observe.
c. The biggest problem with studying dinosaurs is simple. There are no living dinosaurs to observe.
d. The biggest problem with studying dinosaurs, is simply that there are no living dinosaurs to observe.

22. Which of the following would be the best choice for this sentence (reproduced below)?

To gauge behavioral characteristics, scientists cross-examine these (22) finds with living animals that seem similar in order to gain understanding.

a. NO CHANGE
b. finds with living animals to explore potential similarities.
c. finds with living animals to gain understanding of similarities.
d. finds with living animals that seem similar, in order, to gain understanding.

23. Which of the following would be the best choice for this sentence (reproduced below)?

Paleontologists know this sail exists and have ideas for the function of (23) the sail however they are uncertain of which idea is the true function.

a. NO CHANGE
b. the sail however, they are uncertain of which idea is the true function.
c. the sail however they are, uncertain, of which idea is the true function.
d. the sail; however, they are uncertain of which idea is the true function.

24. Which of the following would be the best choice for this sentence (reproduced below)?

Some scientists believe (24) the sail serves to regulate the Spinosaurus' body temperature and yet others believe its used to attract mates.

a. NO CHANGE
b. the sail serves to regulate the Spinosaurus' body temperature, yet others believe it's used to attract mates.
c. the sail serves to regulate the Spinosaurus' body temperature and yet others believe it's used to attract mates.
d. the sail serves to regulate the Spinosaurus' body temperature however others believe it's used to attract mates.

25. Which of the following would be the best choice for this sentence (reproduced below)?

(25) Yet, it's quite possible that the sail could hold a completely unique function.

a. NO CHANGE
b. Yet, it's quite possible,
c. It's quite possible,
d. Its quite possible

26. Which of the following would be the best choice for this sentence (reproduced below)?

While it's (26) plausible, even likely that dinosaurs share many traits with modern animals, there is the danger of over attributing these qualities to a unique, extinct species.

a. NO CHANGE
b. plausible, even likely that, dinosaurs share many
c. plausible, even likely, that dinosaurs share many
d. plausible even likely that dinosaurs share many

27. Which of the following would be the best choice for this sentence (reproduced below)?

(27) For the longest time this image was the prevailing view on dinosaurs, until evidence indicated that they were more likely warm blooded.

a. NO CHANGE
b. For the longest time this was the prevailing view on dinosaurs
c. For the longest time, this image, was the prevailing view on dinosaurs
d. For the longest time this was the prevailing image of dinosaurs

Questions 28–36 are based on the following passage:

Everyone has heard the (28) idea of the end justifying the means; that would be Weston's philosophy. Weston is willing to cross any line, commit any act no matter how heinous, to achieve success in his goal. (29) Ransom is reviled by this fact, seeing total evil in Weston's plan. To do an evil act in order (30) to gain a result that's supposedly good would ultimately warp the final act. (31) This opposing viewpoints immediately distinguishes Ransom as the hero. In the conflict with Un-man, Ransom remains true to his moral principles, someone who refuses to be compromised by power. Instead, Ransom makes it clear that by allowing such processes as murder and lying dictate how one attains a positive outcome, (32) the righteous goal becomes corrupted. The good end would not be truly good, but a twisted end that conceals corrupt deeds.

(33) This idea of allowing necessary evils to happen, is very tempting, it is what Weston fell prey to. (34) The temptation of the evil spirit Un-man ultimately takes over Weston and he is possessed. However, Ransom does not give into temptation. He remains faithful to the truth of what is right and incorrect. This leads him to directly face Un-man for the fate of Perelandra and its inhabitants.

Just as Weston was corrupted by the Un-man, (35) Un-man after this seeks to tempt the Queen of Perelandra to darkness. Ransom must literally (36) show her the right path, to accomplish this, he does this based on the same principle as the "means to an end" argument—that good follows good, and evil follows evil. Later in the plot, Weston/Un-man seeks to use deceptive reasoning to turn the queen to sin, pushing the queen to essentially ignore Melildil's rule to satisfy her own curiosity. In this sense, Un-man takes on the role of a false prophet, a tempter. Ransom must shed light on the truth, but this is difficult; his adversary is very clever and uses brilliant language. Ransom's lack of refinement heightens the weight of Un-man's corrupted logic, and so the Queen herself is intrigued by his logic.

Based on an excerpt from *Perelandra* by C.S. Lewis

28. Which of the following would be the best choice for this sentence (reproduced below)?

Everyone has heard the (28) idea of the end justifying the means; that would be Weston's philosophy.

a. NO CHANGE
b. idea of the end justifying the means; this is Weston's philosophy.
c. idea of the end justifying the means, this is the philosophy of Weston
d. idea of the end justifying the means. That would be Weston's philosophy.

29. Which of the following would be the best choice for this sentence (reproduced below)?

(29) Ransom is reviled by this fact, seeing total evil in Weston's plan.

a. NO CHANGE
b. Ransom is reviled by this fact; seeing total evil in Weston's plan.
c. Ransom, is reviled by this fact, seeing total evil in Weston's plan.
d. Ransom reviled by this, sees total evil in Weston's plan.

30. Which of the following would be the best choice for this sentence (reproduced below)?

To do an evil act in order (30) to gain a result that's supposedly good would ultimately warp the final act.

a. NO CHANGE
b. for an outcome that's for a greater good would ultimately warp the final act.
c. to gain a final act would warp its goodness.
d. to achieve a positive outcome would ultimately warp the goodness of the final act.

31. Which of the following would be the best choice for this sentence (reproduced below)?

(31) This opposing viewpoints immediately distinguishes Ransom as the hero.

a. NO CHANGE
b. This opposing viewpoints immediately distinguishes Ransom, as the hero.
c. This opposing viewpoint immediately distinguishes Ransom as the hero.
d. Those opposing viewpoints immediately distinguishes Ransom as the hero.

32. Which of the following would be the best choice for this sentence (reproduced below)?

Instead, Ransom makes it clear that by allowing such processes as murder and lying dictate how one attains a positive outcome, (32) the righteous goal becomes corrupted.

a. NO CHANGE
b. the goal becomes corrupted and no longer righteous.
c. the righteous goal becomes, corrupted.
d. the goal becomes corrupted, when once it was righteous.

33. Which of the following would be the best choice for this sentence (reproduced below)?

(33) This idea of allowing necessary evils to happen, is very tempting, it is what Weston fell prey to.

a. NO CHANGE
b. This idea of allowing necessary evils to happen, is very tempting. This is what Weston fell prey to.
c. This idea, allowing necessary evils to happen, is very tempting, it is what Weston fell prey to.
d. This tempting idea of allowing necessary evils to happen is what Weston fell prey to.

34. Which of the following would be the best choice for this sentence (reproduced below)?

(34) The temptation of the evil spirit Un-man ultimately takes over Weston and he is possessed.

a. NO CHANGE
b. The temptation of the evil spirit Un-man ultimately takes over and possesses Weston.
c. Weston is possessed as a result of the temptation of the evil spirit Un-man ultimately, who takes over.
d. The temptation of the evil spirit Un-man takes over Weston and he is possessed ultimately.

35. Which of the following would be the best choice for this sentence (reproduced below)?

Just as Weston was corrupted by the Un-man, (35) Un-man after this seeks to tempt the Queen of Perelandra to darkness.

a. NO CHANGE
b. Un-man, after this, would tempt the Queen of Perelandra
c. Un-man, after this, seeks to tempt the Queen of Perelandra
d. Un-man then seeks to tempt the Queen of Perelandra

36. Which of the following would be the best choice for this sentence (reproduced below)?

> Ransom must literally (36) show her the right path, to accomplish this, he does this based on the same principle as the "means to an end" argument—that good follows good, and evil follows evil.

a. NO CHANGE
b. show her the right path. To accomplish this, he uses the same principle as the "means to an end" argument
c. show her the right path; to accomplish this he uses the same principle as the "means to an end" argument
d. show her the right path, to accomplish this, the same principle as the "means to an end" argument is applied

Questions 37–45 are based on the following passage:

> (37) What's clear about the news is today is that the broader the media the more ways there are to tell a story. Even if different news groups cover the same story, individual newsrooms can interpret or depict the story differently than other counterparts. Stories can also change depending on the type of (38) media in question incorporating different styles and unique ways to approach the news. (39) It is because of these respective media types that ethical and news-related subject matter can sometimes seem different or altered. But how does this affect the narrative of the new story?
>
> I began by investing a written newspaper article from the Baltimore Sun. Instantly striking are the bolded Headlines. (40) These are clearly meant for direct the viewer to the most exciting and important stories the paper has to offer. What was particularly noteworthy about this edition was that the first page dealt with two major ethical issues. (41) On a national level there was a story on the evolving Petraeus scandal involving his supposed affair. The other article was focused locally in Baltimore, a piece questioning the city's Ethic's Board and their current director. Just as a television newscaster communicates the story through camera and dialogue, the printed article applies intentional and targeted written narrative style. More so than any of the mediums, news article seems to be focused specifically on a given story without need to jump to another. Finer details are usually expanded on (42) in written articles, usually people who read newspapers or go online for web articles want more than a quick blurb. The diction of the story is also more precise and can be either straightforward or suggestive (43) depending in earnest on the goal of the writer. However, there's still plenty of room for opinions to be inserted into the text.
>
> Usually, all news (44) outlets have some sort of bias, it's just a question of how much bias clouds the reporting. As long as this bias doesn't withhold information from the reader, it can be considered credible. (45) However an over use of bias, opinion, and suggestive language can rob readers of the chance to interpret the news events for themselves.

37. Which of the following would be the best choice for this sentence (reproduced below)?

(37) What's clear about the news today is that the broader the media the more ways there are to tell a story.

a. NO CHANGE
b. What's clear, about the news today, is that the broader the media
c. What's clear about today's news is that the broader the media
d. The news today is broader than earlier media

38. Which of the following would be the best choice for this sentence (reproduced below)?

Stories can also change depending on the type of (38) media in question incorporating different styles and unique ways to approach the news.

a. NO CHANGE
b. media in question; each incorporates unique styles and unique
c. media in question. To incorporate different styles and unique
d. media in question, incorporating different styles and unique

39. Which of the following would be the best choice for this sentence (reproduced below)?

(39) It is because of these respective media types that ethical and news-related subject matter can sometimes seem different or altered.

a. NO CHANGE
b. It is because of these respective media types, that ethical and news-related subject matter, can sometimes seem different or altered.
c. It is because of these respective media types, that ethical and news-related subject matter can sometimes seem different or altered.
d. It is because of these respective media types that ethical and news-related subject matter can sometimes seem different. Or altered.

40. Which of the following would be the best choice for this sentence (reproduced below)?

(40) These are clearly meant for direct the viewer to the most exciting and important stories the paper has to offer.

a. NO CHANGE
b. These are clearly meant for the purpose of giving direction to the viewer
c. These are clearly meant to direct the viewer
d. These are clearly meant for the viewer to be directed

41. Which of the following would be the best choice for this sentence (reproduced below)?

(41) On a national level there was a story on the evolving Petraeus scandal involving his supposed affair.

a. NO CHANGE
b. On a national level a story was there
c. On a national level; there was a story
d. On a national level, there was a story

42. Which of the following would be the best choice for this sentence (reproduced below)?

Finer details are usually expanded on (42) in written articles, usually people who read newspapers or go online for web articles want more than a quick blurb.

a. NO CHANGE
b. in written articles. People who usually
c. in written articles, usually, people who
d. in written articles usually people who

43. Which of the following would be the best choice for this sentence (reproduced below)?

The diction of the story is also more precise and can be either straightforward or suggestive (43) depending in earnest on the goal of the writer.

a. NO CHANGE
b. depending; in earnest on the goal of the writer.
c. depending, in earnest, on the goal of the writer.
d. the goal of the writer, in earnest, depends on the goal of the writer.

44. Which of the following would be the best choice for this sentence (reproduced below)?

Usually, all news (44) outlets have some sort of bias, it's just a question of how much bias clouds the reporting.

a. NO CHANGE
b. outlets have some sort of bias. Just a question of how much
c. outlets have some sort of bias it can just be a question of how much
d. outlets have some sort of bias, its just a question of how much

45. Which of the following would be the best choice for this sentence (reproduced below)?

(45) However an over use of bias, opinion, and suggestive language can rob readers of the chance to interpret the news events for themselves.

a. NO CHANGE
b. However, an over use of bias,
c. However, with too much bias,
d. However, an overuse of bias,

Questions 46–50 are based on the following passage:

Aircraft Engineers

The knowledge of an aircraft engineer is acquired through years of education, and special licenses are required. Ideally, an individual will begin his or her preparation for the profession in high school by taking chemistry, physics, trigonometry, and calculus. Such curricula will aid in one's pursuit of a bachelor's degree in aircraft engineering, which requires several physical and life sciences, mathematics, and design courses.

(47) Some of universities provide internship or apprentice opportunities for the students enrolled in aircraft engineer programs. A bachelor's in aircraft engineering is commonly accompanied by a master's degree in advanced engineering or business administration. Such advanced degrees enable an individual to position himself or herself for executive, faculty, and/or research opportunities. (48) These advanced offices oftentimes require a Professional Engineering (PE) license which can be obtained through additional college courses, professional experience, and acceptable scores on the Fundamentals of Engineering (FE) and Professional Engineering (PE) standardized assessments.

Once the job begins, this line of work requires critical thinking, business skills, problem solving, and creativity. This level of (50) expertise allows aircraft engineers to apply mathematical equations and scientific processes to aeronautical and aerospace issues or inventions. For example, aircraft engineers may test, design, and construct flying vessels such as airplanes, space shuttles, and missile weapons. As a result, aircraft engineers are compensated with generous salaries. In fact, in May 2014, the lowest 10 percent of all American aircraft engineers earned less than $60,110 while the highest paid ten-percent of all American aircraft engineers earned $155,240. In May 2015, the United States Bureau of Labor Statistics (BLS) reported that the median annual salary of aircraft engineers was $107, 830. Conversely, employment opportunities for aircraft engineers are projected to decrease by 2 percent by 2024. This decrease may be the result of a decline in the manufacturing industry. Nevertheless, aircraft engineers who know how to utilize modeling and simulation programs, fluid dynamic software, and robotic engineering tools are projected to remain the most employable.

46. What type of text is utilized in the passage?
    a. Argumentative
    b. Narrative
    c. Biographical
    d. Informative

47. Which of the following would be the best choice for this sentence (reproduced below)?

    (47) Some of universities provide internship or apprentice opportunities for the students enrolled in aircraft engineer programs.

    a. NO CHANGE
    b. Some of universities provided internship or apprentice opportunities
    c. Some of universities provide internship or apprenticeship opportunities
    d. Some universities provide internship or apprenticeship opportunities

48. Which of the following would be the best choice for this sentence (reproduced below)?

(48) These advanced offices oftentimes require a Professional Engineering (PE) license which can be obtained through additional college courses, professional experience, and acceptable scores on the Fundamentals of Engineering (FE) and Professional Engineering (PE) standardized assessments.

a. NO CHANGE
b. These advanced positions oftentimes require acceptable scores on the Fundamentals of Engineering (FE) and Professional Engineering (PE) standardized assessments in order to achieve a Professional Engineering (PE) license. Additional college courses and professional experience help.
c. These advanced offices oftentimes require acceptable scores on the Fundamentals of Engineering (FE) and Professional Engineering (PE) standardized assessments to gain the Professional Engineering (PE) license which can be obtained through additional college courses, professional experience.
d. These advanced positions oftentimes require a Professional Engineering (PE) license which is obtained by acceptable scores on the Fundamentals of Engineering (FE) and Professional Engineering (PE) standardized assessments. Further education and professional experience can help prepare for the assessments.

49. "The knowledge of an aircraft engineer is acquired through years of education." Which statement serves to support this claim?
a. Aircraft engineers are compensated with generous salaries.
b. Such advanced degrees enable an individual to position himself or herself for executive, faculty, or research opportunities.
c. Ideally, an individual will begin his or her preparation for the profession in high school by taking chemistry, physics, trigonometry, and calculus.
d. Aircraft engineers who know how to utilize modeling and simulation programs, fluid dynamic software, and robotic engineering tools will be the most employable.

50. What is the meaning of "expertise" in the marked sentence?
a. Care
b. Skill
c. Work
d. Composition

Part II

Respond to the following prompt. You will have forty-five minutes to plan, write, and edit your response. Your essay should be around 500 words.

Think about a significant event that has occurred in your life. In your essay, describe the event, how you responded to it, and the impact it has had on you.

Social Studies

Canadian Propaganda Poster (1914–1918)

1. The propaganda poster was published during which of the following conflicts?
    a. American Revolutionary War
    b. Cold War
    c. War of 1812
    d. World War I
    e. World War II

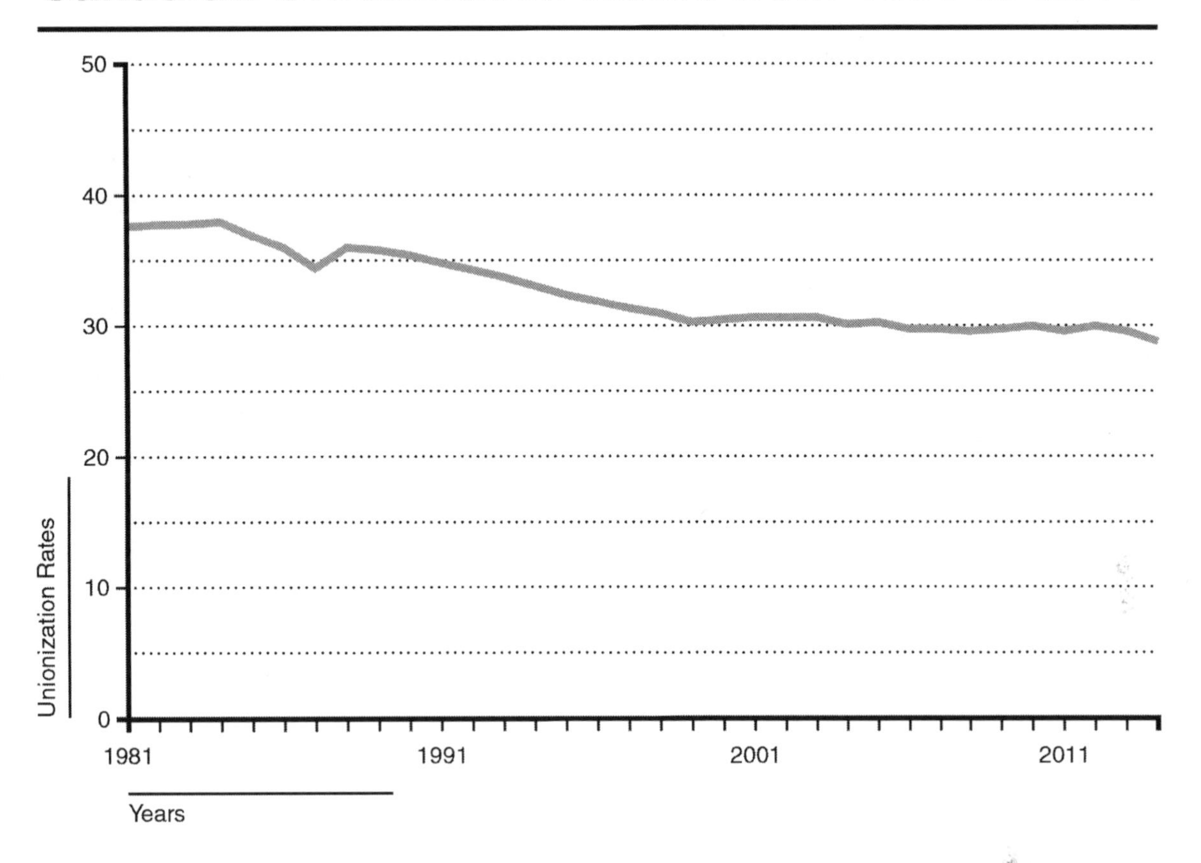

2. Which statement best explains what happened to unionization rates after 1989?

a. The collapse of the Soviet Union triggered a worldwide recession.
b. The United States reneged on its commitments to buy Canadian goods.
c. Free trade weakened labor rights due to increased competition.
d. Labor productivity steeply declined as automation decreased.
e. Canadian manufacturing declined at the same rate as other highly developed countries.

Canadian Trade Balance with the United States from 2014 to 2019 (in billions USD)			
Years	Canadian Exports to U.S.	U.S. Imports to Canada	Canadian Trade Balance
2014	$349.3	$312.8	+ $36.5
2015	$296.3	$280.9	+ $15.5
2016	$277.7	$266.7	+ $11
2017	$299.1	$282.8	+ $16.3
2018	$318.5	$299.7	+ $18.8
2019	$319.4	$292.6	+ $26.8

United States Census Bureau, 2020, https://www.census.gov/foreign-trade/balance/c1220.html

3. Which of the following had the greatest direct effect on the data provided in the table?
   a. General Agreement on Tariffs and Trade
   b. North American Free Trade Agreement
   c. North Atlantic Treaty Organization
   d. Western Hemisphere Travel Initiative
   e. World Trade Organization

4. In which of the following years did greatest trade imbalance occur?
   a. 2015
   b. 2016
   c. 2017
   d. 2018
   e. 2019

Canadian Pacific Railway (1886)

5. What was the most significant consequence of the Canadian Pacific Railway?
   a. It enriched the merchant class.
   b. It led to Canada becoming a confederation.
   c. It attracted immigrants to Eastern Canada.
   d. It facilitated territorial expansion.
   e. It allowed Canada to project military power westward.

Questions 6 and 7 refer to the following map:

Map of Present-Day Canada

6. Which province is entirely landlocked?
   a. British Columbia
   b. Manitoba
   c. Nova Scotia
   d. Ontario
   e. Saskatchewan

7. Which province has the most Arctic tundra?
   a. Newfoundland and Labrador
   b. Northwest Territories
   c. Nunavut
   d. Quebec
   e. Yukon

8. Which term is best defined as a group of people joined by a common culture, language, heritage, history, and religion?
   a. State
   b. Nation
   c. Regime
   d. Government

9. The presidential and parliamentary systems differ in which of the following ways?
   a. The presidential system establishes a separation of powers.
   b. The legislature elects the chief executive in a presidential system.
   c. Voters directly elect the prime minister in a parliamentary system.
   d. The parliamentary system never includes a president.

10. You are the owner of a medium-sized accounting firm in Toronto. You have grown increasingly tired of governmental interference in your business and have decided to use your vote to make a change. You are tired of regulations and taxes. Which platform would you most likely vote for?
   a. Liberal
   b. Democratic
   c. Socialist
   d. Conservative

Questions 11–13 refer to the passage below:

> I come now to the Pyrates that have rose since the Peace of Utrecht; in War Time there is no room for any, because all those of a roving advent'rous Disposition find Employment in Privateers, so there is no Opportunity for Pyrates; like our Mobs in London, when they come to any Height, our Superiors order out the Train Bands, and when once they are raised, the others are suppressed of Course; I take the Reason of it to be, that the Mob go into the tame Army, and immediately from notorious Breakers of the Peace, become, by being put into order, solemn Preservers of it. And should our Legislators put some of the Pyrates into Authority, it would not only lessen their Number, but, I imagine, set them upon the rest, and they would be the likeliest People to find them out, according to the Proverb, set a Thief to catch a Thief. To bring this about, there needs no other Encouragement, but to give all the Effects taken aboard a Pyrate Vessel to the Captors; for in Case of Plunder and Gain, they like it as well from Friends, as Enemies, but are not fond, as Things are carry'd, of ruining poor Fellowes, say the Creoleans, with no Advantage to themselves.

Excerpt from A General History of the Pyrates by Charles Johnson, 1724

11. According to the historian, which of the following groups of people would be the most effective at capturing pirates?
   a. Creolans
   b. Mobs
   c. Pirates
   d. Privateers

12. Which of the following did the historian propose as an incentive for capturing pirates?
    a. Crews that captured pirates received a military promotion.
    b. Crews that captured pirates were allotted territory in the Caribbean.
    c. Crews that captured pirates were given legal permission to plunder merchant ships.
    d. Crews that captured pirates were allowed to keep the treasure they found onboard.

13. According to the historian, what caused the rise in piracy?
    a. The Mobs in London funded piracy to increase their profits.
    b. The government refused to prohibit privateering.
    c. Many sailors were left unemployed after the Peace of Utrecht.
    d. The English Royal Navy stopped patrolling in the Caribbean.

Questions 14–16 refer to the passage below:

> Thus from the first entrance of the Spaniards into New Spain, which hapned on the 18th day of April in the said month of the year 1518, to 1530, the space of ten whole years, there was no end or period put to the Destruction and Slaughters committed by the merciless hands of the Sanguinary and Blood-thirsty Spaniard in the Continent, or space of 450 Miles round about Mexico, and the adjacent or neighboring parts, which might contain four or five spatious Kingdoms, that neither for magnitude or fertility would give Spain her self the pre-eminance. This intire Region was more populous then Toledo, Sevil, Valedolid, Saragoza, and Faventia; and there is not at this day in all of them so many people, nor when they flourisht in their greatest height and splendor was there such a number, as inhabited that Region, which embraceth in its Circumference, four hundred and eighty Miles. Within these twelve years the Spaniards have destroyed in the Said Countinent, by Spears, Fire and Sword, computing Men, Women, Youth, and Children above Four Millions of people in these their Acquests or Conquests (for under that word they mask their Cruel Actions) or rather those of the Turk himself, which are reported of them, tending to the ruin of the Catholick Cause, together with their Invasions and Unjust Wars, contrary to and condemned by Divine as well as Human Laws; nor are they reckoned in this number who perished by their more then Egyptian Bondage and usual Oppressions.

Excerpt from A Brief Account of the Destruction of the Indies by Bartolome de las Casas, 1542

14. According to the friar, how did the Spanish conquer the Amerindians?
    a. The Spanish intentionally spread diseases in the Americas.
    b. The Spanish had superior weaponry, such as firearms and cannons.
    c. The Spanish enslaved the Amerindians on plantations.
    d. The Spanish forcefully converted the Amerindians to Christianity.

15. Which of the following Amerindian civilizations is the friar likely describing?
    a. Aztec Empire
    b. Inca Empire
    c. Iroquois Confederacy
    d. Muisca Confederation

16. Which of the following best explains why a friar accompanied the Spanish explorers?
    a. The Spanish Crown hoped friars would prevent the Spanish explorers from committing war crimes.
    b. Spanish exploration was partially motivated by the desire to spread Catholicism in the Americas.
    c. Friars had more knowledge about navigation and geography as compared to Spanish explorers.
    d. Friars served as the lead negotiators when the Spanish explorers encountered Amerindian tribes.

Questions 17–19 refer to the passage below:

> We sailed to them, and found only one Man advanced in Years, and a Youth; the Man was the greatest Man in the Mechanical Parts of the Mathematicks, I had ever met with; my second Mate was an *Englishman*, an excellent Seaman, as was my Gunner, who had been taken Prisoners at *Campechy*, as well as the Master's Son; they told me the Ship was of *New England*, from a Town called *Boston*. The Owner and the whole Ship's Company came on board the thirtieth; and the Navigator of the Ship, Captain *Shapley*, told me, his Owner was a fine Gentleman, and *Major General* of the largest Colony in *New England*, called the *Maltechusets*; so I received him like a Gentleman, and told him my Commission was to make a Prize of any People seeking a North-west or West Passage into the *South Sea*; but I would look on them as Merchants trading with the Natives for Bevers, Otters and other Furs and Skins, and so for a small Present of Provisions I had no need on, I gave him my Diamond Ring, which cost me twelve Hundred Pieces of Eight (which the modest Gentleman received with difficulty) and having given the brave Navigator *Captain Shapley*, for his fine Charts and Journals, a Thousand Pieces of Eight, and the Owner of the Ship, *Seimor Gibbons*, a quarter Cask of good *Peruan* Wine, and the ten Seamen, each twenty Pieces of Eight, the sixth of *August*, with as much Wind as we could fly before and a Current, we arrived at the first Fall of the River *Parmentiers*.

Excerpt from The Great Probability of a Northwest Passage by Bartholomew de Fonte, 1708

17. According to the passage, what goods did the English merchants primarily trade with Amerindians?
    a. Charts and journals
    b. Diamonds and gold
    c. Furs and skins
    d. Wine and rum

18. Which of the following best describes "Pieces of Eight"?
    a. Pieces of eight were tools that helped explorers read maps.
    b. Pieces of eight were tokens of friendship explorers gave to one another.
    c. Pieces of eight were a raw resource found in the Americas.
    d. Pieces of eight were silver coins that Spain issued as currency.

19. Which of the following best explains why the Spanish explorer was seeking a Northwest Passage?
    a. A Northwest Passage would have reduced the time it took Spanish merchants to reach Asian markets.
    b. A Northwest Passage would have opened more opportunities for fur traders.
    c. A Northwest Passage would have provided Spain with a military advantage in the Americas.
    d. A Northwest Passage would have facilitated Spain's conquest over territory in present-day Canada.

Indian Residential School in the Northwest Territories
Bibliothèque et Archives Canada/PA-042133, Public domain, via Wikimedia Commons,
https://commons.wikimedia.org/wiki/File:R.C._Indian_Residential_School_Study_Time,_Fort_Resolution,_N.W.T.JPG

20. Which statement best describes the primary purpose of the schools depicted in the photograph?
    a. Canada created the schools to suppress Aboriginal culture.
    b. Canada forced Aboriginal children to attend the schools to alleviate poverty.
    c. Canada hoped the schools would increase the Aboriginal literacy rate.
    d. Canada intended the schools to teach Aboriginal children about traditional values.
    e. Canada built the schools as part of its reservation system.

Questions 21 and 22 refer to the following passage:

> By joining these Hurons and Algonquins against their Iroquois enemies, Champlain might make himself the indispensable ally and leader of the tribes of Canada, and at the same time fight his way to discovery in regions which otherwise were barred against him. From first to last it was the policy of France in America to mingle in Indian politics, hold the balance of power between adverse tribes, and envelop in the network of her power and diplomacy the remotest hordes of the wilderness. Of this policy the Father of New France may perhaps be held to have set a rash and premature example. Yet while he was apparently following the dictates of his own adventurous spirit, it became evident, a few years later, that under his thirst for discovery and spirit of knight-errantry lay a consistent and deliberate purpose. That it had already assumed a definite shape is not likely; but his after course makes it plain that, in embroiling himself and his colony with the most formidable savages on the continent, he was by no means acting so recklessly as at first sight would appear.

Excerpt from Pioneers of France in New World by Francis Parkman, 1865

21. According to the passage, how did France typically attempt to subdue the First Nations?
    a. France forged alliances with First Nations to explore regions they otherwise couldn't access.
    b. France deployed a divide-and-conquer to undermine and weaken First Nations.
    c. France offered its assistance in resolving disputes between rival First Nations.
    d. France relied exclusively on its superior weaponry and tactical experience.
    e. France permitted Champlain to act as a knight-errant even when he acted recklessly.

22. Which of the following properly describes the author's characterization of Champlain?
    a. A shrewd leader
    b. A timid explorer
    c. A reckless military general
    d. A foolish knight-errant
    e. A loyal ally

Questions 23–25 refer to the passage below:

> Our system of education turns young people out of the schools able to read, but for the most part unable to weigh evidence or to form an independent opinion. They are then assailed, throughout the rest of their lives, by statements designed to make them believe all sorts of absurd propositions, such as that Blank's pills cure all ills, that Spitzbergen is warm and fertile, and that Germans eat corpses. The art of propaganda, as practiced by modern politicians and governments, is derived from the art of advertisement. The science of psychology owes a great deal to advertisers. In former days most psychologists would probably have thought that a man could not convince many people of the excellence of his own wares by merely stating emphatically that they were excellent. Experience shows, however, that they were mistaken in this. If I were to stand up once in a public place and state that I am the most modest man alive, I should be laughed at; but if I could raise enough money to make the same statement on all the busses and on hoardings along all the principal railway lines, people would presently become convinced that I had an abnormal shrinking from publicity. If I were to go to a small shopkeeper and say: "Look at your competitor over the way, he is getting your business; don't you think it would be a good plan to leave your business and stand up in the middle of the road and try to shoot him before he shoots you?"—if I were to say this, any small shopkeeper would think me mad. But when the Government says it with emphasis and a brass band, the small shopkeepers become enthusiastic, and are quite surprised when they find afterwards that business has suffered. Propaganda, conducted by the means which advertisers have found successful, is now one of the recognized methods of government in all advanced countries, and is especially the method by which democratic opinion is created.

Excerpt from the speech "Free Thought and Official Propaganda" by Bertrand Russell, 1922

23. Which one of the following most accurately describes how propaganda helps governments wage war?
    a. Propaganda convinces people to participate in acts of violence.
    b. Propaganda motivates people to contribute to the war effort in ways they otherwise wouldn't.
    c. Propaganda replaces advertising as the most effective way of shaping public opinion.
    d. Propaganda stimulates the economy through advertising, increasing tax revenue.

24. Governments deployed large-scale propaganda for the FIRST time during which one of the following military conflicts?
    a. Russo-Turkish War
    b. First Sino-Japanese War
    c. Spanish Civil War
    d. World War I

25. According to the passage, what is the relationship between advertising and propaganda?
    a. Propaganda competes with advertising for the public's attention.
    b. Propaganda applies effective advertising techniques to shape public opinion.
    c. Propaganda is a form of advertising that doesn't involve the government.
    d. Propaganda is effective because it's so similar to the types of advertising people enjoy consuming.

26. Which of these choices BEST describes a participatory democracy?
    a. A system in which only the educated and wealthy members of society vote and decide upon the leaders of the country
    b. A system in which groups come together to advance certain select interests
    c. A system that emphasizes everyone contributing to the political system
    d. A system in which one group makes decisions for the population at large

27. Which of the following statements best describes international affairs between World War I and World War II?
    a. A lenient World War I peace treaty for Germany delayed the start of World War II.
    b. The policy of appeasement only encouraged further aggression by Hitler.
    c. A powerful League of Nations fostered increased cooperation and negotiation.
    d. Tensions grew between Germany and Japan.

28. Which of the following is NOT a liberal policy?
    a. Legalization of marijuana
    b. Lowering of taxes
    c. More regulation of the stock exchange
    d. Raising of the minimum wage

Questions 29–32 refer to the passage below:

> By fixing the Reparation payments well within Germany's capacity to pay, we make possible the renewal of hope and enterprise within her territory, we avoid the perpetual friction and opportunity of improper pressure arising out of Treaty clauses which are impossible of fulfillment, and we render unnecessary the intolerable powers of the Reparation Commission.
>
> By a moderation of the clauses relating directly or indirectly to coal, and by the exchange of iron-ore, we permit the continuance of Germany's industrial life, and put limits on the loss of productivity which would be brought about otherwise by the interference of political frontiers with the natural localization of the iron and steel industry.
>
> By the proposed Free Trade Union some part of the loss of organization and economic efficiency may be retrieved, which must otherwise result from the innumerable new political frontiers now created between greedy, jealous, immature, and economically incomplete nationalist States. Economic frontiers were tolerable so long as an immense territory was included in a few great Empires; but they will not be tolerable when the Empires of Germany, Austria-Hungary, Russia, and Turkey have been partitioned between some twenty independent authorities.

Excerpt from The Economic Consequences of the Peace by John Maynard Keynes, 1919

29. Which ONE of the following treaties is the passage most likely referencing?
    a. Munich Agreement
    b. North Atlantic Treaty
    c. Treaty of Versailles
    d. Vienna Convention on the Law of Treaties

30. Why did reparations place an "improper pressure" on Germany?
   a. The reparations prevented Germany from entering into free trade agreements.
   b. The reparations destabilized the German economy and government.
   c. The reparations exacerbated border disputes caused by the dissolution of several large empires.
   d. The reparations placed extreme regulatory pressure on Germany's iron and steel industry.

31. Which ONE of the following best describes an economic benefit of free trade agreements?
   a. They increase international trade by reducing barriers to trade.
   b. They reduce the cost of reparations.
   c. They allow countries to protect domestic iron and steel production.
   d. They facilitate imperialism and the creation of lucrative empires.

32. Which of the following events was partially caused by the failure to resolve the described conflicts?
   a. Cold War
   b. Great Depression
   c. World War I
   d. World War II

33. Which of the following consequences did NOT result from the discovery of the New World in 1492?
   a. Proof that the world was round instead of flat
   b. The deaths of millions of Native Americans
   c. Biological exchange between Europe and the New World
   d. The creation of new syncretic religions

34. Which of the following best describes how culture is transmitted across society?
   a. Culture is almost always transmitted through hierarchical relationships, and it has a trickle-down effect.
   b. Culture is primarily transmitted through religion, economic activities, and government policies.
   c. Cultural exchanges on the internet have given rise to a global popular culture in recent years.
   d. Culture can be transmitted through an endless variety of activities, and the transmission can either be intentional or spontaneous.

35. Regime types fall along a continuum between which two extremes?
   a. Constitutional and non-constitutional
   b. Military and judicial
   c. Federal and communist
   d. Authoritarian and democratic

36. What is the term for the ability of a ruling body to influence the actions, behavior, and attitude of a person or group of people?
   a. Politics
   b. Power
   c. Authority
   d. Legitimacy

37. Which of the following is NOT a shared characteristic sufficient to form a nation?
   a. Culture and traditions
   b. History
   c. Sovereignty
   d. Beliefs and religion

Question 38 is based on the following passage:

> Upon this, one has to remark that men ought either to be well treated or crushed, because they can avenge themselves of lighter injuries, of more serious ones they cannot; therefore the injury that is to be done to a man ought to be of such a kind that one does not stand in fear of revenge.

From Niccolo Machiavelli's The Prince, 1513

38. What advice is Machiavelli giving to the prince?
    a. Lightly injured enemies will overthrow the prince.
    b. Seek to injure everyone you meet.
    c. Hurting people is always the correct course of action.
    d. If you are going to cause an enemy some injury, ensure the injury is fatal.

Question 39 is based on the following passage:

> The creed which accepts as the foundation of morals, Utility, or the Greatest-Happiness Principle, holds that actions are right in proportion as they tend to promote happiness, wrong as they tend to produce the reverse of happiness. By happiness is intended pleasure, and the absence of pain; by unhappiness, pain, and the privation of pleasure.
>
> The utilitarian morality does recognise in human beings the power of sacrificing their own greatest good for the good of others. It only refuses to admit that the sacrifice is itself a good. A sacrifice which does not increase, or tend to increase, the sum total of happiness, it considers as wasted.

From John Stuart Mill's Utilitarianism, 1861

39. What is the meaning of the "Utility"?
    a. Actions should be judged based on the net total of pleasure.
    b. Actions requiring sacrifice can never be valuable.
    c. Actions promoting sacrifice that increase happiness are more valuable than actions that only increase happiness.
    d. Actions can be valuable even if the pain outweighs the pleasure.

Question 40 is based on the following passage:

> The history of all hitherto existing society is the history of class struggles.
>
> Freeman and slave, patrician and plebeian, lord and serf, guildmaster and journeyman, in a word, oppressor and oppressed, stood in constant opposition to one another, carried on an uninterrupted, now hidden, now open fight, that each time ended, either in the revolutionary reconstitution of society at large, or in the common ruin of the contending classes.
>
> Let the ruling classes tremble at a Communistic revolution. The proletarians have nothing to lose but their chains. They have a world to win.
>
> Workingmen of all countries unite!

Karl Marx and Friedrich Engels, The Communist Manifesto, 1848

40. What's the main idea presented in the excerpt?
   a. Working men are morally superior to the ruling class.
   b. Every society will come to an end at some point.
   c. History is defined by class struggle, and working men must now unite and fight the ruling class to gain freedom.
   d. Working men are in the same position as the slave, plebeian, serf, and journeyman.

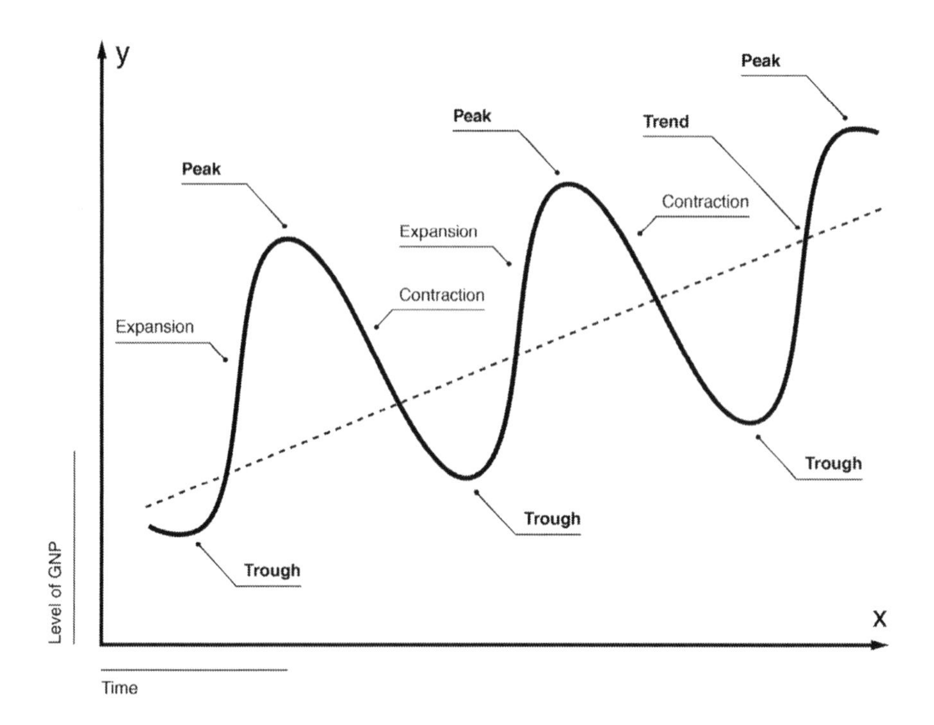

41. Which of the following phases of a business cycle occurs when there is continual growth?
   a. Expansion
   b. Peak
   c. Contraction
   d. Trough

42. Which of the following types of government intervention lowers prices, reassures the supply, and creates opportunity to compete with foreign vendors?
    a. Income redistribution
    b. Price controls
    c. Taxes
    d. Subsidies

43. What type of map would be the most useful for calculating data and differentiating between the characteristics of two places?
    a. Topographic maps
    b. Dot-density maps
    c. Isoline maps
    d. Flow-line maps

44. What accounts for different parts of the Earth experiencing different seasons at the same time?
    a. Differences in the rate of Earth's rotation
    b. Ocean currents
    c. Tilt of the Earth's rotational axis
    d. Elevation

45. Which of the following is NOT a reason why nonrenewable energy sources are used more often than renewables?
    a. Nonrenewable energy is currently cheaper.
    b. Infrastructure was built specifically for nonrenewable sources.
    c. Renewable energy is more difficult and expensive to store for long periods.
    d. Renewable energy cannot be converted into a power source.

46. Which pair of ocean currents or gyres has the most impact on the weather of Canada?
    a. North Atlantic Subtropical Gyre and South Atlantic Subtropical Gyre
    b. Labrador Current and North Pacific Subtropical Gyre
    c. North Pacific Subtropical Gyre and South Pacific Subtropical Gyre
    d. Alaskan Current and Indian Ocean Subtropical Gyre

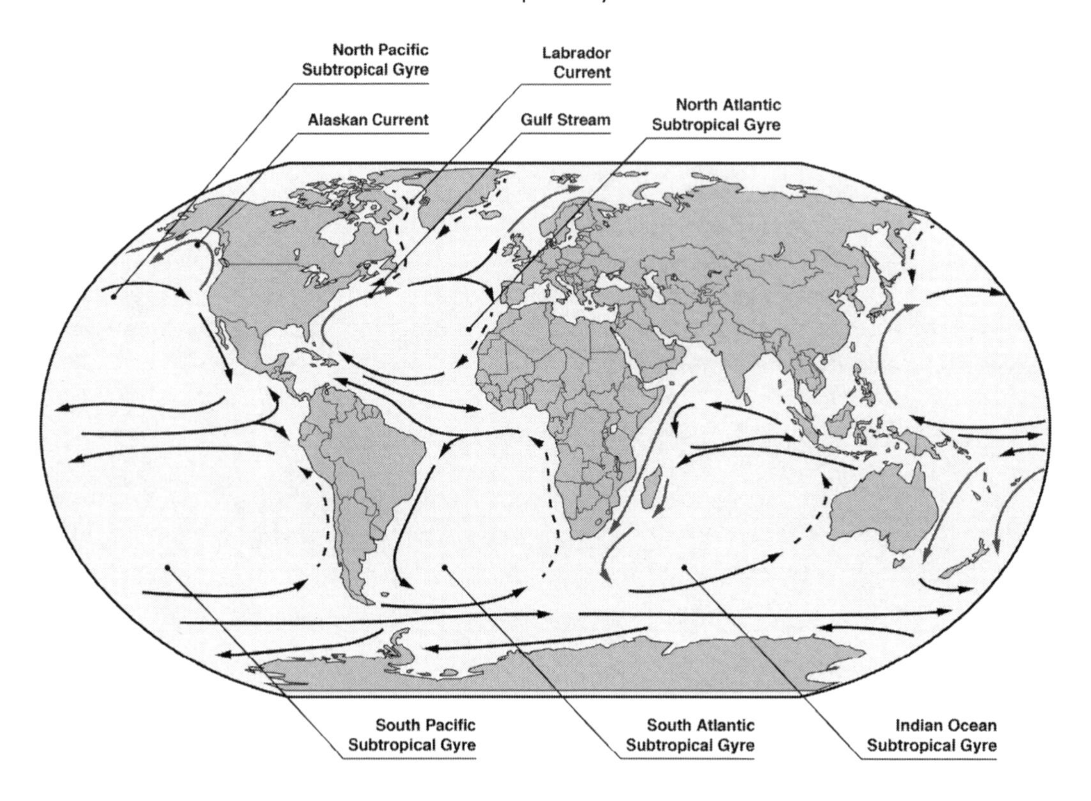

Questions 47–50 are based on the following passage:

In general, orientations on the left of the political spectrum emphasize social and economic equality and advocate for government intervention to achieve it. Orientations on the right of the spectrum generally value the existing and historical political institutions and oppose government intervention, especially in regard to the economy.

Communism is a radical political ideology that seeks to establish common ownership over production and abolish social status and money. Communists believe that the world is split between two social classes—capitalists and the working class (often referred to as the proletariat). Communist politics assert that conflict arises from the inequality between the ruling class and the working class; thus, Communism favors a classless society.

Conservatism is a political ideology that prioritizes traditional institutions within a culture and civilization. Conservatives, in general, oppose modern developments and value stability. Since Conservatism depends on traditional institutions, this ideology differs greatly from country to country. Conservatives often emphasize the traditional family structure and the importance of individual self-reliance. Fiscal Conservatism is one of the most common variants, and in general, the proponents of fiscal Conservatism oppose government spending and public debt.

Progressivism maintains that progress in the form of scientific and technological advancement, social change, and economic development improve the quality of human life. Progressive ideals include the view that the political and economic interests of the ruling class suppress progress, which results in perpetual social and economic inequality.

Libertarianism opposes state intervention on society and the economy. Libertarians advocate for a weak central government, favoring more local rule, and seek to maximize personal autonomy and protect personal freedom. Libertarians often follow a conservative approach to government, especially in the context of power and intervention, but favor a progressive approach to rights and freedom, especially those tied to personal liberty, like freedom of speech.

Liberalism developed during the Age of Enlightenment in opposition to absolute monarchy, royal privilege, and state religion. In general, Liberalism emphasizes liberty and equality, and liberals support freedom of speech, freedom of religion, free markets, civil rights, gender equality, and secular governance. Liberals support government intervention into private matters to further social justice and fight inequality; thus, liberals often favor social welfare organizations and economic safety nets to combat income inequality.

Fascism is a form of totalitarianism that became popular in Europe after World War I. Fascists advocate for a centralized government led by an all-powerful dictator, tasked with preparing for total war and mobilizing all resources to benefit the state. This orientation's distinguishing features include a consolidated and centralized government.

Socialism is closely tied to an economic system. Socialists prioritize the health of the community over the rights of individuals, seeking collective and equitable ownership over the means of production. Socialists tend to be willing to work to elect Socialist policies, like social security, universal health care, unemployment benefits, and other programs related to building a societal safety net.

47. Using the information from the passages above and the introduction about left-axis and right-axis political orientations, which of the following correctly categorizes the orientations mentioned in the passage?

a. Left-axis ideologies: Socialism, Progressivism, Liberalism; Right-axis ideologies: Fascism, Libertarianism, Communism, Conservatism
b. Left-axis ideologies: Socialism, Progressivism, Liberalism, Communism; Right-axis ideologies: Fascism, Libertarianism, Conservatism
c. Left-axis ideologies: Socialism, Progressivism, Libertarianism, Liberalism; Right-axis ideologies: Fascism, Communism, Conservatism
d. Left-axis ideologies: Socialism, Progressivism, Libertarianism, Communism; Right-axis ideologies: Fascism, Liberalism, Conservatism

48. Of the following ideologies, which one advocates for the most radical government intervention to achieve social and economic equality?

a. Socialism
b. Liberalism
c. Libertarianism
d. Fascist

49. Of the following ideologies, which one prioritizes stability and traditional institutions within a culture?
   a. Socialism
   b. Liberalism
   c. Conservatism
   d. Libertarianism

50. Which of the following correctly states one of the biggest differences between Fascism and Libertarianism?
   a. Fascists favor a powerful, centralized government, whereas Libertarians favor powerful local rule.
   b. Fascists prioritize governmental spending on strengthening the military, whereas Libertarians believe governmental involvement in terms of power and intervention should be minimal.
   c. Fascists favor a powerful localized government, whereas Libertarians favor a powerful, centralized government.
   d. Fascists believe governmental involvement in terms of power and intervention should be minimal, whereas Libertarians prioritize governmental spending on strengthening the military.

Science

Passage 1

Questions 1–5 pertain to the following information:

> Worldwide, fungal infections of the lung account for significant mortality in individuals with compromised immune function. Three of the most common infecting agents are *Aspergillus, Histoplasma*, and *Candida*. Successful treatment of infections caused by these agents depends on an early and accurate diagnosis. Three tests used to identify specific markers for these mold species include ELISA (enzyme-linked immunosorbent assay), GM Assay (Galactomannan Assay), and PCR (polymerase chain reaction).
>
> Two important characteristics of these tests include sensitivity and specificity. Sensitivity relates to the probability that the test will identify the presence of the infecting agent, resulting in a true positive result. Higher sensitivity equals fewer false-positive results. Specificity relates to the probability that if the test doesn't detect the infecting agent, the test is truly negative for that agent. Higher specificity equals fewer false-negatives.

Figure 1 shows the timeline for the process of infection from exposure to the pathogen to recovery or death:

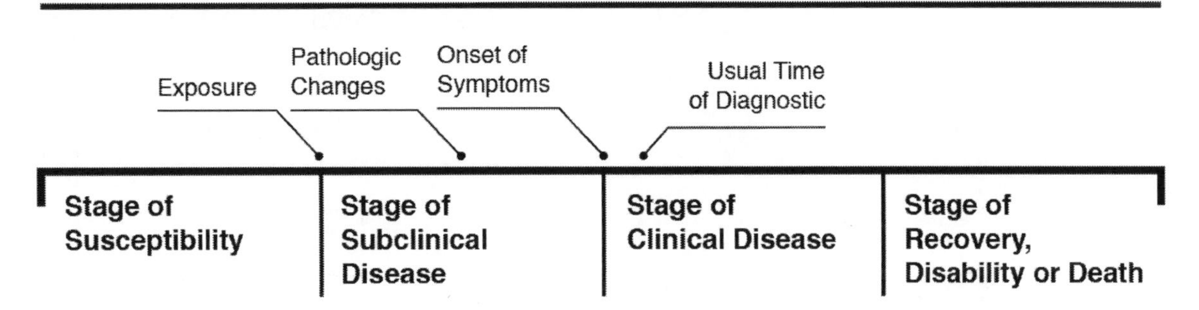

Figure 2 shows the sensitivity and specificity for ELISA, GM assay and PCR related to the diagnosis of infection by *Aspergillus*, *Histoplasma* and *Candida*:

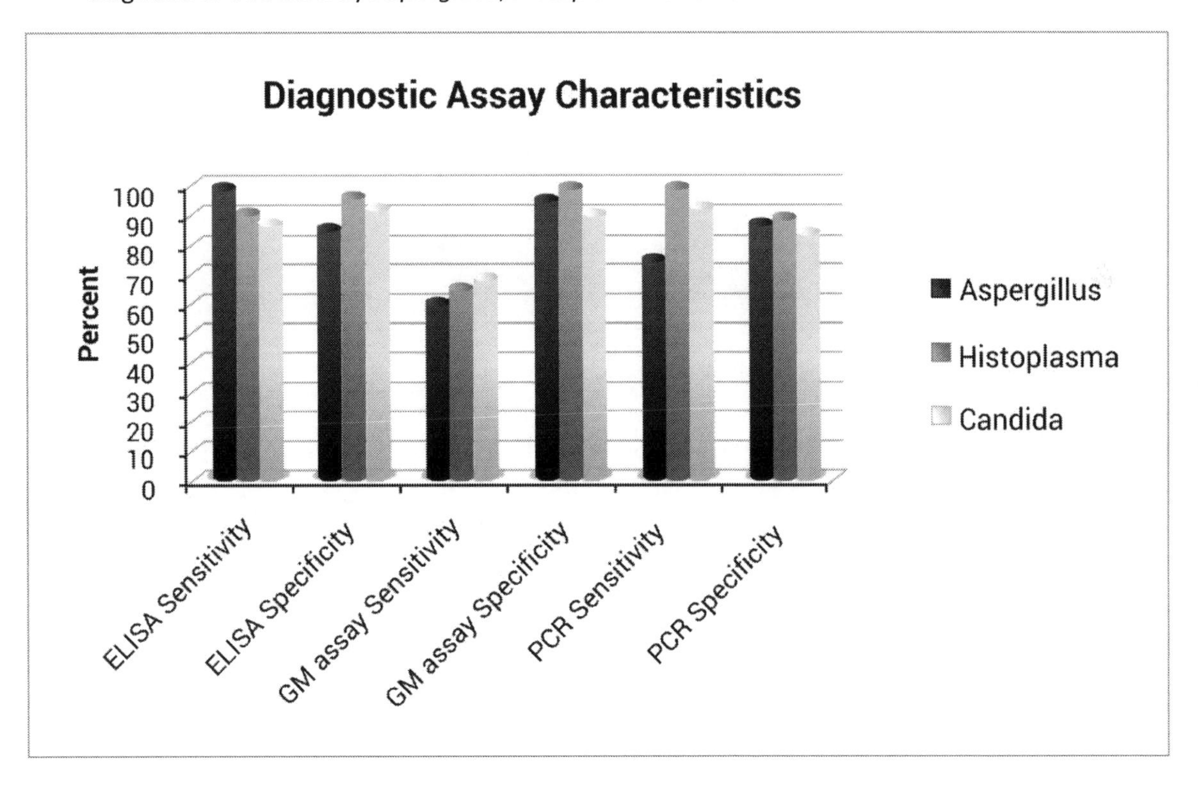

The table below identifies the process of infection in days from exposure for each of the species:

Process of Infection – Days Since Pathogen Exposure			
	Aspergillus	Histoplasma	Candida
Sub-clinical Disease	Day 90	Day 28	Day 7
Detection Possible	Day 118	Day 90	Day 45
Symptoms Appear	Day 145	Day100	Day 120

Figure 3 identifies the point at which each test can detect the organism. Time is measured in days from the time an individual is exposed to the pathogen:

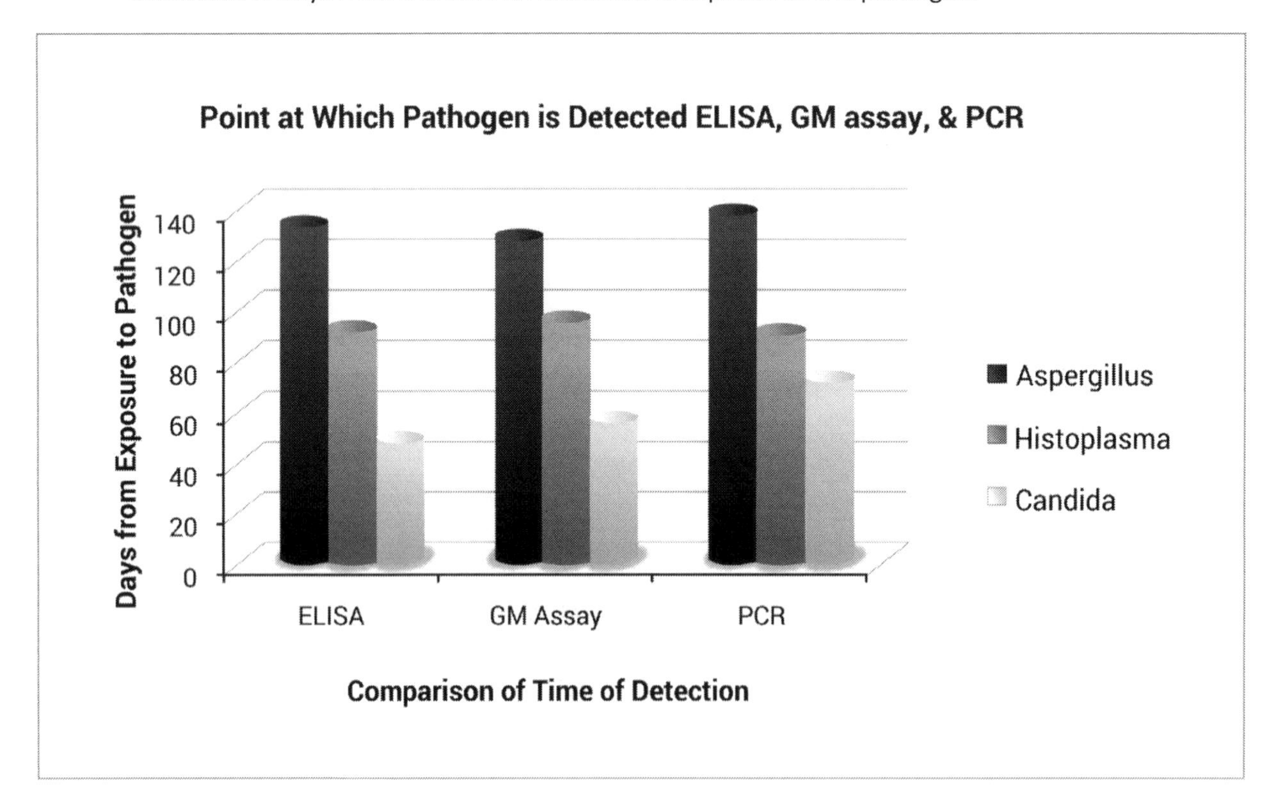

1. Which of the following statements is supported by Figure 2?
   a. For *Candida*, the GM assay will provide the most reliable results.
   b. ELISA testing for *Aspergillus* is the most specific of the three tests.
   c. PCR is the most sensitive method for testing *Histoplasma*.
   d. True positive rates were greater than 75% for all three testing methods.

2. In reference to the table and Figure 3, which pathogen can be detected earlier in the disease process, and by which method?
   a. *Candida* by PCR testing
   b. *Aspergillus* by ELISA testing
   c. *Candida* by GM assay
   d. *Histoplasma* by PCR testing

3. In reference to Figure 2, which statement is correct?
   a. There is a 20% probability that ELISA testing will NOT correctly identify the presence of *Histoplasma.*
   b. When GM assay testing for *Candida* is conducted, there is a 31% probability that it will NOT be identified if the organism is present.
   c. The probability that GM assay testing for *Aspergillus* will correctly identify the presence of the organism is 99%.
   d. The false-negative probabilities for each of the three testing methods identified in Figure 2 indicate that the organism will be detected when present less than 70% of the time.

4. Physicians caring for individuals with suspected *Histoplasma* infections order diagnostic testing prior to instituting treatment. PCR testing results will not be available for 10 days. GM assay results can be obtained more quickly. The physicians opt to wait for the PCR testing. Choose the best possible rationale for that decision.
   a. The treatment will be the same regardless of the test results.
   b. The individual was not exhibiting any disease symptoms.
   c. The probability of PCR testing identifying the presence of the organism is greater than the GM assay.
   d. The subclinical disease phase for *Histoplasma* is more than 100 days.

5. Referencing the data in Figures 2 and 3, if ELISA testing costs twice as much as PCR testing, why might it still be the best choice to test for *Candida*?
   a. ELISA testing detects the presence of *Candida* sooner than PCR testing.
   b. ELISA testing has fewer false-positives than PCR testing.
   c. There is only a 69% probability that PCR testing will correctly identify the presence of *Candida.*
   d. PCR testing is less sensitive than ELISA testing for *Candida.*

Passage 2

Questions 6–12 pertain to the following information:

Scientists disagree about the cause of Bovine Spongiform Encephalopathy (BSE), also known as "mad cow disease." Two scientists discuss different explanations about the cause of the disease.

<u>Scientist 1</u>

Mad cow disease is a condition that results in the deterioration of brain and spinal cord tissue. This deterioration manifests as sponge-like defects or holes that result in irreversible damage to the brain. The cause of this damage is widely accepted to be the result of an infectious type of protein, called a prion. Normal prions are located in the cell wall of the central nervous system and function to preserve the myelin sheath around the nerves. Prions are capable of turning normal proteins into other prions by a process that is still unclear, thereby causing the proteins to be "refolded" in abnormal and harmful configurations. Unlike viruses and bacteria, the harmful prions possibly don't contain DNA or RNA, based on the observation of infected tissues in the laboratory that remain infected after immersion in formaldehyde or exposure to ultraviolet light. The transformation from normal to abnormal protein structure and function in a given individual is thought to occur as the result of proteins that are genetically weak or abnormally prone to mutation, or through transmission from another host through food, drugs or organ transplants from infected animals. The abnormal prions also don't trigger an immune response. After prions accumulate in large enough numbers, they form damaging conglomerations that result in the sponge-like holes in tissues, which eventually cause the loss of proper brain function and death.

Figure 1 depicts formation of abnormal prions that results from the abnormal folding of amino acids:

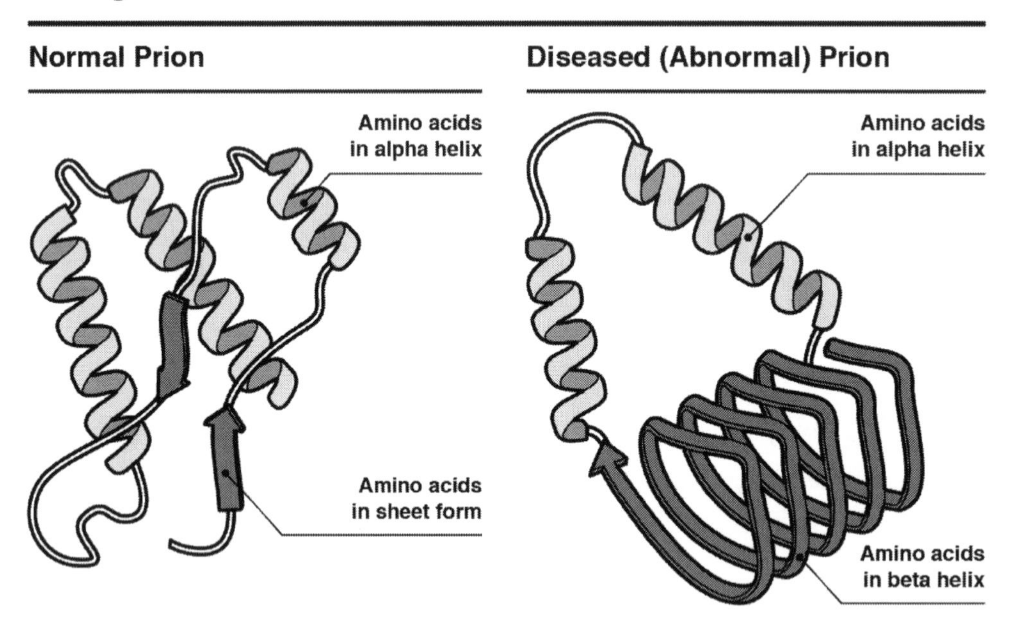

<u>Scientist 2</u>

The degeneration of brain tissue in animals afflicted with mad cow disease is widely considered to be the result of prion proteins. This theory fails to consider other possible causes, such as viruses. Recent studies have shown that infected tissues often contain small particles that match the size and density of viruses. In order to demonstrate that these viral particles are the cause of mad cow disease, researchers used chemicals to inactivate the viruses. When the damaged, inactivated viruses were introduced into healthy tissue, no mad cow disease symptoms were observed. This result indicates that viruses are likely the cause of mad cow disease. In addition, when the infected particles from an infected animal are used to infect a different species, the resulting particles are identical to the original particles. If the infecting agent was a protein, the particles would not be identical because proteins are species-specific. Instead, the infective agent is viewed as some form of a virus that has its own DNA or RNA configuration and can reproduce identical infective particles.

6. Which statement below best characterizes the main difference in the scientists' opinions?
    a. The existence of species-specific proteins
    b. Transmission rates of mad cow disease
    c. The conversion process of normal proteins into prions
    d. The underlying cause of mad cow disease

7. Which of the following statements is INCORRECT?
    a. Scientist 2 proposes that viruses aren't the cause of mad cow disease because chemicals inactivated the viruses.
    b. Scientist 1 suggests that infectious proteins called prions are the cause of mad cow disease.
    c. Scientist 1 indicates that the damaging conglomerations formed by prions eventually result in death.
    d. Scientist 2 reports that infected tissues often contain particles that match the size profile of viruses.

8. Which of the following is true according to Scientist 1?
    a. Normal proteins accumulate in large numbers to produce damaging conglomerations.
    b. Prions can change normal proteins into prions.
    c. Species-specific DNA sequences of infected tissues indicate that proteins cause mad cow disease.
    d. Prions are present only in the peripheral nervous system of mammals.

9. Which of the following statements would be consistent with the views of BOTH scientists?
    a. Resulting tissue damage is reversible.
    b. The infecting agent is composed of sheets of amino acids in an alpha helix configuration.
    c. Species-specific DNA can be isolated from infected tissue.
    d. Cross-species transmission of the illness is possible.

10. How does the *conglomeration* described in the passage affect function?
    a. Synapses are delayed
    b. Sponge-like tissue formations occur
    c. Space-occupying lesions compress the nerves
    d. The blood supply to surrounding tissues is decreased

11. What evidence best supports the views of Scientist 2?
    a. Species-specific DNA is present in the infected particles.
    b. Prions are present in the cell membrane.
    c. Prions can trigger an immune response.
    d. The infected particles were inactivated and didn't cause disease.

12. Which of the following statements is supported by this passage?
    a. Scientist 1 favors the claim that viruses are the cause of mad cow disease.
    b. Prions are a type of infectious virus.
    c. The process that results in the formation of the abnormal prion is unclear.
    d. Mad cow disease is caused by normal proteins.

Passage 3

Questions 13–17 pertain to the following information:

Scientists have long been interested in the effect of sleep deprivation on overeating and obesity in humans. Recently, scientists discovered that increased levels of the endocannabinoid 2-Arachidonoylglycerol (2-AG) in the human body is related to overeating. The endocannabinoids play an important role in memory, mood, the reward system, and metabolic processes including glucose metabolism and generation of energy. The endocannabinoid receptors CB1-R and CB2-R are protein receptors located on the cell membrane in the brain, the spinal cord and, to a lesser extent, in the peripheral neurons and the organs of the immune system. The two principal endogenous endocannabinoids are AEA (Anandamide) and 2-Arachidonoylglycerol (2-AG). The endocannabinoids can affect the body's response to chronic stress, mediate the pain response, decrease GI motility, and lessen the inflammatory response in some cancers.

Figure 1 identifies the chemical structure of the endogenous cannabinoids including 2-AG:

Figure 1:
Chemical Structure of Common Endogenous Cannabinoids

The Five-Best known Endocannabinoids Showing the Common 19 - C Backbone Structure and specific R-group Constituents

Anandamide | **2 - Arachidonoyl- glycerol** | **Noladin Ether**

N-arachidonoyl-dopamine | **Virodhamine** | **EC backbone structure**

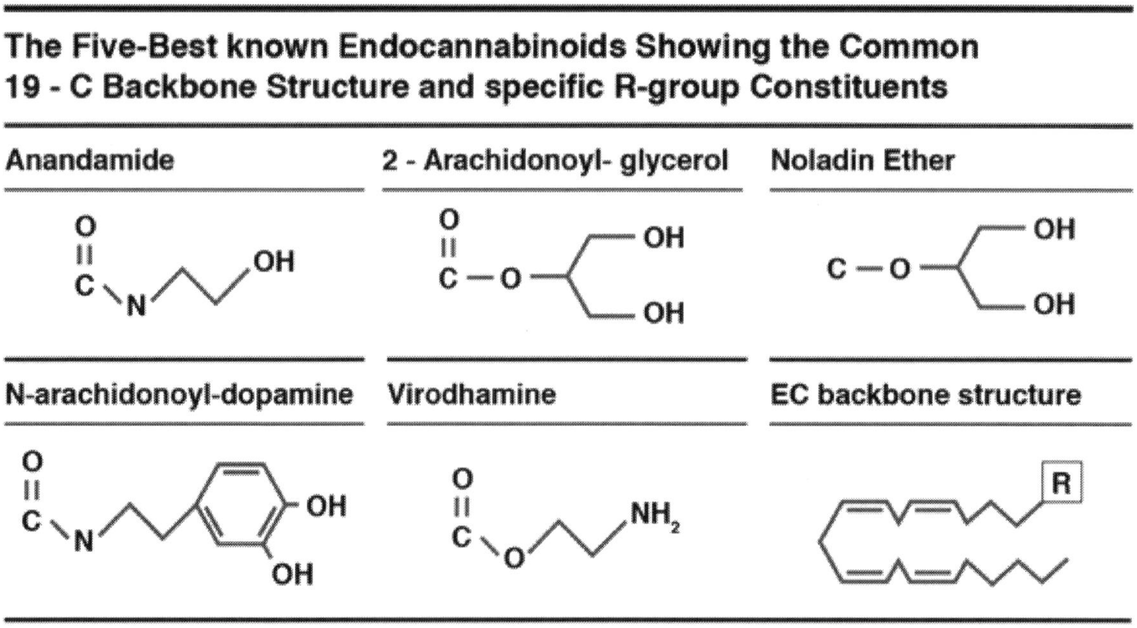

Recent research has also examined the relationship between sleep deprivation and the levels of 2-AG present in blood, as these conditions relate to obesity. The circadian fluctuations of 2-AG are well-known. Levels normally increase in late afternoon and evening. This physiological increase is thought to contribute to late-day snacking behaviors even after adequate calories have been consumed. The relationship between sleep deprivation and 2-AG appears to relate to the effect of 2-AG on the stress response, represented by sleep deprivation in this study. In order to examine this relationship, university scientists conducted an experiment to identify the influence of injections of 2-AG and sleep deprivation on overeating in a population of non-obese male and female participants that ranged in age from 20–40 years old. To accomplish this, human research subjects (participants) were allowed to eat their favorite junk foods in addition to consuming sufficient calories each day. All of the participants were injected daily with a solution of either sterile normal saline or 2-AG. Daily weight gain was recorded for the three treatment groups that included: participants A–E who received sterile normal saline injections, participants F–J who received 2-AG injections, and participants K–O who received 2-AG injections and were limited to 4.5 hours of sleep each night for 7 nights. The results of the three trials are shown below.

Figure 2 identifies the daily weight gain (in grams) of participants receiving sterile normal saline injections:

Daily Weight Gain for Patients Receiving Sterile Normal Saline Injections

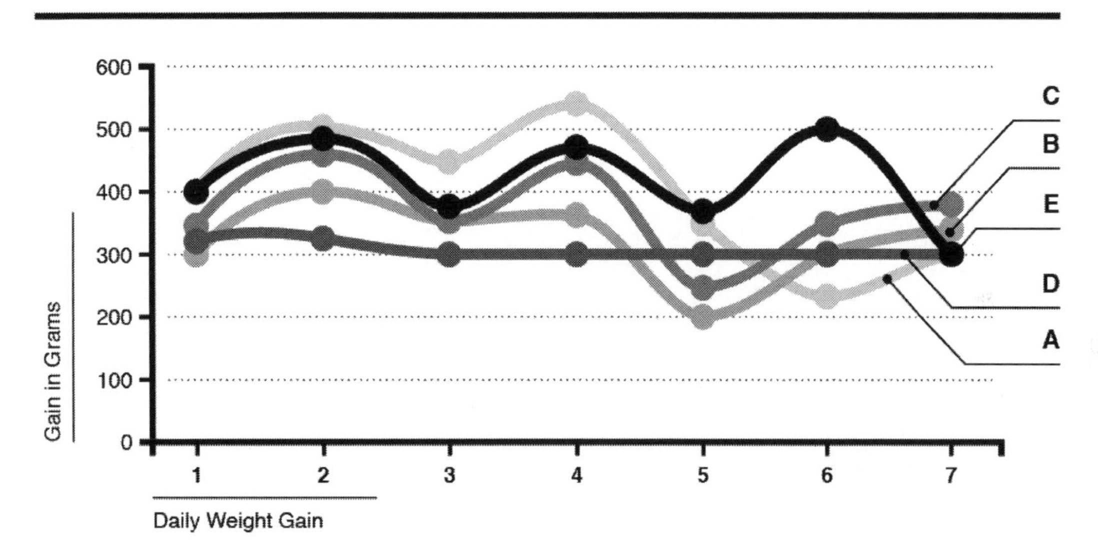

Figure 3 identifies the daily weight gain for participants receiving 2-AG injections:

Figure 3:

Daily Weight Gain for Participants Receiving Daily 2-AG Injections

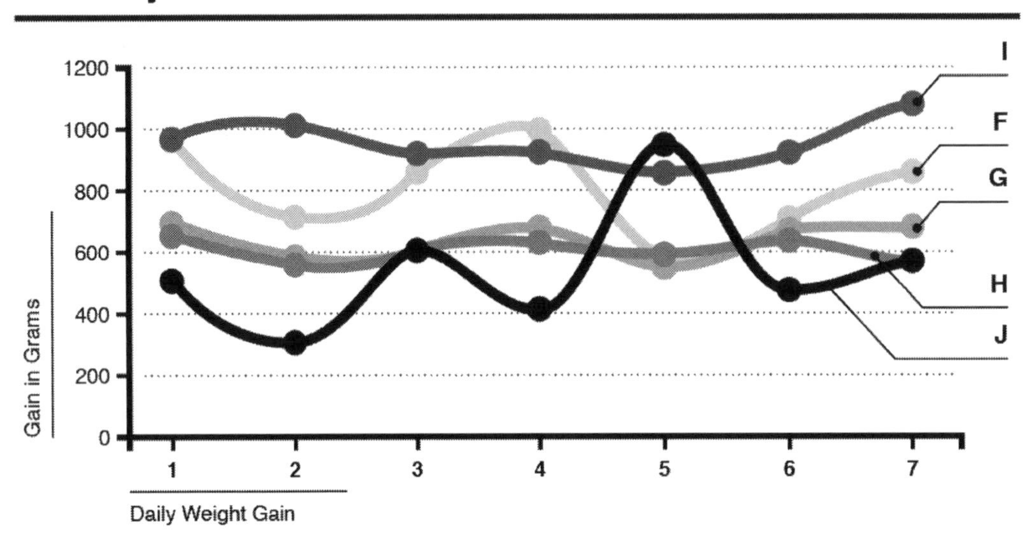

Figure 4 identifies the daily weight gain for participants receiving daily injections of 2-AG who were also limited to 4.5 hours sleep per night for 7 consecutive nights:

Figure 4:

Daily Weight Gain for Participants Receiving Daily 2-AG Injections Who Were Limited to 4.5 Hours of Sleep Per Night for 7 Consecutive Nights

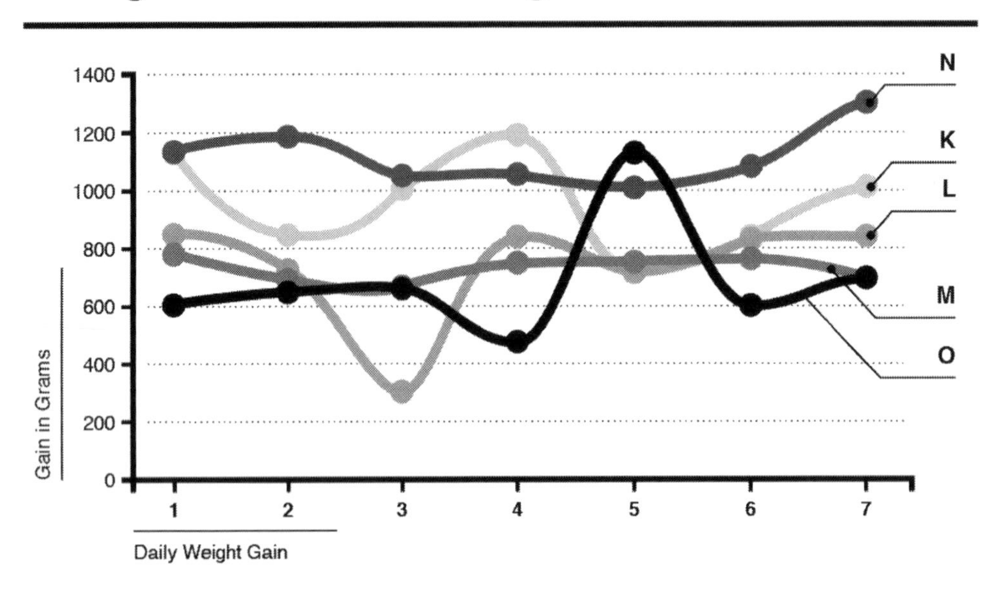

Figure 5 identifies the participants' average daily weight gain by trial:

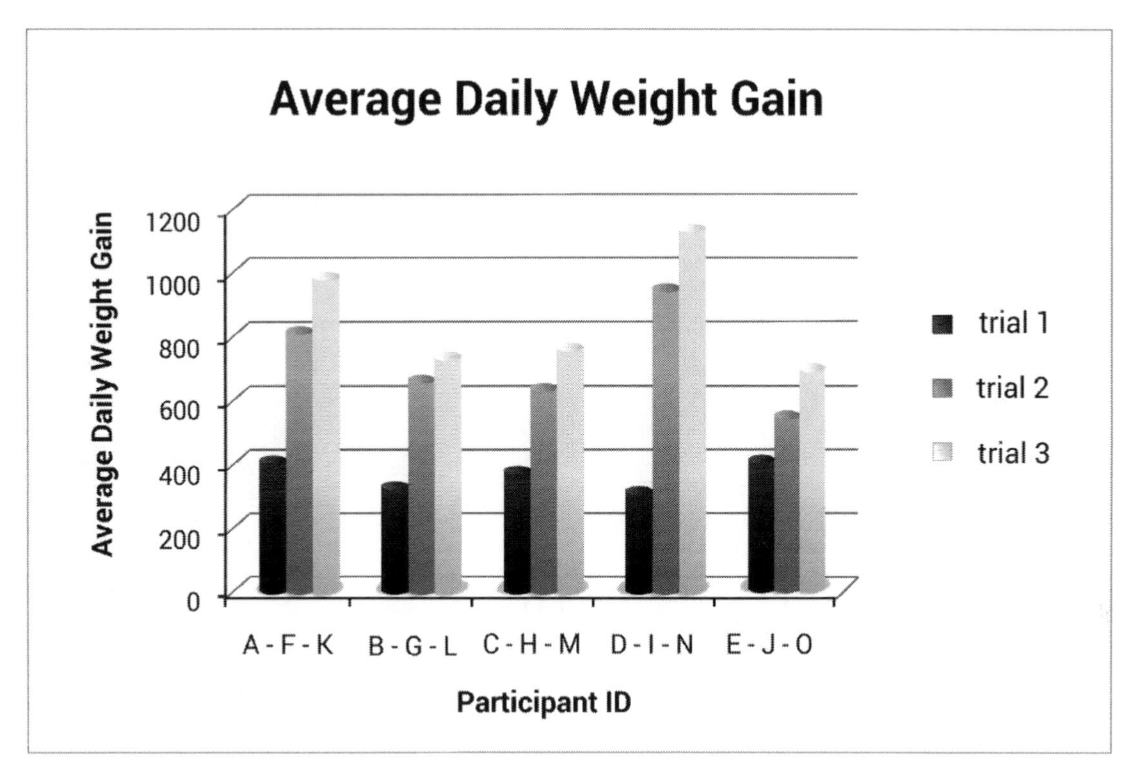

13. What was the main hypothesis for this study?
    a. 2-AG injections combined with sleep deprivation will result in weight gain.
    b. 2-AG injections will increase food intake beyond satiety.
    c. Sleep deprivation will result in weight gain.
    d. The placebo effect of the sterile normal saline will influence eating behavior.

14. Do the study results support the hypothesis? Choose the best answer.
    a. No, participants in trials 1 and 3 all gained weight.
    b. Yes, participants in trial 1 gained more weight daily than participants in trial 3.
    c. No, the average weight gain of participants in trial 2 and trial 3 was the same.
    d. Yes, all trial 3 participants gained more weight than trial 1 participants.

15. Describe the study results for participants D and H.
    a. Participant H gained more than one pound each day.
    b. Weight gain for each participant was inconsistent with the study hypothesis.
    c. There was significant fluctuation in the daily weight gain for both participants.
    d. Participant D's average daily weight was two times participant H's average daily weight gain.

16. According to the researchers, which of the following best describes the influence of sleep deprivation on eating behaviors?
    a. The total number of sleep hours is unrelated to the degree of body stress.
    b. Sleep deprivation stimulates the release of endogenous cannabinoids that may increase food intake.
    c. Deprivation of any variety triggers the hunger response.
    d. Sleep deprivation increases eating behaviors in the early morning hours.

17. According to the passage, how does 2-AG influence eating behaviors?
    a. Circadian fluctuations result in increased levels of 2-AG in the afternoon and evening.
    b. Endogenous cannabinoids like 2-AG increase gastric motility, which stimulates the hunger response.
    c. The sedation that results from the presence of 2-AG limits food intake.
    d. Endogenous cannabinoids block the opioid system, which decreases food-seeking behaviors.

Passage 4

Questions 18–22 pertain to the following passage:

A national wholesale nursery commissioned research to conduct a cost/benefit analysis of replacing existing fluorescent grow lighting systems with newer LED lighting systems. LEDs (light-emitting diodes) are composed of various semi-conductor materials that allow the flow of current in one direction. This means that LEDs emit light in a predictable range, unlike conventional lighting systems that give off heat and light in all directions. The wavelength of light of a single LED is determined by the properties of the specific semi-conductor. For instance, the indium gallium nitride system is used for blue, green, and cyan LEDs. As a result, growing systems can be individualized for the specific wavelength requirements for different plant species. In addition, LEDs don't emit significant amounts of heat compared to broadband systems, so plant hydration can be controlled more efficiently.

Figure 1 identifies the visible spectrum with the wavelength expressed in nanometers:

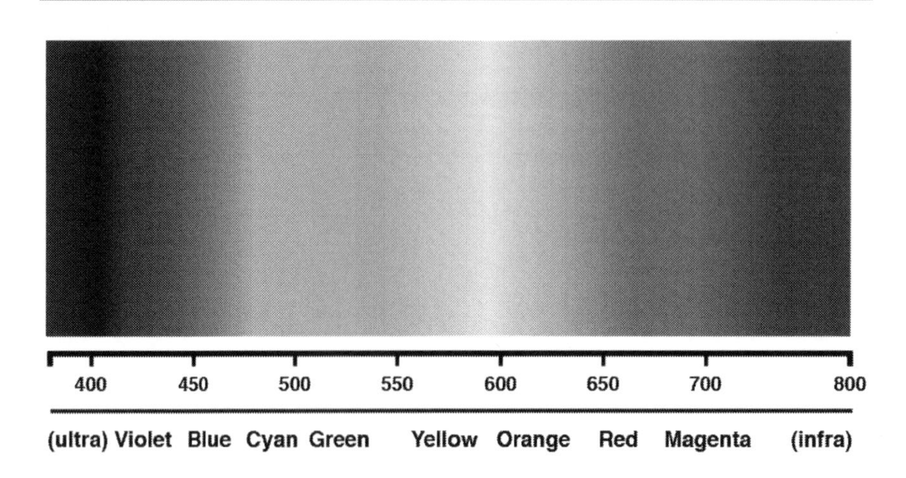

Figure 2 identifies the absorption rates of different wavelengths of light:

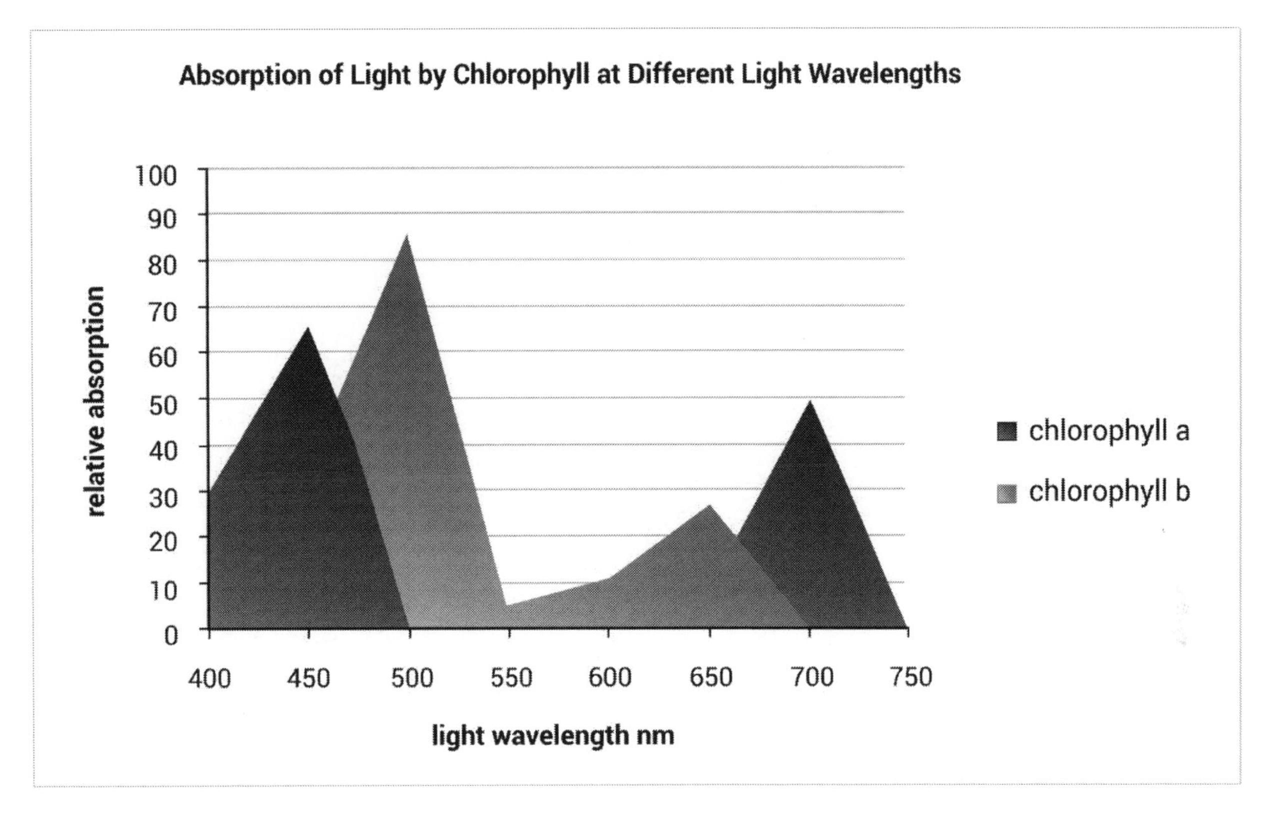

Researchers conducted three trials and hypothesized that LEDs would result in greater growth rates than conventional lighting or white light. They also hypothesized that using a combination of red, blue, green, and yellow wavelengths in the LED lighting system would result in a greater growth rate than using red or blue wavelengths alone. Although green and yellow wavelengths are largely reflected by the plant (Figure 2), the absorption rate is sufficient to make a modest contribution to plant growth. Fifteen Impatiens walleriana seed samples were planted in the same growing medium. Temperature, hydration, and light intensity were held constant. Plant height in millimeters was recorded as shown in the following figures.

Figure 3 identifies the plant growth rate in millimeters with light wavelengths of 440 nanometers:

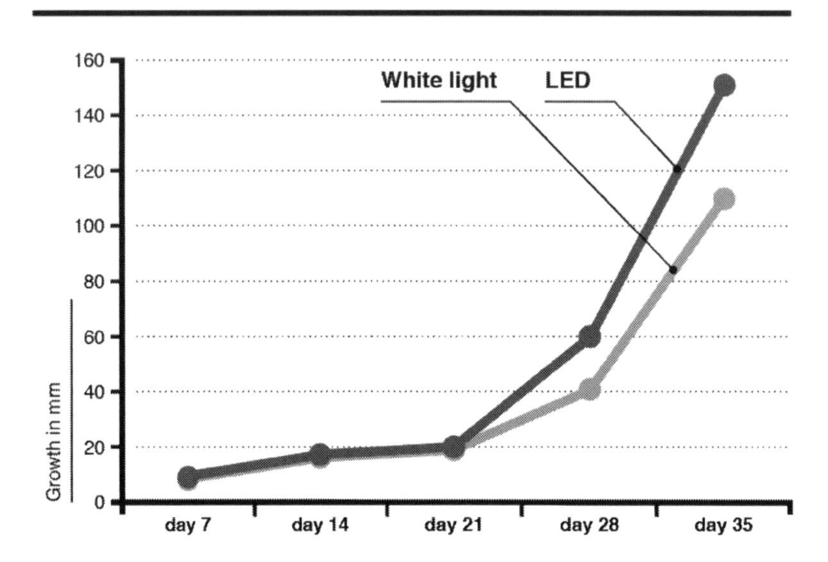

Figure 4 identifies the plant growth rate in mm with light wavelengths of 650 nanometers:

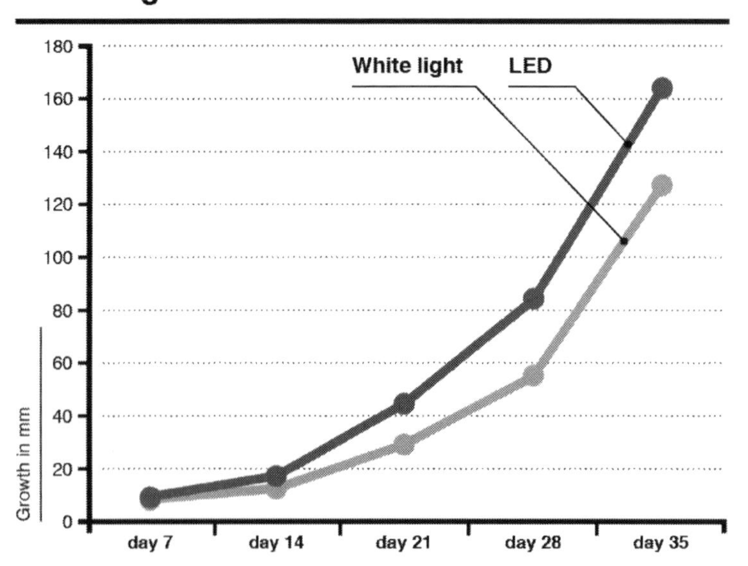

Figure 5 identifies the plant growth rate in millimeters with combined light wavelengths of 440, 550, and 650 nanometers:

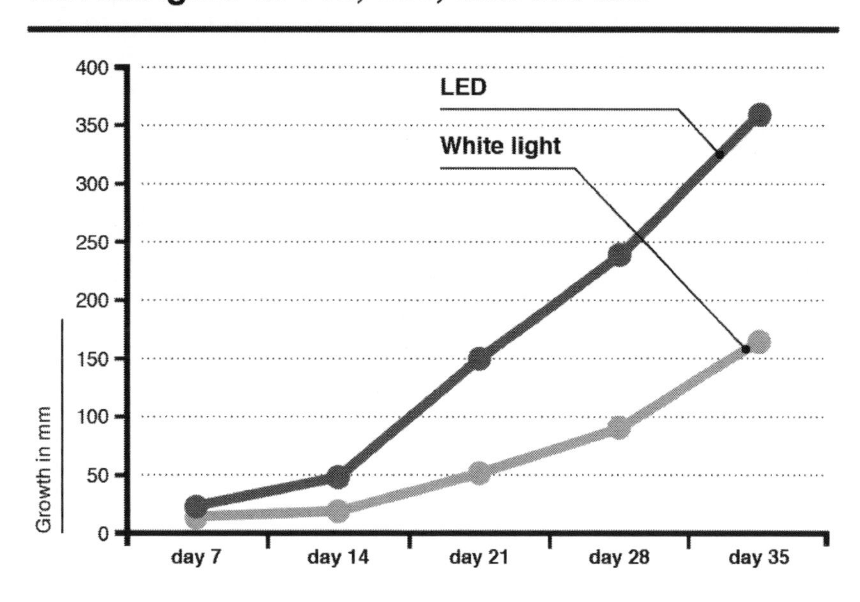

Figure 6 identifies average daily plant growth rate in millimeters:

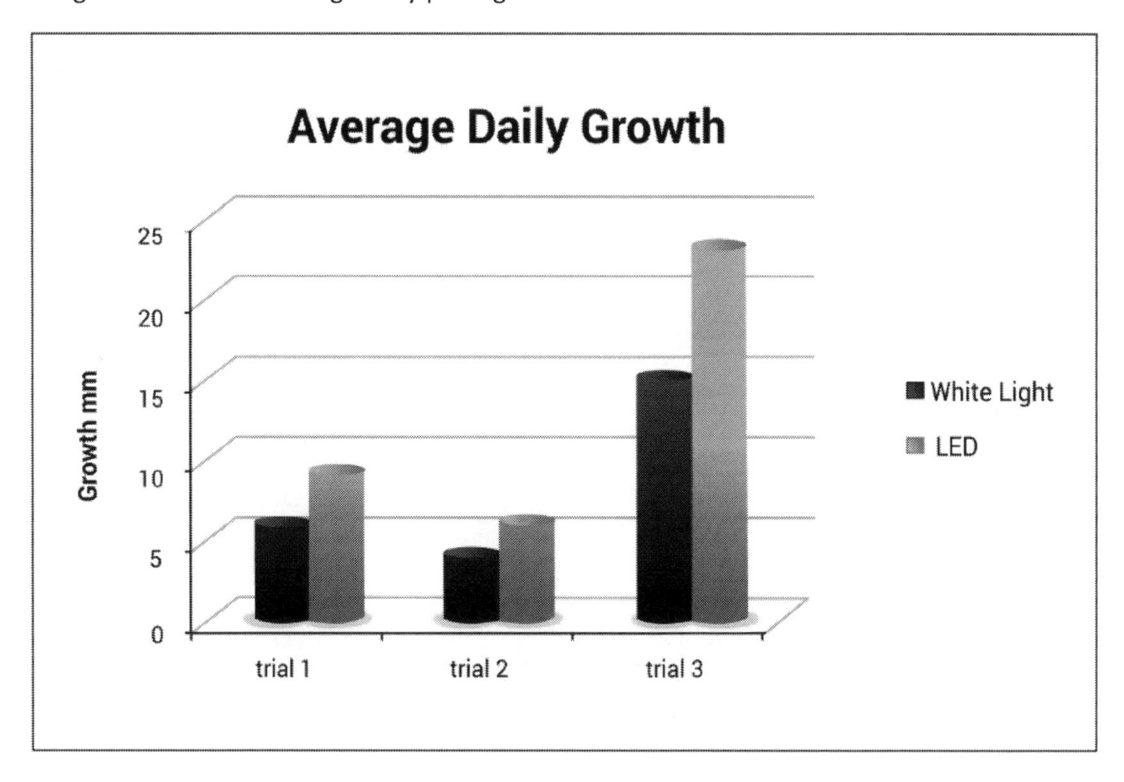

18. If the minimum plant height required for packaging a plant for sale is 150 millimeters, based on plant growth, how much sooner will the LED plants be packaged compared to the white light plants?
   a. 14 days
   b. 21 days
   c. 35 days
   d. 42 days

19. Plants reflect green and yellow light wavelengths. Do the results of the three trials support the view that plants also absorb and use green and yellow light wavelengths for growth?
   a. Yes, green and yellow light wavelengths were responsible for plant growth in trial 3.
   b. No, white light alone was responsible for measurable plant growth.
   c. Yes, the growth rates in trial 3 were greater than the rates in trials 1 and 2.
   d. No, only the red and blue wavelengths were effective in stimulating plant growth.

20. When did the greatest rate of growth occur for both groups in trial 1 and trial 2?
   a. From 7 days to 14 days
   b. From 28 days to 35 days
   c. From 21 days to 28 days
   d. From 14 days to 21 days

21. If an LED lighting system costs twice as much as a white light system, based only on the average daily growth rate as noted above, would it be a wise investment?
   a. No, because multiple different semi-conductors would be necessary.
   b. Yes, growth rates are better with LEDs.
   c. No, the LED average daily growth rate was not two times greater than the white light rate.
   d. Yes, LEDs use less electricity and water.

22. If the researchers conducted an additional trial, trial 4, to measure the effect of green and yellow wavelengths on plant growth, what would be the probable result?
   a. The growth rate would exceed trial 1.
   b. The growth rate would equal trial 3.
   c. The growth rate would be the same as trial 2.
   d. The growth rate would be less than trial 1 or trial 2.

Passage 5

Questions 23–28 pertain to the following passage:

> Mangoes are a tropical fruit that grow on trees native to Southern Asia called the *Mangifera*. Mangoes are now grown in most frost-free tropical and subtropical locations around the world. India and China harvest the greatest numbers of mangoes. A major problem the mango industry faces each year is the destruction of fruit after harvest. This destruction is the result of spoilage or rotting that occurs during long shipping and storage times.
>
> To prevent the spoilage of mangoes, fruits are stored and shipped in climate-controlled containers. Ideally, mangoes should be stored at around 5 °C, which is about the same temperature as a home refrigerator. Although storage at 5 °C is highly effective at

preventing spoilage, the monetary costs associated with maintaining this temperature during long shipping times are prohibitive.

Fruit companies spend large amounts of money to learn about the underlying cause of spoilage and possible methods to prevent loss of their product. Anthracnose, an infection that causes mango decay, is caused by *Colletotrichum,* a type of fungus that has been identified as a major contributor to mango spoilage. This fungus, which may remain dormant on green fruit, grows on the surface of the mango and can penetrate the skin and cause spoilage. The infection first appears during the flowering period as small black dots that progress to dark brown or black areas as ripening occurs. Humidity and excessive rainfall increase the severity of this infection. Previous studies established that colony sizes smaller than 35 millimeters after 4 weeks of travel resulted in acceptable amounts of spoilage.

Currently, several additional pre-treatment measures aimed at prevention are employed to slow decay of the fruit from the harvest to the marketplace. Industry researchers examined the individual and collective benefits of two of these processes, including post-harvest hot water treatment and air cooling at varied transport temperatures in order to identify optimum post-harvest procedures.

Table 1 identifies the observed mango decay in millimeters at 5 °C, 7.5 °C, and 10 °C with two pre-treatment processes over time measured in days since harvest of the fruit.

Table 1: Days Since Harvest

	2	4	6	8	10	12	14	16	18	20	22	24	26	28
5° C														
Water	1	4	7	9	11	12	14	19	22	23	25	27	28	30
Air	0	2	3	6	8	9	11	12	13	15	16	17	18	19
7.5°C														
Water	2	3	4	5	6	8	9	11	12	13	14	15	16	27
Air	0	2	5	6	7	9	10	15	22	23	24	27	32	39
10°C														
Water	2	3	5	7	8	11	12	14	22	35	42	44	47	62
Air	1	2	4	6	7	9	10	15	19	23	27	29	35	44

Figure 1 identifies the observed mango decay of fruit stored at 5 °C measured in millimeters, with two pretreatment processes, over time measured in days since harvest of the fruit:

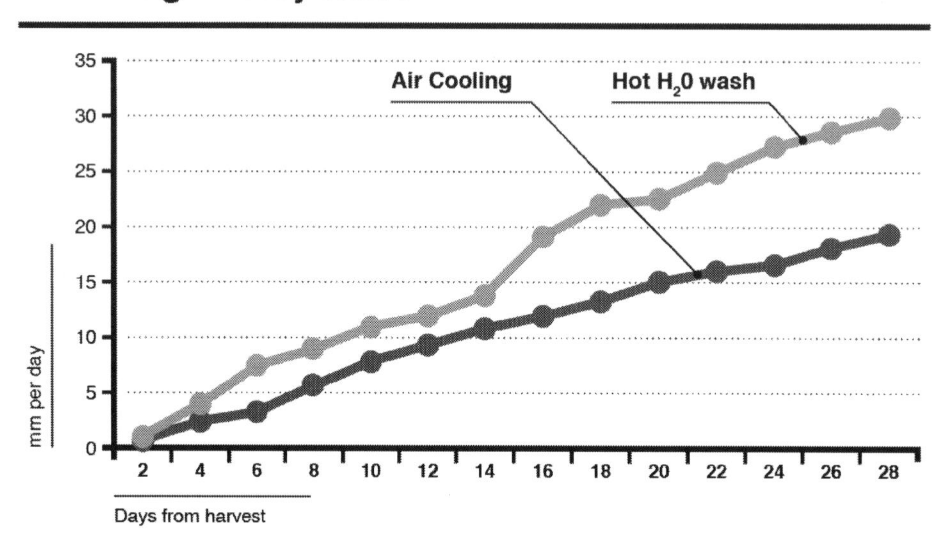

Figure 2 identifies the observed mango decay of fruit stored at 7.5 °C, measured in millimeters, with two pretreatment processes, over time measured in days since harvest of the fruit:

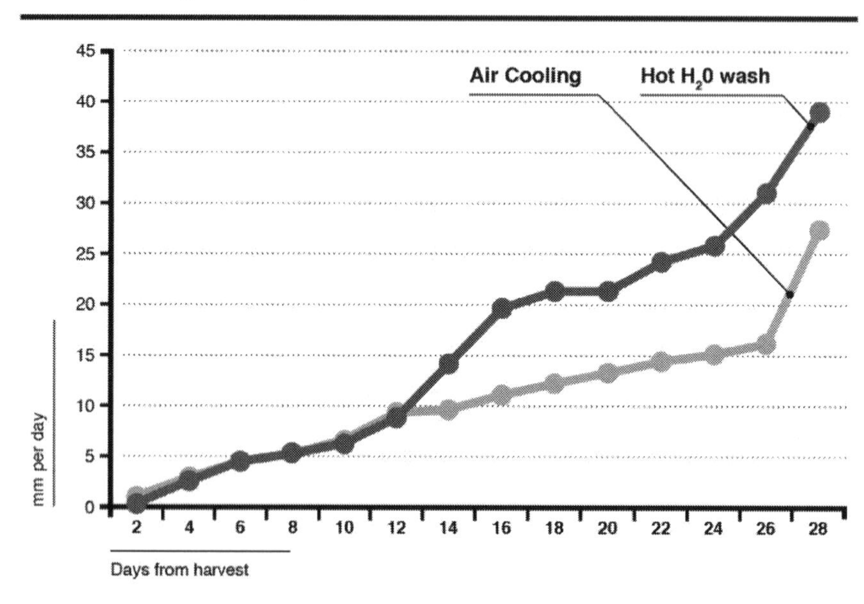

Figure 3 identifies the observed mango decay of fruit stored at 10 °C measured in millimeters, with two pretreatment processes, over time measured in days since harvest of the fruit:

Figure 3:
10 °C Mango Decay Rates

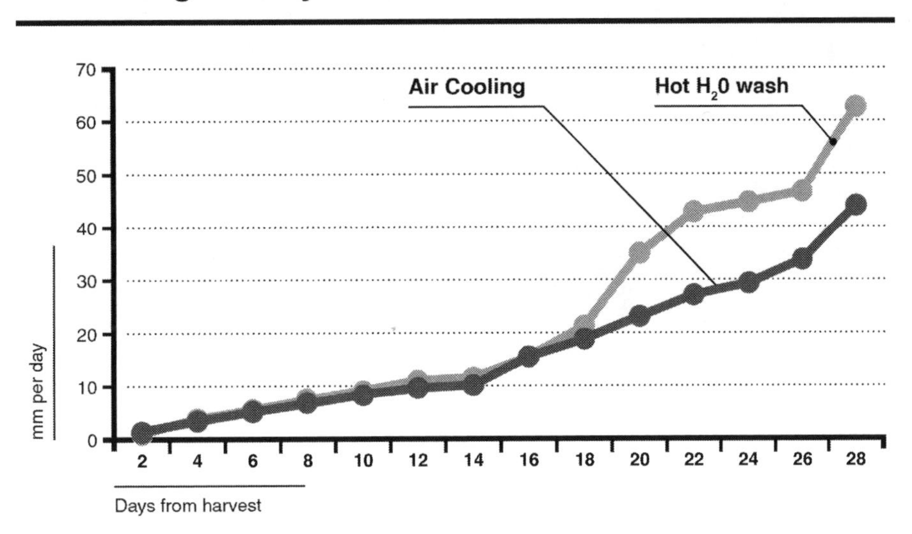

Figure 4 identifies fruit decay measured in millimeters at 5 °C, 7.5 °C, and 10 °C with the combined pre-treatments over 28 days:

Figure 4:
Mango Decay Rates at 5 °C, 7.5 °C, and 10 °C with the Combined Pre-Treatments (Air Cooling & Water Bath)

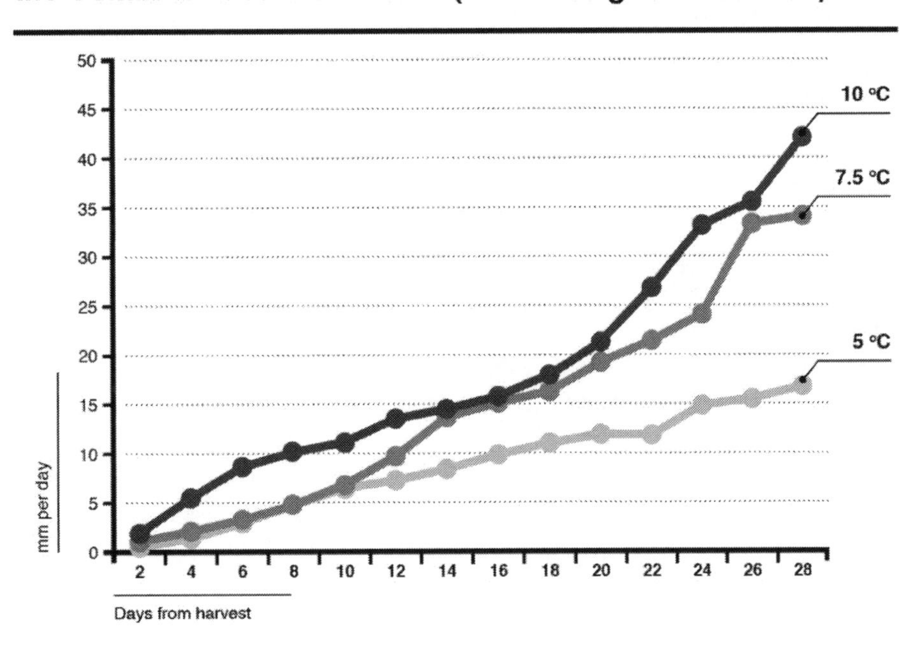

23. According to the passage above, which of the following statements is false?
    a. The optimal temperature for storing mangos is 5 °C.
    b. *Anacardiaceae Magnifera* is responsible for mango spoilage.
    c. Storing fruit at 5 °C is costly.
    d. Long distance shipping is a critical factor in mango spoilage.

24. If the mangoes were shipped from India to the U.S., and the trip was expected to take 20 days, which model would be best according to the data in Table 1?
    a. The 10 °C model, because fungal levels were acceptable for both pre-treatments.
    b. The 5 °C, model because it's more cost-effective.
    c. The 7.5 °C model, because this temperature is less expensive to maintain, and the fungal levels were acceptable.
    d. No single model is better than the other two models.

25. According to Figures 1–3 above, the largest one-day increase in fruit decay occurred under which conditions?
    a. Air cooling at 10 °C
    b. Hot water wash at 10 °C
    c. Air cooling at 7.5 °C
    d. Hot water wash at 7.5 °C

26. Which pre-treatment method reached unacceptable fungal levels first?
    a. Hot water wash at 7.5 °C
    b. Air cooling at 5 °C
    c. Hot water wash at 10 °C
    d. Air cooling at 10 °C

27. The researchers were attempting to identify the best shipping conditions for mangoes for a 28-day trip from harvest to market. Referencing Figures 1–4, which conditions would be the most cost-effective?
    a. Air cooling pre-treatment at 5 °C
    b. Air cooling and hot water wash pre-treatment at 5 °C
    c. Hot water wash pre-treatment at 7.5 °C
    d. Air cooling at 10 °C

28. Shipping mangoes at 5 °C is costly. According to the researchers' findings, is shipping mangoes at 5 °C more cost effective than 7.5 °C for trips lasting more than 28 days when combined air cooling and hot water wash treatments are applied?
    a. Yes, shipping at 7.5 °C combined with both pre-treatments resulted in an unacceptable fungal infection rate.
    b. Yes, fungal infection rates were below 35 mm for both pre-treatments 5 °C.
    c. No, air cooling pre-treatment was acceptable at 10 °C, and it's less expensive to ship fruit at 10 °C.
    d. No, hot water wash rates were lower than air cooling at 5 °C.

Passage 6

Questions 29–34 pertain to the following passage:

Scientists recently discovered that circadian rhythms help regulate sugar consumption by brown adipose tissue. The results of this study suggest that circadian rhythms and fat cells work together to warm the body in preparation for early morning activities involving cold weather. A circadian rhythm refers to life processes controlled by an internal "biological clock" that maintains a 24-hour rhythm. Sleep is controlled by one's circadian rhythm. To initiate sleep, the circadian rhythm stimulates the pineal gland to release the hormone melatonin, which causes sleepiness. Importantly, the circadian rhythm discerns when to begin the process of sleep based on the time of day. During the daytime, sunlight stimulates special cells within the eye, photosensitive retinal ganglion cells, which, in turn, allow the "biological clock" to keep track of how many hours of sunlight there are in a given day.

Brown adipose tissue (BAT) is a type of fat that plays an important role in thermogenesis, a process that generates heat. In humans and other mammals, there are two basic types of thermogenesis: shivering thermogenesis and non-shivering thermogenesis. Shivering thermogenesis involves physical movements, such as shaky hands or clattering teeth. Heat is produced as a result of energy being burned during physical activity. Non-shivering thermogenesis doesn't require physical activity; instead, it utilizes brown adipose tissue to generate heat. Brown fat cells appear dark because they contain large numbers of mitochondria, the organelles that burn sugar to produce energy and heat.

Researchers know that brown adipose tissue (BAT) is essential for maintaining body temperature. A new discovery in humans has shown that circadian rhythms cause BAT to consume more sugar in the early-morning hours. This spike in sugar consumption causes more heat to be produced in BAT. Scientists propose that our human ancestors could have benefited from extra body heat during cold hunts in the morning.

Perhaps more significantly, these new findings may suggest a role for BAT in the prevention of Type 2 Diabetes. Two important questions remain; to what degree does BAT affect blood glucose levels, and is it possible to increase BAT in a given individual? The demonstrated increase in sugar consumption and heat production of BAT is thought to be related to insulin-sensitivity. To examine the first question, researchers conducted three trials to examine the relationship between brown fat and blood glucose levels at different points in the day. PET scanning was used to estimate total body brown fat in 18 non-diabetic participants. Total body brown fat expressed as a proportion of total body fat (either 5%, 10%, or 20%) was the basis for group selection. The researchers hypothesized that the blood glucose levels would be inversely related to the percentage of BAT. Resulting data is included below.

Figure 1 identifies the circadian cycle of blood glucose:

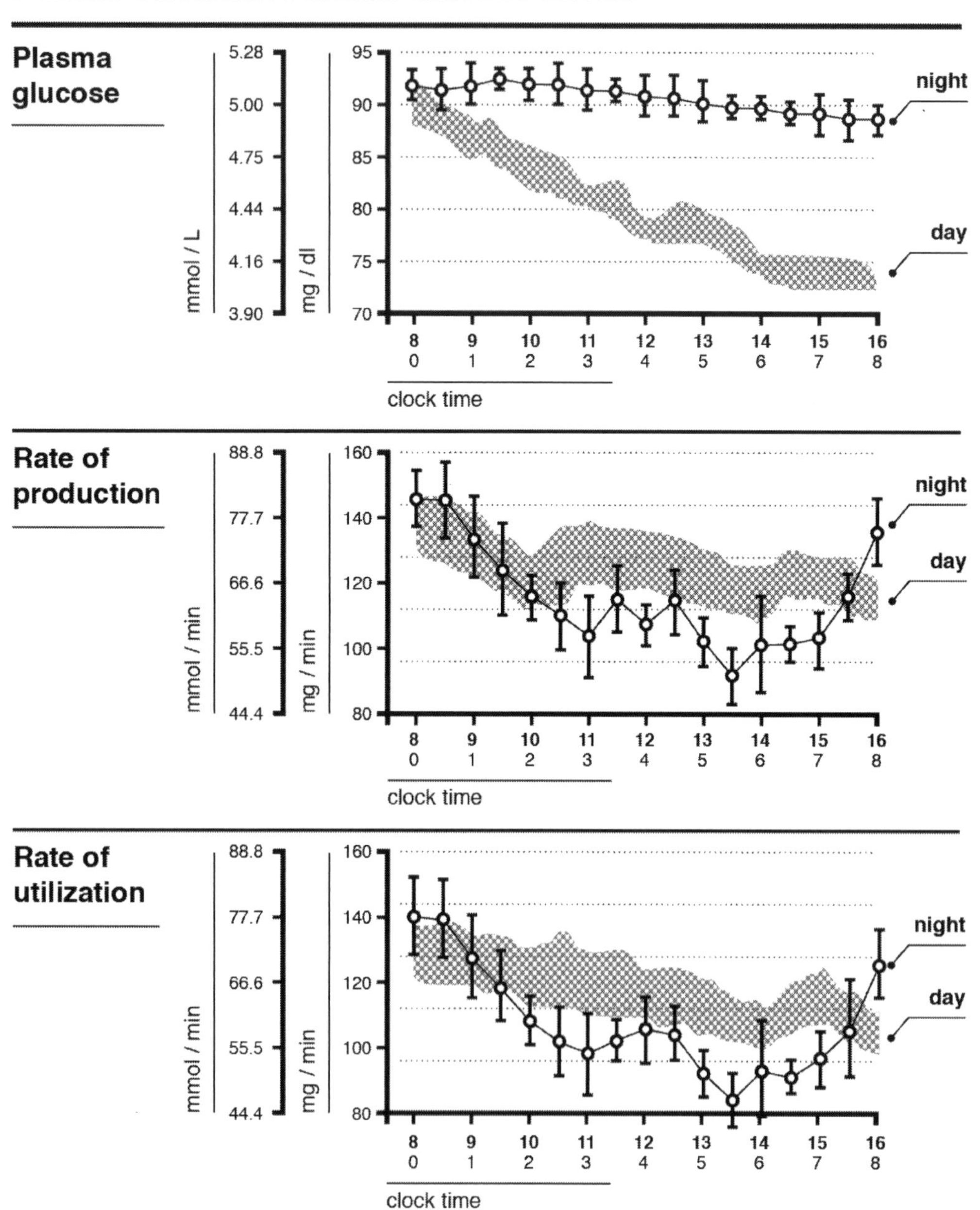

Figure 2 identifies the resulting blood glucose measurements for participants with 5% total brown body fat:

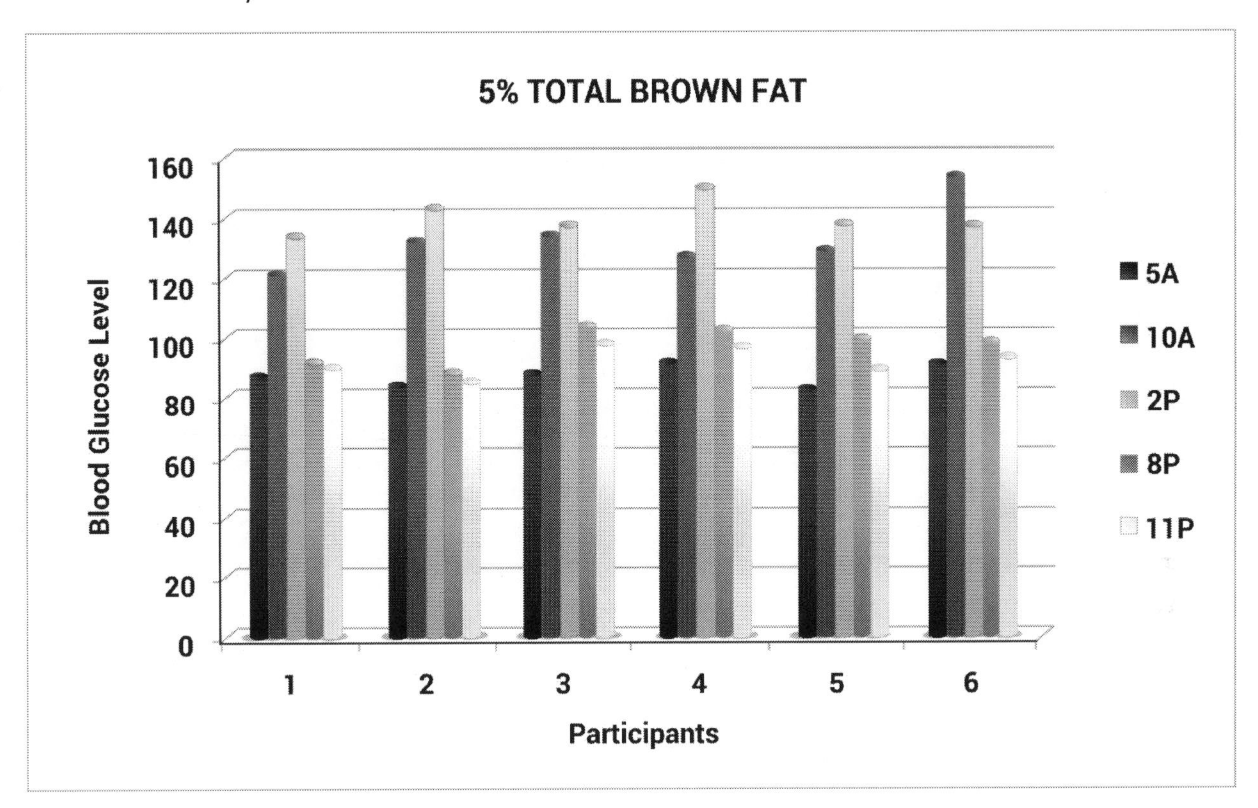

Figure 3 identifies the resulting blood glucose measurements for participants with 10% total brown body fat:

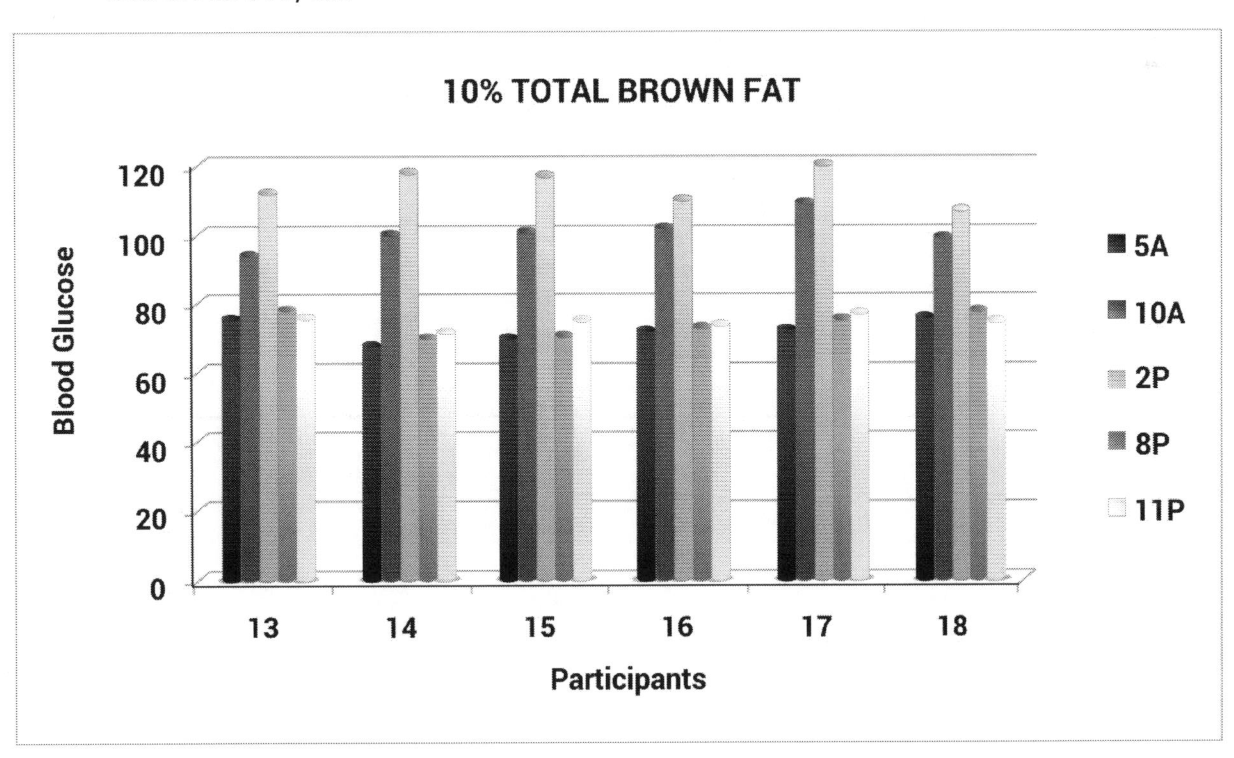

Figure 4 identifies the resulting blood glucose measurements for participants with 20% total brown body fat:

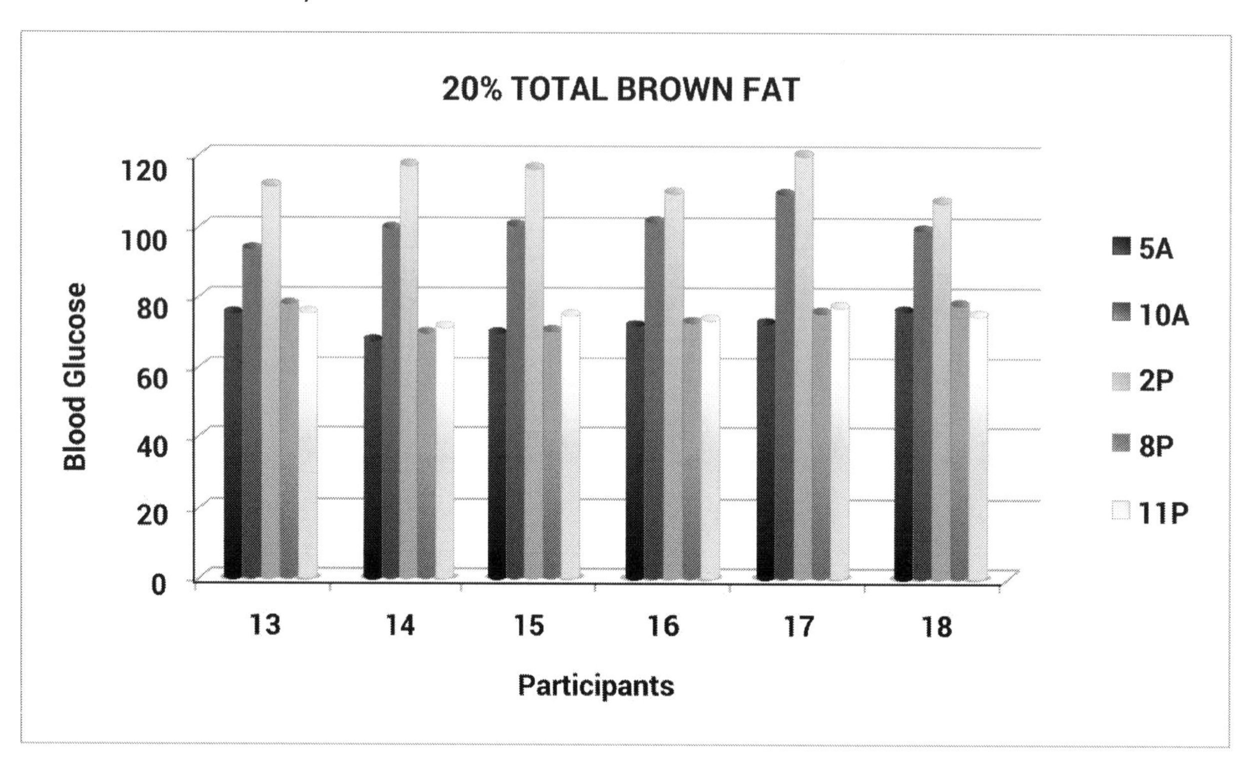

Figure 5 identifies the average blood glucose measurements for the three trials:

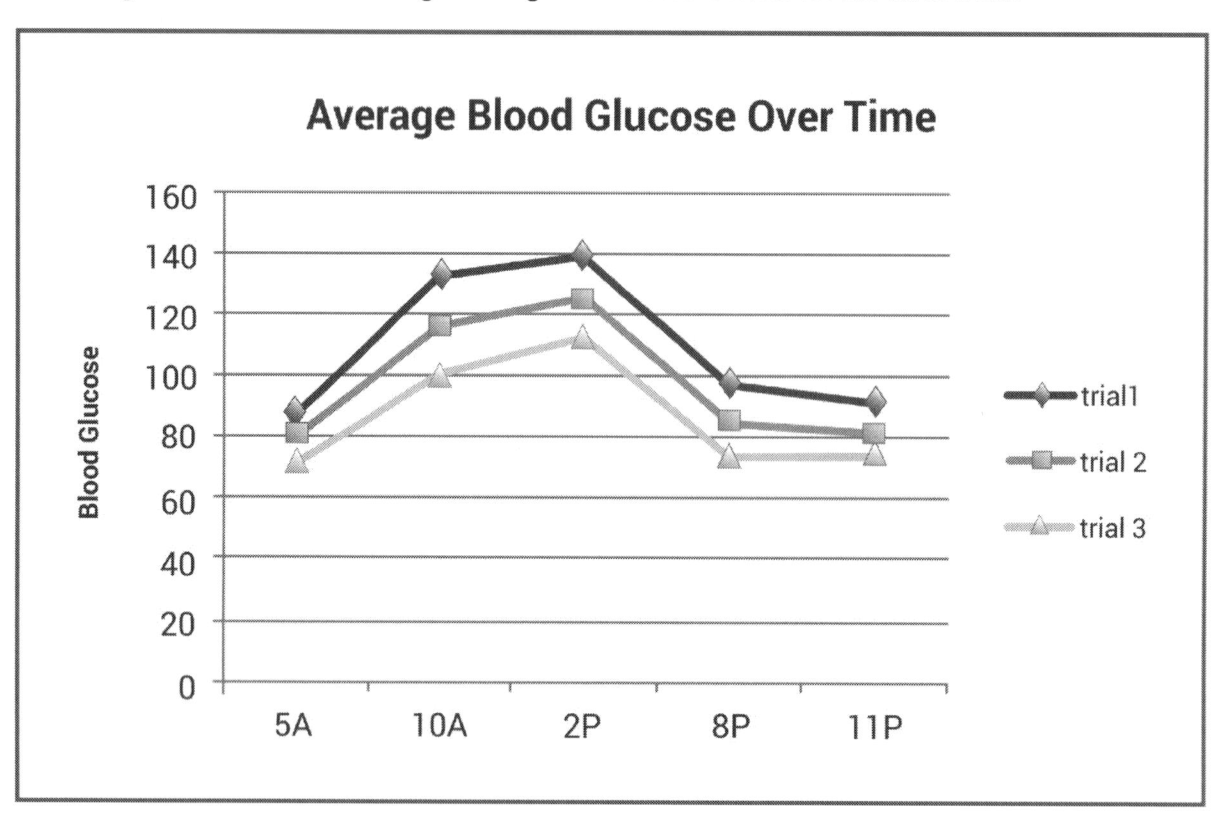

29. Which of the following describes the relationship of the research results in Figure 5?
    a. Positive correlation among the three trials
    b. Curvilinear relationship
    c. Weak negative relationship
    d. No demonstrated relationship

30. Which of the following statements concerning mitochondria is INCORRECT?
    a. Mitochondrial function is diminished in the presence of elevated blood glucose levels.
    b. Mitochondria are responsible for the color of brown fat.
    c. Mitochondria are capable of reproduction in response to energy needs.
    d. Mitochondria are responsible for binding oxygen in mature red blood cells.

31. According to Figure 5, participants' blood sugars were highest at what time of day?
    a. 5 a.m., because heat is generated early in the morning
    b. 8 p.m., because the participants ate less for dinner than lunch
    c. 11 p.m., because brown adipose tissue is not active at night
    d. 2 p.m., because the effects of the early-morning activity of the brown adipose tissue had diminished

32. Circadian rhythms control sleep by doing which of the following?
    a. Stimulating the pineal gland to release ganglia
    b. Stimulating the release of melatonin
    c. Suppressing shivering during cold mornings
    d. Instructing brown adipose tissue to release sugar

33. Which Participant in trial 1 had the highest average blood sugar for the group?
    a. 1
    b. 3
    c. 4
    d. 6

34. Is the data in Figure 5 consistent with the daytime plasma glucose trend in Figure 1?
    a. Yes, Figure 5 blood glucose readings declined from a morning to afternoon.
    b. No, blood glucose readings peaked at 2 p.m..
    c. Yes, morning glucose readings were higher in group 1.
    d. No, nighttime levels fluctuated between 100 and 110.

Passage 7

Questions 35–40 pertain to the following passage:

> A biome is a major terrestrial or aquatic environment that supports diverse life forms. Freshwater biomes—including lakes, streams and rivers, and wetlands—account for 0.01% of the Earth's fresh water. Collectively, they are home to 6% of all recognized species. Standing water bodies may vary in size from small ponds to the Great Lakes. Plant life in lakes is specific to the zone of the lake that provides the optimal habitat for a specific species, based on the depth of the water as it relates to light. The photic layer is the shallower layer where light is available for photosynthesis. The aphotic layer is deeper, and the levels of sunlight are too low for photosynthesis. The benthic layer is the bottom-most layer, and its inhabitants are nourished by materials from the photic

layer. Light-sensitive cyanobacteria and microscopic algae are two forms of phytoplankton that exist in lakes. As a result of nitrogen and phosphorous from agriculture and sewage run-off, algae residing near the surface can multiply abnormally so that available light is diminished to other species. Oxygen supplies may also be reduced when large numbers of algae die.

Recently, concerns have been raised about the effects of agriculture and commercial development on the quality of national freshwater bodies. In order to estimate the effect of human impact on freshwater, researchers examined plant life from the aphotic layer of three freshwater lakes of approximately the same size located in three different environments. Lake A was located in a remote forested area of western Montana. Lake B was located in central Kansas. Lake C was located in a medium-size city on the west coast of Florida. The researchers hypothesized that the microscopic algae and cyanobacteria populations from Lake A would approach appropriate levels for the size of the lake. They also hypothesized that the remaining two samples would reveal abnormal levels of the phytoplankton. In addition, the researchers measured the concentration of algae at different depths at four different times in another lake identified as having abnormal algae growth. These measurements attempted to identify the point at which light absorption in the photic layer was no longer sufficient for the growth of organisms in the aphotic layer. Resulting data is identified below.

Figure 1 illustrates the zones of the freshwater lake:

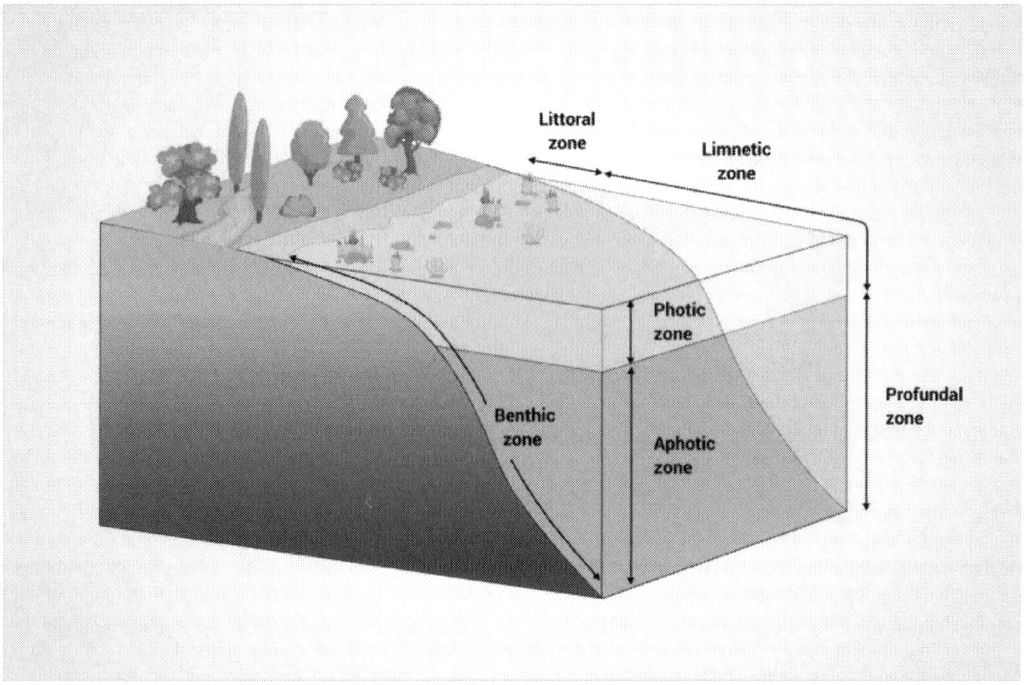

Figure 2 identifies algae and cyanobacteria levels in parts per million for Lake A over six measurements:

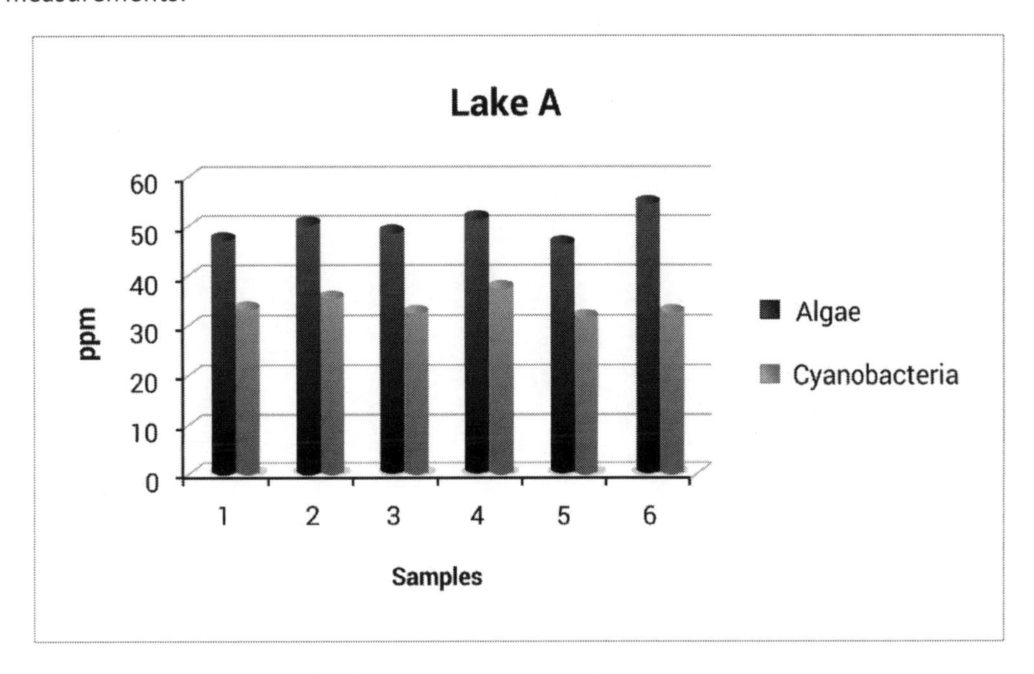

normal: Algae 50 p.p.m. Cyanobacteria 35 p.p.m.

Figure 3 identifies algae and cyanobacteria levels in parts per million for Lake B over six measurements:

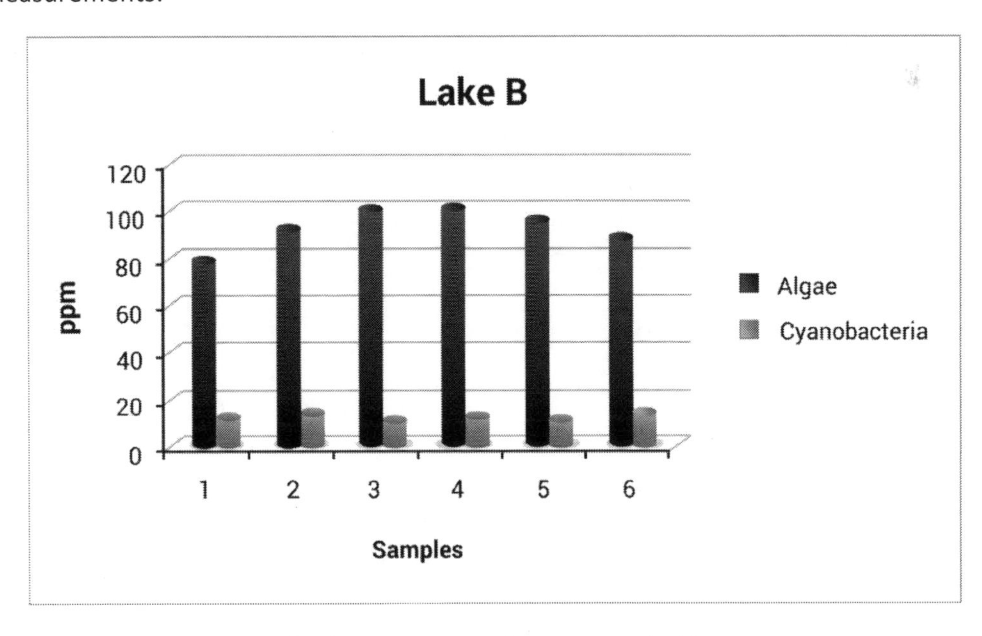

normal: Algae 50 p.p.m. Cyanobacteria 35 p.p.m.

Figure 4 identifies algae and cyanobacteria levels in parts per million for Lake C over six measurements:

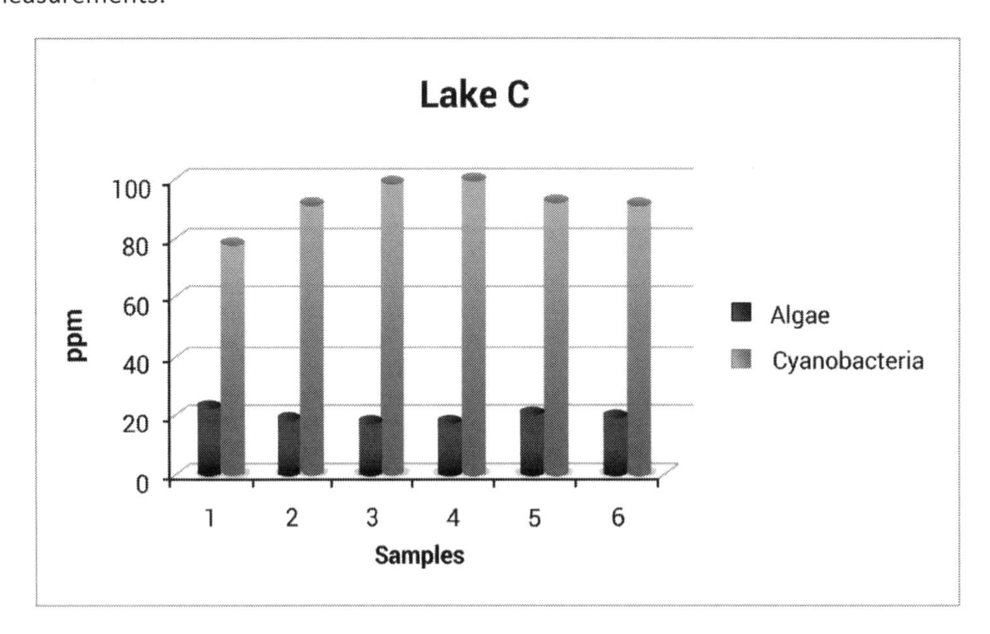

normal: Algae 50 p.p.m. Cyanobacteria 35 p.p.m.

Figure 5 identifies cyanobacteria levels at different depths over time:

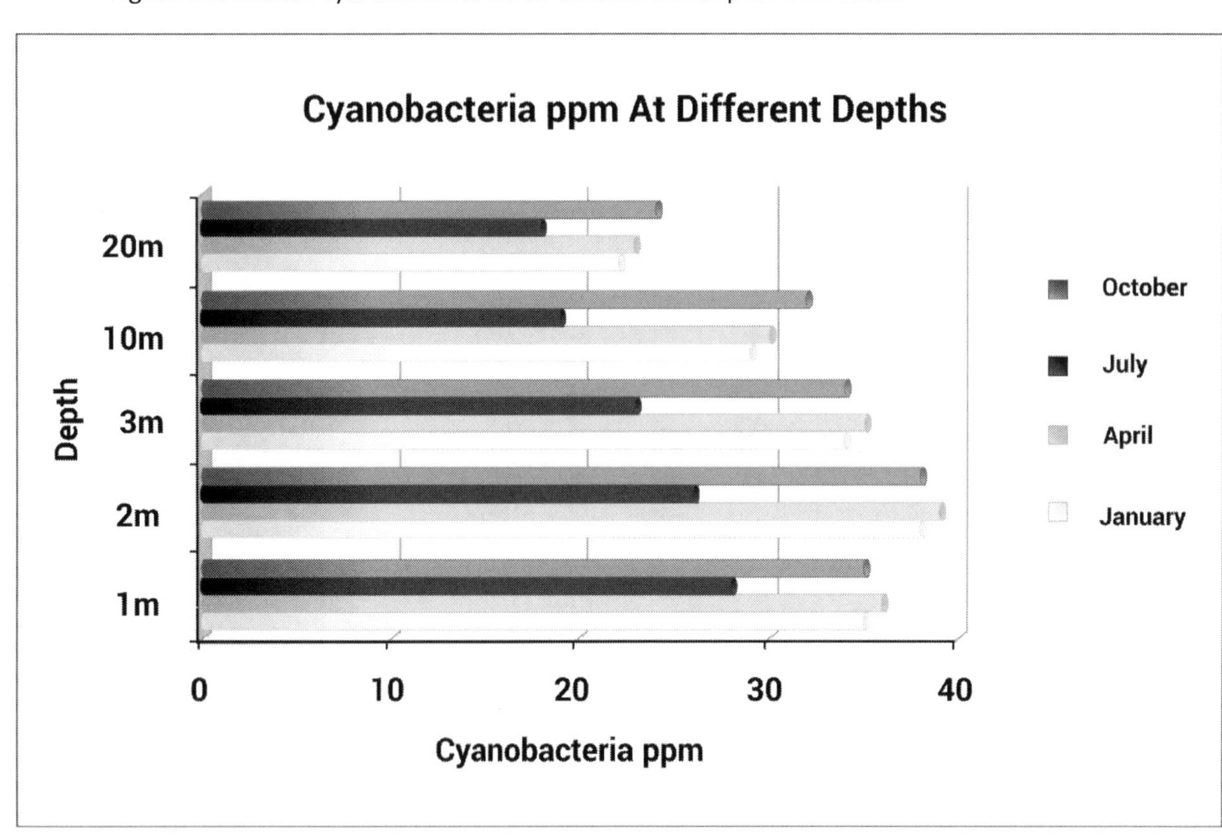

35. Based on Figure 2, was the researchers' hypothesis confirmed?
    a. No, the phytoplankton levels were not elevated in the first trial.
    b. Yes, the phytoplankton levels were raised above normal in each sample.
    c. No, the Lake A numbers were normal.
    d. Yes, algae levels were above normal in Lake C.

36. In Lake B, cyanobacteria were decreased and algae were increased. Which of the following is a possible explanation for this finding?
    a. The overgrowth of algae decreased the light energy available for cyanobacteria growth.
    b. Lake B experienced severe flooding, causing the water levels in the lake to rise above normal.
    c. Agricultural chemical residue depleted the food source for cyanobacteria.
    d. Cyanobacteria cannot survive in the cold winter weather in Lake B.

37. What common factor might explain the results for Lake B and Lake C?
    a. Population concentration
    b. Average humidity of the locations
    c. Average heat index
    d. Excess nitrogen and phosphorous in the ground water

38. As algae levels increase above normal, what happens to organisms in the aphotic level?
    a. Growth is limited but sustained.
    b. Species eventually die due to decreased oxygenation.
    c. Cyanobacteria increase to unsafe levels.
    d. Aerobic bacteria multiply.

39. Referencing Figures 2 and 3, which environment would favor organisms in the benthic layer of the corresponding lake?
    a. Figure 4, because the cyanobacteria are protective.
    b. Figure 3, because increased numbers of Algae provide more light.
    c. Figure 4, because cyanobacteria are able to survive.
    d. Figure 3, because the levels of both species are normal.

40. Which of the following statements is supported by the data in Figure 5?
    a. Algae growth is greater in July than April.
    b. Cyanobacteria can't exist at 20 meters in this lake.
    c. There's insufficient light in the aphotic layer at 3 meters to support algae growth.
    d. Cyanobacteria growth rates are independent of algae growth at 1 meter.

Passage 8

Questions 41–45 pertain to Passage 8:

Scientists use the scientific method to investigate a theory or solve a problem. It includes four steps: observation, hypothesis, experiment, and conclusion. Observation occurs when the scientist uses one of their senses to identify what they want to study. A hypothesis is a conclusive sentence about what the scientist wants to research. It generally includes an explanation for the observations, can be tested experimentally, and predicts the outcome. The experiment includes the parameters for the testing that will occur. The conclusion will state whether or not the hypothesis was supported.

Scientist A would like to know how sunlight affects the growth of a plant. She says that more sunlight will cause the plant to grow faster. She sets up her experimental groups and tests her hypothesis over 11 days.

Figure 1 below shows the experimental data Scientist A collected over 11 days:

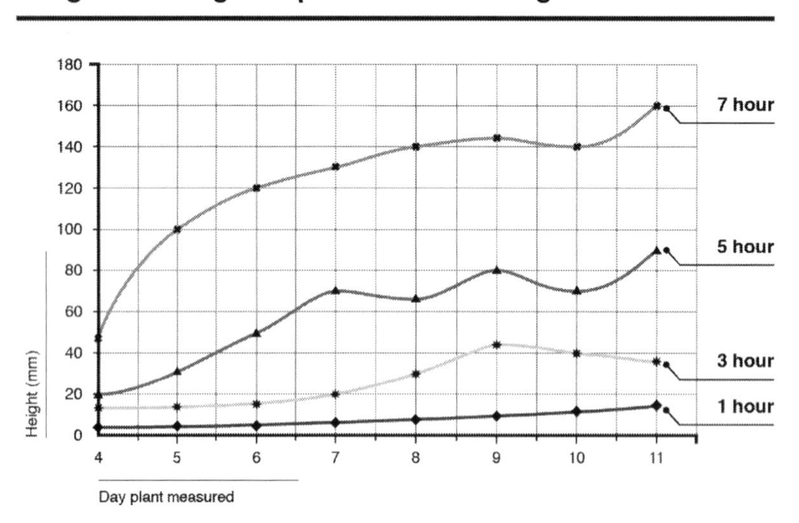

Figure 2 represents the process of photosynthesis that occurs in plants:

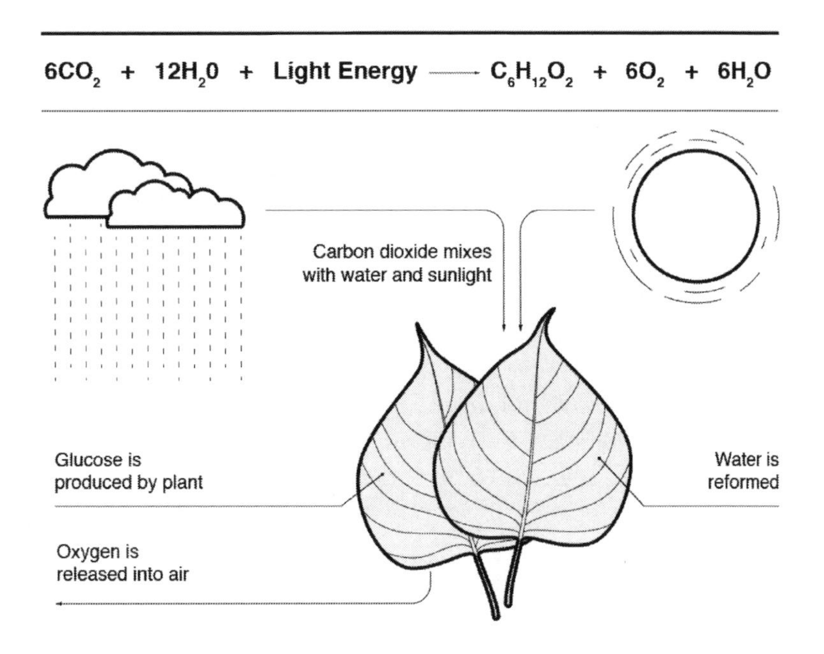

41. What is her hypothesis?
    a. More sunlight will cause the plant to grow faster.
    b. She will test her theory over 11 days.
    c. How sunlight affects plant growth.
    d. Plants do not grow well with one hour of sunlight per day.

42. How many experimental groups does she have?
    a. 1
    b. 3
    c. 4
    d. 11

43. What type of chart is represented in the first figure?
    a. Bar graph
    b. Line graph
    c. Pie chart
    d. Pictogram

44. What part of the photosynthesis reaction is provided directly by sunlight?
    a. Light energy
    b. H_2O
    c. CO_2
    d. Glucose

45. What should her conclusion be based on her experimental data?
    a. 5 hours of sunlight is optimal for plant growth.
    b. Plants should only be measured for 11 days.
    c. Less sunlight is better for plant growth.
    d. Providing plants with more sunlight makes them grow bigger.

Passage 9

Questions 46–50 pertain to Passage 9:

The periodic table contains all known 118 chemical elements. The first 98 elements are found naturally while the remaining were synthesized by scientists. The elements are ordered according to the number of protons they contain, also known as their atomic number. For example, hydrogen has an atomic number of one and is found in the top left corner of the periodic table, whereas radon has an atomic number of 86 and is found closer on the right side of the periodic table, several rows down. The rows are called periods and the columns are called groups. The elements are arranged by similar chemical properties.

Each chemical element represents an individual atom. When atoms are linked together, they form molecules. The smallest molecule contains just two atoms, but molecules can also be very large and contain hundreds of atoms. In order to find the mass of a molecule, the atomic mass of each individual atom in the molecule must be added together.

Figure 1 below depicts the trends and commonalities between the elements that can be seen in the periodic table:

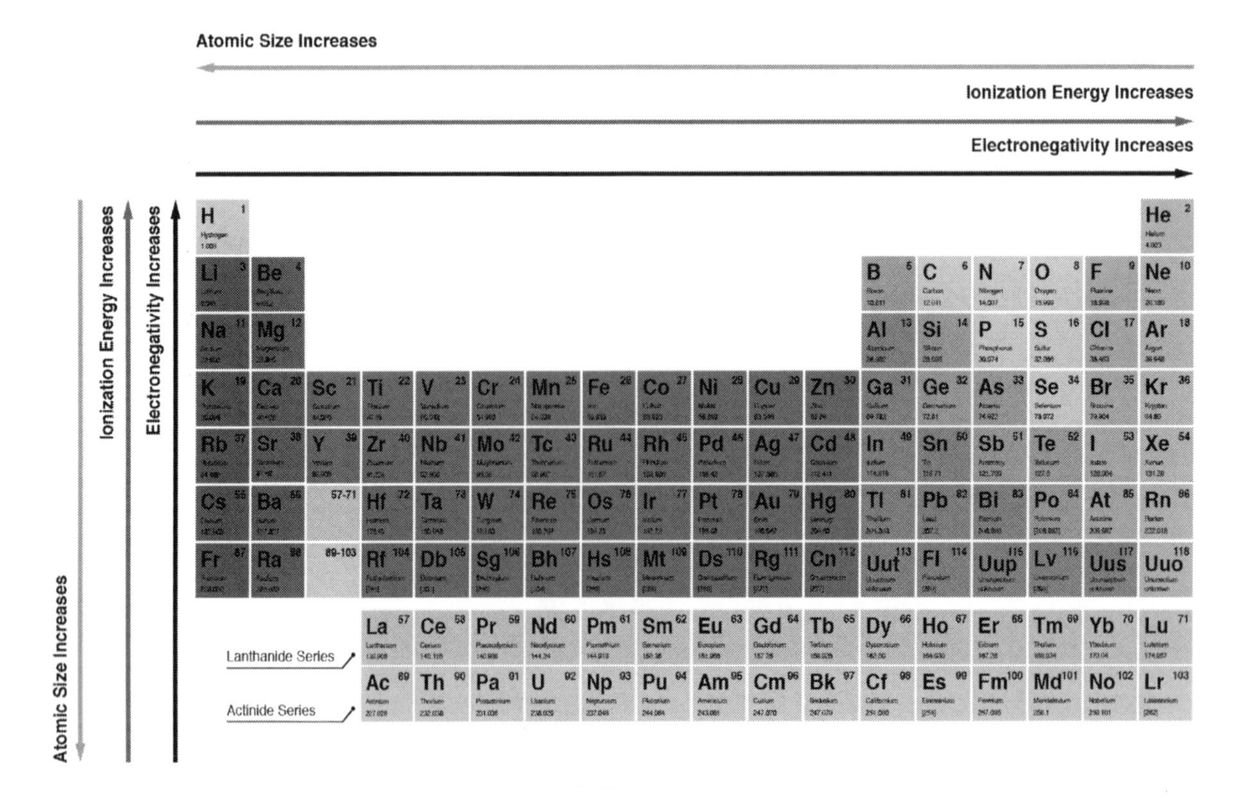

Figure 2 shows what the information in each element's box represents:

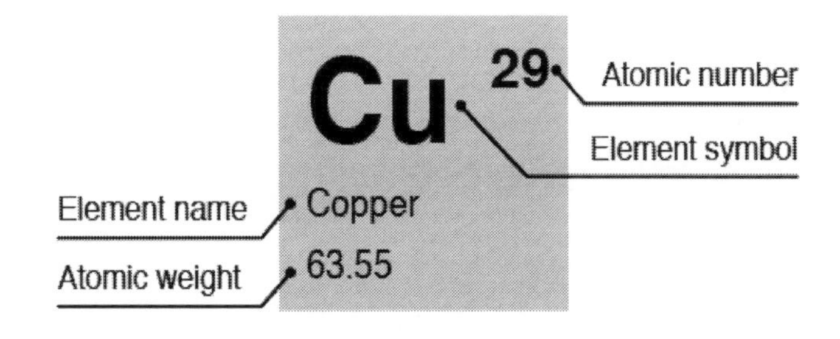

Figure 3 shows the periodic table with color coding according to the groups and periods:

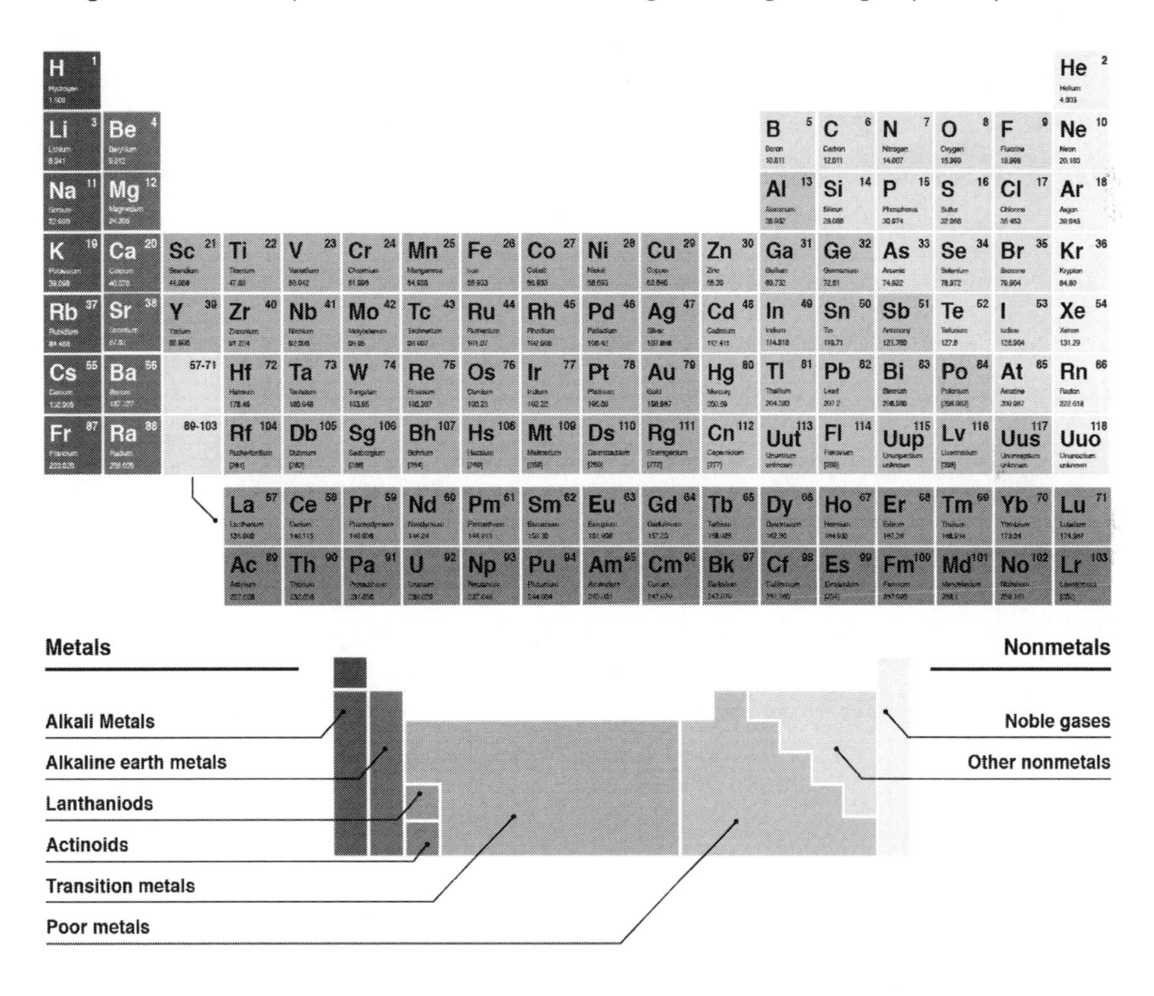

46. What is the atomic mass of NaCl?

a. 23
b. 58.5
c. 35.5
d. 71

47. Which of the following elements is most electronegative?
    a. Ununoctium (Uuo)
    b. Francium (Fr)
    c. Hydrogen (H)
    d. Helium (He)

48. Which of the following elements has the lowest ionization energy?
    a. Potassium (K)
    b. Gallium (Ga)
    c. Sodium (Na)
    d. Aluminum (Al)

49. Which element has the fewest number of protons?
    a. Radon (Rn)
    b. Boron (B)
    c. Nitrogen (N)
    d. Hydrogen (H)

50. Scientist A needs a noble gas for her experiment. Which of these elements should she consider using?
    a. Nitrogen (N)
    b. Radon (Rn)
    c. Copper (Cu)
    d. Boron (B)

Answer Explanations

Mathematics

Calculator Questions

1. D: This problem can be solved using basic arithmetic. Xavier starts with 20 apples, then gives his sister half, so 20 divided by 2.

$$\frac{20}{2} = 10$$

He then gives his neighbor 6, so 6 is subtracted from 10.

$$10 - 6 = 4$$

Lastly, he uses ¾ of his apples to make an apple pie, so to find remaining apples, the first step is to subtract ¾ from one and then multiply the difference by 4.

$$\left(1 - \frac{3}{4}\right) \times 4 = ?$$

$$\left(\frac{4}{4} - \frac{3}{4}\right) \times 4 = ?$$

$$\left(\frac{1}{4}\right) \times 4 = 1$$

2. C: The product of two irrational numbers can be rational or irrational. Sometimes, the irrational parts of the two numbers cancel each other out, leaving a rational number. For example, $\sqrt{2} \times \sqrt{2} = 2$ because the roots cancel each other out. Technically, the product of two irrational numbers can be complex because complex numbers can have either the real or imaginary part (in this case, the imaginary part) equal zero and still be considered a complex number. However, Choice *D* is incorrect because the product of two irrational numbers is not an imaginary number so saying the product is complex *and* imaginary is incorrect.

3. B: The car is traveling at a speed of five meters per second. On the interval from one to three seconds, the position changes by ten meters. By making this change in position over time into a rate, the speed becomes ten meters in two seconds or five meters in one second.

4. C: The number negative four is classified as a real number because it exists and is not imaginary. It is rational because it does not have a decimal that never ends. It is an integer because it does not have a fractional component. The next classification would be whole numbers, for which negative four does not qualify because it is negative. Choices *D* and *E* are wrong because -4 is not considered an irrational number because it does not have a never-ending decimal component.

5. B: Since $850 is the price *after* a 20% discount, $850 represents 80% of the original price. To determine the original price, set up a proportion with the ratio of the sale price (850) to original price (unknown) equal to the ratio of sale percentage (where x represents the unknown original price):

$$\frac{850}{x} = \frac{80}{100}$$

To solve a proportion, cross multiply the numerators and denominators and set the products equal to each other: $(850)(100) = (80)(x)$. Multiplying each side results in the equation $85{,}000 = 80x$.

To solve for x, divide both sides by 80: $\frac{85{,}000}{80} = \frac{80x}{80}$, resulting in $x = 1062.5$. Remember that x represents the original price. Subtracting the sale price from the original price ($\$1062.50 - \$850$) indicates that Frank saved $212.50.

6. D: $0.45

List the givens.

$$Store\ coffee = \$1.23/lbs$$

$$Local\ roaster\ coffee = \$1.98/1.5\ lbs$$

Calculate the cost for 5 lbs of store brand.

$$\frac{\$1.23}{1\ lbs} \times 5\ lbs = \$6.15$$

Calculate the cost for 5 lbs of the local roaster.

$$\frac{\$1.98}{1.5\ lbs} \times 5\ lbs = \$6.60$$

Subtract to find the difference in price for 5 lbs.

```
 $6.60
-$6.15
 $0.45
```

7. C: The first step in solving this problem is expressing the result in fraction form. Separate this problem first by solving the division operation of the last two fractions. When dividing one fraction by another, invert or flip the second fraction and then multiply the numerator and denominator.

$$\frac{7}{10} \times \frac{2}{1} = \frac{14}{10}$$

Next, multiply the first fraction with this value:

$$\frac{3}{5} \times \frac{14}{10} = \frac{42}{50}$$

In this instance, you can find the decimal form by converting the fraction into $\frac{x}{100}$, where x is the number from which the final decimal is found. Multiply both the numerator and denominator by 2 to get the fraction as an expression of $\frac{x}{100}$.

$$\frac{42}{50} \times \frac{2}{2} = \frac{84}{100}$$

In decimal form, this would be expressed as 0.84.

8. E: To find the average of a set of values, add the values together and then divide by the total number of values. In this case, include the unknown value of what Dwayne needs to score on his next test, in order to solve it.

$$\frac{78 + 92 + 83 + 97 + x}{5} = 90$$

Add the unknown value to the new average total, which is 5. Then multiply each side by 5 to simplify the equation, resulting in:

$$78 + 92 + 83 + 97 + x = 450$$

$$350 + x = 450$$

$$x = 100$$

Dwayne would need to get a perfect score of 100 in order to get an average of at least 90.

Test this answer by substituting back into the original formula:

$$\frac{78 + 92 + 83 + 97 + 100}{5} = 90$$

9. D: 1, 2, 3, 4, 6, 12. A given number divides evenly by each of its factors to produce an integer (no decimals). The number 5, 7, 8, 9, 10, 11 (and their opposites) do not divide evenly into 12. Therefore, these numbers are not factors.

10. E: The augmented matrix that represents the system of equations has dimensions 4×3 because there are three equations with three unknowns. The coefficients of the variables make up the first three columns, and the last column is made up of the numbers to the right of the equal sign. This system can be solved by reducing the matrix to row-echelon form, where the last column gives the solution for the unknown variables.

11. B: There are two zeros for the function: $x = 0, -2$.

The zeros can be found several ways, but this particular equation can be factored into:

$$f(x) = x(x^2 + 4x + 4) = x(x + 2)(x + 2)$$

By setting each factor equal to zero and solving for x, there are two solutions. On a graph these zeros can be seen where the line crosses the x-axis.

12. D: This problem involves a composition function, where one function is plugged into the other function. In this case, the $f(x)$ function is plugged into the $g(x)$ function for each x-value. The composition equation becomes:

$$g(f(x)) = 2^3 - 3(2^2) - 2(2) + 6$$

Simplifying the equation gives the answer:

$$g(f(x)) = 8 - 3(4) - 2(2) + 6 = 8 - 12 - 4 + 6 = -2$$

13. D: This system of equations involves one quadratic function and one linear function, as seen from the degree of each equation. One way to solve this is through substitution. Solving for y in the second equation yields $y = x + 2$. Plugging this equation in for the y of the quadratic equation yields:

$$x^2 - 2x + x + 2 = 8$$

Simplifying the equation, it becomes:

$$x^2 - x + 2 = 8$$

Setting this equal to zero and factoring, it becomes:

$$x^2 - x - 6 = 0 = (x - 3)(x + 2)$$

Solving these two factors for x gives the zeros $x = 3, -2$. To find the y-value for the point, each number can be plugged in to either original equation. Solving each one for y yields the points $(3, 5)$ and $(-2, 0)$.

14. D: The expression is simplified by collecting like terms. Terms with the same variable and exponent are like terms, and their coefficients can be added.

15. A: Finding the product means distributing one polynomial onto the other. Each term in the first must be multiplied by each term in the second. Then, like terms can be collected. Multiplying the factors yields the expression:

$$20x^3 + 4x^2 + 24x - 40x^2 - 8x - 48$$

Collecting like terms means adding the x^2 terms and adding the x terms. The final answer after simplifying the expression is:

$$20x^3 - 36x^2 + 16x - 48$$

16. E: Finding the zeros for a function by factoring is done by setting the equation equal to zero, then completely factoring. Since there was a common x for each term in the provided equation, that would be factored out first. Then the quadratic that was left could be factored into two binomials, which are $(x + 1)(x - 4)$. Setting each factor equal to zero and solving for x yields three zeros.

17. A: 13 nurses

Using the given information of 1 nurse to 25 patients and 325 patients, set up an equation to solve for number of nurses (N):

$$\frac{N}{325} = \frac{1}{25}$$

Multiply both sides by 325 to get N by itself on one side.

$$\frac{N}{1} = \frac{325}{25} = 13\ nurses$$

18. E: The equation can be solved by factoring the numerator into $(x + 6)(x - 5)$.

Since that same factor exists on top and bottom, that factor $(x - 5)$ cancels.

This leaves the equation $x + 6 = 11$.

Solving the equation gives the answer $x = 5$. When this value is plugged into the equation, it yields a zero in the denominator of the fraction. Since this is undefined, there is no solution.

19. D: This problem can be solved by using unit conversion. The initial units are miles per minute. The final units need to be feet per second. Converting miles to feet uses the equivalence statement 1 mile = 5,280 feet. Converting minutes to seconds uses the equivalence statement 1 minute = 60 seconds. Setting up the ratios to convert the units is shown in the following equation:

$$\frac{72\ miles}{90\ minutes} * \frac{1\ minute}{60\ seconds} * \frac{5280\ feet}{1\ mile} = 70.4 \text{ feet per second.}$$

The initial units cancel out, and the new units are left.

20. B: The formula can be manipulated by dividing both the length, *l*, and the width, *w*, on both sides. The length and width will cancel on the right, leaving height by itself.

21. B: The domain is all possible input values, or *x*-values. For this equation, the domain is every number greater than or equal to zero. There are no negative numbers in the domain because taking the square root of a negative number results in an imaginary number.

22. D: This problem can be solved by setting up a proportion involving the given information and the unknown value. The proportion is:

$$\frac{21\ pages}{4\ nights} = \frac{140\ pages}{x\ nights}$$

Solving the proportion by cross-multiplying, the equation becomes $21x = 4 * 140$, where $x = 26.67$.

Since it is not an exact number of nights, the answer is rounded up to 27 nights. Twenty-six nights would not give Sarah enough time.

23. D: The slope from this equation is 50, and it is interpreted as the cost per gigabyte used. Since the *g*-value represents number of gigabytes and the equation is set equal to the cost in dollars, the slope relates these two values. For every gigabyte used on the phone, the bill goes up 50 dollars.

24. C: Graphing the function $y = \cos(x)$ shows that the curve starts at $(0, 1)$, has an amplitude of 2, and a period of 2π. This same curve can be constructed using the sine graph, by shifting the graph to the left $\frac{\pi}{2}$ units. This equation is in the form:

$$y = \sin\left(x + \frac{\pi}{2}\right)$$

25. B: Start by squaring both sides to get $1 + x = 16$. Then subtract 1 from both sides to get $x = 15$.

No Calculator Questions

26. A: This inverse of a function is found by switching the x and y in the equation and solving for y. In the given equation, solving for y is done by adding 5 to both sides, then dividing both sides by 3. This answer can be checked on the graph by verifying the lines are reflected over $y = x$.

27. B: The zeros of this function can be found by using the quadratic formula:

$$x = \frac{-b \pm \sqrt{b^2 - 4ac}}{2a}$$

Identifying a, b, and c can also be done from the equation because it is in standard form. The formula becomes:

$$x = \frac{0 \pm \sqrt{0^2 - 4(1)(4)}}{2(1)} = \frac{\sqrt{-16}}{2}$$

Since there is a negative underneath the radical, the answer is a complex number:

$$x = \pm 2i$$

28. E: Setting up a proportion is the easiest way to represent this situation. The proportion becomes $\frac{20}{x} = \frac{40}{100}$, where cross-multiplication can be used to solve for x. The answer can also be found by observing the two fractions as equivalent, knowing that twenty is half of forty, and fifty is half of one hundred.

29. C: Division with scientific notation can be solved by grouping the first terms together and grouping the tens together. The first terms can be divided, and the tens terms can be simplified using the rules for exponents. The initial expression becomes $0.4 * 10^4$. This is not in scientific notation because the first number is not between 1 and 10. Shifting the decimal and subtracting one from the exponent yields $4.0 * 10^3$.

30. A: The relative error can be found by finding the absolute error and making it a percent of the true value. The absolute error is $36 - 35.75 = 0.25$. This error is then divided by 36—the true value—to find 0.7%.

31. B: The y-intercept of an equation is found where the x-value is zero. Plugging zero into the equation for x allows the first two terms to cancel out, leaving -4.

32. A: The equation is *even* because:

$$f(-x) = f(x)$$

Plugging in a negative value will result in the same answer as when plugging in the positive of that same value. The function:

$$f(-2) = \frac{1}{2}(-2)^4 + 2(-2)^2 - 6 = 8 + 8 - 6 = 10$$

yields the same value as:

$$f(2) = \frac{1}{2}(2)^4 + 2(2)^2 - 6 = 8 + 8 - 6 = 10$$

33. C: The equation $x = 3$ is not a function because it does not pass the vertical line test. This test is made from the definition of a function, where each x-value must be mapped to one, and only one, y-value. This equation is a vertical line, so the x-value of 3 is mapped with an infinite number of y-values.

34. A: The function can be factored to identify the zeros. First, the term $3x$ is factored out to the front because each term contains $3x$. Then, the quadratic is factored into $(x + 3)(x - 2)$.

35. C: A die has an equal chance for each outcome. Since it has six sides, each outcome has a probability of $\frac{1}{6}$. The chance of a 1 or a 2 is therefore $\frac{1}{6} + \frac{1}{6} = \frac{1}{3}$.

36. A: The slope is given by:

$$m = \frac{y_2 - y_1}{x_2 - x_1} = \frac{0 - 4}{0 - (-3)} = -\frac{4}{3}$$

37. B: The table shows values that are increasing exponentially. The differences between the inputs are the same, while the differences in the outputs are changing by a factor of 2. The values in the table can be modeled by the equation $f(x) = 2^x$.

38. C: The formula for continually compounded interest is $A = Pe^{rt}$. Plugging in the given values to find the total amount in the account yields the equation:

$$A = 2000e^{0.05*8} = 2983.65$$

39. E: The sample space is made up of $8 + 7 + 6 + 5 = 26$ balls. The probability of pulling each individual ball is $^1/_{26}$. Since there are 7 yellow balls, the probability of pulling a yellow ball is $^7/_{26}$.

40. B: For the first card drawn, the probability of a King being pulled is $\frac{4}{52}$. Since this card isn't replaced, if a King is drawn first the probability of a King being drawn second is $\frac{3}{51}$. The probability of a King being drawn in both the first and second draw is the product of the two probabilities: $\frac{4}{52} \times \frac{3}{51} = \frac{12}{2652}$. This fraction, when divided by 12, equals $\frac{1}{221}$.

		3	8	0

41. 380 miles. To find the total driving time, the total distance of 350 miles can be divided by the constant speed of 60 miles per hour. This yields a time of 5.8333 hours, which is then rounded. Once the driving time is computed, the break times need to be found. If Gary takes a break for 10 minutes every 100 miles, he will take 3 breaks on his trip. This will yield a total of 30 minutes of break time. Since the answer is needed in minutes, 5.8333 can be converted to minutes by multiplying by 60, giving a driving time of 350 minutes. Adding the break time of 30 minutes to the driving time of 350 minutes gives a total travel time of 380 minutes.

			6	6

42. 66 Cookies. If the sixth graders bought 45% of the cookies, the number they bought is found by multiplying 0.45 by 500. They bought 225 cookies. The number of cookies left is:

$$500 - 225 = 275$$

During the second lunch, the seventh graders bought 40% of the cookies, which is found by multiplying 0.40 by the remaining 275 cookies. The seventh graders bought 110 cookies. This leaves 165 cookies to sell to the eighth graders. If they bought 60% of the remaining cookies, then they bought 99 cookies. Subtracting 99 from 165 cookies leaves Kelly with 66 cookies remaining after the three lunches.

			2	0

43. 20. Because these are 2 parallel lines cut by a transversal, the angle with a measure of 45 degrees is equal to the measure of angle 6. Angle 6 and the angle labeled $5x + 35$ are supplementary to one another. The sum of these angles should be 180, so the following equation can be generated:

$$5x + 35 + 45 = 180$$

Solving for x, the sum of 35 and 45 is 80, which is then subtracted from 180 to yield 100. Dividing 100 by 5 gives the value of x, which is 20.

			1	0

44. 10. 8 squared is 64, and 6 squared is 36. These should be added together to get:

$$64 + 36 = 100$$

Then, the last step is to find the square root of 100 which is 10.

			2	4

45. 24; The long division would be completed as follows:

```
    24
36|864
  - 72↓
   144
```

			1	8
-	-	-	-	-
	.	.	.	.
	0	0	0	0
	1	1	1	1
	2	2	2	2
	3	3	3	3
	4	4	4	4
	5	5	5	5
	6	6	6	6
	7	7	7	7
	8	8	8	8
	9	9	9	9

46. 18; If Ray will be 25 in three years, then he is currently 22. The problem states that Lisa is 13 years younger than Ray, so she must be 9. Sam's age is twice that, which means that the correct answer is 18.

	5	0	.	7
-	-	-	-	-
	.	.	.	.
	0	0	0	0
	1	1	1	1
	2	2	2	2
	3	3	3	3
	4	4	4	4
	5	5	5	5
	6	6	6	6
	7	7	7	7
	8	8	8	8
	9	9	9	9

47. 50.7; The values for the missing sides must first be found before the perimeter can be calculated. The missing side that is the hypotenuse of the right triangle can be calculated using the Pythagorean Theorem as follows:

$$11^2 + 4^2 = x^2$$

$$121 + 16 = x^2$$

$$137 = x^2$$

$$x = 11.7$$

The other missing side is equal to the value of the length of the larger rectangle less than the value of the side of the square $12 - 4 = 8$. Then, all the sides can be added together to find the perimeter:

$$12 + 6 + 8 + 5 + 4 + 4 + 11.7 = 50.7$$

			1	5
-	-	-	-	-
	.	.	.	.
	0	0	0	0
	1	1	1	1
	2	2	2	2
	3	3	3	3
	4	4	4	4
	5	5	5	5
	6	6	6	6
	7	7	7	7
	8	8	8	8
	9	9	9	9

48. 15; Follow the *order of operations* in order to solve this problem. Solve the parentheses first, and then follow the remainder as usual.

$$(6 \times 4) - 9$$

This equals $24 - 9$, or 15.

	8	7	.	5

49. 87.5; For an even number of total values, the *median* is calculated by finding the *mean* or average of the two middle values once all values have been arranged in ascending order from least to greatest. In this case, $(92 + 83) \div 2$ would equal the median 87.5.

				6

50. 6; The formula for the perimeter of a rectangle is $P = 2L + 2W$, where P is the perimeter, L is the length, and W is the width. The first step is to substitute all of the data into the formula:

$$36 = 2(12) + 2W$$

Simplify by multiplying 2×12:

$$36 = 24 + 2W$$

Simplifying this further by subtracting 24 on each side, which gives:

$$36 - 24 = 24 - 24 + 2W$$

$$12 = 2W$$

Divide by 2:

$$6 = W$$

The width is 6 cm. Remember to test this answer by substituting this value into the original formula:

$$36 = 2(12) + 2(6)$$

Reading

1. B: Mr. Button's wife is about to have a baby. The passage begins by giving the reader information about traditional birthing situations. Then, we are told that Mr. and Mrs. Button decide to go against tradition to have their baby in a hospital. The next few passages are dedicated to letting the reader know how Mr. Button dresses and goes to the hospital to welcome his new baby. There is a doctor in this excerpt, as Choice *C* indicates, and Mr. Button does put on clothes, as Choice *D* indicates. However, Mr. Button is not going to the doctor's office nor is he about to go shopping for new clothes. Choice *E* is incorrect; the passage tells us that Mr. Button is already president of his company.

2. E: The tone of the above passage is nervous and excited. We are told in the fourth paragraph that Mr. Button "arose nervously." We also see him running without caution to the doctor to find out about his wife and baby—this indicates his excitement. We also see him stuttering in a nervous yet excited fashion as he asks the doctor if it's a boy or girl.

3. C: Dedicated. Mr. Button is dedicated to the task before him. Choice *A*, numbed, Choice *B*, chained, Choice *D*, moved, and Choice *E*, exonerated, all could grammatically fit in the sentence. However, they are not synonyms with *consecrated* like Choice *C* is.

4. D: Giving readers a visual picture of what the doctor is doing. The author describes a visual image—the doctor rubbing his hands together—first and foremost. The author may be trying to make a comment about the profession; however, the author does not "explain the detail of the doctor's profession" as Choice *B* suggests.

5. D: To introduce the setting of the story and its characters. We know we are being introduced to the setting because we are given the year in the very first paragraph along with the season: "one day in the summer of 1860." This is a classic structure of an introduction of the setting. We are also getting a long explanation of Mr. Button, what his work is, who is related to him, and what his life is like in the third paragraph.

6. B: The narrator offers a metaphor of the taxi driver as a fisherman. The driver is watching the merging of traffic as a fisherman would watch the merging, or "confluence," of two river currents. The other choices are depictions of actions that aren't represented in the passage.

7. E: In relation to the last two paragraphs, the rest of the passage serves primarily to highlight the accuracy and depth of Aomame's knowledge, specifically of music and history. We see Aomame in the first paragraph, "settled…back," listening to the music while in a taxi. The second, third, and fourth paragraphs depict her detailed knowledge of Janáček's *Sinfonietta*, Czechoslovakia during the First World War, and then Japan around 1926. The last two paragraphs are more about Aomame's personality and family life.

8. B: The narrator's perspective is that of an intrigued storyteller. The passage seems to be an excerpt from a narrative story, and the narrator is certainly intrigued by the main character, Aomame. The narrator describes her actions and thoughts—going into detail about what she loves (history and sport) and her personality and background.

9. A: The statement serves as imagery to symbolize the blissful ignorance of the upcoming World War II in Eastern Europe, as the imagery is followed with Aomame thinking "about the vicissitudes of history," or the downturn of events that World War II brought with it. Using this same imagery, one could say the "carefree winds" turned into a raging storm. The figurative language of simile, synecdoche, rhetorical question, and personification are not represented in the statement.

10. C: The setting of the story is located in Japan. Although Czechoslovakia (former Czech Republic) and Germany are mentioned, we have one context clue that gives us information as to the setting. In the fourth paragraph about Japan, the narrator gives us this statement: "It was the beginning of a terrible, dark time in this country, too." The words "this country" tells us that Japan is the country in which the narrator is telling the story.

11. B: The first paragraph explains the difficulties facing the group of travelers. Language and making a living are listed in the second sentence, so Choices *A* and *C* are both incorrect. Extra fees and exorbitant rates are described in the first paragraph's final sentence; therefore, Choice *D* is incorrect. Although the men were later separated from their wives and children, that separation was due to a problem with logistics. There was no prohibition specifically against bringing their wives and children. The entire group was legally barred from leaving the country. Choice *E* is incorrect; we can see how the travelers were treated during their travels, especially in the second paragraph. Thus, Choice *B* is the correct answer.

12. A: We can see that this text records parts of history, so this is most likely a historical memoir or the author's personal journal retelling an event that happened. We see a brief summary of the event before the passage begins.

13. D: The ship hired at Boston in Lincoln-shire is the group's first attempt to flee their homeland, and it's described in the second paragraph. After the group reached an agreement with the shipmaster, he betrayed them to the government, which led to government officials arresting the group. The women didn't become sick or get separated until their second attempt at fleeing, so Choices *A* and *B* are both incorrect. Following the first shipmaster's betrayal, the group did go to jail, but their imprisonment occurred after their first attempt to leave went wrong; therefore, Choice *C* is incorrect. There is no shipwreck in this passage, so Choice *E* is incorrect. Thus, Choice *D* is the correct answer.

14. A: After the men boarded the ship, armed government officials arrived on the shore. Rather than risk everyone getting caught, the Dutch shipmaster raised the anchor, hoisted the sails, and took off. So, even though the men were moved to tears by the separation, there was nothing they could do. The government refused to let some of the group's member out of prison, but that's not what caused the separation; therefore, Choice *B* is incorrect. Choices *C, D,* and *E* are factually incorrect based on the information contained in the third paragraph. Thus, Choice *A* is the correct answer.

15. C: Ships charged the group of travelers extra fees because the government outlawed their departure. The first paragraph says that the government barred them from the ports and havens, so they needed to find a secret passage out of the country. As helping the group was illegal, ships charged them extra fees in exchange for taking on that risk. The passage doesn't state why the group is being persecuted; therefore, Choice *A* is incorrect. Choice *B* is incorrect because the passage doesn't mention whether the voyage was particularly long or dangerous. While the government does seize property from the group, it's unclear whether the group is particularly wealthy; therefore, Choice *D* is incorrect. Choice *E* is incorrect because nothing is mentioned about the group being rowdy. Thus, Choice *C* is the correct answer.

16. B: The speaker is referring to the current war, Choice *B*. The lines say: "I set and sew—my heart aches with desire— / That pageant terrible, that fiercely pouring fire / On wasted fields, and writhing grotesque things / Once men" (lines 8–11). Pageant is a metaphor for the battlefield in this poem. After she mentions the word "pageant," the speaker then goes on to describe the "pageant," or war, by describing the fire, fields, and dying men.

17. C: The speaker is referring to soldiers in these lines, so Choice *C* is correct. If we look at the surrounding context clues, the word "war" is mentioned before the description of these men. The word "tred" refers to a sort of weary marching.

18. E: The speaker effectively contrasts the idea between activeness and passiveness in the sense that the speaker views sewing as something passive and longs to do something active in order to help out in the war. Let's look at the poem for proof of this contrast. In the first stanza, the dull act of sewing is set in contrast with the horror that happens in her dreams. In the second stanza, the massacre in the war is contrasted by her, again "sitting and sewing" and yearning to help. Finally, in the third stanza, her "useless seam" is contrasted with the soldiers dying in the mud. The stark contrast of the comforts and ease of sewing to the pain and suffering of soldiers in active duty is a central theme in this poem.

19. C: Poetic lines that can be seen in this poem are rhyming couplets. Rhyming couplets are two pairs of lines that end with rhyming words. We see this all the way through, except for the last line in each stanza, which ends with the word "sew." Here are some examples of the end rhymes of the rhyming couplets: "seems" and "dreams," "things" and "flings," "rain" and "slain."

20. C: A stanza is a group of lines that make up a repeating metrical unit of a poem. We see that there are three stanzas, and each stanza consists of seven lines each. The seven lines have three rhyming couplets (six lines) and end with a seventh line that has no end rhyme. However, the last word in every unit is "sew," which makes a cohesive repetition in the poem.

21. B: Readers should carefully focus their attention on the beginning of the passage to answer this series of questions. Even though the sentences may be worded a bit differently, all but one statement is true. It presents a false idea that descriptive details are not necessary until the climax of the story. Even if one does not read the passage, he or she probably knows that all good writing begins with descriptive details to develop the main theme the writer intends for the narrative.

22. C: This choice allows room for the fact that not all people who attempt to write a play will find it easy. If the writer follows the basic principles of narrative writing described in the passage, however, writing a play does not have to be an excruciating experience. None of the other options can be supported by points from the passage.

23. A: Choice *A* is true based on the sentence that reads, "Next, use dialogue to reveal the attitudes and personalities of each of the characters who have a key part in the unfolding story." Choice *B* is false because there is no drama with predictable progression. Choice *C* is incorrect based on the information that claims an image is like using a thousand words. Choice *D* contradicts the point that the protagonist should experience self-discovery. Finally, Choice *E* is incorrect because all drama suggests some challenge for the characters to experience.

24. B: To suggest that a ten-minute play is accessible does not imply any timeline, nor does the passage mention how long a playwright spends with revisions and rewrites. So, Choice *A* is incorrect. Choice *B* is correct because of the opening statement that reads, "Learning how to write a ten-minute play may seem like a monumental task at first; but, if you follow a simple creative writing strategy, similar to writing a narrative story, you will be able to write a successful drama." None of the remaining choices are supported by points in the passage.

25. E: Note that the only element not mentioned in the passage is the style feature that is part of a narrative writer's tool kit. It is not to say that ten-minute plays do not have style. The correct answer denotes only that the element of style was not illustrated in this particular passage.

26. C: The author contrasts two different viewpoints, then builds a case showing preference for one over the other. Choice *A* is incorrect because the introduction does not contain an impartial definition, but rather another's opinion. Choice *B* is incorrect. There is no puzzling phenomenon given, as the author doesn't mention any peculiar cause or effect that is in question regarding poetry. Choice *D* does contain another's viewpoint at the beginning of the passage; however, to say that the author has no stake in this argument is incorrect; the author uses personal experiences to build their case. Finally, Choice *E* is incorrect because there is no description of the history of poetry offered within the passage.

27. B: Choice *B* accurately describes the author's argument in the text—that poetry is not irrelevant. While the author does praise—and even value—Buddy Wakefield as a poet, he or she never heralds Wakefield as a genius. Eliminate Choice *A*, as it is an exaggeration. Not only is Choice *C* an exaggerated statement, but the author never mentions spoken word poetry in the text. Choice *D* is wrong because this statement contradicts the writer's argument. Choice *E* can also be eliminated, because the author mentions how performance actually *enhances* poetry and that modern technology is one way poetry remains vital.

28. D: *Exiguously* means not occurring often, or occurring rarely, so Choice *D* would LEAST change the meaning of the sentence. Choice *A*, *indolently*, means unhurriedly, or slow, and does not fit the context of the sentence. Choice *B*, *inaudibly*, means quietly or silently. Choice *C*, *interminably*, means endlessly, or all the time, and is the opposite of the word *exiguously*. Choice *E*, *impecunious,* means impoverished or destitute, and does not fit within the context of the sentence.

29. E: The author of the passage tries to insist that performance poetry is a subset of modern poetry, and therefore prove that modern poetry is not "dying," but thriving on social media for the masses. Choice *A* is incorrect, as the author is not refusing any kind of validation. Choice *B* is incorrect; the author's insistence is that poetry will *not* lose popularity. Choice *C* mimics the topic but compares two different genres, while the author makes no comparison in this passage. Choice *D* is incorrect as well; again, there is no cause or effect the author is trying to prove.

30. B: The author's purpose is to disprove Gioia's article claiming that poetry is a dying art form that only survives in academic settings. In order to prove his argument, the author educates the reader about new developments in poetry (Choice *A*) and describes the brilliance of a specific modern poet (Choice *C*), but these are used to serve as examples of a growing poetry trend that counters Gioia's argument. Choice *D* is incorrect because it contradicts the author's argument. Choice *E* is incorrect because the passage uses the performance as a way to convey the author's point; it's not the focus of the piece. It's also unclear if the author was actually present at the live performance.

31. C: Because the details in Choice *A* and Choice *B* are examples of how an emotionally intelligent leader operates, they are not the best choice for the definition of the term *emotional intelligence*. They are qualities observed in an EI leader. Choice *C* is true as noted in the second sentence of the passage: Emotional Intelligence (EI) includes developing the ability to know one's own emotions, to regulate impulses and emotions, and to use interpersonal communication skills with ease while dealing with other people. It makes sense that someone with well-developed emotional intelligence will have a good handle on understanding their emotions, be able to regulate impulses and emotions, and use interpersonal communication skills. Choice *D* is not a definition of EI. Choice *E* is the opposite of the definition of EI, so both Choice *D* and Choice *E* are incorrect.

32. C: Choice *E* can be eliminated immediately because of the signal word "obviously." Choice *A* can be eliminated because it does not reflect an accurate fact. Choices *B* and *D* do not support claims about how to be a successful leader.

33. E: The qualities of an unsuccessful leader possessing a transactional leadership style are listed in the passage. Choices *A* and *B* are incorrect because these options reflect the qualities of a successful leader. Choices *C* and *D* are definitely not characteristics of a successful leader; however, they are not presented in the passage and readers should do their best to ignore such options.

34. D: Even though some choices may be true of successful leaders, the best answer must be supported by sub-points in the passage. Therefore, Choices *A* and *C* are incorrect. Choice *B* is incorrect because uncompromising transactional leadership styles squelch success. Choice *E* is never mentioned in the passage.

35. E: To support Choice *E*, the idea that a leader can develop emotional intelligence if desired, the passage says, "There are ways to develop emotional intelligence for the person who wants to improve his or her leadership style."

36. D: In this passage, Juliet says, "that which we call a rose by any other name would smell as sweet," meaning that regardless of Romeo's name, he is still Romeo. She would still love him and see him as perfection, no matter his title. She isn't saying that she is special here or that Romeo shouldn't love her. While the titles of Capulet and Montague are what keep them apart, she is not saying that here. Her main point is that their names are meaningless when it comes to their love.

37. C: Romeo is comparing Juliet's beauty to the sun. He is essentially saying that the moon pales in comparison to Juliet's beauty and should be envious of how she shines. The sun is rising and shining, but that is not why the moon is envious. He says the moon is sick with grief because Juliet is so much more beautiful. While the moon is paler than the sun, this doesn't best describe Romeo's intention with this statement.

38. A: Romeo has seen Juliet and is desperate to be with her. Lines like "it is my love O that she knew she were" and "That I might touch that cheek!" indicate that he desires to touch and be with Juliet. He does not demonstrate any sadness or remorsefulness in these lines. While he may respect Juliet, he is more enamored with her than respectful in his tone. He is neither tentative nor hesitant, as he is describing her beauty with desire and love.

39. C: Romeo compares Juliet to a winged messenger of heaven because he hears her voice over his head. This, along with the set direction, suggests that she is above him on the balcony, and he is in the garden. The other lines do not suggest anything about Romeo's proximity to Juliet in the play.

40. B: Romeo uses the word *as* to compare Juliet to the winged messenger of heaven. This comparison using *like* or *as* is a simile. An example of a metaphor would be "Juliet is the sun" because it does not use the words *like* or *as*. An example of personification would be when Romeo describes the moon as envious. Imagery and theme are not demonstrated in this example.

Writing

1. C: Choice *C* correctly uses *from* to describe the fact that dogs are related to wolves. The word *through* is incorrectly used here, so Choice *A* is incorrect. Choice *B* makes no sense. Choice *D* unnecessarily changes the verb tense in addition to incorrectly using *through*.

2. B: Choice *B* is correct because the Oxford comma is applied, clearly separating the specific terms. Choice *A* lacks this clarity. Choice *C* is correct but too wordy since commas can be easily applied. Choice *D* doesn't flow with the sentence's structure.

3. D: Choice *D* correctly uses the question mark, fixing the sentence's main issue. Thus, Choice *A* is incorrect because questions do not end with periods. Choice *B*, although correctly written, changes the meaning of the original sentence. Choice *C* is incorrect because it completely changes the direction of the sentence, disrupts the flow of the paragraph, and lacks the crucial question mark.

4. A: Choice *A* is correct since there are no errors in the sentence. Choices *B* and *C* both have extraneous commas, disrupting the flow of the sentence. Choice *D* unnecessarily rearranges the sentence.

5. D: Choice *D* is correct because the commas serve to distinguish that *artificial selection* is just another term for *selective breeding* before the sentence continues. The structure is preserved, and the sentence can flow with more clarity. Choice *A* is incorrect because the sentence needs commas to avoid being a run-on. Choice *B* is close but still lacks the required comma after *selection*, so this is incorrect. Choice *C* is incorrect because the comma to set off the aside should be placed after *breeding* instead of *called*.

6. B: Choice *B* is correct because the sentence is talking about a continuing process. Therefore, the best modification is to add the word *to* in front of *increase*. Choice *A* is incorrect because this modifier is missing. Choice *C* is incorrect because with the additional comma, the present tense of *increase* is inappropriate. Choice *D* makes more sense, but the tense is still not the best to use.

7. A: The sentence has no errors, so Choice *A* is correct. Choice *B* is incorrect because it adds an unnecessary comma. Choice *C* is incorrect because *advantage* should not be plural in this sentence without the removal of the singular *an*. Choice *D* is very tempting. While this would make the sentence more concise, this would ultimately alter the context of the sentence, which would be incorrect.

8. C: Choice *C* correctly uses *on to*, describing the way genes are passed generationally. The use of *into* is inappropriate for this context, which makes Choice *A* incorrect. Choice *B* is close, but *onto* refers to something being placed on a surface. Choice *D* doesn't make logical sense.

9. D: Choice *D* is correct, since only proper names should be capitalized. Because the name of a dog breed is not a proper name, Choice *A* is incorrect. In terms of punctuation, only one comma after *example* is needed, so Choices *B* and *C* are incorrect.

10. D: Choice *D* is the correct answer because "rather" acts as an interrupting word here and thus should be separated by commas. Choices B and C use commas unwisely, breaking the flow of the sentence.

11. B: Since the sentence can stand on its own without *Usually*, separating it from the rest of the sentence with a comma is correct. Choice *A* needs the comma after *Usually*, while Choice *C* uses commas incorrectly. Choice *D* is tempting but changing *turn* to past tense goes against the rest of the paragraph.

12. A: In Choice *A*, the dependent clause *Sometimes in particularly dull seminars* is seamlessly attached with a single comma after *seminars*. Choice *B* contain too many commas. Choice *C* does not correctly combine the dependent clause with the independent clause. Choice *D* introduces too many unnecessary commas.

13. D: Choice *D* rearranges the sentence to be more direct and straightforward, so it is correct. Choice *A* needs a comma after *on*. Choice *B* introduces unnecessary commas. Choice *C* creates an incomplete sentence, since *Because I wasn't invested in what was going on* is a dependent clause.

14. C: Choice *C* is fluid and direct, making it the best revision. Choice *A* is incorrect because the construction is awkward and lacks parallel structure. Choice *B* is clearly incorrect because of the unnecessary comma and period. Choice *D* is close, but its sequence is still awkward and overly complicated.

15. B: Choice *B* correctly adds a comma after *person* and cuts out the extraneous writing, making the sentence more streamlined. Choice *A* is poorly constructed, lacking proper grammar to connect the sections of the sentence correctly. Choice *C* inserts an unnecessary semicolon and doesn't enable this section to flow well with the rest of the sentence. Choice *D* is better but still unnecessarily long.

16. D: This sentence, though short, is a complete sentence. The only thing the sentence needs is an em-dash after "Easy." In this sentence the em-dash works to add emphasis to the word "Easy" and also acts in place of a colon, but in a less formal way. Therefore, Choice *D* is correct. Choices *A* and *B* lack the crucial comma, while Choice *C* unnecessarily breaks the sentence apart.

17. C: Choice *C* successfully fixes the construction of the sentence, changing *drawing* into *to draw*. Keeping the original sentence disrupts the flow, so Choice *A* is incorrect. Choice *B*'s use of *which* offsets the whole sentence. Choice *D* is incorrect because it unnecessarily expands the sentence content and makes it more confusing.

18. B: Choice *B* fixes the homophone issue. Because the author is talking about people, *their* must be used instead of *there*. This revision also appropriately uses the Oxford comma, separating and distinguishing *lives, world, and future*. Choice *A* uses the wrong homophone and is missing commas. Choice *C* neglects to fix these problems and unnecessarily changes the tense of *applies*. Choice *D* fixes the homophone but fails to properly separate *world* and *future*.

19. C: Choice *C* is correct because it fixes the core issue with this sentence: the singular *has* should not describe the plural *scientists*. Thus, Choice *A* is incorrect. Choices *B* and *D* add unnecessary commas.

20. D: Choice *D* correctly conveys the writer's intention of asking if, or *whether*, early perceptions of dinosaurs are still influencing people. Choice *A* makes no sense as worded. Choice *B* is better, but *how* doesn't coincide with the context. Choice *C* adds unnecessary commas.

21. A: Choice *A* is correct, as the sentence does not require modification. Choices *B* and *C* implement extra punctuation unnecessarily, disrupting the flow of the sentence. Choice *D* incorrectly adds a comma in an awkward location.

22. B: Choice *B* is the strongest revision, as adding *to explore* is very effective in both shortening the sentence and maintaining, even enhancing, the point of the writer. To explore is to seek understanding in order to gain knowledge and insight, which coincides with the focus of the overall sentence. Choice *A* is not technically incorrect, but it is overcomplicated. Choice *C* is a decent revision, but the sentence could still be more condensed and sharpened. Choice *D* fails to make the sentence more concise and inserts unnecessary commas.

23. D: Choice *D* correctly applies a semicolon to introduce a new line of thought while remaining in a single sentence. The comma after *however* is also appropriately placed. Choice *A* is a run-on sentence. Choice *B* is also incorrect because the single comma is not enough to fix the sentence. Choice *C* adds commas around *uncertain* which are unnecessary.

24. B: Choice *B* not only fixes the homophone issue from *its*, which is possessive, to *it's*, which is a contraction of *it is*, but also streamlines the sentence by adding a comma and eliminating *and*. Choice *A* is incorrect because of these errors. Choices *C* and *D* only fix the homophone issue.

25. A: Choice *A* is correct, as the sentence is fine the way it is. Choices *B* and *C* add unnecessary commas, while Choice *D* uses the possessive *its* instead of the contraction *it's*.

26. C: Choice *C* is correct because the phrase *even likely* is flanked by commas, creating a kind of aside, which allows the reader to see this separate thought while acknowledging it as part of the overall sentence and subject at hand. Choice *A* is incorrect because it seems to ramble after *even* due to a missing comma after *likely*. Choice *B* is better but inserting a comma after *that* warps the flow of the writing. Choice *D* is incorrect because there must be a comma after *plausible*.

27. D: Choice *D* strengthens the overall sentence structure while condensing the words. This makes the subject of the sentence, and the emphasis of the writer, much clearer to the reader. Thus, while Choice *A* is technically correct, the language is choppy and over-complicated. Choice *B* is better but lacks the reference to a specific image of dinosaurs. Choice *C* introduces unnecessary commas.

28. B: Choice B correctly joins the two independent clauses. Choice A is decent, but "that would be" is too verbose for the sentence. Choice *C* incorrectly changes the semicolon to a comma. Choice *D* splits the clauses effectively but is not concise enough.

29. A: Choice *A* is correct, as the original sentence has no error. Choices *B* and *C* employ unnecessary semicolons and commas. Choice *D* would be an ideal revision, but it lacks the comma after *Ransom* that would enable the sentence structure to flow.

30. D: By reorganizing the sentence, the context becomes clearer with Choice *D*. Choice *A* has an awkward sentence structure. Choice *B* offers a revision that doesn't correspond well with the original sentence's intent. Choice *C* cuts out too much of the original content, losing the full meaning.

31. C: Choice *C* fixes the disagreement between the singular *this* and the plural *viewpoints*. Choice *A*, therefore, is incorrect. Choice *B* introduces an unnecessary comma. In Choice *D*, *those* agrees with *viewpoints*, but neither agrees with *distinguishes*.

32. A: Choice *A* is direct and clear, without any punctuation errors. Choice *B* is well-written but too wordy. Choice *C* adds an unnecessary comma. Choice *D* is also well-written but much less concise than Choice *A*.

33. D: Choice *D* rearranges the sentence to improve clarity and impact, with *tempting* directly describing *idea*. On its own, Choice *A* is a run-on. Choice *B* is better because it separates the clauses, but it keeps an unnecessary comma. Choice *C* is also an improvement but still a run-on.

34. B: Choice *B* is the best answer simply because the sentence makes it clear that Un-man takes over and possesses Weston. In Choice *A*, these events sounded like two different things, instead of an action and result. Choices *C* and *D* make this relationship clearer, but the revisions don't flow very well grammatically.

35. D: Changing the phrase *after this* to *then* makes the sentence less complicated and captures the writer's intent, making Choice *D* correct. Choice *A* is awkwardly constructed. Choices *B* and *C* misuse their commas and do not adequately improve the clarity.

36. B: By starting a new sentence, the run-on issue is eliminated, and a new line of reasoning can be seamlessly introduced, making Choice *B* correct. Choice *A* is thus incorrect. While Choice *C* fixes the run-on via a semicolon, a comma is still needed after *this*. Choice *D* contains a comma splice. The independent clauses must be separated by more than just a comma, even with the rearrangement of the second half of the sentence.

37. C: Choice *C* condenses the original sentence while being more active in communicating the emphasis on changing times/media that the author is going for, so it is correct. Choice *A* is clunky because it lacks a comma after *today* to successfully transition into the second half of the sentence. Choice *B* inserts unnecessary commas. Choice *D* is a good revision of the underlined section, but not only does it not fully capture the original meaning, it also does not flow into the rest of the sentence.

38. B: Choice *B* clearly illustrates the author's point, with a well-placed semicolon that breaks the sentence into clearer, more readable sections. Choice *A* lacks punctuation. Choice *C* is incorrect because the period inserted after *question* forms an incomplete sentence. Choice *D* is a very good revision but does not make the author's point clearer than the original.

39. A: Choice *A* is correct: while the sentence seems long, it actually doesn't require any commas. The conjunction "that" successfully combines the two parts of the sentence without the need for additional punctuation. Choices *B* and *C* insert commas unnecessarily, incorrectly breaking up the flow of the sentence. Choice *D* alters the meaning of the original text by creating a new sentence, which is only a fragment.

40. C: Choice *C* correctly replaces *for* with *to*, the correct preposition for the selected area. Choice *A* is not the answer because of this incorrect preposition. Choice *B* is unnecessarily long and disrupts the original sentence structure. Choice *D* is also too wordy and lacks parallel structure.

41. D: Choice *D* is the answer because it inserts the correct punctuation to fix the sentence, linking the dependent and independent clauses. Choice *A* is therefore incorrect. Choice *B* is also incorrect since this revision only adds content to the sentence while lacking grammatical precision. Choice *C* overdoes the punctuation; only a comma is needed, not a semicolon.

42. B: Choice *B* correctly separates the section into two sentences and changes the word order to make the second part clearer. Choice *A* is incorrect because it is a run-on. Choice *C* adds an extraneous comma, while Choice *D* makes the run-on worse and does not coincide with the overall structure of the sentence.

43. C: Choice *C* is the best answer because of how the commas are used to flank *in earnest*. This distinguishes the side thought (*in earnest*) from the rest of the sentence. Choice *A* needs punctuation. Choice *B* inserts a semicolon in a spot that doesn't make sense, resulting in a fragmented sentence and lost meaning. Choice *D* is unnecessarily elaborate and leads to a run-on.

44. A: Choice *A* is correct because the sentence contains no errors. The comma after *bias* successfully links the two halves of the sentence, and the use of *it's* is correct as a contraction of *it is*. Choice *B* creates a sentence fragment, while Choice *C* creates a run-on. Choice *D* incorrectly changes *it's* to *its*.

45. D: Choice *D* correctly inserts a comma after *However* and fixes *over use* to *overuse*—in this usage, it is one word. Choice *A* is therefore incorrect, as is Choice *B*. Choice *C* is a good revision but does not fit well with the rest of the sentence.

46. D: This passage is informative (*D*) because it is nonfiction and factual. The passage's intent is not to state an opinion, discuss an individual's life, or tell a story. Thus, the passage is not argumentative (*A*), biographical (*C*), or narrative (*B*).

47. D: To begin, *of* is not required here. *Apprenticeship* is also more appropriate in this context than *apprentice opportunities*, *apprentice* describes an individual in an apprenticeship, not an apprenticeship itself. Both of these changes are needed, making (*D*) the correct answer.

48. D: To begin, the selected sentence is a run-on, and displays confusing information. Thus, the sentence does need revision, making (*A*) wrong. The main objective of the selected section of the passage is to communicate that many positions (*positions* is a more suitable term than *offices,* as well) require a PE license, which is gained by scoring well on the FE and PE assessments. This must be the primary focus of the revision. It is necessary to break the sentence into two, to avoid a run-on. Choice *B* fixes the run-on aspect, but the sentence is indirect and awkward in construction. It takes too long to establish the importance of the PE license. Choice *C* is wrong for the same reason and it is a run on. Choice *D* is correct because it breaks the section into coherent sentences and emphasizes the main point the author is trying to communicate: the PE license is required for some higher positions, it's obtained by scoring well on the two standardized assessments, and college and experience can be used to prepare for the assessments in order to gain the certification.

49. C: Any time a writer wants to validate a claim, he or she ought to provide factual information that proves or supports that claim: "beginning his or her preparation for the profession in high school" supports the claim that aircraft engineers undergo years of education. For this reason, Choice *C* is the correct response. However, completing such courses in high school does not guarantee that aircraft engineers will earn generous salaries (*A*), become employed in executive positions (*B*), or stay employed (*D*).

50. B: Choice *B* is correct because "skill" is defined as having certain aptitude for a given task. (*C*) is incorrect because "work" does not directly denote "critical thinking, business skills, problem solving, and creativity." (*A*) is incorrect because the word "care" doesn't fit into the context of the passage, and (*D*), "composition," is incorrect because nothing in this statement points to the way in which something is structured.

Social Studies

1. D: Based on the context provided in the title, it can be inferred that Britain is the old lion and Canada is the young lion in the propaganda poster. Additionally, the title states that it was disseminated between 1914 and 1918. During World War I (1914–1918), Canadians fought alongside British forces because Canada was a dominion within the British Empire. Thus, Choice *D* is the correct answer. Both the American Revolutionary War (1775–1783) and War of 1812 (1812–1815) predated the propaganda poster by a century, so Choice *A* and Choice *C* are incorrect. Canada was an independent country during World War II (1939–1945) and the Cold War (1947–1991), so Britain wouldn't have sought to enlist Canadian soldiers. Therefore, Choice *B* and Choice *E* are both incorrect.

2. C: The graph illustrates a consistent decline in unionization rates after 1989. The United States and Canada entered into the Canada–United States Free Trade Agreement in 1988, and the North American Free Trade Agreement (NAFTA) superseded it in 1994. Free trade generated tremendous economic growth, but it also severely undermined Canadian labor rights since NAFTA effectively created a North American labor market. Thus, Choice *C* is the correct answer. The Soviet Union collapsed in the late 1980s, but its dissolution didn't have a direct impact on Canadian labor rights. Therefore, Choice *A* is incorrect. The United States committed itself to buying far more Canadian goods when it entered into NAFTA, so Choice *B* is incorrect. Automation has very arguably contributed to a decline in unionization rates, but Choice *D* is incorrect because automation has increased and boosted labor productivity since 1989. NAFTA undermined Canadian manufacturing but not at the same rate as other highly developed countries like the United States. Choice *C* states a more direct effect on labor, so Choice *E* is incorrect.

3. B: The table provides statistics related to trade between the United States and Canada, and both countries were signatories of the North American Free Trade Agreement (NAFTA). From 1994 until 2020, NAFTA lowered trade barriers between the United States, Canada, and Mexico. Thus, Choice *B* is the correct answer. The General Agreement on Tariffs and Trade (GATT) played a major role in promoting international free trade in the postwar era, but NAFTA had a far more direct impact on Canadian exports to the United States and American imports to Canada. Therefore, Choice *A* is incorrect. The United States and Canada are both founding members of the North Atlantic Treaty Organization (NATO); however, NATO is a military alliance, so Choice *C* is incorrect. The Western Hemisphere Travel Initiative is an American framework for border security; therefore, Choice *D* is incorrect. The World Trade Organization (WTO) replaced GATT in 1994, and the WTO has worked on several trade disputes between the United States and Canada. However, NAFTA had a more direct impact because it set the framework for all trade between the two countries, so Choice *E* is incorrect.

4. E: A trade balance is the difference between a country's exports and imports. According to the table, Canada had a positive trade balance with the United States in all six years, meaning it exported more Canadian goods than it imported American goods. The relevant trade balances for this question were: $15.5 billion in 2015 (Choice *A*), $11 billion in 2016 (Choice *B*), $16.3 billion in 2017 (Choice *C*), $18.8 billion in 2018 (Choice *D*), and $26.8 billion in 2019 (Choice *E*). Thus, Choice *E* is the correct answer.

5. D: The Canadian Pacific Railway convinced British Columbia and Prince Edward Island to join the confederation, and the transportation of immigrants and critical supplies spurred rapid population growth in Western Canada. Thus, Choice *D* is the correct answer. The Canadian Pacific Railway indisputably enriched Canadian merchants by creating a larger and more unified marketplace; however, territorial expansion represents a broader and more significant effect, so Choice *A* is incorrect. While support for a transcontinental railroad helped build support for the confederation, it wasn't the most important factor. Additionally, construction didn't begin until after the confederation's establishment.

Therefore, Choice *B* is incorrect. Choice *C* is incorrect because the railway had a much greater impact on attracting immigrants to Western Canada from Eastern Canada as well as foreign countries. The Canadian Pacific Railway did transport Canadian law enforcement agents and military units into Western Canada, but this only relates to one aspect of expansion. So Choice *E* is incorrect.

6. E: Landlocked means being completely surrounded by land, so the correct answer is a province without a coastline. Saskatchewan borders the Northwest Territories, Alberta, Manitoba, and the United States. Thus, Choice *E* is the correct answer. British Columbia is bordered by the Pacific Ocean, so Choice *A* is incorrect. Manitoba borders Hudson Bay, so Choice *B* is incorrect. Nova Scotia borders the Atlantic Ocean, so Choice *C* is incorrect. Ontario borders both the Great Lakes and Hudson Bay, so Choice *D* is incorrect.

7. C: The northernmost region of Canada is mostly Arctic tundra due to its proximity to the Arctic Circle and frigid temperatures. Nunavut is the northernmost Canadian province, and it consists almost entirely of tundra. Thus, Choice *C* is the correct answer. The northern tip of Newfoundland and Labrador is covered in tundra, but the rest of the province is much more temperate. Therefore, Choice *A* is incorrect. The Northwest Territories have significant amounts of tundra; however, the Northwest Territories have a warmer climate than Nunavut, and some areas are covered in boreal forests. Therefore, Choice *B* is incorrect. Northern Quebec has tundra-like conditions, but the vast majority of Quebec isn't tundra. So Choice *D* is incorrect. Yukon's coast on the Arctic Ocean features tundra-like conditions, but the rest of the province has a subarctic climate with warm summers. Therefore, Choice *E* is incorrect.

8. B: A nation is defined as a group of people who have common traits, such as heritage, history, language, culture, and religion. It has nothing to do with borders, sovereignty, power, people in office, or the rules by which a government operates (all of which are found in the other answer choices of state, government, and regime).

9. A: The presidential system establishes a separation of powers. In the presidential system, voters directly elect the chief executive, and the presidential system establishes a separation of powers between different branches of government. In contrast, in a parliamentary system, parliament elects the chief executive, and the increased collaboration and dependency creates a more responsive government. Choices *B* and *C* confuse how the executive is elected in each system. Choice *D* is incorrect because many parliamentary systems include a president, though the status of head of state is often purely ceremonial.

10. D: If you're the owner, you want the candidate who is going to get the government most fully out of your business. That means the liberal, democratic, and socialist platforms are out because they advocate for greater government involvement. The conservative platform, however, wants less government regulation and would therefore be your choice.

11. C: The historian argues granting pirates with legal authority to capture their fellow pirates because it takes a "Thief to catch a Thief." Thus, Choice *C* is the correct answer. Creolans are mentioned as people pirates would only rob when it was financially advantageous to do so; therefore, Choice *A* is incorrect. The historian only references the Mobs in London in a metaphor to support enlisting pirates as law enforcement officers, so Choice *B* is incorrect. Choice *D* is the second-best answer choice. During the seventeenth and eighteenth centuries, governments occasionally enlisted former pirates as privateers, which were essentially state-sanctioned pirates. However, Choice *C* more directly addresses this phenomenon, so Choice *D* is incorrect.

12. D: The historian argued that pirates would be motivated to capture fellow pirates if they were allowed to keep what they found aboard. The relevant line appears at the end of the passage—"To bring this about, there needs no other Encouragement, but to give all the Effects taken aboard a Pyrate Vessel to the Captors; for in Case of Plunder and Gain, they like it as well from Friends, as Enemies." Thus, Choice *D* is the correct answer. Military promotions, territorial allotments, and plundering merchant ships are not mentioned anywhere in the passage, so Choices *A, B,* and *C* are all incorrect. Choice *C* is the second-best answer because governments often tasked privateers with capturing pirates, and privateers were also generally allowed to attack enemy merchant ships. However, Choice *C* is too broad because it doesn't specify what types of merchant ships could be plundered.

13. C: In the first sentence, the historian mentions that piracy increased after the Peace of Utrecht because sailors could no longer find employment as privateers. Thus, Choice *C* is the correct answer. During the seventeenth and eighteenth centuries, European criminal syndicates funded piracy and shared in the profits, but the historian doesn't claim that the Mobs in London are funding piracy. So, Choice *A* is incorrect. Choice *B* is incorrect because the historian doesn't describe or support a prohibition on privateering; in fact, the historian is arguing for the government to hire pirates as privateers. The Royal Navy is not mentioned in the passage, so Choice *D* is incorrect.

14. B: At the end of the passage, the friar mentions how the Spaniards destroyed the Amerindian civilizations they encountered by using "Spears, Fire and Sword." Fire is a reference to firearms and cannons, which played a major role in European colonization of the Americas. Thus, Choice *B* is the correct answer. Choice *A* is the second-best answer choice. Diseases also facilitated European conquest of the Americas; however, for the most part, Europeans didn't intentionally spread disease. In addition, the friar doesn't mention diseases in the passage, so Choice *A* is incorrect. Spanish colonizers did enslave Amerindians, and the friar alludes to this practice by mentioning "Egyptian Bondage" at the end of the passage. However, slavery was a consequence of European colonization, not a tool of conquest, and weaponry is more strongly supported by the passage. Therefore, Choice *C* is incorrect. Likewise, Choice *D* is a true statement about Spanish practices in the Americas, but Spain conquered Amerindian tribes through force before converting them. So, Choice *D* is incorrect.

15. A: The friar mentions fighting that took place in Mexico between 1518 and 1530. The Aztec Empire was located in Mexico, and the Spanish *conquistador* Hernán Cortés's campaign against the Aztecs lasted from 1519 to 1521. Thus, Choice *A* is the correct answer. The Inca Empire held territory in present-day Peru, Bolivia, Argentina, Chile, and Colombia, and Francisco Pizarro didn't conquer the empire until 1533. So, Choice *B* is incorrect. The Iroquois Confederacy was primarily located in the present-day United States and Canada, so Choice *C* is incorrect. Spain conquered the Muisca Confederation, but it held territory in present-day Colombia, not Mexico. Therefore, Choice *D* is incorrect.

16. B: Spain partially explored and colonized the Americas to spread Catholicism. After the Protestant Reformation, the Catholic Church instituted a Counter-Reformation and established new religious orders to convert new followers in Africa, Asia, and the Americas. As such, the Spanish Crown regularly included friars in expeditions to the Americas. Thus, Choice *B* is the correct answer. Catholic clergymen weren't intended to prevent war crimes, and many Catholic friars actively contributed to the abuse of Amerindians. So, Choice *A* is incorrect. While friars were well educated, they generally didn't know more about navigation and geography than explorers. Therefore, Choice *C* is incorrect. Choice *D* is incorrect because the friars didn't have an official role in negotiating with Amerindians.

17. C: The Spanish explorers meet English merchants from New England, and the fur trade was common in that region. In the middle of the passage, the Spanish explorer describes how the English merchants traded with Amerindians for furs and skins. Thus, Choice *C* is the correct answer. Choice *A* is incorrect because the English navigator Captain Shapley provides charts and journals to the Spanish explorer, not Amerindians. The Spanish explorer gives the major general of Massachusetts a diamond ring as a gift, and gold isn't mentioned in the passage. So, Choice *B* is incorrect. Similarly, the Spanish explorer gives Peruvian wine as a gift to the English, and rum isn't mentioned in the passage. Therefore, Choice *D* is incorrect.

18. D: The Spanish explorer references "Pieces of Eight" several times in the passage to describe the price of his diamond ring and gifts to the English sailors. Spain minted pieces of eight out of the raw silver it imported from the Americas. Pieces of eight functioned as Spain's currency, and they were also known as the *Spanish dollar*. These silver coins rose to the status of a global currency during the sixteenth century, largely due to the high demand for silver in China. Thus, Choice *D* is the correct answer. Pieces of eight weren't tools used to read maps, so Choice *A* is incorrect. Although the Spanish explorer used pieces of eight as tokens of friendship in the passage, this wasn't their primary function. So, Choice *B* is incorrect. Choice *C* is the second-best answer. Pieces of eight were the product of a raw resource (silver) in the Americas. However, Choice *D* is a more accurate description because it correctly characterizes pieces of eight as Spanish currency. As such, Choice *C* is incorrect.

19. A: Europeans initially sailed west seeking a faster path to Asian markets. After European rulers realized Columbus had arrived in the Americas, they continued to search for this route. When these attempts also failed, European explorers focused on a passage to Asia in the North Atlantic Ocean, which is commonly referred to as the Northwest Passage. The Spanish explorer in this passage describes his mission as locating the Northwest Passage. Thus, Choice *A* is the correct answer. Several European explorers established settlements to capitalize on the fur trade while searching for a Northwest Passage, but this was not the goal. The fur trade is only mentioned in the passage to describe the European sailors' relationship to Amerindians. So, Choice *B* is incorrect. Choices *C* and *D* are both incorrect because the search for a Northwest Passage was about finding a quicker route to Asia, not obtaining a military advantage or conquering territory.

20. A: Canada explicitly created the residential schools in order to acculturate Aboriginal children and undermine Aboriginal culture. The schools were highly coercive and abusive, distinguishing them from the rest of the Canadian education system. In recent years, Canada has officially apologized for this policy on several occasions. Thus, Choice *A* is the correct answer. Poverty alleviation and increased literacy rates were both factors in Canada's decision to create the schools, but in comparison to acculturation, they were mere side benefits. Therefore, Choice *B* and Choice *C* are both incorrect. Canada intended the schools to teach Aboriginal children about Western values, not traditional values, so Choice *D* is incorrect. Reservations and residential schools were both part of the Indian Act, 1876, but the location of the schools isn't directly related to the schools' primary purpose. Therefore, Choice *E* is incorrect.

21. B: The first two sentences of the passage outline France's divide-and-conquer strategy for subduing First Nations. In effect, France allied itself with the weaker First Nations to conquer the regional hegemon, and once this was completed, France could either maintain diplomatic relations or turn against its former allies. Thus, Choice *B* is the best answer. France did forge some alliances for the purposes of exploration, and the passage expresses how some people criticized Champlain for appearing to do so. However, the passage explicitly shows how the divide-and-conquer strategy was more typical, so Choice *A* is not the best answer. Likewise, France did sometimes act as a mediator between rival First

Nations to garner diplomatic power, but this was only part of its divide-and-conquer strategy. Therefore, Choice *C* is not the best answer. France did not exclusively rely on its superior weaponry, and the Iroquois very arguably held a significant tactical advantage over the French for much of this period. So Choice *D* is incorrect. Champlain's critics accused him of being a foolish knight-errant, but he wasn't one in reality, so Choice *E* is incorrect.

22. A: The passage concludes the passage with statements about how Champlain deceived everyone by appearing reckless while actually acting with great deliberation, which is a shrewd move. Thus, Choice *A* is the correct answer. The author describes Champlain as being the opposite of timid; therefore, Choice *B* is incorrect. Champlain isn't described as a general, and the author explains how Champlain was really being deliberate. So Choice *C* is incorrect. The author describes Champlain as initially appearing to act as a foolish knight-errant, but the passage concludes by characterizing Champlain as consistent and deliberate. Therefore, Choice *D* is incorrect. While the passage describes Champlain positioning himself as an indispensable ally to First Nations, his divide-and-conquer strategy inherently involved deception since the overriding goal was to make France the region's dominant power. So Choice *E* is incorrect.

23. B: The passage describes how propaganda can shape public opinion, which was invaluable for governments seeking to increase civilian compliance with rations and drafts. Additionally, the passage describes how this tactic can effectively persuade people to adopt extreme positions, like calls to violence. Thus, Choice *B* is the correct answer. Choice *A* is the second-best answer choice. Although successful wartime propaganda encourages citizens to commit violence against enemy forces, Choice *B* is the better answer because it's broader. While Choice *A* only mentions violence, Choice *B* references the entire war effort, which includes everything from violence to electoral support. Therefore, Choice *A* is incorrect. Choice *C* is incorrect because the passage never claims propaganda replaced advertising. Similarly, Choice *D* is incorrect because the passage doesn't mention propaganda's economic benefit.

24. D: Governments first deployed large-scale propaganda during World War I. Propaganda was a critical part of the governments' total war strategy, which called for the mobilization of every possible resource for the war effort. In order to fight this unprecedented global conflict, governments had to convince the public to sacrifice their food, goods, and lives to the war effort like never before. Thus, Choice *D* is the correct answer. Propaganda was used in the Russo-Turkish War (1877–1878) and First Sino-Japanese War (1894–1895), but it was not widespread and orchestrated by the government. During World War I, nearly every government created official propaganda departments for the first time in history. So, Choices *A* and *B* are incorrect. The Spanish, German, and Soviet governments all published a significant amount of propaganda. However, World War I (1914–1918) occurred several decades before the Spanish Civil War (1936–1939). Therefore, Choice *C* is incorrect.

25. B: At the beginning of the passage, Russell states that propaganda is "derived from the art of advertisement." The passage then goes on to describe how advertising was groundbreaking in its ability to bend public opinion toward the author's message. At the end of the passage, Russell states that propaganda mimics advertisers' successful tactics, and as a result, it is the "method by which democratic opinion is created." Thus, Choice *B* is the correct answer. Choice *A* is incorrect because propaganda doesn't compete for advertising; it's a form of advertising. Governments create propaganda, so Choice *C* is incorrect. Choice *D* is the second-best answer choice. Although effective propaganda is similar to advertising, the passage doesn't mention anything about the public enjoying advertising of any type. Therefore, Choice *D* is incorrect.

26. C: A participatory democracy in its truest form is a system in which everyone participates in the political system. Choice *A* describes an elite democracy. Choice *B* is a pluralist democracy—one where interest groups and advocacy for certain issues dominates the government. Choice *D* is the exact opposite of a democracy.

27. B: Eager to avoid another global conflict, European leaders tried to appease Hitler by letting him occupy Austria and Czechoslovakia. This policy failed because it only emboldened Hitler, and he invaded Poland in 1939. Rather than receiving leniency after World War I, Germany was forced to sign a humiliating peace treaty. Furthermore, the League of Nations failed to prevent conflict because it lacked any real power. This encouraged continued aggression from Italy, Germany, and Japan, which culminated in World War II.

28. B: Liberals tend to want governmental interference when it comes to inequality and the marketplace. They want strong federal power, but only to protect certain safeguards they deem necessary. Legalization of marijuana certainly falls into their belief in rights of privacy. Regulation of the stock exchange is also in their wheelhouse. The raising of minimum wage is something they have supported for a long time. Lowering of taxes, however, is generally not pushed for by liberal voters.

29. C: The passage is referencing the Treaty of Versailles. The first paragraph mentions reparations owed by Germany. The Treaty of Versailles forced Germany to assume full responsibility for causing World War I and pay reparations to the victorious Allied Powers. Thus, Choice *C* is the correct answer. None of the other answer choices is a treaty that involved German reparations. The Munich Agreement (1938) was part of Western Europe's policy of appeasement toward Adolf Hitler, and it allowed Nazi Germany to annex the Sudetenland in the present-day Czech Republic. So, Choice *A* is incorrect. The North Atlantic Treaty (1949) established the North Atlantic Treaty Organization, a military alliance between the United States and Western Europe. Therefore, Choice *B* is incorrect. The Vienna Convention on the Law of Treaties (1969) codified a series of international laws in regard to treaties, so Choice *D* is incorrect.

30. B: Reparations placed improper pressure on Germany because it was an unsustainable economic and political burden. World War I decimated Germany's economic infrastructure, and the country was already severely in debt. Consequently, reparations were a major financial burden, and the financial crisis deepened during the Great Depression. The German public considered the Treaty of Versailles to be a national humiliation, and reparations further undermined the post-war government's legitimacy in the eyes of the public. Thus, Choice *B* is the correct answer. Choice *A* is incorrect because the treaty didn't place any prohibitions on free trade agreements. Although numerous border disputes erupted after World War I, German reparations were not the reason why, so Choice *C* is incorrect. Choice *D* is the second-best answer. As referenced in the passage's second paragraph, the Treaty of Versailles included some clauses related to Germany's coal, iron, and steel industries. However, those clauses weren't directly related to reparations, so Choice *D* is incorrect.

31. A: Free trade agreements seek to increase international trade by limiting or eliminating tariffs and subsidies for domestic industries. Overall, international trade increased dramatically after the signing of the General Agreement on Tariffs and Trade (1947) and formation of the World Trade Organization (1995). Thus, Choice *A* is the correct answer. Although Keynes issued proposals to reduce German reparations and establish a Free Trade Union, they are separate proposals. Reparations aren't directly related to free trade agreements. As such, Choice *B* is incorrect. Choice *C* is incorrect because free trade agreements generally prohibit countries from subsidizing or protecting domestic industries. Free trade agreements don't facilitate imperialism, so Choice *D* is incorrect.

32. D: The passage is describing the Treaty of Versailles and the aftermath of World War I. The Treaty of Versailles exacerbated Germany's economic woes and political dysfunction, which led to the rise of

Adolf Hitler's Nazi Party. Additionally, many border disputes weren't adequately resolved after World War I, and those conflicts contributed to World War II. Thus, Choice *D* is the correct answer. The Cold War (1946–1991) occurred several decades after World War I, and it was triggered by ideological and geopolitical disputes between the United States and Soviet Union. So, Choice *A* is incorrect. Although the Great Depression (1929–1945) deepened the unresolved conflicts mentioned in the passage, it wasn't caused by those conflicts; rather, the Great Depression was caused by financial speculation, unsustainable national debts, and falling productivity. So, Choice *B* is incorrect. The Treaty of Versailles marked the end of World War I, so it could not have been the cause of the war. As such, Choice *C* is incorrect.

33. A: Most scholars already knew the world was round by 1492. On the other hand, the arrival of Europeans in North and South America introduced deadly diseases that killed millions of native peoples. Europeans had developed immunity to diseases such as smallpox, while Native Americans had not. In addition, Europeans introduced a number of new plants and animals to the New World, but they also adopted many new foods as well, including potatoes, tomatoes, chocolate, and tobacco. Finally, Europeans tried to convert Native Americans to Christianity, but Indians did not completely give up their traditional beliefs. Instead, they blended Christianity with indigenous and African beliefs to create new syncretic religions.

34. D: Culture can be transmitted in nearly endless ways, ranging from governmental policies to entertainment consumption. Furthermore, culture can be transmitted intentionally or spontaneously. For example, powerful institutions can sometimes unilaterally shift the culture to achieve a goal, but other times cultural change is a natural byproduct of social interactions that spirals in an unforeseen direction. Thus, Choice *D* is the correct answer. Culture is not always transmitted through hierarchical relationships, so Choice *A* is incorrect. Religion, economic activities, and government policies play a powerful role in cultural development, but culture can be transmitted in other important ways, such as through social interactions and digital networks. Therefore, Choice *B* is incorrect. Choice *C* is a true statement, but it does not describe how culture is transmitted across society, so it's incorrect.

35. D: Governmental regimes fall along a continuum between total authoritarianism and complete direct democracy. Most countries are neither totally authoritarian, nor a complete direct democracy, but they do all fall along this continuum. China and Iran would be towards the authoritarian end of the spectrum and the United States, United Kingdom, and Mexico would be towards the democratic end.

36. B: Choice *B* is correct, as power is the ability of a ruling body to influence the actions, behavior, and attitude of a person or group of people. Choice *A* is incorrect, as politics is the process of governance typically exercised through the enactment and enforcement of laws over a community, most commonly a state. Although closely related to power, Choice *C* is incorrect, because authority refers to a political entity's justification to exercise power. Legitimacy is synonymous with authority, so Choice *D* is also incorrect.

37. C: Choice *C* is correct. There are no definitive requirements to be a nation; rather, the nation only needs a group bound by some shared characteristic. Examples include language, culture, religion, homeland, ethnicity, and history. Choice *C* isn't a requirement to be a nation, though it is required to be a state.

38. D: Choice *D* is correct. Machiavelli was an Italian diplomat, politician, and historian, and *The Prince* is his best-known political treatise. The excerpt instructs the Prince that if he injures a man, then he must ensure the injury is "of such a kind that one does not stand in fear of revenge." Choices *B* and *C* contradict the first sentence of the excerpt, which says that men "ought either to be well treated or crushed." Choice *A* is close, but the selection goes too far, assuming revenge will result in overthrowing the Prince.

39. A: Choice *A* is correct. In the excerpt, "utility" is defined as actions that are "right in proportion as they tend to promote happiness, wrong as they tend to produce the reverse of happiness." The excerpt then explains that happiness is measured by pleasure, and the reverse is pain. Therefore, the author calls for actions to be evaluated based on the net total of pleasure. Choice *D* contradicts the definition provided in the excerpt. The excerpt doesn't support Choice *C*, as there's no evidence that pleasure-generating sacrifices merit special status. Choice *B* is incorrect because sacrifice can still be valuable if it leads to more pleasure than pain.

40. C: Choice *C* is correct. Karl Marx, a philosopher, social scientist, historian, and revolutionary, is considered the father of communism. All the answer choices contain true statements or reasonable assumptions from the passage; however, Choice *C* best articulates the main idea—society is the history of class struggle, and working men must unite and fight a revolutionary battle like their historical ancestors.

41. A: Choice *A* is correct. A business cycle is when the gross domestic product (GDP) moves downward and upward over a long-term growth trend, and the four phases are expansion, peak, contraction, and trough. An expansion is the only phase where employment rates and economic growth continually grow. Contraction is the opposite of expansion. The peak and trough are the extreme points on the graph.

42. D: Choice *D* is correct. The government can intervene in the economy through income redistribution, taxes, subsidies, and price controls, but only subsidies lower prices, reassure the supply, and create opportunity to compete with foreign vendors.

43. C: Choice *C* is correct. Isoline maps are used to calculate data and differentiate between the characteristics of two places. In an isoline map, symbols represent values, and lines can be drawn between two points to determine differences. The other answer choices are maps with different purposes. Topographic maps display contour lines, which represent the relative elevation of a particular place. Dot-density maps and flow-line maps are types of thematic maps. Dot-density maps illustrate the volume and density of a characteristic of an area. Flow-line maps use lines to illustrate the movement of goods, people, or even animals between two places.

44. C: Choice *C* is correct. The tilt of the Earth's rotation causes the seasons due to the difference in direct exposure to the Sun. For example, the northern hemisphere is tilted directly toward the Sun from June 22 to September 23, which creates the summer in that part of the world. Conversely, the southern hemisphere is tilted away from the Sun and experiences winter during those months. Choice *A* is factually incorrect—the rate of Earth's rotation is constant. Choice *B* and *D* are factors in determining climate, but differences in climate don't cause the seasons.

45. D: Choice *D* is correct. Nonrenewable energy resources are oil, natural gas, and coal, collectively referred to as fossil fuels. Nonrenewable energy is more widely used due to its abundance and relatively cheap price. In addition, countries have tailored their existing infrastructure to nonrenewable energy. Currently, the technology to store renewable energies for long periods is either nonexistent or expensive. Choice *D* is correct because it's inaccurate. Renewable energy can be converted into a power source, but the issue is scale of use. For example, Canada converts renewable resources to derive about sixteen percent of the country's energy.

46. B: Choice *B* is correct. Ocean currents dramatically impact the climate by storing heat from the Sun and transporting the warmth around the globe. The evaporation of ocean water increases the temperature and humidity in the nearby landmasses. A gyre is a system of circulating currents. Countries are most impacted by the currents and gyres closest to their shores. The question stem asks what currents have the most impact on Canada. According to the map, the North Pacific Gyre, Labrador Current, Alaskan Current, and Gulf Stream impact Canada. Choice *B* is the only answer with a pair of those currents or gyres.

47. B: As stated in the passage, left-axis political orientations typically favor social and economic equality and advocate for government intervention to achieve these goals. Communism, Socialism, Liberalism, and Progressivism are examples of political orientations on the left side of the spectrum that are discussed in the passage. Right-axis political orientations generally value the existing and historical political institutions and oppose government intervention, especially in regard to the economy. Libertarianism, Conservatism, and Fascism are the three orientations from the passage that are considered right-axis orientations. It is true that Libertarians have some progressive ideals; however, Libertarianism is still usually considered a right-axis orientation because the political ideals are conservative.

48. A: Choice *A* is correct. On the political spectrum, ideologies on the left side of the axis emphasize socioeconomic equality and advocate for government intervention, while ideologies on the right axis seek to preserve society's existing institutions and oppose government intervention. Therefore, the answer will be the farthest left on the axis, making Choice *A* correct.

49. C: Choice *C* is correct, as it most closely corresponds to the provided definition. Conservatism prioritizes traditional institutions. In general, conservatives oppose modern developments and value stability. Socialism and liberalism both feature the desire to change the government to increase equality. Libertarianism is more concerned with establishing a limited government to maximize personal autonomy.

50. A: Fascists advocate for a strong, consolidated, centralized government led by an all-powerful dictator. They believe a key role of this centralized government is to prepare for war. While Libertarians still tend to maintain conservative approach to government, especially in the context of power and military intervention, they favor a weaker central government with powerful local rule instead.

Science

1. C: There is a 99% probability of PCR testing identifying *Histoplasma*. GM assay was more specific for identifying *Aspergillus,* 95% to 85%. True positive is defined by sensitivity. The sensitivity of GM assay testing is less than 70%.

2. D: *Histoplasma* is detectable 90 days from exposure. PCR testing is able to detect *Histoplasma* 91 days from exposure—one day after sufficient organisms exist for detection. *Candida* is detectable 45 days from exposure. PCR testing is able to detect *Candida* 72 days from exposure—27 days after a sufficient number of organisms exist for detection. *Aspergillus* is detectable 118 days from exposure. ELISA testing is able to detect *Aspergillus* 134 days from exposure—16 days after a sufficient number of organisms exist for detection. *Candida* is detectable 45 days from exposure. GM assay testing is able to detect *Candida* 56 days from exposure—11 days after a sufficient number of organisms exist for detection.

3. B: The probability that the GM assay will identify *Candida* is 69%. Therefore, there's a 31% probability that it won't be identified. ELISA sensitivity and specificity for *Histoplasma* are both greater than 80%. False-negative probabilities are represented by the specificity of a given testing method. The sensitivity and specificity for GM assay testing for *Aspergillus* is 9% and 96% respectively. All testing methods had greater than 90% specificity for the organisms.

4. C: The sensitivity of PCR testing for *Histoplasma* is 99%, and the test can identify the organism one day after it reaches a detectable colony size. The sensitivity for GM assay testing for *Histoplasma* is 65%. If physicians rely on GM assay testing, they may determine that the individual doesn't have the *Histoplasma* infection. Treatment will depend on the presence or absence of the infection as indicated by testing. Waiting for PCR testing is based on the sensitivity of the test, not the individual's current symptoms. The subclinical phase of *Histoplasma* is 28 days.

5. A: ELISA testing detects *Candida* three days after the organism is present in sufficient numbers to be recognized. PCR detects the organism more than three weeks after it is first detectable. ELISA testing sensitivity for *Candida* is 87% and PCR testing is 92%. However, the ability to identify the presence of the organism earlier in the process of infection (allowing early intervention) outweighs the differences in the probability of identifying the presence of the organism. There's a 92% probability that PCR testing will identify the presence of *Candida*. PCR testing is more sensitive than ELISA: 92% versus 87%.

6. D: The main difference in the scientists' opinions is related to the cause of mad cow disease. The existence of species-specific proteins was used by Scientist 2 to support viral infection as the cause of the disease. Transmission rates of the disease and the conversion of normal proteins to prions were not debated in the passage.

7. A: Scientist 2 proposed that viruses were the cause of mad cow disease because chemicals inactivated the viruses. The remaining choices are correct.

8. B: According to Scientist 1, abnormal prions are capable of "refolding" normal proteins in harmful prions. Abnormal proteins accumulate to produce the damaging conglomerations. Scientist 2 didn't find species-specific DNA and used this fact to support viruses as the cause of mad cow disease. According to Scientist 1, prions are located in the central nervous system, not the peripheral nervous system.

9. D: Mad cow disease can be spread between animal species and from animals to humans through consumption of diseased animal products. The resulting damage to the central nervous system is irreversible and will eventually cause the death of the animal. Scientist 2 would not agree that the infecting agent contained amino acids, as they form proteins, and Scientist 2 believes that a virus causes the disease. Scientist 2 demonstrated that the infected tissue of animals that were infected by a different species didn't contain species-specific DNA, which would have been the expected outcome if the infecting agent were a protein.

10. C: The accumulated masses of abnormal prions eventually form sponge-like holes in the brain and spinal cord that result in death. The passage doesn't mention the effects of the synapses, nerves, or blood supply.

11. D: The absence of disease resulting from the inactivated viral particles best supports the views of Scientist 2. There were no species-specific DNA sequences found in the infected particles. Scientist 2 didn't support the existence of prions as the cause of mad cow disease.

12. C: The actual process of "refolding" the normal protein into the abnormal protein isn't clear from this passage. Scientist 1 claims that prions cause the disease. Prions are an abnormal protein, not a virus. Scientist 1 claims that mad cow disease is caused by abnormal proteins.

13. C: The main hypothesis for this study involved the influence of 2-AG levels combined with sleep deprivation on eating behaviors. The combination of the two conditions, not each one separately, constitutes the main hypothesis. The passage didn't discuss a placebo effect in the normal saline injection group.

14. D: The study results support the hypothesis because the participants who received 1-AG injections and were sleep deprived gained more weight than participants who received sterile normal saline injections. The remaining choices do not support the hypothesis.

15. A: Participant H gained more than 1 pound (450g) per day. There was little fluctuation in the day-to-day weight gain for each participant. Participant H in trial 2 gained more weight than participant D in trial 1.

16. B: Sleep deprivation increases the levels and duration of action of 2-AG, an endogenous cannabinoid, especially in the late afternoon. The stress effect increases with the degree of sleep deprivation. The passage doesn't discuss a relationship between sleep deprivation and the hunger response. Eating behaviors are increased in late afternoon as a result of the extended duration of 2-AG action.

17. A: Circadian fluctuations increase the levels of 2-AG during the afternoon and evening. This increase is believed to stimulate food intake beyond the point of satiety. Endogenous cannabinoids decrease gastric motility. 2-AG may have a calming effect on mood, but food intake is still increased in the presence of afternoon and evening levels of 2-AG. Endogenous cannabinoids work with the opioid system to mediate the pain response, not food-seeking behaviors.

18. A: In trial 3, the plants grown with the combined-wavelength LED's reached 150 millimeters by day 21. The plants grown with white light reached 160 millimeters by day 35.

19. C: In trial 3, with LED lighting that included green and yellow wavelengths, plant growth was greater than trial 1 or trial 2 with either blue or red wavelengths. However, from the available information, it can only be said that green and yellow wavelengths *contributed to* plant growth in trial 3, but not that green and yellow wavelengths *alone* were responsible for plant growth in trial 3. There was plant growth in all lighting conditions.

20. B: In trial 1, from day 28 to day 35, white light growth increased by 71 millimeters, and red light increased by 78 millimeters. In trial 2, from day 28 to day 35, white light growth increased by 71 millimeters, and blue light increased by 78 millimeters.

21. C: The average daily growth with LED lighting was not twice the white light average daily growth. LED systems did result in better growth rates and they do require less water and electricity. However, the question is based on recorded average daily growth, and that rate was not double the white light rate.

22. D: The passage says that green and yellow wavelengths are reflected by the plant. Therefore, it's expected that those wavelengths would result in slower growth than the blue or red wavelengths, which are absorbed.

23. B: *Anacardiaceae Magnifera* is the genus and family name for the mango. The *Colletotrichum* fungus causes the spoilage. The remaining choices are correct.

24. C: According to Table 1, at 20 days, the fungal level at 7.5 °C was the same as the fungal level at 5 °C. Because the 7.5 °C temperature is less expensive than the 5 °C temperature, the 7.5 °C model is best. The 10 °C model is less expensive than the 7.5 °C, but fungal levels are greater. Only the 7.5 °C and 5 °C models had acceptable fungal levels at 20 days.

25. B: The hot water wash pre-treatment fungal level increased by 15 millimeters from day 26 to day 28 at 10 °C. It was the single largest one-day increase across the trials. Air cooling at 10 °C increased by 5 millimeters from day 14 to day 16. Air cooling at 7.5 °C increased by 7 millimeters from day 26 to day 28. Hot water wash at 7.5 °C increased 11 millimeters from day 26 to day 28.

26. C: The hot water wash fungal level at 10 °C reached 35 millimeters on day 20. The maximum fungal level for air cooling at 5 °C was 19 millimeters, and at 10 °C, 35 millimeters on day 26. The maximum fungal level for hot water at 7.5 °C was 36 millimeters on day 28.

27. C: The fungal levels were acceptable with the hot water wash at 7.5 °C, and the 7.5 °C temperature is less expensive to maintain. Air cooling and hot water wash pre-treatment at 5 °C resulted in acceptable fungal levels, but the 5 °C temperature is costlier. Fungal levels were not acceptable at 28 days at 10 °C.

28. B: Shipping mangoes at 5 °C is costly, but for the 28-day trip, the fungal levels were only acceptable in the 5 °C model. Air cooling fungal rates at 5 °C were lower than the hot water wash rates, but each was acceptable. Fungal rates at 7.5 °C and 10 °C were unacceptable.

29. A: The correlation was positive, because when one variable increased, the other increased, and when one variable decreased, the other decreased. There are two forms for a curvilinear relationship. In one curvilinear relationship, when variable 1 increases, a second variable increases as well, but only to a certain point, and then variable 2 decreases as variable 1 continues to increase. In the other form, variable 1 increases while variable 2 decreases to a certain point, after which both variables increase. In a negative relationship, high values for one variable are associated with low values for the second variable.

30. D: Mitochondrial activity is suppressed by elevated blood glucose levels. Mitochondria use sugar to produce cellular energy, and the presence of large numbers of mitochondria in BAT gives BAT a brownish color. Mitochondria contain DNA and can reproduce additional mitochondria when additional energy is required. In the body, mature red blood cells are the only cells that don't contain mitochondria.

31. D: Blood sugars for all groups identified in Figure 5 were highest at 2 p.m.

32. B: Circadian rhythms control sleep by stimulating the release of melatonin from the pineal gland. Ganglia are nerve cells, not hormones, that affect sleep. BAT doesn't release sugar; it utilizes sugar for heat production. Shivering on cold mornings is a desirable form of thermogenesis but isn't associated with sleep.

33. D: The average blood sugar for participant 6 was 115. Participant 1 was 105, participant 3 was 112, and participant 4 was 82.

34. B: The daytime blood glucose levels in Figure 1 decreased as the day progressed. The blood glucose levels in Figure 5 peaked for the day at 2 p.m. Night blood glucose levels didn't reach 100. Group I's levels are irrelevant to the question.

35. A: Based only on Figure 2, the researchers' hypothesis wasn't confirmed. Subsequent trials confirmed the hypothesis.

36. A: Increased algae levels can block sunlight, limiting growth of species inhabiting lower zones. The passage doesn't identify the effects of rainfall or cold temperatures on phytoplankton growth, so Choices *B* and *D* are incorrect. The passage identifies the effect of phosphorous and nitrogen residue on algae growth, but not as a food source for cyanobacteria.

37. D: The passage identifies freshwater contamination by phosphorous and nitrogen as the most common cause of algae overgrowth. Population density would be more common in Florida than Kansas.

38. B: As algae levels increase above normal, organisms in the aphotic level plants don't receive adequate light for normal growth and oxygen levels are decreased, resulting in the death of oxygen-dependent species.

39. C: Algae block the sunlight, which limits growth.

40. A: Algae growth was greater in July, which limited the amount of light reaching the lower zones of the lake, decreasing the levels of cyanobacteria. Cyanobacteria existed in less-than-normal concentrations at 20 meters, but there were measurable levels of the organisms. Algae growth at 3 meters wasn't measured. The passage states that cyanobacteria growth is associated with algae growth, not independent of algae growth.

41. A: The hypothesis is the sentence that describes what the scientist wants to research with a conclusive expected finding. Choice *A* describes how she believes sunlight will affect plant growth. Choice *B* includes details about the experiment. Choice *C* is not a conclusive theory. Choice *D* describes the data that she found after conducting the experiment.

42. C: Looking at Figure 1, four experimental groups are shown on the graph for which data were collected: plants that received 1 hour of sunlight, 3 hours of sunlight, 5 hours of sunlight, and 7 hours of sunlight. Choices *A* and *B* could be describing two of the experimental groups and how much sunlight they received. Choice *D* describes how many days' data was collected.

43. B: After the data was collected, it was compiled into a line graph. The data points were collected, and then a line was drawn between the points. Data is represented by horizontal or vertical bars in bar graphs, Choice *A*. Pie charts are circular charts, with the data being represented by different wedges of the circle, Choice *C*. Pictograms use pictures to describe their subject, Choice *D*.

44. A: Looking at Figure 2, the sun provides light energy that drives forward the process of photosynthesis, which is how plants make their own source of energy and nutrients. Choices *B* and *C* are found in the environment around the plants. They combine with light energy to make the photosynthesis reaction work. Choice *D* is a product of photosynthesis.

45. D: Looking at the Figure 1, the experimental group that received 7 hours of sunlight every day grew taller than any of the other groups that received less sunlight per day. Therefore, it is reasonable to conclude that more sunlight makes plants grow bigger. Choice *A* is not a reasonable conclusion because it did not have the tallest plants. The scientist decided to measure the plants only for 11 days, but that does not describe a conclusion for the experiment, Choice *B*. Choice *C* is the opposite of the correct conclusion and does not have evidence to support it.

46. B: The atomic mass of a molecule can be found by adding the atomic mass of each component together. Looking at Figure 2, the atomic mass of each element is found below its symbol. The atomic mass of Na is 23, Choice *A*, and the atomic mass of Cl is 35.5, Choice *C*. The sum of those two components is 58.5, Choice *B*. Choice *D* is equal to two Cl atoms joined together.

47. D: Figure 1 shows the trends of the periodic table. Looking at the black arrows representing electronegativity, it is shown that electronegativity increases going towards the top row of the table and also increases going towards the right columns of the table. Therefore, the most electronegative element would be found in the top right corner of the table, which is where the element Helium is found. Choices *A* and *B* are found at the bottom of the table. Choice *D* is found on the left side of the table.

48. A: Looking at Figure 1, the ionization energy increases as it goes right and up on the periodic table. Of the choices given, potassium (K) is found in the first column, fourth row. Each of the other choices is found above or to the right of potassium, so they would tend to have higher ionization energy. Since we're looking for the element with the lowest ionization energy, Choice *A*, potassium, is correct.

49. D: The atomic number of an element represents the number of protons. Looking at Figure 2, the atomic number is located at the top of the box, above the element's symbol. Hydrogen (H) has an atomic number of 1 and has the least number of protons of any other element in the periodic table. Radon (Rn), Choice *A*, has 86 protons. Boron (B), Choice *B*, has 5 protons. Nitrogen (N), Choice *C*, has 7 protons.

50. B: Looking at Figure 3, the elements are color coded in periods and groups according to their similar properties. Noble gases are located in the right most column of the table. Radon (Rn) is the only one of the element choices marked as a noble gas and would be the right choice for Scientist A. Nitrogen (N) and Boron (B), Choices *A* and *D*, are nonmetals. Copper (Cu), Choice *C*, is a transition metal.

Dear GED Test-Taker,

We would like to start by thanking you for purchasing this study guide for your GED Canada exam. We hope that we exceeded your expectations.

Our goal in creating this study guide was to cover all of the topics that you will see on the test. We also strove to make our practice questions as similar as possible to what you will encounter on test day. With that being said, if you found something that you feel was not up to your standards, please send us an email and let us know.

We have study guides in a wide variety of fields. If you're interested in one, try searching for it on Amazon or send us an email.

Thanks Again and Happy Testing!
Product Development Team
info@studyguideteam.com

FREE Test Taking Tips DVD Offer

To help us better serve you, we have developed a Test Taking Tips DVD that we would like to give you for FREE. **This DVD covers world-class test taking tips that you can use to be even more successful when you are taking your test.**

All that we ask is that you email us your feedback about your study guide. Please let us know what you thought about it – whether that is good, bad or indifferent.

To get your **FREE Test Taking Tips DVD**, email freedvd@studyguideteam.com with "FREE DVD" in the subject line and the following information in the body of the email:

a. The title of your study guide.

b. Your product rating on a scale of 1-5, with 5 being the highest rating.

c. Your feedback about the study guide. What did you think of it?

d. Your full name and shipping address to send your free DVD.

If you have any questions or concerns, please don't hesitate to contact us at freedvd@studyguideteam.com.

Thanks again!

Manufactured by Amazon.ca
Bolton, ON